Advances in Isotope Geochemistry

Series editor

Jochen Hoefs, Göttingen, Germany

More information about this series at http://www.springer.com/series/8152

Horst Marschall · Gavin Foster
Editors

Boron Isotopes

The Fifth Element

Springer

Editors
Horst Marschall
Geology & Geophysics Department
Woods Hole Oceanographic Institution
Woods Hole, MA
USA

and

Institut für Geowissenschaften
Goethe Universität Frankfurt
Frankfurt am Main
Germany

Gavin Foster
National Oceanography Centre
Southampton
University of Southampton
Southampton
UK

ISSN 2364-5105 ISSN 2364-5113 (electronic)
Advances in Isotope Geochemistry
ISBN 978-3-319-64664-0 ISBN 978-3-319-64666-4 (eBook)
https://doi.org/10.1007/978-3-319-64666-4

Library of Congress Control Number: 2017947859

Printed on acid-free paper

This Springer imprint is published by Springer Nature
The registered company is Springer International Publishing AG
The registered company address is: Gewerbestrasse 11, 6330 Cham, Switzerland

Contents

Boron Isotopes in the Earth and Planetary Sciences—A Short History and Introduction

1

Horst R. Marschall and Gavin L. Foster

Me fifth element – supreme being. Me protect you.
Leeloo

Abstract

This volume on boron isotope geochemistry contains chapters reviewing the low- and high-temperature geochemistry, marine chemistry, and cosmochemistry of boron isotopes. It covers theoretical aspects of B isotope fractionation, experiments and atomic modeling, as well as all aspects of boron isotope analyses in geologic materials by the full range of solution and in situ methods. The book provides guidance for researchers on the analytical and theoretical end, and introduces the various scientific applications and research fields in which boron isotopes play a growing role today. This chapter provides a brief history of boron isotope research and analytical development and provides an overview of the other chapters of the volume "Boron Isotopes—The Fifth Element" in the series *Advancements in Isotope Geochemistry*.

Keywords

Boron isotopes · History of boron · Geochemistry

1.1 Introduction

Boron is a moderately volatile, lithophile metalloid with a low atomic mass and two stable isotopes (^{10}B and ^{11}B). The ^{11}B/^{10}B ratio in terrestrial materials is approximately 4, but shows a variation of ~100 ‰ in nature (Palmer and Swihart 1996; Foster et al. 2016).The large B isotopic fractionation at low temperatures has led to the use of the B stable isotope system for studies on processes acting on the Earth's surface (e.g. Hemming and Hanson 1992; Barth 1998) and for the detection of fluid-rock interaction

H.R. Marschall (✉)
Department Geology and Geophysics, Woods Hole Oceanographic Institution, Woods Hole, MA 02543, USA
e-mail: marschall@em.uni-frankfurt.de

H.R. Marschall
Institut für Geowissenschaften, Goethe Universität Frankfurt, Altenhöferallee 1, 60438 Frankfurt am Main, Germany

G.L. Foster
Ocean and Earth Science, National Oceanography Centre Southampton, University of Southampton, Southampton SO14 3ZH, UK
e-mail: Gavin.Foster@noc.soton.ac.uk

H. Marschall and G. Foster (eds.), *Boron Isotopes*, Advances in Isotope Geochemistry,
https://doi.org/10.1007/978-3-319-64666-4_1

processes, such as in seafloor hydrothermal systems or enrichment processes in subduction zones (e.g. Spivack and Edmond 1987; Ishikawa and Nakamura 1994; Scambelluri and Tonarini 2012). The pH-dependent speciation of B in water and hydrous fluids leads to pH-dependent boron isotope fractionation between water and solids (Kakihana et al. 1977; Wunder et al. 2005; Klochko et al. 2006). Consequently, the boron isotopic composition of biogenic carbonate can be employed as a paleo-pH proxy for seawater during, for example, glacial-interglacial cycles and periods of the more distant geological past (e.g., Hönisch and Hemming 2005; Martínez-Botí et al. 2015).

The strong enrichment of B in the crust and the significant difference in B isotopic composition among continental crust, modern seawater and the depleted mantle make boron a powerful tracer for the secular evolution of the ocean-crust-mantle system (e.g., Leeman and Sisson 1996; Marschall et al. 2017). In cosmochemistry, boron isotope ratios are employed to investigate the timing of condensation and accretion, and in particular irradiation processes in the early solar system that led to the production of short-lived ^{10}Be and the subsequent ingrowth of radiogenic ^{10}B (e.g., McKeegan et al. 2000; Hoppe et al. 2001; MacPherson et al. 2003; Gounelle et al. 2013).

1.2 A Short History of Boron Isotope Analyses

1.2.1 The Discovery of Boron Stable Isotopes

The invention of the mass spectrometer by F.W. Aston and J.J. Thomson at Cambridge University (Thomson 1913; Aston 1919) and by A. J. Dempster at the University of Chicago (Dempster 1918) led to the discovery of stable isotopes in 1920 for which F. W. Aston was awarded the Noble Price in chemistry in 1922. Boron was among the first elements investigated with the new instrument and the discovery of its two stable isotopes was already reported in one of the first papers (Aston 1920). Soon, the masses of ^{10}B and ^{11}B were determined from ionized BF_3 (Aston 1927), and the results were already within 0.02‰ of the currently accepted values.

1.2.2 Natural Abundances and Variations

Other workers used acid titration of BCl_3 and BBr_3 solution to determine the atomic mass of boron and, hence, the natural abundances of the two stable isotopes (Baxter and Scott 1921). Discrepancies among different workers led to the consideration that natural isotopic abundances could vary depending on the source of boron, and attempts to demonstrate this experimentally occurred as early as 1925 (Briscoe and Robinson 1925). However the δ^{11}B values determined for sassolite from Italy, colemanite from California and boracite from Turkey by Briscoe and Robinson (1925) ranged from +54 to +208‰.[1] These values are much higher and show a much larger spread than any values determined for natural boron minerals more recently, suggesting that fractionation occured in the laboratory during sample preparation. The results and other contrasting reports of the 1920s for natural variations of boron isotope abundance ratios were questioned in the 1930s and explained as due to experimental errors. The quest for the determination of *the* atomic mass of boron (and hence its natural isotopic abundance ratio) continued.

Photometric analysis of boron from BF_3 and from boron oxide in the 1930s led to the establishment of the atomic mass of boron closer to the presently accepted value, and it was assumed that natural variations would not be resolvable analytically with the methods available at the time (Elliott 1930a, b; Jenkins and McKellar 1932; Ornstein and Vreeswijk 1933). These early pioneers quickly recognised something that modern analysts know all too well: the isotopic composition of boron was difficult to measure

[1] δ^{11}B values are calculated from the reported ^{11}B/^{10}B ratios relative to the certified ratio of SRM951.

precisely. In this case, it was suspected that the widely used BF_3 method caused instrumental fractionation and memory effects, leading to the use of other boron compounds instead, such as trimethyl boroxine, $B_3O_3(CH_3)_3$ (Inghram 1946).

The lighter boron isotope, ^{10}B, has a very large cross section for thermal neutrons that is only rivaled by Cd and some of the rare earths (Sears 1992). The cross section of ^{11}B is approximately six orders of magnitude smaller than that of ^{10}B; the efficacy of boron as a neutron absorber in nuclear reactors, therefore, depends strongly on its isotopic composition. Following the discovery of nuclear fission in the 1930s and 1940s there was a growing interest in the natural isotopic abundance of boron. Indeed, such investigations played an important role in the discovery of the neutron itself (Chadwick 1933). The high neutron absorption of boron was also investigated early on in the context of neutron irradiation treatment of cancer (Zahl and Cooper 1941; Conger 1953). However, the natural isotope abundances became less important for technical applications once efficient isotope enrichment methods were developed (e.g., Beams and Haynes 1936; Yates 1938).

Nevertheless, the debate on the natural isotope abundances of boron continued, and some authors still suspected that natural variations of boron isotope ratios could exist (e.g. Aston 1931). This hypothesis was encouraged especially after Urey and Greiff (1935) had predicted variations in the atomic masses of light elements in nature; they argued for natural stable isotope fractionation based on a theoretical discussion of mass-dependent fractionation among molecules. This renewed the interest in the natural variation of the abundances of boron isotopes (Thode et al. 1948). However, most of these early studies did not include controlled sample preparation strategies from sampling of natural material to the stage of sample introduction into the mass spectrometer. Instead, commercially available, industrially processed boric acid was used for which the mine sources and original mineralogy were known (Thode et al. 1948). The possibility (and likelihood) of isotope fractionation during chemical processing was not discussed, and indeed all published boron isotope ratios from this early period are suspiciously enriched in the heavy isotope with $\delta^{11}B$ ranging from +56 to +94‰ (see Footnote 1) (Inghram 1946; Thode et al. 1948).

The debate continued with improvements in mass spectrometric methods, but subsequent papers reporting full preparation procedures for analyzed minerals found no resolvable variation for a large selection of borates and borosilicates (Osberghaus 1950; Parwel et al. 1956). This also includes the first analyses of borosilicates (axinite and tourmaline) and of seawater, but unfortunately, no actual values were reported by Parwel et al. (1956). Analytical methods initially included the use of single-focussing mass spectrometers, but then moved to double-focussing instruments, and employed a range of boron compounds, such as boron hydride, trimethyl borane $(CH_3)_3B$, and trimethyl boroxine (Inghram 1946; Lehmann and Shapiro 1959; Abernathey 1960), as well as BF_3 and BCl_3 (Inghram 1946; Thode et al. 1948; Osberghaus 1950; Bentley 1960), all of which resulted in very high $^{11}B/^{10}B$ ratios and hence high estimates for the atomic mass of boron.

1.2.3 TIMS and the Establishment of Standards

The debate was resolved in the 1960s with the introduction of isotope reference materials that could be used to correct for instrumental mass fractionation, and the use of thermal-ionization mass spectrometry (TIMS), in which boron compounds with higher masses (and so with reduced machine induced mass fractionation) were analyzed (McMullen et al. 1961). Ions of boron compounds were analyzed in P-TIMS, most importantly $Na_2BO_2^+$, but also $K_2BO_2^+$, which show less instrumental fractionation than BF_3 due to the much smaller relative mass difference between the ^{10}B- and ^{11}B-bearing compounds, respectively (McMullen et al. 1961; Finley et al. 1962; Shima 1962, 1963; Agyei and McMullen 1968). Other alkali-borate compounds were also used later on, including $Rb_2BO_2^+$ and $Li_2BO_2^+$ (Gensho and Honda 1971), and $Cs_2BO_2^+$ (Ramakumar et al. 1985; Spivack and Edmond 1986). In particular the cesium borate method produced very precise data, owing to the low instrumental fractionation of masses 309/308. In the early 1980s, TIMS employing the analysis of negative ions (BO_2^-) was developed, which enabled the analyses of much smaller samples compared to P-TIMS (Zeininger and Heumann 1983; Duchateau and De Bièvre 1983).

The first modern studies on minerals, rocks and waters included the investigation of borates and borosilicates, which contain boron as a major component, but also meteorites, igneous and sedimentary rocks and seawater, all of which contain B as a trace element (Finley et al. 1962; Shima 1962, 1963; Agyei and McMullen 1968). The possibility of interlaboratory comparison was given with the establishment of an internationally distributed boron isotope standard, a boric acid distributed by NIST (then the NBS) in 1970, named standard reference material 951 (Catanzaro et al. 1970). This material was established as the primary standard for boron isotopes, and boron isotope ratios of terrestrial materials are since reported in delta notation relative to NIST-SRM951:

$$\delta^{11}B = \left(\frac{^{11}B/^{10}B_{\text{sample}}}{^{11}B/^{10}B_{\text{SRM951}}} - 1 \right) \cdot 1000 \quad (1.1)$$

A number of secondary reference materials have since been established, as listed in this book (see Foster et al. 2017). These materials include a range of silicate glasses, silicate minerals, boro-silicates, borates, waters, rocks, and carbonates and cover a large compositional and structural range (e.g., Kasemann et al. 2001, 2009; Tonarini et al. 2003; Rosner and Meixner 2004; Brand et al. 2014).

1.2.4 Plasma Mass Spectrometry and Interlaboratory Comparison

Inductively-coupled plasma mass spectrometry (ICP-MS) was first utilized for the determination of boron isotopes in the late 1980s using quadropole mass spectrometers for the analysis of waters, borates and borosilicates (Gregoire 1987; Porteous et al. 1995; Al-Ammar et al. 2000). This was followed by the use of magnetic sector-field instruments (Gäbler and Bahr 1999) and eventually the employment of multi-collector ICP-MS, which led to improvements in precision and accuracy to the sub-permil level (Lécuyer et al. 2002; Aggarwal et al. 2003; Foster 2008).

The growing number of laboratories and analytical techniques necessitated the execution of interlaboratory comparison studies. The first two such intercomparison studies involved the majority of labs worldwide that had been working on boron isotopes at the time (Gonfiantini et al. 2003; Aggarwal et al. 2009, with 27 and 28 different labs, respectively). The outcomes of these studies were sobering: in both studies half of the laboratories did not release their analyses; and the results from the labs that did showed interlaboratory discrepancies that were much larger than the uncertainties reported by individual labs. For example, most laboratories reported sub-permil uncertainties, whereas reported values would vary by up to 15‰ between labs for the same reference material with non-systematic off-sets among different labs (Gonfiantini et al. 2003; Aggarwal et al. 2009). This demonstrated that analyses of boron isotope ratios in geologic materials and even in simple boric acid solutions is still challenging, and that reported uncertainties did not reflect accuracies of the measurements—at least for a number of laboratories.

A more encouraging outcome was presented by a more recent intercomparison study on boron isotopes in seawater and carbonates, which included only four different labs (Foster et al.

2013). This study showed agreement of reported $\delta^{11}B$ values for seawater that agreed within the reported uncertainties (2SD ≈ 0.4‰), and only a slightly larger discrepancy for the analysed carbonates (1.5‰).

1.2.5 The Development of in Situ Techniques

It was realised early on that geological samples can show small-scale heterogeneity in boron isotopes that bear a wealth of information on the physical or geochemical evolution of their formation environments. These heterogeneities are inaccessible through bulk methods and require in situ analytical methods instead.

The development of in situ analytical methods followed the development of the bulk methods, which provided boron isotope reference values for solid materials (glasses and minerals) that could be used to correct for instrumental mass fractionation. The first published B isotope analyses by secondary-ion mass spectrometry ("ionprobe") were completed in the late 1970s and early 1980s (Phinney et al. 1979; Christie et al. 1981; Shimizu and Hart 1982), but it took ten years before quantitative B isotope data became available for a larger number of samples, mostly produced by the group of Chaussidon et al. (Chaussidon and Alabrède 1992; Chakraborty et al. 1993; Chaussidon and Jambon 1994; Chaussidon and Robert 1995; Chaussidon et al. 1997) and studies by Smith and Yardley (1996) and Peacock and Hervig (1999). And it was not until the early 2000s that SIMS was applied to boron isotope geochemistry and cosmochemistry in a more widespread manner, as the number of installed facilities grew larger (e.g., Kasemann et al. 2001; Williams et al. 2001; Hoppe et al. 2001; Nakano and Nakamura 2001; Sugiura et al. 2001; Rose et al. 2001; Gurenko and Schmincke 2002; Kobayashi et al. 2004; Altherr et al. 2004). In situ analyses of boron isotopes by laser-ablation ICP-MS has also been completed more recently with a growing number of laboratories applying this method (e.g., le Roux et al. 2004; Tiepolo et al. 2006; Fietzke et al. 2010; Hou et al. 2010). Advantages and limitations of modern, state-of-the-art boron isotope analyses by various mass-spectrometric methods are discussed in this book (Foster et al. 2017) and in earlier reviews (You 2004; Aggarwal and You 2016).

1.2.6 Theoretical and Experimental Boron Isotope Fractionation

Boron isotope fractionation was demonstrated experimentally and calculated theoretically by Kakihana et al. (1977) and applied to pH-dependent B isotope fractionation between carbonates and seawater by Hemming and Hanson (1992). Equilibrium stable isotope fractionation is dominated by differences in the bond strength of the element of interest to its bonding partners. Boron does not form B^{3+} ions, but is almost exclusively bonded to O in nature with a strongly covalent character of the B–O bonds (Hawthorne et al. 1996). It forms $B(OH)_3$, $B(OH)_4^-$ or other complex ions in fluids. Fractionation effects are notably large where the coordination polyhedra of B between two phases differ. Boron in minerals, melts and fluids is either trigonally (3-fold) or tetrahedrally (4-fold) coordinated to oxygen, and therefore, displays large isotopic fractionation between different phases (Kakihana et al. 1977). The higher bond strength of the trigonal coordination leads to a fractionation of ^{11}B into the trigonal sites, and a relative enrichment of the light isotope ^{10}B in the phase with tetrahedrally coordinated B.

In the early 2000s various laboratories started to conduct experiments on boron isotope fractionation at a range of pressures and temperatures that were compared to the theoretical predictions. These experimentally derived fractionation factors were used to calibrate temperature- and pH-dependent fractionation laws for specific mineral–fluid (± melt) systems (Williams et al. 2001; Hervig et al. 2002; Williams and Hervig 2005; Wunder et al. 2005). Experiments on boron sorption onto organic and inorganic surfaces demonstrated the importance of boron speciation in surface complexes (Lemarchand et al. 2007), and carbonate–water B isotope fractionation experiments improved the accuracy of B isotopes as a tool of paleo-pH indicator for seawater from fossil biogenic carbonate (Hemming et al. 1995; Sanyal et al. 2000; Klochko et al. 2006). Theoretical predictions also became more sophisticated with the application of *ab initio* calculations (Rustad and Bylaska 2007; Kowalski et al. 2013). These theoretical studies have accompanied and informed the gradual improvement in the analysis of boron in natural materials.

1.3 The Fifth Element

The last compendium that summarized the geochemistry of boron, with some coverage of its isotope geochemistry, was published more than 20 years ago (Grew and Anovitz 1996), and we have since seen significant progress in analytical techniques, applications and scientific insight in the isotope geochemistry of boron. The present volume, therefore, provides a unique resource for students and professionals alike, including those who will use it as an introduction into a new field and those who use it as a reference in their ongoing research.

This introductory chapter is followed by a review and outlook by Foster et al. (2017, Chap. 2) of the laboratory methods employed to analyze the boron isotopic compositions of geologic materials, which may include waters, glasses, minerals or biogenic carbonates. Major obstacles in the accurate and precise analysis of boron isotopes are evaluated that arise from sample preparation, chemical purification and mass spectrometric methods. The treatment of reference materials and matrix bias are discussed, as well as recent improvements in precision and accuracy. Contamination and unintended isotope fractionation in the laboratory are highlighted as major sources of inaccuracy.

In Chap. 3, Kowalski and Wunder (2017) discuss boron isotope fractionation in vapor–fluid–melt–mineral systems from a theoretical standpoint together with a review of fractionation experiments that have been reported. The current state of atomistic modeling of boron isotope fractionation among fluid and solid phases is discussed and the experimental work conducted at a range of pressure-temperature conditions is summarized and critically evaluated. The authors then explore the feedback between the two approaches and identify areas in the geosciences where further experimental data is sorely needed.

These three general chapters of the volume are followed by a block of four Chaps. (4–7) on the low-temperature geochemistry of boron isotopes in the surface environments of the Earth, such as during weathering and in riverine and marine environments, and by a block of four Chaps. (8–11) on the high-temperature geochemistry and cosmochemistry of boron isotopes.

Chapter 4 provides a detailed discussion of boron incorporation into calcium carbonate on the molecular and crystallographic level. Branson (2017) discusses the adsorption and structural incorporation of boron into inorganic carbonate (calcite and aragonite) and their dependence on seawater chemical composition, temperature, pH and abundances of boron, as well as on growth rates and structure of the carbonate. The relevant experimental work is reviewed and discussed in the context of isotope fractionation among aqueous dissolved boron species and boron adsorbed to $CaCO_3$ growth surfaces and incorporated in the mineral structure. The chapter then proceeds to discuss boron isotope fractionation between seawater and biogenic carbonate and the vital effects that offset various species from the inorganic fractionation line. The chapter lays out the framework on which the following two chapter discuss the specific application of boron isotope signatures in foraminifera and corals, respectively.

Chapter 5 discusses the systematics of boron isotopes in foraminifera and their use as a pH proxy for paleo-seawater. Rae (2017) reviews calibration studies for benthic and planktic species using culturing experiments and the results from natural specimen sampled in the seawater column and in sediment cores. His chapter provides an overview of the applications of boron isotopes in foraminifera in studies of past seawater pH and atmospheric CO_2 levels and discusses current obstacles and future opportunities in these fields.

Chapter 6 reviews the use of boron isotopes in scleractinian corals for the reconstruction of seawater pH. Emphasis is placed on the pH offset between seawater and the calcification fluid generated by the organism itself to stabilize the aragonite skeleton. McCulloch et al. (2017) describe how the investigation of boron isotopes in corals has led to the detailed study of, and new insight into, the process of calcification and pH regulation by different species. The chapter highlights the complexities of pH variations in the calcifying fluid in scleractinian corals and, therefore, of their boron isotopic compositions. Both parameters may depend on more than just temperature, seawater pH and availability of light, but are also influenced by seawater chemistry including the level of dissolved inorganic carbon, as well as vital processes that produce different responses for example between symbiont and asymbiont species.

The use of boron isotopes as a tracer for rock weathering and erosion is reviewed in Chap. 7. Gaillardet and Lemarchand (2017) discuss boron isotope fractionation in subaerial weathering processes including isotopic signatures recorded in soil, vegetation, groundwater and by rivers. Fractionation mechanisms during chemical weathering and the formation of secondary minerals are examined, as well as the influence of precipitation and of the vegetation cycle. The riverine transport is estimated with its effects on the global budget of the ocean, including fluctuations caused by changes in climate and vegetation. The anthropogenic influence, such as land use, on the boron cycle is also explored.

The boron isotopic composition of oceanic basalts and the Earth's mantle are discussed in Chap. 8. Marschall (2017) also reviews the available data on altered oceanic crust, serpentinized abyssal peridotites, and oceanic sediments, as well as hydrothermal vent fluids. The fractionation of boron isotopes during low- and high-temperature alteration of the oceanic crust is discussed based on these data. The chapter also reviews the Earth boron budget and isotopic composition of the crust–mantle system and its major reservoirs, and provides an overview of our current knowledge of the secular evolution of seawater over geological time.

The boron isotope systematics of subduction zones is evaluated in Chap. 9. De Hoog and Savov (2017) review the boron isotope variations observed in fore-arc materials and magmas erupted along active convergent margins and show that boron and its isotope ratio varies systematically with the physical geometry of subduction zones and with a range of regularly employed geochemical tracers. The boron budget of the slab entering subduction zones is summarized, followed by a discussion of boron isotope fractionation processes related to slab dehydration and the role of the serpentinized mantle wedge as an important boron reservoir.

In Chap. 10 Trumbull and Slack (2017) elucidate the geochemistry of boron isotopes in the continental crust with a discussion of prograde metamorphism, of various types of granite and pegmatites, and of hydrothermal systems including ore deposits. They show via a compilation of global granite data that S-type granites are on average isotopically lighter then the mantle, which is in agreement with their interpretation as derived from melting of sedimentary rocks. In contrast, I-type granites, which may be to a large degree derived from fractionated or remelted mantle-derived magmas or magmatic rocks, are on average isotopically heavier than the mantle. This agrees with the isotopically heavy composition of basaltic magmas erupted along modern subduction zones. The authors show that this S-type/I-type dichotomy also holds for the hydrothermal ore deposits that are associated with the respective granitic magmatism.

Finally, Chap. 11 reviews the cosmochemistry of boron isotopes. Liu and Chaussidon (2017) summarize our knowledge of the boron isotopic composition of the solar system and its components, as well as that of other stars of the galaxy and of the interstellar medium as derived from spectroscopy. The possible sources of the two isotopes of boron and their nucleosynthesis are discussed along with the meteoritic record of boron in the early solar system. The isotopic record of live ^{10}Be in the early solar system as recorded in early condensates is reviewed, which points not only to spallation processes during this phase, but also to boron isotopic heterogeneity in the planetary disc. Liu and Chaussidon (2017) close their chapter with a number of major unresolved questions of boron isotope cosmochemistry that concern for example the origin of the boron isotopic composition of the solar system near a $^{11}B/^{10}B$ ratio of 4, which is shared by other stars of the galaxy, and the origin of the small-scale isotopic heterogeneity observed among primitive meteorites and individual components of meteorites.

Acknowledgements We would like to thank series editor Jochen Hoefs for inviting us to contribute this volume on boron isotopes to the Springer series *Advances in Isotope Geochemistry*. We are also grateful to Springer editors Annett Büttner and Chris Bendall for their support at various stages of the project. James Rae is thanked for editorial handling of Chaps. 1 and 2. Preparation of this review chapter was financially supported by a WHOI Independent Study Award from the Andrew W. Mellon Foundation to HRM.

References

Abernathey RM (1960) Isotopic analysis of boron as trimethyl borate, vol 14503. U. S. Atomic Energy Commision, IDO, Idaho Falls, Idaho

Aggarwal SK, You CF (2016) A review on the determination of isotope ratios of boron with mass spectrometry. Mass Spectrom Rev 9999:1–21

Aggarwal JK, Sheppard D, Mezger K, Pernicka E (2003) Precise and accurate determination of boron isotope ratios by multiple collector ICP-MS: origin of boron in the Ngawha geothermal system, New Zealand. Chem Geol 199:331–342

Aggarwal J, Böhm F, Foster G, Halas S, Hönisch B, Jiang SY, Košler J, Liba A, Rodushkin I, Sheehan T, Shen JJS, Tonarini S, Xie Q, You CF, Zhao ZQ, Zuleger E (2009) How well do non-traditional stable isotope results compare between different laboratories: results from interlaboratory comparison of boron isotope measurements. J Analyt Atom Spectrom 24:825–831

Agyei EK, McMullen CC (1968) A study of the isotopic abundance of boron from various sources. Can J Earth Sci 5:921–927

Al-Ammar AS, Reitznerová E, Barnes RM (2000) Improving boron isotope ratio measurement precision with quadropole inductively coupled plasma-mass spectrometry. Spectrochim Acta B 55:1861–1867

Altherr R, Topuz G, Marschall H, Zack T, Ludwig T (2004) Evolution of a tourmaline-bearing lawsonite eclogite from Elekdag area (Central Pontides, N Turkey): evidence for infiltration of slab-derived -rich fluids during exhumation. Contrib Mineral Petrol 148:409–425

Aston FW (1919) A positive ray spectrograph. Phil Mag 38:707–714

Aston FW (1920) The mass-spectra of chemical elements (Part 2.). Phil Mag 40:628–634

Aston FW (1927) Bakerian Lecture. A new mass-spectrograph and the whole number rule. Proc Royal Soc A 115:487–514

Aston FW (1931) The isotopic constitution and atomic weights of selenium, bromine, boron, tungsten, antimony, osmium, ruthenium, tellurium, germanium, rhenium and chlorine. Proc Royal Soc 132: 487–498

Barth S (1998) $^{11}B/^{10}B$ variations of dissolved boron in a freshwater-seawater mixing plume (Elbe Estuary, North Sea). Marine Chem. 62:1–14

Baxter GP, Scott AF (1921) The atomic weight of boron. Science 54:524–525

Beams JW, Haynes FB (1936) The separation of isotopes by centrifuging. Phys Rev 50:491–492

Bentley PG (1960) Isotopic analysis of boron in boron trifluoride by mass spectrometery and measurement of natural boron 10 concentration. J Sci Instrum 37:323–328

Brand WA, Coplen TB, Vogl J, Rosner M, Prohaska T (2014) Assessment of international reference materials for isotope-ratio analysis (IUPAC Technical Report). Pure Appl Chem 86:425–467

Branson O (2017) Boron incorporation into marine $CaCO_3$. In: Marschall HR, Foster GL (eds) Boron isotopes—The fifth element, advances in isotope geochemistry, vol 7. Springer, Heidelberg

Briscoe HVA, Robinson PL (1925) A redetermination of the atomic weight of boron. J Chem Soc Transact 127:696–720

Catanzaro EJ, Champion CE, Garner EL, Marinenko G, Sappenfield KM, Shields WR (1970) Boric acid: isotopic and assay standard reference materials. NBS (US) Spec Publ 260–17:1–71

Chadwick J (1933) Bakerian Lecture. The neutron. Proc Royal Soc A 142:1–25

Chakraborty S, Dingwell DB, Chaussidon M (1993) Chemical diffusivity of boron in melts of haplogranitic composition. Geochim Cosmochim Acta 57:1741–1751

Chaussidon M, Albarède F (1992) Secular boron isotope variations in the continental crust: an ion microprobe study. Earth Planet Sci Lett 108:229–241

Chaussidon M, Jambon A (1994) Boron content and isotopic composition of oceanic basalts: geochemical and cosmochemical implications. Earth Planet Sci Lett 121:277–291

Chaussidon M, Robert F (1995) Nucleosynthesis of ^{11}B-rich boron in the pre-solar cloud recorded in meteoritic chondrules. Nature 374:337–339

Chaussidon M, Robert F, Mangin D, Hanon P, Rose EF (1997) Analytical procedures for the measurement of boron isotope composition by ion microprobe in meteorites and mantle rocks. Geostand Newsl 21:7–17
Christie WH, Eby RE, Warmack RJ, Landau L (1981) Determination of boron and lithium in nuclear materials by secondary ion mass spectrometry. Anal Chem 53:13–17
Conger AD (1953) The effect of boron enrichment on slow neutron-irradiated tissues. Genetics 38:128–133
De Hoog CJ, Savov IP (2017) Subduction zones, dehydration, metasomatism, mud and serpentinite volcanoes, and arc magmatism. In: Marschall HR, Foster GL (eds) Boron isotopes—The fifth element, advances in isotope geochemistry, vol. 7. Springer, Heidelberg
Dempster AJ (1918) A new method of positive ray analysis. Phys Rev 11:316–325
Duchateau NL, De Bièvre P (1983) Boron isotopic measurements by thermal ionization mass spetrometry using the negative BO_2^- ion. Internat. J. Mass Spectrom 54:289–297
Elliott A (1930a) Determination of the abundance ratios of isotopes from band spectra. Nature 126:845–846
Elliott A (1930b) Isotope effect in the spectrum of boron monoxide: intensity measurements and structure of the β-bands. Nature 126:203–204
Fietzke J, Heinemann A, Taubner I, Böhm F, Erez J, Eisenhauer A (2010) Boron isotope ratio determination in carbonates *via* LA-MC-ICP-MS using soda-lime glass standards as reference material. J Analyt Atom Spectrom 25:1953–1957
Finley HO, Eberle AR, Rodden CJ (1962) Isotopic boron composition of certain boron minerals. Geochim Cosmochim Acta 26:911–914
Foster GL (2008) Seawater pH, pCO_2 and $[CO_3^{2-}]$ variations in the Caribbean Sea over the last 130kyr: a boron isotope and B/Ca study of planktic foraminifera. Earth Planet Sci Lett 271:254–266
Foster GL, Hönisch B, Paris G, Dwyer GS, Rae JWB, Elliott T, Gaillardet J, Hemming NG, Louvat P, Vengosh A (2013) Interlaboratory comparison of boron isotope analyses of boric acid, seawater and marine $CaCO_3$ by MC-ICPMS and NTIMS. Chem Geol 358:1–14
Foster GL, Lécuyer C, Marschall HR (2016) Boron stable isotopes. In: White WM (ed) Encyclopedia of geochemistry, encyclopedia earth science series. Springer, Berlin, pp 1–6
Foster GL, Marschall HR, Palmer MR (2017) Boron isotope analysis of geologic materials. In: Marschall HR, Foster GL (eds) Boron isotopes—The fifth element, advances in isotope geochemistry, vol 7. Springer, Heidelberg
Gäbler HE, Bahr A (1999) Boron isotope ratio measurements with a double-focusing magnetic sector ICP mass spectrometer for tracing anthropogenic input into surface and ground water. Chem Geol 156:323–330
Gaillardet J, Lemarchand D (2017) Boron isotopes in riverine systems and the weathering environment. In: Marschall HR, Foster GL (eds) Boron isotopes—The fifth element, Advances in Isotope Geochemistry, vol. 7. Springer, Heidelberg
Gensho R, Honda M (1971) Measurement of the isotopic ratio of boron. J Mass Spectrom Soc Japan 19:134–143. (Japanese with English abstract)
Gonfiantini R, Tonarini S, Gröning M, Adorni-Braccesi A, Al-Ammar AS, Astner M, Bächler S, Barnes RM, Basset RL, Cocherie A, Deyhle A, Dini A, Ferrara G, Gaillardet J, Grimm J, Guerrot C, Krähenbühl U, Layne G, Lemarchand D, Meixner A, Northington DJ, Pennisi M, Reitznerová E, Rodushkin I, Sugiura N, Surberg R, Tonn S, Wiedenbeck M, Wunderli S, Xiao Y, Zack T (2003) Intercomparison of boron isotope and concentration measurements. Part II: evaluation of results. Geostand Newsl 27:41–57
Gounelle M, Chaussidon M, Rollion-Bard C (2013) Variable and extreme irradiation conditions in the early solar system inferred from the initial abundance of ^{10}Be in Isheyevo CAIs. Astrophys J Lett 763:L33
Gregoire DC (1987) Determination of boron isotope ratios in geological materials by inductively coupled plasma mass spectrometry. Anal Chem 59:2479–2484
Grew ES, Anovitz LM (1996) Boron: mineralogy, petrology and geochemistry, reviews in mineralogy, vol 33. Mineralogical Society of America, Washington, D.C., 864p
Gurenko AA, Schmincke HU (2002) Orthopyroxene-bearing tholeiites of the Iblean Plateau (Sicily): constraints on magma origin and evolution from glass inclusions in olivine and orthopyroxene. Chem Geol 183:305–331
Hawthorne FC, Burns PC, Grice JD (1996) The crystal chemistry of boron. In: Grew E, LM A (eds) Boron: mineralogy, petrology and geochemistry, reviews in mineralogy, vol 33. Mineralogical Society of America, Washington, D.C., pp 41–116
Hemming NG, Hanson GN (1992) Boron isotopic composition and concentration in modern marine carbonates. Geochim Cosmochim Acta 56:537–543
Hemming NG, Reeder RJ, Hanson GN (1995) Mineral-fluid partitioning and isotopic fractionation of boron in synthetic calcium carbonate. Geochim Cosmochim Acta 59:371–379
Hervig RL, Moore GM, Williams LB, Peacock SM, Holloway JR, Roggensack K (2002) Isotopic and elemental partitioning of boron between hydrous fluid and silicate melt. Am Mineral 87:769–774
Hönisch B, Hemming NG (2005) Surface ocean pH response to variations in pCO_2 through two full glacial cycles. Earth Planet Sci Lett 236:305–314
Hoppe P, Goswami JN, Krähenbühl U, Marti K (2001) Boron in chondrules. Meteor Planet Sci 36: 1331–1343
Hou KJ, Li YH, Xiao YK, Liu F, Tian YR (2010) In situ boron isotope measurements of natural geological materials by LA-MC-ICP-MS. Chinese Sci Bull 55:3305–3311
Inghram MG (1946) The isotopic constitution of tungsten, silicon, and boron. Phys Rev 70:653–660

Ishikawa T, Nakamura E (1994) Origin of the slab component in arc lavas from across-arc variation of B and Pb isotopes. Nature 370:205–208
Jenkins FA, McKellar A (1932) Mass ratio of the boron isotopes from the spectrum of BO. Phys Rev 42:464–487
Kakihana H, Kotaka M, Satoh S, Nomura M, Okamoto M (1977) Fundamental studies on the ion exchange separation of boron isotopes. Bull Chem Soc Japan 50:158–163
Kasemann S, Meixner A, Rocholl A, Vennemann T, Rosner M, Schmitt AK, Wiedenbeck M (2001) Boron and oxygen isotopic composition of certified reference materials NIST SRM 610/612 and reference materials JB-2 and JR-2. Geostand Newsl 25:405–416
Kasemann SA, Schmidt DN, Bijma J, Foster GL (2009) In situ boron isotope analysis in marine carbonates and its application for foraminifera and palaeo-pH. Chem Geol 260:138–147
Klochko K, Kaufmann AJ, Yao W, Byrne RH, Tossell JA (2006) Experimental measurement of boron isotope fractionation in seawater. Earth Planet Sci Lett 248:276–285
Kobayashi K, Tanaka R, Moriguti T, Shimizu K, Nakamura E (2004) Lithium, boron, and lead isotope systematics of glass inclusions in olivines from Hawaiian lavas: evidence for recycled components in the Hawaiian plume. Chem Geol 212:143–161
Kowalski P, Wunder B (2017) Boron-isotope fractionation among solids-fluids-melts: experiments and atomic modeling. In: Marschall HR, Foster GL (eds) Boron isotopes—The fifth element, advances in isotope geochemistry, vol 7. Springer, Heidelberg
Kowalski PM, Wunder B, Jahn S (2013) Ab initio prediction of equilibrium boron isotope fractionation between minerals and aqueous fluids at high *P* and *T*. Geochim Cosmochim Acta 101:285–301
le Roux PJ, Shirey SB, Benton L, Hauri EH, Mock TD (2004) In situ, multiple-multiplier, laser ablation ICP-MS measurement of boron isotopic composition ($\delta^{11}B$) at the nanogram level. Chem Geol 203:123–138
Lécuyer C, Grandjean P, Reynard B, Albarède F, Telouk P (2002) $^{11}B^{10}B$ analysis of geological materials by ICP-MS Plasma 54: application to the boron fractionation between brachiopod calcite and seawater. Chem Geol 186:45–55
Leeman WP, Sisson VB (1996) Geochemistry of boron and its implications for crustal and mantle processes. In: Grew ES, Anovitz LM (eds) Boron: mineralogy, petrology and geochemistry, reviews in mineralogy, Vol 33. Mineralogical Society of America, Washington, D.C., pp 645–695
Lehmann WJ, Shapiro I (1959) Isotopic composition of boron and its atomic weight. Nature 183:1324
Lemarchand D, Schott J, Gaillardet J (2007) How surface complexes impact boron isotope fractionation: evidence from Fe and Mn oxides sorption experiments. Earth Planet Sci Lett 260:277–296
Liu MC, Chaussidon M (2017) The cosmochemistry of boron isotopes. In: Marschall HR, Foster GL (eds) Boron isotopes—The fifth element, advances in isotope geochemistry, vol 7. Springer, Heidelberg
MacPherson GJ, Huss GR, Davis AM (2003) Extinct ^{10}Be in Type A calcium-aluminum-rich inclusions from CV chondrites. Geochim Cosmochim Acta 67:3165–3179
Marschall HR (2017) Boron isotopes in the ocean floor realm and the mantle. In: Marschall HR, Foster GL (eds) Boron isotopes—The fifth element, advances in isotope geochemistry, vol 7. Springer, Heidelberg
Marschall HR, Wanless VD, Shimizu N, Pogge von Strandmann PAE, Elliott T, Monteleone BD (2017) The boron and lithium isotopic composition of mid-ocean ridge basalts and the mantle. Geochimica et Cosmochimca Acta 207:102–138. doi:10.1016/j.gca.2017.03.028
Martnez-Bot MA, Marino G, Foster GL, Ziveri P, Henehan MJ, Rae JWB, Mortyn PG, Vance D (2015) Boron isotope evidence for oceanic carbon dioxide leakage during the last deglaciation. Nature 518:219–222
McCulloch MT, D'Olivio JP, Falter J, Georgiou L, Holcomb M, Montagna P, Trotter J (2017) Boron isotopes in corals. In: Marschall HR, Foster GL (eds) Boron isotopes—The fifth element, advances in isotope geochemistry, vol 7. Springer, Heidelberg
McKeegan KD, Chaussidon M, Robert F (2000) Incorporation of short-lived ^{10}Be in a calciu-aluminium-rich inclusion from the Allende meteorite. Science 289:1334–1337
McMullen CC, Cragg CB, Thode HG (1961) Absolute ratio of $^{11}B/^{10}B$ in Searles Lake borax. Geochim Cosmochim Acta 23:147–150
Nakano T, Nakamura E (2001) Boron isotope geochemistry of metasedimentary rocks and tourmalines in a subduction zone metamorphic suite. Phys Earth Planet Inter 127:233–252
Ornstein LS, Vreeswijk JA (1933) Messung des Intensitätsverhältnisses der Bor-Isotope B_{10} und B_{11}. Zeitschr Phys 80:57–58
Osberghaus O (1950) Die Isotopenhäufigkeit des Bors. Massenspektrometrische Untersuchung der Elektronenstoßprodukte von BF_3 und BCl_3. Zeitschr Phys 128:366–377
Palmer MR, Swihart GH (1996) Boron isotope geochemistry: an overview. In: Grew ES, Anovitz LM (eds) Boron: mineralogy, petrology and geochemistry, Reviews in Mineralogy, 1st edn, Vol 33. Mineralogical Society of America, Washington, DC, pp 709–740
Parwel A, von Ubisch H, Wickman FE (1956) On the variations in the relative abundance of boron isotopes in nature. Geochim Cosmochim Acta 10:185–190
Peacock SM, Hervig RL (1999) Boron isotopic composition of subduction-zone metamorphic rocks. Chem Geol 160:281–290
Phinney D, Whitehead B, Anderson D (1979) Li, Be, and B in minerals of a refractory-rich Allende inclusion.

In: Proceedings 10th Lunar and Planetary Science Conference, pp 885–905
Porteous NC, Walsh JN, Jarvis KE (1995) Measurement of boron isotope ratios in groundwater studies. Analyst 120:1397–1400
Rae JWB (2017) Boron isotopes in foraminfera. In: Marschall HR, Foster GL (eds) Boron isotopes—the fifth element, advances in isotope geochemistry, vol 7. Springer, Heidelberg
Ramakumar KL, Parab AR, Khodade PS, Almaula AI, Chitambar SA, Jain HJ (1985) Determination of isotopic composition of boron. Radioanal Nucl Chem Lett 94:53–62
Rose E, Shimizu N, Layne G, Grove TL (2001) Melt production beneath Mt. Shasta from boron data in primitive melt inclusions. Science 293:281–283
Rosner M, Meixner A (2004) Boron isotopic composition and concentration of ten geological reference materials. Geostand Geoanal Res 28:431–441
Rustad JR, Bylaska EJ (2007) Ab initio calculation of isotopic fractionation in $BH(OH)_3$(aq) and $BHO-H_4^-$(aq). J Am Chem Soc 129:2222–2223
Sanyal A, Nugent M, Reeder RJ, Bijma J (2000) Seawater pH control on the boron isotopic composition of calcite: evidence from inorganic calcite precipitation experiments. Geochim Cosmochim Acta 64:1551–1555
Scambelluri M, Tonarini S (2012) Boron isotope evidence for shallow fluid transfer across subduction zones by serpentinized mantle. Geology 40:907–910
Sears VF (1992) Neutron scattering lengths and cross sections. Neutr News 3:26–37
Shima M (1962) Boron in Meteorites. J Geophys Res 67:4521–4523
Shima M (1963) Geochemical study of boron isotopes. Geochim Cosmochim Acta 27:911–913
Shimizu N, Hart SR (1982) Isotope fractionation in secondary ion mass spectometry. J Appl Phys 53:1303–1311
Smith MP, Yardley BWD (1996) The boron isotopic composition of tourmaline as a guide to fluid processes in the southwestern England orefield: an ion microprobe study. Geochim Cosmochim Acta 60:1415–1427
Spivack AJ, Edmond JM (1986) Determination of boron isotope ratios by thermal ionization mass spectrometry of the dicesium metaborate cation. Anal Chem 58:31–35
Spivack AJ, Edmond JM (1987) Boron isotope exchange between seawater and the oceanic crust. Geochim Cosmochim Acta 51:1033–1043
Sugiura N, Shuzou Y, Ulyanov A (2001) Beryllium-boron and aluminium-magnesium chronology of calcium-aluminium-rich inclusions in CV chondrites. Meteor Planet Sci 36:1397–1408
Thode HG, Macnamara J, Lossing FP, Collins CB (1948) Natural variations in the isotopic content of boron and its chemical atomic weight. J Am Chem Soc 70:3008–3011
Thomson JJ (1913) Bakterian Lecture: rays of positive electricity. Proc Royal Soc A 89:1–20
Tiepolo M, Bouman C, Vannucci R, Schwieters J (2006) Laser ablation multicollector ICP-MS determination of δ^{11}B in geological samples. Appl Geochem 21:788–801
Tonarini S, Pennisi M, Adorno-Braccesi A, Dini A, Ferrara G, Gonfiantini R, Wiedenbeck M, Gröning M (2003) Intercomparison of boron isotope and concentration measurements. Part I: selection, preparation and homogeneity tests of the intercomparison materials. Geostand Newsl 27:21–39
Trumbull RB, Slack JF (2017) Continental crust, granites, pegmatites, high-*T* crustal metamorphism, and ore deposits. In: Marschall HR, Foster GL (eds) Boron isotopes—The fifth element, advances in isotope geochemistry, vol 7. Springer, Heidelberg
Urey HC, Greiff LJ (1935) Isotopic exchange equilibria. J Am Chem Soc 57:321–327
Williams LB, Hervig RL (2005) Lithium and boron isotopes in illite-smectite: the importance of crystal size. Geochim Cosmochim Acta 69:5705–5716
Williams LB, Hervig RL, Holloway JR, Hutcheon I (2001) B isotope geochemistry during diagenesis. Part I. Experimental determination of fractionation during illitization of smectite. Geochim Cosmochim Acta 65:1769–1782
Wunder B, Meixner A, Romer RL, Wirth R, Heinrich W (2005) The geochemical cycle of boron: constraints from boron isotope partitioning experiments between mica and fluid. Lithos 84:206–216
Yates EL (1938) The separation of isotopes for the investigation of nuclear transmutations. Proc Royal Soc 168:148–158
You CF (2004) Thermal ionization mass spectrometry techniques for boron isotopic analysis: a review. In: de Groot PA (ed) Handbook of Stable Isotope Analytical Techniques, vol 1. Elsevier, Amsterdam, pp 142–152
Zahl PA, Cooper FS (1941) Physical and biological considerations in the use of slow neutrons for cancer therapy. Radiology 37:673–682
Zeininger H, Heumann KG (1983) Boron isotope ratio measurements by negative ionization mass spectrometry. Internat J Mass Spectrom 48:377–380

Boron Isotope Analysis of Geological Materials

2

Gavin L Foster, Horst R Marschall and Martin R Palmer

Abstract

Over the last twenty years applications of the boron isotope system have expanded from the analysis of boron-rich phases (e.g., tourmaline, borates) to include other materials with low B concentrations (e.g., carbonates, basaltic glass). The accurate and precise determination of the boron isotopic composition of geological materials is however a difficult task, particularly for those where boron is present in low-concentration. For solution methods, this difficulty arises principally from the near ubiquitous level of boron contamination in most standard clean laboratories, the light mass of the element, the occurrence of only two stable isotopes, and the large mass difference between them. For in situ approaches, such as secondary-ion mass spectrometry, additional difficulties arise from the restricted availability of well-characterized reference materials, from surface contamination, from limited precision in low-concentration samples, and limitations in reproducibility in high-concentration samples that may partly arise from small-scale heterogeneities in the analyzed materials. Nevertheless, a variety of novel techniques, strategies and methodologies have been developed over the past two decades to meet these challenges. We describe here some of these developments and focus on those that we feel are going to play a major role in the growing use of the boron isotope system in the earth and planetary sciences in decades to come.

Keywords

Analytical methods · Sample digestion · Sample purification

G.L. Foster (✉) · M.R. Palmer
Ocean and Earth Science, National Oceanography Centre Southampton, University of Southampton, Southampton SO14 3ZH, UK
e-mail: Gavin.Foster@noc.soton.ac.uk

H.R. Marschall
Department of Geology & Geophysics, Woods Hole Oceanographic Institution, Woods Hole, MA 02540, USA

H.R. Marschall
Institut für Geowissenschaften, Goethe Universität Frankfurt, Altenhöferallee 1, 60438 Frankfurt am Main, Germany

H. Marschall and G. Foster (eds.), *Boron Isotopes*, Advances in Isotope Geochemistry,
https://doi.org/10.1007/978-3-319-64666-4_2

2.1 Introduction

Boron has two naturally occurring stable isotopes: ^{11}B and ^{10}B that occur approximately in a 4:1 ratio in terrestrial materials. The large relative mass difference between the two isotopes ($\sim 10\%$), and the behavior of boron in aqueous solutions, results in variations in the isotopic composition of boron in Earth and extra-terrestrial materials providing numerous and unique insights into low- and high-temperature geochemical processes. As with other stable isotope systems, boron isotope variations are expressed in delta notation with reference to a standard: in this case, a boric acid (NIST SRM 951; Catanzaro et al. 1970; Brand et al. 2014). Despite its utility, the applicability of the boron isotope system is often restricted by the analytical challenges involved in its precise and accurate measurement. Some of the difficulties are common to other dual isotope systems (e.g., Li and Cu isotopes), whilst others, such as the volatility of boron in acidic solutions and the near ubiquitous level of blank contamination in most clean laboratories, are unique to the boron isotope system. Despite these complications boron isotope measurement of geological materials has a long history (e.g., McMullen et al. 1961; Marschall and Foster 2017). In recognition of the important information that can be gained from boron isotope studies, analytical techniques have evolved considerably, and there are now many dissolution, purification and mass spectrometric methods available, each with its own advantages and disadvantages. Several publications reviewed the methods in use by the late 1990s (e.g., Aggarwal and Palmer 1995; Swihart 1996), so here we will focus only on those methods that are most commonly used today (also see the recent review of analytical techniques by Aggarwal and You 2016). In the following sections, the required purification methods and the analytical methodology are discussed briefly in turn, and the more commonly used analytical techniques are summarized in Table 2.1. The $\delta^{11}B$ values of commonly used standards and reference materials for boron isotope research are shown in Table 2.2. Note that many of the standards listed in Table 2.2 are not commercially available; we therefore recommend that the interested reader contact the lead author of the appropriate study to enquire about availability.

2.2 Digestion and Purification Methods for Boron Isotope Analysis of Geologic Materials

For the most accurate bulk sample isotopic analysis the sample is first dissolved and then boron is generally purified from the sample matrix to remove isobaric interferences and enhance ionization. This requirement applies to all analyses by Inductively Coupled Plasma Mass Spectrometry (ICPMS) and for most, but not all, Thermal Ionization Mass Spectrometry (TIMS) techniques.

2.2.1 Sample Digestion

A number of procedures are available for digestion of a geological sample. These include: pyro-hydrolysis, alkali fusion and acid digestion. Pyro-hydrolysis is time consuming and is perhaps most suitable for boron-rich materials such as tourmalines (Spivack and Edmond 1986). Acid digestion is currently popular due to the small volumes of reagents needed and hence lower levels of contamination. For carbonates cold HNO_3 (or HCl; e.g., Foster et al. 2013) is sufficient to completely dissolve a sample, whereas silicates are most commonly dissolved for geochemical analysis by heating the sample together with high concentrations of HF (to volatilise Si in the sample) and HNO_3 (to retain the dissolved cations in solution), with additional acids (e.g., $HClO_4$) being employed for particularly refractory/organic-rich samples. This creates problems for boron analyses of silicates because BF_3 is highly volatile and can easily be lost from solution. Hence, studies employing the acid dissolution technique typically add mannitol to the dissolution cocktail to form a boron-mannitol complex that is stable in acid solution (Nakamura et al. 1992). Alternatively, by using the alkali fusion method the

Table 2.1 Summary of boron isotope measurement techniques

Analytical technique	Acronym	Chemical purification commonly performed	Sample types	Sample size (ng)	Precision (‰)	Example references
Positive ion thermal ionisation mass spectrometry	PTIMS	Cation and anion (typically Amberlite IRA743) exchange resin	Carbonates, silicates, waters	20–1000	0.1–0.3	He et al. (2013), Trotter et al. (2011)
Negative ion thermal ionisation mass spectrometry	NTIMS	None or cation and anion (typically Amberlite IRA743) exchange resin	Carbonates, waters	1–10	0.3–1.0	Hemming and Hönisch (2007), Kasemann et al. (2009), Foster et al. (2013), Clarkson et al. (2015)
Total evaporation negative ion thermal ionisation mass spectrometry	TE-NTIMS	None	Carbonates, waters	0.3–1	1–2	Foster et al. (2006), Ni et al. (2010), Liu et al. (2013)
High-resolution inductively coupled plasma mass spectrometry	HR-ICPMS	Microsublimation or cation and/or anion (typically Amberlite IRA743) exchange resin	Carbonates, waters	3–5	0.5–0.7	Misra et al. (2014)
Multi-collector inductively coupled plasma mass spectrometry	MC-ICPMS	Microsublimation or cation and/or anion (typically Amberlite IRA743) exchange resin	Carbonates, silicates, waters	5–50	0.2–0.3	Foster et al. (2013), Louvat et al. (2011), Foster (2008)
Laser ablation multi-collector inductively coupled plasma mass spectrometry	LA-MC-ICPMS	None required-in situ technique	Carbonate, silicates	0.1–0.3	0.5–1.75	Fitzke et al. (2010), Thil et al. (2016)
Secondary ion mass spectrometry	SIMS	None required-in situ technique	Carbonate, silicates	0.001–0.00001	0.5–3	Chaussidon et al. (1997), Marschall and Monteleone (2015), Kasemann et al. (2009), Liu et al. (2010)

aluminosilicate matrix can be broken down by high temperature fusion of the sample with a fluxing agent, typically K_2CO_3, followed by dissolution of the fusion cake in pure water (Tonarini et al. 1997).

For all digestion methods any further chemical treatments prior to purification (e.g., the conversion of the dissolved silicate sample to chloride) must be done with care to avoid any isotopic fractionation arising from the volatility

Table 2.2 Summary of boron isotope reference materials and commonly used standards

Description	Material	$\delta^{11}B_{951}$	Uncertainty	Suitable for:					References
				PTIMS	NTIMS	ICPMS	SIMS	LA-ICPMS	
NIST SRM 951 (951a)	Boric acid	0		x	x	x			Brand et al. (2014) and references therein
IRMM-011	Boric acid	0.16/ −0.375		x	x	x			Brand et al. (2014) and references therein
ERM-AE101	Boric acid (aqueous)	−122.96		x	x	x			Brand et al. (2014) and references therein
		−123.01	0.41						Brand et al. (2014) and references therein
ERM-AE120	Boric acid (aqueous)	−20.2	0.6	x	x	x			Vogl and Rosner (2011)
ERM-AE121	Boric acid (aqueous)	19.9	0.6	x	x	x			Vogl and Rosner (2011)
ERM-AE122	Boric acid (aqueous)	39.7	0.6	x	x	x			Vogl and Rosner (2011)
NRC NASS-5	Seawater	39.89		x	x	x			Brand et al. (2014) and references therein
OSIL IAPSO	Seawater	39.64	0.42	x	x	x			Brand et al. (2014) and references therein
Natural seawater	Seawater	39.61	0.2	x	x	x			From Foster et al. (2010)
IAEA-B-1	Seawater	38.6	1.66	x	x	x			Brand et al. (2014) and references therein
		38.76	0.79						Brand et al. (2014) and references therein
IAEA-B-2	Groundwater	13.8	0.79	x	x	x			Brand et al. (2014) and references therein
		14.38	1.2						Brand et al. (2014) and references therein
IAEA-B-3	Groundwater	−21.4	0.89	x	x	x			Brand et al. (2014) and references therein
		−20.82	0.86						Brand et al. (2014) and references therein

(continued)

Table 2.2 (continued)

Description	Material	$\delta^{11}B_{951}$	Uncertainty	Suitable for:					References
				PTIMS	NTIMS	ICPMS	SIMS	LA-ICPMS	
IAEA-B-4	Tourmaline (schorl)	−8.7	0.18	x	x	x	x	x	Brand et al. (2014) and references therein
IAEA-B-5	Basalt	−3.8	2	x	x	x			Brand et al. (2014) and references therein
IAEA-B-6	Obsidian	−1.8	1.5	x	x	x	x	x	Brand et al. (2014) and references therein
IAEA-B-7	Limestone	9.7	5.9	x	x	x			Brand et al. (2014) and references therein
IAEA-B-8	Clay	−5.1	0.87	x	x	x			Brand et al. (2014) and references therein
H112566	Tourmaline (schorl)	−12.5	1.5	x	x	x	x	x	Leeman and Tonarini (2001)
H108796	Tourmaline (dravite)	−6.6	0.6	x	x	x	x	x	Leeman and Tonarini (2001)
H98144	Tourmaline (elbaite)	−10.4	0.3	x	x	x	x	x	Leeman and Tonarini (2001)
GOR-128-G	Komatiitic glass	13.55	0.21	x	x	x	x	x	Rosner and Meixner (2004)
GOR-132-G	Komatiitic glass	7.11	0.97	x	x	x	x	x	Rosner and Meixner (2004)
StHs6/80-G	Andesitic glass	−4.48	0.29	x	x	x	x	x	Rosner and Meixner (2004)
TB	Shale rock powder	−12.6	1.3	x	x	x			Rosner and Meixner (2004)
Am-21805	Amphibole	−0.4	0.9	x	x	x	x	x	Pabst et al. (2012)
Phe-80-3	Phengite (white mica)	−13.5	0.7	x	x	x	x	x	Pabst et al. (2012)
NIST SRM 610	Silicate glass	−0.52	0.53	x	x	x	x	x	Brand et al. (2014) and references therein
NIST SRM 612	Silicate glass	−0.51	0.52	x	x	x	x	x	Brand et al. (2014) and references therein
GSJ JB-2	Basalt rock powder	7.24	0.33	x	x	x	x	x	Brand et al. (2014) and references therein
GSJ JB-3	Basalt rock powder	5.9	0.8	x	x	x			Rosner and Meixner (2004)
GSJ JGb-1	Gabbro rock powder	4.0	2.8	x	x	x			Rosner and Meixner (2004)

(continued)

Table 2.2 (continued)

Description	Material	$\delta^{11}B_{951}$	Uncertainty	Suitable for:					References
				PTIMS	NTIMS	ICPMS	SIMS	LA-ICPMS	
GSJ JA-1	Andesite rock powder	5.3	0.3	x	x	x			Rosner and Meixner (2004)
GSJ JR-2	Rhyolite rock powder	2.9	0.2	x	x	x			Rosner and Meixner (2004)
GSJ JR-2e	Rhyolite	2.57	1.06	x	x	x	x	x	Brand et al. (2014) and references therein
GSJ JCp-1	Coral powder	24.24		x	x	x			Brand et al. (2014) and references therein
NIST RM 1547	Peach leaves	41.09	1.12	x	x	x			Brand et al. (2014) and references therein
NIST RM 8433	Corn bran	8.3	1.69	x	x	x			Brand et al. (2014) and references therein
BCR-679	White cabbage	−23.8	1.15	x	x	x			Brand et al. (2014) and references therein
BIG-D	Boric acid (aqueous)	14.56	0.61	x	x	x			Foster et al. (2013)
BIG-E	Boric acid (aqueous)	24.96	0.31	x	x	x			Foster et al. (2013)
UMD	Boric acid (aqueous)	−12.87	0.29	x	x	x			Foster et al. (2013)
IC_A	Deep ocean carbonate	15.96	1.50	x	x	x			Foster et al. (2013)
IC_B/UWC-1	Inorganic calcite rhomb	8.99	1.64	x	x	x	x	x	Foster et al. (2013)
		8.08	1.16						Kasemann et al. (2009)
IC_C/PS69/318-1	Cold water octocoral	15.90	0.65	x	x	x	x	x	Foster et al. (2013)
		16.07	0.67						Kasemann et al. (2009)
IC_D/ M93-TB-FC-1	Aragonitic *Porites* (sp) coral	25.23	0.67	x	x	x	x	x	Foster et al. (2013)
		24.8							Kasemann et al. (2009)
IC_1	*Porites cylindrica* coral	24.86	0.89	x	x	x			Foster et al. (2013)
IC_2	*Acropora nobilis* coral	24.03	1.81	x	x	x			Foster et al. (2013)
IC_3	*Porites cylindrica* coral	23.75	1.35	x	x	x			Foster et al. (2013)

(continued)

Table 2.2 (continued)

Description	Material	$\delta^{11}B_{951}$	Uncertainty	Suitable for:					References
				PTIMS	NTIMS	ICPMS	SIMS	LA-ICPMS	
IC_4	*Acropora nobilis* coral	22.72	0.92	x	x	x			Foster et al. (2013)
IC_5	*Acropora nobilis* coral	22.04	1.37	x	x	x			Foster et al. (2013)
IC_6	Planktic foraminifers (*Globigerinoides sacculifer*)	20.16	2.27	x	x	x			Foster et al. (2013)
IC_7	Planktic foraminifers (*Globigerinoides sacculifer*)	20.87	1.73	x	x	x			Foster et al. (2013)
IC_8	Planktic foraminifers (*Globigerinoides sacculifer*)	19.99	2.24	x	x	x			Foster et al. (2013)
IC_9	Planktic foraminifers (*Globigerinoides sacculifer*)	20.78	2.15	x	x	x			Foster et al. (2013)
IC_{10}	Brachiopod (*Terebratalia transversa*)	16.95	2.86	x	x	x			Foster et al. (2013)
IC_{11}	Brachiopod (*Terebratalia transversa*)	15.38	1.60	x	x	x			Foster et al. (2013)
KW-T6	*Tridacna gigas* bivalve	15.53	0.64	x	x	x	x	x	Kasemann et al. (2001)

Uncertainty is typically at a 95% confidence level but the reader should refer to the cited reference. For Vogl and Rosner (2011) quoted values are the mean of gravimetric determination and two measurements, uncertaity is 2sd of these. For the data from Foster et al. (2010) uncertainty is mean and 2sd of 28 seawater samples. For standards from Foster et al. (2013) the quoted value is the mean of all laboratories and the uncertainty is the 2sd about this mean. For the data from Kasemann et al. (2009) the quoted values are the mean and uncertainty from their Table 2.2

of boron. The alkali fusion method was developed with this limitation in mind, especially with regard to silicates where the formation of a K_2CO_3 fusion cake avoids the need for HF for dissolution (e.g., Tonarini et al. 1997). Regardless of the sample type, all reagents must have very low boron concentrations ($\ll$1 ng/g) to keep the procedural blank contamination sufficiently low. To this end, most clean laboratories carrying out boron isotope analysis are fitted with boron-free HEPA filters to reduce the blank contamination during sample handling (e.g., Rosner et al. 2005).

Once a sample is dissolved, there are two principal methods currently in use to purify boron prior to analysis. These are discussed in turn below.

2.2.2 Ion Exchange

For carbonate samples, boron purification from the relatively simple $CaCO_3$ matrix typically involves the use of the boron specific ion-exchange resin Amberlite 743 (Kiss 1988) either in a one-step (Foster 2008) or two step procedure (Lemarchand et al. 2002; McCulloch et al. 2014). The Amberlite 743 resin contains a hydrophobic styrene backbone and tertiary amine group that only absorbs boron in the borate form ($B(OH)_4^-$; Yoshimura et al. 1998). In the one-step method of Foster (2008), the dissolved $CaCO_3$ sample is added to a 2 M Na acetate–0.5 M acetic acid buffer to ensure that the loading pH is ~5. This pH is sufficient to ensure quantitative retention of boron on the column and low enough to prevent Ca precipitation prior to loading. In the two-step process (Foster et al. 2013; Lemarchand et al. 2002; McCulloch et al. 2014), dissolved cations (e.g., Ca, Mg) are first removed using a cationic exchange resin (e.g., Dowex AG50X8; Lemarchand et al. 2002; McCulloch et al. 2014). The pH of the solution from this first step is then increased to ~9 using NH_4OH without fear of forming precipitates and is then further purified using Amberlite 743 (Lemarchand et al. 2002; McCulloch et al. 2014). Following rinses with MQ water, and in some methods with $NaNO_3$ as well to remove the matrix (Foster et al. 2013; Lemarchand et al. 2002), quantitative elution of boron from Amberlite 743 is typically achieved using 0.1–0.5 M HNO_3. Experiments by Lemarchand et al. (2002) illustrate that relatively large isotopic fractionations can result if recovery is not close to 100%, although this is relatively easy to achieve and monitor (e.g., Foster et al. 2013; Lemarchand et al. 2002). The volume of HNO_3 required to elute all the boron depends on its molarity, resin mesh size, column volume and column geometry. Due to the volatility of boron in acidic solutions further concentration of the eluant via a dry-down step is generally avoided. Hence, column design has to be optimized to recover boron in sufficiently low quantities of HNO_3 such that final boron concentration is not diluted beyond that required for precise measurement by mass spectrometry. Hence, for measurement of $CaCO_3$ samples that generally have low boron contents (10–50 μg/g B; Hemming and Hansen, 1992), the columns are typically small volume (<50 μl) and reagent volumes are also small (<1 ml HNO_3). Thus blank contamination from purification is low: typically <1 ng and in some instances <50 pg (Foster et al. 2013; Henehan et al. 2013).

Purification of boron contained within aluminosilicate geological samples is more complex. This is partly because rock samples contain a more diverse suite of cations that must be removed from the sample, but the complexities of purification largely arise because of the more aggressive procedures that are required to release boron from the rock matrix (see above). Regardless of whether the rock is dissolved or fused, the boron must then be purified from a concentrated solution containing high concentrations of a wide range of dissolved species. Hence, the purification method has typically involved at least three separate ion-exchange columns (Nakamura et al. 1992; Tonarini et al. 1997). This is time-consuming and problematic in terms of controlling potential contamination of the sample from the large volumes of reagents required and prolonged sample processing. More recently, Romer and Meixner (2014) have developed an improved two-step method and Wei et al. (2013) developed a single column procedure using Bio-Rad AG MP-1 (an anion exchange resin) for separation of rock samples that had undergone acid dissolution. This latter technique greatly simplifies the purification stage but, in its current incarnation, requires large volumes of 24 M HF ($\sim$24 ml per sample) in nine separate aliquots during the column chemistry. Recent tests by one of us (Palmer and Cooper, in prep) however suggest the approach is well-suited to miniaturisation allowing for a reduction in blank and HF volume.

2.2.3 Purification by Non Exchange Resin Methodologies

A number of methods exist to purify boron that do not rely on anion/cation exchange resins. The earliest of these, methyl borate distillation, exploits the volatility of trimethyl borate to separate boron from more refractory species such as Si (Aggarwal and Palmer 1995; Spivack and Edmond 1986). More recently, a method initially used as a final purification step to remove organic contamination (Gaillardet et al. 2001; Lemarchand et al. 2002) known as "microsublimation" is being used to separate boron from $CaCO_3$ and other matrices (Misra et al. 2014; Pi et al. 2014; Wang et al. 2010). The microsublimation approach exploits the volatility of boron and its ability to sublime from a solid at low temperatures ($\sim$75–100 °C). In this separation method $CaCO_3$ samples are dissolved in a small volume (<50 μl) of HNO_3 or HCl with pH <2 and loaded onto the center of the cap of a 5 ml Savillex Teflon fin legged screw top beaker (Misra et al. 2014; Wang et al. 2010). The inverted beaker with sample loaded on the lid is then placed on a hotplate at a relatively well-controlled temperature ideally with wells or digestion blocks to ensure an even heating of the vial (Liu et al. 2013; Wang et al. 2010). The distillation process is then carried out for >12 h and the purified boron is simply recovered from the drop that forms in the conical end of the inverted beaker. The potential for fractionation if recovery is not quantitative, as with the column procedures above, is large (Gaillardet et al. 2001; Misra et al. 2014; Wang et al. 2010) and incomplete recovery can occur through loss of sample through the screw-top seal of the distillation vial or because complexes can form in the residue that bind to boron (Gaillardet et al. 2001). The extent of recovery when applied to $CaCO_3$ purification is checked by analyzing the B/Ca ratio of what was loaded on the lid and the B concentration of the final distillate and residue (Misra et al. 2014), although the accuracy of these approaches limits these assessments of B concentration to ±1–12% and significant fractionation occurs if B recovery is <90% (Misra et al. 2014). Due to the limited handling and small volume of reagents used the total procedural blanks, provided B-clean reagents are used, tend to be smaller for this method than for column-based approaches (e.g., <10 pg).

2.3 Analytical Methods for Boron Isotope Analysis of Geological Materials

Boron isotopes are commonly measured using a variety of analytical techniques. Figure 2.1 provides an up-to-date summary of the state of the art with regard to the precision and sample size required for each of the techniques and Table 2.2 lists some of the commonly used reference materials. In the following we attempt to briefly summarize the relevant methodologies, paying particular attention to their strengths and weaknesses.

2.3.1 The Isotopic Analysis of Boron in Solution Mode

2.3.1.1 Positive Ion Thermal Ionization Mass Spectrometry

The first boron isotope measurement technique suitable for high precision analysis of geological materials was Positive ion Thermal Ionization Mass Spectrometry (PTIMS; e.g., Ramakumar et al. 1985). Initial attempts using a sodium-borate complex only achieved a precision of around ±2‰ (at 95% confidence; e.g., Swihart et al. 1986) due to the relatively light mass of the $Na_2BO_2^+$ ion (mass 88 and 89 for ^{10}B and ^{11}B, respectively). Better precision was achieved with using Cs-borate complexes ($Cs_2BO_2^+$; masses 308 and 309) due its heavier mass minimizing instrumental fractionation (±0.4‰ at 95% confidence; Ramakumar et al. 1985). However the relatively low ion yield from these heavy alkali-borate complexes, the requirement for dynamic runs due to the close spacing of masses 308 and 309, and the potential high level of B-contamination during purification of the boron prior to mass spectrometric analysis, typically resulted in sample size being relatively large for these early PTIMS methods (5000 ng; Spivack and Edmond 1986). Since the initial development of the Cs-borate TIMS method there has been a series of improvements to sample size and precision. For example, improved signal intensity and reproducibility was achieved by coating the filament with graphite (Xiao et al. 1988) and changes to the sample dissolution method and boron purification process (Nakamura et al. 1992; Tonarini et al. 1997). In addition, the introduction of a specially designed double collector package allowed for static multi-collection of both boron isotopes (Nakano and Nakamura 1998) that further improved precision and reduced the required sample size, such that it is now possible to achieve accuracy and precision of around ±0.3‰ (at 95% confidence) on sample sizes as small as 20 ng of boron (Deyhle 2001; He et al. 2013). These latest methods therefore offer the potential for achieving precise data on small samples using PTIMS.

2.3.1.2 Negative Ion Thermal Ionization Mass Spectrometry

Negative ion thermal ionization mass spectrometry (NTIMS) approaches to the measurement of

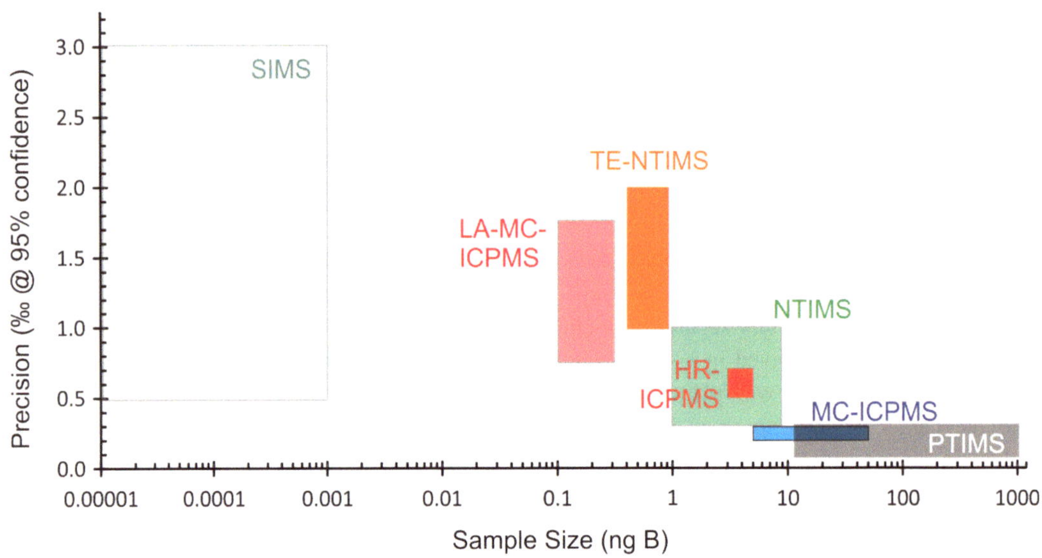

Fig. 2.1 Summary of the sample size and reported precision (at 95% confidence) for the commonly used techniques for boron isotope analysis from a selection of contributions. *Sources* Secondary ion mass spectrometry (SIMS; Marschall and Monteleone 2015); laser ablation multi-collector inductively-coupled plasma mass spectrometry (LA-MC-ICPMS; Fietzke et al. 2010); total evaporation negative ion thermal ionization mass spectrometry (TE-NTIMS; Foster et al. 2006; Ni et al. 2010); high-resolution ICPMS (Misra et al. 2014); negative ion thermal ionization mass spectrometry (NTIMS; Hönisch and Hemming 2007; Kasemann et al. 2009); multi-collector ICPMS (MC-ICPMS; Foster et al. 2013); positive ion thermal ionization mass spectrometry (PTIMS; He et al. 2013; Trotter et al. 2011). Note that across the techniques as sample size increases precision tends to improve

boron were first developed by Zeininger and Heumann (1983). This approach, where $^{10}B^{16}O_2^-$ and $^{11}B^{16}O_2^-$ are measured statically at masses 42 and 43, has a very high ion yield when an activator (e.g., lanthanum nitrate or boron-free seawater) is used, meaning samples as small as 1 ng can be measured to relatively high precision (±0.4–0.7‰ at 95% confidence; Hemming and Hanson 1992; Kasemann et al. 2001; Clarkson et al. 2015; Farmer et al. 2016). Hemming and Hanson (1994) noted that a complex ionic electrolyte like acidified seawater or dissolved calcium carbonate also resulted in very high ion yields with NTIMS. Hemming and Hanson (1992) therefore proposed that for carbonates and natural waters no boron purification was necessary provided the same boron-free seawater was added to samples (dissolved carbonate or natural water) and standards (boric acid NIST SRM 951) to ensure some similarity in matrix. The principal difficulties of this approach are: (i) the accurate correction of machine induced mass fractionation of the two relatively light isotopologues of BO_2^- at masses 42 and 43; (ii) interference of minor borate isotopologues (e.g., $^{10}B^{16}O^{17}O^-$ on mass 43, and CNO^- from organic contamination on mass 42). Whilst the interference of CNO^- can be adequately monitored using CN^- and the other borate isotopologues are typically minor, it is the accurate correction of machine-induced bias that has proven most difficult to overcome. Several approaches exist that all require a high level of operator skill and care, from using a linear regression to correct for the time-dependent fractionation relating to ionization (e.g., Kasemann et al. 2001), to total evaporation (TE-NTIMS) whereby the entire sample is ionized and the $^{11}B/^{10}B$ ratio is determined from the total ^{11}B measured/total ^{10}B measured (Foster et al. 2006; Liu et al. 2013). The degree of instrumental mass fractionation and the precision of NTIMS analyses are improved if care is taken to control the filament temperature during warm-up of the filament (up to ~970 °C) and during the data acquisition (approximately 950–1000 °C) (Hemming and Hanson 1994; Hemming and Hönisch 2007). For some, but not all (e.g., Foster et al. 2013; Kasemann et al. 2009), NTIMS methodologies, the accuracy of $CaCO_3$ analysis remains an issue and biases of 0.5 to 2.7‰ are evident and appear to be related to sample matrix in some way (e.g., Foster et al. 2013; Farmer et al. 2016).

2.3.1.3 Inductively Coupled Plasma Mass Spectrometry

The measurement of boron isotopes by ICPMS also has a relatively long history, with the first study by Gregoire (1987). These early attempts however were associated with high uncertainty (±3‰ at 95% confidence; Gregoire 1987) making the methodology unsuitable for most geochemical applications. The last 20 years or so have seen a dramatic growth in the use of multi-collector inductively coupled plasma mass spectrometry (MC-ICPMS; e.g., Walder and Freedman 1992; Halliday et al. 1998). This new technology was rapidly adopted for boron isotope measurement with early applications meeting some success (e.g., Aggarwal et al. 2003; Lecuyer et al. 2002). However the challenges of boron isotope analysis by MC-ICPMS (and ICPMS) are also manifold: (i) due to the volatility of boron in acidic solutions wash out between samples can be very long (Al-Amar et al. 2000); (ii) $^{40}Ar^{4+}$ and/or $^{20}Ne^{2+}$ from the ionization gas interfere on ^{10}B (Wang et al. 2010); (iii) mass fractionation is large although typically relatively stable (up to 16%; Foster 2008); (iv) due to its light mass, boron has very low transmission across the plasma interface leading to poor sensitivity; (v) matrix effects mean solutions must be purified first in order to obtain accurate and precise data (e.g., Guerrot et al. 2011; Pi et al. 2014). Improved analytical and abundance sensitivity, and mass fractionation stability with the most recent generations of MC-ICPMS has overcome many of these problems (Foster 2008). Purification is typically achieved using one of the options discussed in Sect. 2.2 and several strategies are also available to overcome the wash-out problem, including: (a) the addition of small quantities of NH_3 gas to ensure spray chamber pH is high enough that boron is present in the non-volatile borate form (Foster 2008); (b) protracted wash-outs and the use of dilute solutions (Wang et al. 2010; McCulloch et al. 2014); (c) the use of a mixed 0.3 M HF/0.4 M HNO_3 reagent for the introduction of samples, standards, blanks and wash (Misra et al. 2014); (d) the use of direct injection where the volume of wetted parts is small and hence the boron wash-out is relatively quick (Louvat et al. 2011). Regardless of which strategy is used to overcome the wash out issue, the most precise $\delta^{11}B$ data are obtained when sample-standard bracketing is as fast as possible (i.e. <5 min between sample and standard), wash out is stable and machine-blank is <2% of analyte signal, and the MC-ICPMS is optimized for mass fractionation stability rather than maximum signal (e.g., Foster 2008; Guerrot et al. 2011; Louvat et al. 2011; Wang et al. 2010). Until recently, analytical sensitivity placed a lower limit on sample size ($\sim$10 ng of B), however improved electronics (10^{12} Ω resistors for faraday cup amplifiers; e.g., John and Adkins 2010) now allow a further reduction in sample sizes (<5 ng of B) making laboratory blank contamination, rather than total available boron, the limiting factor.

2.3.2 The Isotopic Analysis of Boron in Situ

Many geological materials show small-scale chemical and isotopic heterogeneities that contain a wealth of relevant information but are inaccessible by bulk methods, as these only reveal average compositions of the dissolved sample mass. This is where in situ methods of boron isotope analyses become indispensable, and the two most common techniques are discussed here.

2.3.2.1 Secondary Ion Mass Spectrometry (SIMS)

Secondary-ion mass spectrometry (SIMS) employs a beam of focused ions to sputter and ionize small amounts of material from the surface of a solid sample. The secondary ions created in this process are accelerated by a homogenous high-voltage electric field into a mass spectrometer. The mass spectrometer may employ a time-of-flight mass separation system (e.g., Lyon et al. 2007) or a quadrupole filter, but more typically for boron isotope analysis consists of an electrostatic filter and a sector magnet (e.g., Straub and Layne 2002). More recently, the NanoSIMS, which enables higher lateral resolution than standard ion probes, has been employed to analyze B isotopes in meteorites (Liu et al. 2010).

SIMS is a method that requires introduction of the sample in solid form into the mass spectrometer. Typically, these include silicate minerals, glass, and biogenic carbonates that may contain trace concentrations of boron or are boro-silcates or borates that contain boron as a major component. Ideally, the samples should be very well polished to provide a flat, smooth surface. This is important, because the sample surface is part of the extraction lens for the secondary ion beam, and surface topography leads to distortion of the electric field that may result in irreproducible boron isotope fractionation. SIMS has also been applied to analyze boron isotopes of water samples and clay suspensions. These liquid materials were dried on Si wafers or glass slides, and the dry residues were analyzed by ion probe (Rose-Koga et al. 2006; Williams et al. 2001).

The prominent advantage of SIMS compared to any alternative analytical method lies in the very high spatial resolution of the method. This follows from the gentle sputtering process with slow depth progress (compared with laser-ablation methods) and from the ability to focus the primary ion beam to spot sizes of 5–100 μm, depending on required signal intensity and B concentration in the sample. In NanoSIMS, spot diameters below 1 μm are possible. The sputter rate depends on the material and the primary beam density and is typically on the order of 0.1–5 nm/s for beam currents of 1–40 nA in carbonates, silicates and silicate glasses (e.g., Marschall and Monteleone 2015; Rollion-Bard et al. 2007). This translates to a pit depth of 0.1–5 μm for typical B isotope analyses that take 10–30 min total analysis time for a single spot. These small volumes of sputtered material make SIMS the ideal method for resolving spatial heterogeneity, for instance in foraminifera and coral skeletons (e.g., Rollion-Bard et al. 2003; Blamart et al. 2007) and for material with limited volume, such as melt inclusions (e.g., Schmitt et al. 2002). The small analyzed volume also translates to a very small mass boron required for a SIMS isotope analysis, which is several orders of magnitude lower than for any other method, and may range from approximately 10 femto grams to several pico grams (Fig. 2.1).

For high-concentration samples (≥ 20 µg/g B, depending on the instrument), the analytical precision of SIMS depends on the stability of the mass spectrometer, most importantly the stability of the magnet, the detection system (electron multiplier = ion counter; or Faraday cups), and the primary beam. Most published B isotope analyses from magnetic-sector field SIMS have been completed using a single collector and magnet cycling between the collection of $^{10}B^+$ and $^{11}B^+$. Two electron multipliers in multi-collection mode have also been used, but precision was not significantly better compared to mono-collection (Kobayashi et al. 2004). More recently, the B isotope ratio of tourmaline (ca. 30,000 µg/g B) has been analyzed in multi-collection mode using two Faraday cups, which reduced the total analysis time to 2 min per spot at a precision of ±0.8‰ (2SE; Büttner et al. 2016). A $^{16}O^-$ primary beam was used in all cases, including small- and large-radius spectrometers and NanoSIMS. Fluctuations of the primary beam intensity that are on the same time scale as the integration cycles of the measurement will directly affect internal precision in single-collection mode and cannot be corrected. Magnet stability has been improved over the years through better electronic hardware, laminated magnets, software improvements, and the introduction of a nuclear magnetic resonance magnet control. The combined instrument stability translates to an internal precision of typically between 0.4 and 1.0‰ (2SE) for a $^{11}B/^{10}B$ analysis of a high-concentration sample (Büttner and Kasemann 2007; Drivenes et al. 2015; Marschall and Monteleone 2015). Internal precision refers to the standard deviation among the analyzed ^{11}B–^{10}B cycles in a single measurement divided by the square root of the number of cycles in that measurement.

For low-concentration analysis, the internal precision depends on the number of counted ions and can be predicted from counting statistics (Fitzsimons et al. 2000). The total counts of the measurement depend on the boron concentration in the sample, the primary beam current (or beam density and the analyzed area), the composition of the analyzed sample, and the sensitivity of the instrument.

A way to express instrument sensitivity is through the useful ion yield, which depicts the fraction of ions counted at the conditions of the measurement relative to the amount present in the sputtered sample volume. For boron, this value ranges from approximately 0.15–0.3% for small-radius instruments to 1.8% for the large-radius instruments (Hervig et al. 2006; Marschall and Monteleone 2015). That means that >98% of the boron present in the sample is currently wasted, mostly because of poor ionization during the sputter process. Large-radius magnet SIMS instruments now achieve precisions of ±3‰ (2σ) in samples with 1 µg/g boron in a single-spot 30-min analysis (Marschall and Monteleone 2015).

Strong or systematic deviations of the precision of low-concentration analysis from the precision predicted from counting statistics may be due to small-scale sample heterogeneity, or they may be a sign of an analytical problem, for example caused by insufficient instrument stability or by surface contamination of the sample. Surface contamination is a challenge in SIMS particularly for boron analysis (e.g., Chaussidon et al. 1997; Marschall and Ludwig 2004; Shaw et al. 1988). Several measures have been applied to reduce and quantify surface contamination: (i) avoid boron-bearing lubricants and polishing materials during sample preparation; (ii) produce a sample surface that is as smooth as possible without roughness, cracks, pits, or other topography; (iii) clean sample mounts (or thin sections) in ultra-sonic bath with B-free de-ionized water immediately before gold-coating and introduction into the air-lock system of the mass spectrometer; (iv) use sufficient pre-sputter time to remove remaining surface contamination prior to signal integration; (v) raster the beam over a larger area during pre-sputter and reduce the rastered area (or fixed spot) during collection of the analysis signal; (vi) use a field aperture to block the margin of the secondary beam, so that only the central, low-contamination part of the sample is analyzed (Chaussidon et al. 1997; Liu et al. 2009; Marschall and Ludwig 2004; Marschall and Monteleone 2015). The best way to quantify any remaining contamination is by the analysis of a material that contains as little B as possible. Silica glass Herasil-102 was suggested for this purpose, which contains ≤ 1 ng/g boron (Marschall and Ludwig 2004).

Mass interferences are not a problem for B isotope analysis by SIMS, and a mass resolving power (m/Δm at 10% peak height) of approximately 1200 is sufficient to suppress the principle interferences ($^{9}BeH^{+}$ and $^{30}Si^{3+}$ on $^{10}B^{+}$; $^{10}BH^{+}$ on $^{11}B^{+}$). Machine drift is typically slow (≤ 0.3‰/h) and can be corrected for by completing regular analysis of well-characterized (ideally internationally distributed) reference materials or standards. Reproducibility of the measurements is quantified by repeated analyses of a homogenous material, which may include the reference materials. Reported reproducibility (2 SD) for SIMS boron isotope analysis ranges from 0.4 to 4.7‰, but is typically on the order of 1.5‰, as summarized by Marschall and Monteleone (2005) (see also Büttner and Kasemann 2007; Büttner et al. 2016; Kasemann et al. 2009; Nakano and Nakamura 2001; Trumbull et al. 2008).

Accuracy is the most difficult aspect to quantify in any measurement, and B isotopes by SIMS are no exception. Instrumental mass fractionation of $^{11}B/^{10}B$ in SIMS is typically between 2 and 6%, and it may vary among different instruments, from session to session on a particular instrument, and even within an analytical session (e.g., Chaussidon et al. 1997). The primary B isotope standard (NIST SRM951) is boric acid, a white crystalline powder that is water soluble and widely used for TIMS and ICP-MS analyses, but for most applications it is not suitable for SIMS.

The mass fractionation of SIMS may depend on the chemical composition of the analyzed material. This so-called "chemical matrix effect" requires that reference materials used for correction of instrumental mass fractionation need to be similar to the analyzed samples, or at least it needs to be demonstrated through the analysis of a range of materials that the matrix effect is negligible or can be quantified for data correction. A range of silicate glasses has been distributed internationally and has been characterized for their B isotopic compositions. They are homogeneous at the level of current analytical precision, and range from komatiitic to basaltic to andesitic to rhyolitic in composition (e.g., Jochum et al. 2006; see Table 2.2).

The matrix effect for SIMS boron isotope analyses among this wide range of naturally occurring glass compositions was analytically not resolvable (less than 0.5‰) in a study by Marschall and Monteleone (2015). This had also been demonstrated previously for a similar set of basaltic to rhyolitic glasses in two other SIMS labs at a level of ≤ 1.3 and ≤ 2.1‰, respectively by Rosner et al. (2008), showing that matrix effects are not limiting the accuracy of B isotope analyses of natural silicate glasses by SIMS. In contrast, the widely distributed NIST glasses of the 61X series (e.g., SRM610, SRM612) are not suitable for SIMS boron isotope analyses as a reference for instrumental mass fractionation correction; they consistently produce fractionation factors that are 2–4‰ different from all glasses with natural compositions (Rosner et al. 2008; Gurenko and Kamenetsky 2011). These glasses are, therefore, generally avoided for SIMS boron isotope work in modern studies.

Chaussidon et al. (1997) investigated boric acid solution and seawater evaporated on a Si wafer, silicate glasses, and tourmaline, and found less than 1.6‰ matrix effect among all these materials. In an earlier study, the same group had reported a strong effect of tourmaline chemistry on instrumental mass fractionation, amounting to approximately 15‰ between the Fe–Mg-rich types (schorl-dravite) on the one hand and the

Li-rich elbaite on the other. However, the reported relationship is mostly anchored on one elbaite sample, and this set of tourmalines was only used in that particular lab. A set of three tourmaline reference materials (dravite, schorl and elbaite) has subsequently been established and is now widely distributed through the Harvard museum (Leeman and Tonarini 2001). Small (<2‰) or negligible matrix effects among various tourmaline compositions have been reported by a number of labs (e.g., Büttner and Kasemann 2007; Drivenes et al. 2015; Ludwig et al. 2011; Nakano and Nakamura 2001). However, recently McGregor et al. (2013) reported difference in instrumental mass fractionation of approximately 5‰ among the Harvard reference tourmalines, in addition to a matrix effect of 4–5‰ for the B-rich silicate prismatine depending on its Fe content. It is, therefore, recommended that a range of reference materials should be analyzed in each laboratory and during each analytical session (Table 2.2).

Few other minerals have been investigated for their matrix effect in SIMS (e.g., Pabst et al. 2012), but the growing number of laboratories that specialize on B isotope studies may lead to the establishment of a wider range of reference materials. For the analysis of carbonates, a characterized standard of inorganic calcite and a Porites coral sample are in use for mass fractionation correction; the latter has been analyzed by NTIMS and MC-ICP-MS (Kasemann et al. 2009; Table 2.2).

2.3.2.2 Laser Ablation Inductively Coupled Mass Spectrometry

The rapid growth in the adoption of MC-ICPMS for isotopic analysis stems partly from the utility of the plasma interface, as it is able to efficiently ionize both wet and dry aerosols (Halliday et al. 1998). This allows the direct coupling of the MC-ICPMS with a laser ablation (LA) system for high spatial resolution sample introduction. Shortly after the first applications of solution mode MC-ICPMS for boron isotope analysis (e.g., Lecuyer et al. 2002), LA-MC-ICPMS was applied to the in situ determination of $\delta^{11}B$ (le Roux et al. 2004). As with solution-mode MC-ICPMS, there has been a resurgence in recent years, following the enhanced analytical sensitivity and mass fractionation stability of the latest generation of MC-ICPMS. Several studies now report precise (<1‰ at 95% confidence) $\delta^{11}B$ in high boron matrices like tourmaline (e.g., Hou et al. 2010; Martin et al. 2015; Ribeiro da Costa et al. 2014) and low boron matrices such as biogenic $CaCO_3$ with a spatial resolution of <100 μm (Fietzke et al. 2010; Yang et al. 2015). Although, as with the other approaches discussed here, demonstrating accuracy is challenging and requires the use of an appropriate standards and reference materials (e.g., Devulder et al. 2015; Table 2.2).

The LA-MC-ICPMS introduction system is very flexible allowing for a diverse range of sample sizes and types to be analysed, and is relatively rapid (1 sample every 2–5 min). However, despite this utility the sample/spot size of LA-MC-ICPMS is still larger than is achievable with SIMS. Focusing of the laser beam on the surface of the sample is not thought to be critical (Thil et al. 2016) so a high degree of polish on the sectioned sample is not necessary. As with all boron isotopic analytical techniques, blank contamination can however be an issue and many studies do a pre-analysis ablation to clean the surface of the sample immediately prior to analysis (e.g., Fietzke et al. 2010).

For most applications NIST 610 (356 μg/g B) or NIST 612 (34.7 μg/g B) are used to correct for machine induced isotopic fractionation (Table 2.2) through a sample-standard bracketing routine (e.g., Fietzke et al. 2010). The importance of matrix matching sample and standard is currently unclear, with some studies reporting that there is no need (e.g., Fietzke et al. 2010; Hou et al. 2010) but others suggesting corrections of >2‰ are required if NIST glasses are used for sample-standard bracketing of $CaCO_3$ matrixes (e.g., Thil et al. 2016). This debate is well illustrated for tourmaline, for which Mikova et al. (2014) proposed that variations in major element chemistry were sufficient to impart significant matrix effects, whereas Ribeiro da Costa et al. (2014) reported no significant matrix dependence for 18 tourmaline samples covering a $\delta^{11}B$ range of −24.5 to +3.2‰ as determined by PTIMS (Palmer and Slack 1989). Indeed, in Ribeiro da Costa et al. (2014) the average deviation between solution and laser ablation was only 0.7% (maximum deviation of 2.3%; 15 of the 18 analyses differed by <1‰), despite the differences in sample size: the TIMS technique involved analysis of the total B extracted from multiple grains of tourmaline sometimes showing evidence of optical and isotopic zoning, whereas the LA-MC-ICPMS technique only analyses a small portion of individual tourmaline grains (using a 25 μm spot rastered over 100–200 μm).

Despite some difficulties, the advantage of the high spatial resolution and relatively quick throughput makes LA-MC-ICPMS analysis of $\delta^{11}B$ an attractive technique, albeit one where accuracy and precision approaching the bulk methods is a significant challenge.

2.4 Summary and Outlook

The last 20 years or so have seen the interest and application of stable isotope systems to key problems in the Earth sciences grow at a phenomenal rate. The stable isotopes of boron have a long history in this regard but analytical difficulties have always made the analysis of the $^{11}B/^{10}B$ ratio difficult and hard won. There are, however, a wealth of sample preparation, digestion, purification and analytical techniques available that allow for the analysis of boron at a wide range of scales and from a plethora of sample types (Table 2.1; Fig. 2.1). Given the ever wider proliferation of MC-ICPMS technology and the suitability of this technique for both bulk and in situ analysis, along with the installation of large-radius ion probes with high transmission in a number of laboratories worldwide, we are in no doubt that the years to come will continue to see a growth in the application and utility of boron isotopes. This, along with the continued growth in interest in boron isotopes in general, will lead to the establishment of new standards and quality control materials for in situ and solid B isotope analyses.

Acknowledgements The authors wish to thank Martin Rosner and Simone Kasemann for their thorough reviews of this chapter that greatly improved it. James Rae is acknowledged for his efficient editorial handling.

References

Aggarwal JK, Palmer MR (1995) Boron isotope analysis a review. Analyst 120:1301–1307

Aggarwal SK, You C-F (2016) A review of the determination of isotope ratios of boron with mass spectrometry. Mass Spectrom Rev 9999:1–21

Aggarwal JK, Sheppard D, Mezger K, Pernicka E (2003) Precise and accurate determination of boron isotope ratios by multi collector ICP-MS: origin of boron in the Ngawha geothermal system, New Zealand. Chem Geol 199:331–342

Al-Ammar A, Gupta RJ, Barnes RM (2000) Elimination of boron memory effect in inductively coupled plasma-mass spectrometry by ammonia gas injection into the spray chamber during analysis. Spectrochim Acta B 55:629–635

Blamart D, Rollion-Bard C, Meibom A, Ciuf JP, Juillet-Leclerc A, Dauphin Y (2007) Correlation of boron isotopic composition with ultrastructure in the deep-sea coral *Lophelia pertusa*: implications for biomineralization and paleo pH. Geochem Geophys Geosyst 8:Q12001

Brand WA, Coplen TB, Vogl J, Rosner M, Prohaska T (2014) Assessment of international reference materials for isotope-ratio analysis (IUPAC Technical Report). Pure Appl Chem 86(3):425–467

Büttner SH, Kasemann S (2007) Deformation-controlled cation diffusion in tourmaline: a microanalytical study on trace elements and boron isotopes. Am Miner 92:1862–1874

Büttner SH, Reid W, Glodny J, Wiedenbeck M, Chuwa G, Moloto T, Gucsik A (2016) Fluid sources in the Twangiza-Namoya Gold Belt (Democratic Republic of Congo): Evidence from tourmaline and fluid compositions, and from boron and Rb–Sr isotope systematics. Precambr Res 280:161–178

Catanzaro EJ, Champion CE, Garner EL, Marinenko G, Sappenfield KM, Shields WR (1970) Boric assay; isotopic, and assay standard reference materials, p 70

Chaussidon M, Robert F, Mangin D, Hanon P, Rose EF (1997) Analytical procedures for the measurement of boron isotope composition by ion microprobe in meteorites and mantle rocks. Geostand Newslett 21:7–17

Clarkson MO, Kasemann SA, Wood RA, Lenton TM, Daines SJ et al. (2015) Ocean acidification and the Permo-Triassic mass extinction. Science 348:229–232

Devulder V, Gerdes A, Vanhaecke F, Degryse P (2015) Validation of the determination of the B isotopic composition in Roman glasses with laser ablation multi-collector inductively coupled plasma-mass spectrometry. Spectrochim Acta Part B 105:116–120

Deyhle A (2001) Improvemnts of boron isotope analysis by positive thermal ionization mass spectrometry using static multicollection of $Cs_2BO_2^+$ ions. Int J Mass Spectrom 206:79–89

Drivenes K, Larsen RB, Müller A, Sørensen BE, Wiedenbeck M, Raanes MP (2015) Late-magmatic immiscibility during batholith formation: assessment of B isotopes and trace elements in tourmaline from the Land's End granite, SW England. Contrib Miner Petrol 169:56

Farmer JR, Honisch B, Uchikawa J (2016) Single laboratory comparison of MC-ICP-MS and N-TIMS boron isotope analyses in marine carbonates. Chem Geol 447:173–182

Fietzke J, Heinemann A, Taubner I, Bohm F, Erez J, Eisenhauer A (2010) Boron isotopic ratio determination in carbonates via LA-MC-ICP-MS using soda-lime glass standards as reference materials. J Anal At Spectrom 25:1953–1957

Fitzsimons ICW, Harte B, Clark RM (2000) SIMS stable isotope measurement: counting statistics and analytical precision. Mineral Mag 64:59–83

Foster GL (2008) Seawater pH, pCO_2 and $[CO_3^{2-}]$ variations in the Caribbean Sea over the last 130 kyr: A boron isotope and B/Ca study of planktic foraminifera. Earth Planet Sci Lett 271:254–266

Foster GL, Ni Y, Haley B, Elliott T (2006) Accurate and precise isotopic measurement of sub-nanogram sized samples of foraminiferal hosted boron by total evaporation NTIMS. Chem Geol 230:161–174

Foster GL, Pogge von Strandmann PAE, Rae JWB (2010) Boron and magnesium isotopic composition of seawater. Geochem Geophys Geosyst 11:Q08018. doi:10.1029/2010GC003201

Foster GL, Honisch B, Paris G, Dwyer GS, Rae JWB et al (2013) Interlaboratory comparison of boron isotope analysis of boric acid, seawater and marine $CaCO_3$ by MC-ICPMS and NTIMS. Chem Geol 358:1–14

Gaillardet J, Lemarchand D, Gopel C, Manhes G (2001) Evaporation and sublimation of boric acid: application for boron purification from organic rich solutions. Geostand Newslett 25:67–75

Gregoire DC (1987) Determination of boron isotope ratios in geological materials by inductively coupled plasma mass spectrometry. Anal Chem 58:2478–2484

Guerrot C, Millot R, Robert M, Negrel P (2011) Accurate and high-precision determination of boron isotopic ratios at low concentration by MC-ICP-MS (Neptune). Geostand Geoanal Res 35:275–284

Gurenko AA, Kamenetsky VS (2011) Boron isotopic composition of olivine-hosted melt inclusions from Gorgona komatiites, Colombia: New evidence supporting wet komatiite origin. Earth Planet Sci Lett 312:201–212

Halliday AN, Lee DC, Christensen JN, Rehkamper M, Yi W et al (1998) Applications of multiple collector-ICPMS to cosmochemistry, geochemistry, and paleoceanography. Geochim Cosmochim Acta 62:919–940

He M, Xiao Y, Jin Z, Ma Y, Xiao J et al (2013) Accurate and precise determination of boron isotopic ratios at low concentration by positive thermal ionization mass spectrometry using static multicollection of $CS_2BO_2^+$. Anal Chem 85:6248–6253

Hemming NG, Hanson GN (1992) Boron isotopic composition and concentration in modern marine carbonates. Geochim Cosmochim Acta 56:537–543

Hemming NG, Hanson GN (1994) A procedure for the isotopic analysis of boron by negative thermal ionization mass spectrometry. Chem Geol 114:147–156

Hemming NG, Hönisch B (2007) Boron isotopes in marine carbonate sediments and the pH of the ocean. Dev Mar Geol 1:717–734

Henehan MJ, Rae JWB, Foster GL, Erez J, Prentice KC et al (2013) Calibration of the boron isotope proxy in the planktonic foraminifera *Globigerinoides ruber* for use in palaeo-CO_2 reconstruction. Earth Planet Sci Lett 364:111–122

Hervig RL, Mazdad FK, Williams P, Guan Y, Huss GR, Leshin LA (2006) Useful ion yields for Cameca IMS 3f and 6f SIMS: limits on quantitative analysis. Chem Geol 227:83–99

Hou K-J, Li Y-H, Liu F, Tian Y-R (2010) In situ boron isotope measurements of natural geological materials by LA-MC-ICPMS. Chin Sci Bull 55:3305–3311

Jochum KP, Stoll B, Herwig K, Willbold M, Hofmann A et al (2006) MPI-DING reference glasses for in situ microanalysis: new reference values for element concentrations and isotope ratios. Geochem Geophys Geosyst 7

John SG, Adkins JF (2010) Analysis of dissolved iron isotopes in seawater. Mar Chem 119:12

Kasemann S, Meixner A, Rocholl A, Vennemann T, Rosner M et al (2001) Boron and oxygen isotope composition of certified reference materials NIST SRM 610/612 and reference materials JB-2 and JR-2. Geostand Newslett 25:405–416

Kasemann S, Schmidt DN, Bijma J, Foster GL (2009) In situ boron isotope analysis of marine carbonates and its application for foraminifera and palaeo-pH. Chem Geol 260:138–147

Kiss E (1988) Ion-exchange separation and spectrophotometric determination of boron in geological materials. Anal Chim Acta 211:243–256

Kobayashi K, Tanaka R, Moriguti T, Shimizu K, Nakamura E (2004) Lithium, boron, and lead isotope systematics of glass inclusions in olivines from Hawaiian lavas: evidence for recycled components in the Hawaiian plume. Chem Geol 212:143–161

le Roux PJ, Shirley SB, Benton L, Hauri EH, Mock TD (2004) In situ, multiple-multiplier, laser ablation ICP-MS measurement of boron isotopic composition (δ11B) at the nanogram level. Chem Geol 203:123–138

Lecuyer C, Grandjean P, Reynard B, Albarede F, Telouk P (2002) $^{11}B/^{10}B$ analysis of geological

materials by ICP-MS Plasma 54: Application to the boron fractionation between brachiopod calcite and seawater. Chem Geol 186:45–55
Leeman WP, Tonarini S (2001) Boron isotopic analysis of proposed borosilicate mineral reference samples. Geostand Geoanal Res 25:399–403
Lemarchand D, Gaillardet J, Gopel C, Manhes G (2002) An optimized procedure for boron separation and mass spectrometry analysis for river samples. Chem Geol 182:323–334
Liu MC, McKeegan KD, Goswami JN, Marhas KK, Sahijpal S et al (2009) Isotopic records in CM hibonites: implications for timescales of mixing of isotope reservoirs in the solar nebula. Geochim Cosmochim Acta 73:5051–5079
Liu MC, Nittler LR, Alexander CMO, Lee T (2010) Lithium-beryllium-boron isotopic compositions in meteoritic hibonite: implications for origin of ^{10}Be and early solar system irradiation. Astrophys J Lett 719:L99–L103
Liu Y-W, Aciego SM, Wanamaker AD, Sell BK (2013) A high-throughput system for boron microsublimation and isotope analysis by total evaporation thermal ionization mass spectrometry. Rapid Commun Mass Spectrom 27:1705–1714
Louvat P, Bouchez J, Paris G (2011) MC-ICP-MS Isotope measurments with direct injection nebulisation (d-DIHEN): optimisation and application to boron in seawater and carbonate samples. Geostand Geoanal Res 35:75–88
Ludwig T, Marschall HR, Pogge von Strandmann PAE, Shabaga BM, Fayek M, Hawthorne FC (2011) A secondary ion mass spectrometry (SIMS) re-evaluation of B and Li isotopic compositions of Cu-bearing elbaite from three global localities. Mineral Mag 75:2485–2494
Lyon IC, Tizard JM, Henkel T (2007) Evidence for lithium and boron from star-forming regions implanted in presolar SiC grains. Meteorit Planet Sci 42
MacGregor JR, Grew ES, de Hoog JCM, Harley SL, Kowalski PM et al (2013) Boron isotopic composition of tourmaline, prismatine, and grandidierite from granulite facies paragneisses in the Larsemann Hills, Prydz Bay, East Antarctica: evidence for a non-marine evaporite source. Geochim Cosmochim Acta 123:261–283
Marschall HR, Ludwig T (2004) The low-boron contest: minimising surface contamination and analysing boron concentration at the ng/g-level by secondary ion mass spectrometry. Mineral Petrol 81:265–278
Marschall HR, Monteleone BD (2015) Boron isotope analysis of silicate glass with very low boron concentrations by secondary ion mass spectrometry. Geostand Geoanal Res 39:31–46
Marschall HR, Foster GL (2017) Boron isotopes in the earth and planetary sciences—a short history and introduction. In: Marschall HR, Foster GL (eds) Boron isotopes—The fifth element, Advances in Isotope Geochemistry, vol 7, Springer, Heidelberg, p 1–11
Martin C, Ponzevera E, Harlow G (2015) In situ lithium and boron isotope determinations in mica, pyroxene, and serpentine by LA-MC-ICP-MS. Chem Geol 412:107–116
McCulloch MT, Holcomb M, Rankenburg K, Trotter J (2014) Rapid, high-precision measurments of boron isotopic compositions in marine carbonates. Rapid Commun Mass Spectrom 28:2704–2712
McMullen CC, Gragg CB, Thode HG (1961) Absolute ratio B11/B10 in Searles Lake borax. Geochim Cosmochim Acta 23:147–149
Mikova J, Kosler J, Wiedenbeck M (2014) Matrix effects during laser ablation MC ICP-MS analysis of boron isotopes in tourmaline. J Anal At Spectrom 29:903–914
Misra S, Owen R, Kerr J, Greaves M, Elderfield H (2014) Determination of δ11B by HR-ICP-MS from mass limited samples: application to natural carbonates and water samples. Geochim Cosmochim Acta 140:531–552
Nakamura E, Ishikawa T, Birck JL, Allègre CJ (1992) Precise boron isotopic analysis of natural rock samples using a boron-mannitol complex. Chem Geol 94:193–204
Nakano T, Nakamura E (1998) Static multi-collection of $Cs_2BO_2^+$ ions for precise boron isotope analysis with positive thermal ionization mass spectrometry. Int J Mass Spectrom 176:13–21
Nakano T, Nakamura E (2001) Boron isotope geochemistry of metasedimentary rocks and tourmalines in a subduction zone metamorphic suite. Phys Earth Planet Inter 127:233–252
Ni Y, Foster GL, Elliott T (2010) The accuracy of δ11B measurements of foraminifers. Chem Geol 274:187–195
Pabst S, Zack T, Savov IP, Ludwig T, Rost D et al (2012) The fate of subducted oceanic slabs in the shallow mantle: insights from boron isotopes and light element composition of metasomatized blueschists from the Mariana forearc. Lithos 132–133:162–179
Palmer MR, Slack JF (1989) Boron isotopic composition of tourmaline from massive sulfide deposits and tourmalinites. Contrib Miner Petrol 103:434–451
Pi J, You CF, Chung CH (2014) Micro-sublimation separation of boron in rock samples for isotopic measurements by MC-ICPMS. J Anal At Spectrom 29:861–867
Ramakumar KL, Parab AR, Khodade PS, Almaula AI, Chitambar SA, Jain HC (1985) Determination of isotopic composition of boron. J Radioanal Nucl Chem 94:53–61
Ribeiro da Costa I, Mourao C, Recio C, Guimaraes F, Antunes IM et al (2014) Tourmaline occurrences within the Penamacor-Monsanto granitic pluton and host-rocks (Central Portugal): genetic implications of crystal-chemical and isotopic features. Contrib Miner Petrol 167:993–1016
Rollion-Bard C, Chaussidon M, France-Lanord C (2003) pH control on oxygen isotopic composition of symbiotic corals. Earth Planet Sci Lett 215:275–288

Rollion-Bard C, Vigier N, Spezzaferri S (2007) In situ measurements of calcium isotopes by ion microprobe in carbonates and application to foraminifera. Chem Geol 244:679–690

Romer RL, Meixner A (2014) Lithium and boron isotopic fractionation in sedimentary rocks during metamorphism—the role of rock composition and protolith mineralogy. Geochem Cosmocimica Acta 128:158–177

Rose-Koga EF, Sheppard SMF, Chaussidon M, Carigan J (2006) Boron isotopic composition of atmospheric precipitations and liquid–vapour fractionations. Geochim Cosmochim Acta 70:1603–1615

Rosner M, Meixner A (2004) Boron isotopic composition and concentration of ten geological reference materials. Geostand Geoanal Res 28:431–441

Rosner M, Romer RL, Meixner A (2005) Air handling in clean laboratory environments: the reason for anomalously high boron background levels. Anal Bioanal Chem 382:120–124

Rosner M, Wiedenbeck M, Ludwig T (2008) Composition-induced variations in SIMS instrumental mass fractionation during boron isotope ratio measurements of silicate glasses. Geostand Geoanal Res 32:27–38

Schmitt AK, Kasemann S, Meixner A, Rhede D (2002) Boron in central Andean ignimbrites: implications for crustal boron cycles in an active continental margin. Chem Geol 183:333–347

Shaw DM, Higgins MD, Truscott MG, Middleton TA (1988) Boron contamination in polished thin sections of meteorites: implications for other trace-element studies by alpha-track image or ion microprobe. Am Miner 73:894–900

Spivack AJ, Edmond JM (1986) Determination of boron isotope ratios by thermal ionisation mass spectrometry of the dicesium metaborate cation. Anal Chem 58:31–35

Straub SM, Layne GD (2002) The systematics of boron isotopes in Izu arc front volcanic rocks. Earth Planet Sci 198: 25–39

Swihart GH (1996) Instrumental techniques for boron isotope analysis. In: Grew ES, Anovitz LM (eds) Boron mineralogy, petrology, and geochemistry. Mineralogical Society of America Reviews in Mineralogy, pp 845–862

Swihart GH, Moore PB, Callis EL (1986) Boron isotopic composition of marine and nonmaine evaporite borates. Geochim Cosmochim Acta 50:1297–1301

Thil F, Blamart D, Assailly C, Lazareth CE, Leblanc T et al (2016) Development of laser ablation multi-collector inductively coupled plasma mass spectrometry for boron isotopic measurements in marine biocarbonates: new improvements and application to a modern *Porites* coral. Rapid Commun Mass Spectrom 30:359–371

Tonarini S, Pennisi M, Leeman WP (1997) Precise boron isotopic analysis of complex silicate (rock) samples using alkali carbonate fusion and ion-exchange separation. Chem Geol 142:129–137

Trotter J, Montagna P, McCulloch MT, Silenzi S, Reynaud S et al (2011) Quantifying the pH "vital effect" in temperate zooxanthellate coral *Cladocora caespitosa*: validation of the boron seawater pH proxy. Earth Planet Sci Lett 303:163–173

Trumbull RB, Krienitz M-S, Gottesmann B, Wiedenbeck M (2008) Chemical and boron-isotope variations in tourmalines from an S-type granite and its source rocks: the Erongo granite and tourmalinites in the Damara Belt, Namibia. Contrib Miner Petrol 155:1–18

Vogl J, Rosner M (2011) Production and certification of a unique set of isotope and delta reference materials for boron isotope determination in geochemical, environmental and industrial materials. Geostand Geoanal Res 36:161–175

Walder AJ, Freedman PA (1992) Isotopic ratio measurement using a double focusing magnetic sector mass analyser with an inductively coupled plasma as an ion source. J Anal Mass Spectrom 7:571–575

Wang B-S, You C-F, Huang K-F, Wu S-F, Aggarwal SK et al (2010) Direct separation of boron from Na- and Ca-rich matrices by sublimation for stable isotope measurement by MC-ICP-MS. Talanta 82:1378–1384

Wei GJ, Wei JX, Liu Y, Ke T, Ren ZY et al (2013) Measurement of high-precision boron isotope of silicate materials by a single column purification method and MC-ICP-MS. J Anal At Spectrom 28:606–612

Williams LB, Hervig RL, Weiser ME, Hutcheon I (2001) The influence of organic matter on the boron isotope geochemistry of the gulf coast sedimentary basin, USA. Chem Geol 174

Xiao Y-K, Beary ES, Fassett JD (1988) An improved method for the high-precision isotopic measurement of boron by thermal ionization mass spectrometry. Int J Mass Spectrom Ion Processes 85:203–213

Yang S-Y, Jiang S-Y, Palmer MR (2015) Chemical and boron isotopic compositions of tourmaline from the Nyalam leucogranites, South Tibetan Himalaya: implications for their formation from B-rich melt to hydrothermal fluids. Chem Geol 419:102–113

Yoshimura K, Miyazaki Y, Ota F, Matsuoka S, Hirofumi S (1998) Complexation of boric acid with the *N*-methyl-D-glucamine group in solution and in cross-linked polymer. J Chem Soc, Faraday Trans 94:683–689

Zeininger H, Heumann KG (1983) Boron isotopic ratio measurement by negative thermal ionization mass spectrometry. Int J Mass Spectrom Ion Phys 48:377–380

Boron Isotope Fractionation Among Vapor–Liquids–Solids–Melts: Experiments and Atomistic Modeling

3

Piotr M. Kowalski and Bernd Wunder

Abstract

A quantitative understanding of the principle factors that govern their geochemical behavior is required to employ boron and its isotopes as geochemical tracers of any vapor-, liquid- or melt-mediated process in the Earth's interior. Feedback between experiments and computational predictions are required to gain insight into the processes driving isotope partitioning. This chapter comprises methods and results of selected experimental studies and first principles atomistic modeling techniques aimed at determining and predicting temperature-, pressure-, and pH-dependent B-isotope fractionation among B-bearing geomaterials.

3.1 Introduction

One of the fundamental tasks in Geosciences is unraveling the transport of Earth's materials and their interactions with the dynamic, evolving Earth system. For any sophisticated description of geochemical cycles, knowledge of the distribution of elements and isotopes among geomaterials is a prerequisite to understand the underlying mechanisms and laws that govern their interactions. The light element boron is highly fluid-mobile and tends to strongly partition into the fluid phase during fluid-rock interaction processes (e.g., Mottl and Holland 1978). The two stable isotopes, ^{10}B and ^{11}B, differ in their atomic mass by 10%. Consequently, boron isotopes strongly fractionate during geological processes, thereby leading to natural $\delta^{11}B$-variations of about 100‰ (Barth 1993; Palmer and Swihart 2002). Therefore, B isotopes have great potential for modeling geochemical cycles and are widely used as geochemical tracers to unravel processes operating at the surface, as well as in the deeper crust and mantle (e.g., Leeman and Sisson 2002; Palmer and Swihart 2002).

Arc magmas, for example, are typically enriched in B relative to fresh mid-ocean ridge basalts and mantle (e.g., Ryan and Langmuir

P.M. Kowalski (✉)
Forschungszentrum Jülich, Institute of Energy and Climate Research (IEK-6), 52425 Jülich, Germany
e-mail: p.kowalski@fz-juelich.de

B. Wunder
Helmholtz-Zentrum Potsdam, GFZ German Research Centre for Geosciences, 14473 Potsdam, Germany
e-mail: wunder@gfz-potsdam.de

H. Marschall and G. Foster (eds.), *Boron Isotopes*, Advances in Isotope Geochemistry,
https://doi.org/10.1007/978-3-319-64666-4_3

1993). In addition, the boron concentration in arc magmas often systematically decreases with increasing distance to the underlying old and cold subducting plates (e.g., Rosner et al. 2003). This indicates continuous dehydration within these slabs during the breakdown of (OH, B)-bearing minerals (e.g., micas, serpentine) with increasing depth. Furthermore, ^{11}B preferentially fractionates into hydrous fluids relative to the co-existing minerals, and arc magmas are successively depleted in the heavy ^{11}B with progressing subduction (e.g., Ishikawa and Nakamura 1994). Both observations indicate that B is transported from subducted sediments and/or altered oceanic crust into the magma's source region by a fluid or melt. As those reservoirs show different B concentrations and $\delta^{11}B$-values (Palmer and Swihart 2002), B and its isotopes are ideal geochemical tracers for tracking the source and to quantify the mass flux from the slab into the mantle wedge and finally to the surface. Details of a large body of work on this topic are presented in this book (De Hoog and Savov 2017).

Another important natural process, which strongly fractionates B isotopes, is the interaction of seawater with various solid marine components (e.g., biogenic carbonates, clays, oxides) at near surface conditions. The isotopic composition of seawater is governed by the B isotope exchange between the two main aqueous B-species, $B(OH)_3$ and $[B(OH)_4]^-$, which can be expressed as:

$$^{10}B(OH)_3 + \left[^{11}B(OH)_4\right]^- \leftrightarrow {}^{11}B(OH)_3 + \left[^{10}B(OH)_4\right]^- . \quad (3.1)$$

This reaction in turn controls the isotopic composition of geomaterials in contact with seawater through fractionation during surface adsorption and/or a structural exchange mechanism. As the B-isotopic fractionation factor α with

$$\alpha = \left(^{11}B(OH)_3/\left[^{10}B(OH)_4\right]^-\right) / \left(\left[^{11}B(OH)_4\right]^-/^{10}B(OH)_3\right) \quad (3.2)$$

strongly depends on pH, paleo-pH values of seawater can be reconstructed from the B-isotopic composition of biogenic carbonates in marine sediments (e.g., Hemming and Hanson 1992; Spivack et al. 1993; Gaillardet and Allègre 1995; Sanyal et al. 1995; Palmer et al. 1998). Three chapters within this book (Branson 2017; Rae 2017; McCulloch 2017).

The use of such carbonate archives is a key to the understanding of past and present climate changes (e.g., Lea 2003), whereas the first example is used to expand geochemical models in subduction-zone regimes (e.g., Marschall et al. 2007; Konrad-Schmolke and Halama 2014). Both, either require the assumption of isotope equilibrium or quantitative knowledge of pH-, temperature-, and pressure-dependent equilibrium B-isotope fractionation among the various phases of interest in order to identify disequilibrium.

The establishment of B-isotope fractionation data has progressed via two different routes, namely by laboratory experiments and by computational modeling. In the first part of this chapter, we present methods and results from selected experimental studies, designed to determine B-isotopic fractionation data at low to high temperatures in the systems vapor–liquids–solids–melts. In the second part, we present the modeling methodology behind the first-principles simulation approach to the ab initio prediction of equilibrium isotope fractionation between phases at high temperatures and pressures recently developed by Kowalski and Jahn (2011) and Kowalski et al. (2013). For various systems and conditions, B-isotopic fractionation data are calculated and compared with results from the experimental studies.

3.2 Notations

The B-isotope composition of a phase is generally given as per mil variations relative to the B isotopic standard NIST 951 boric acid according to

$$\delta^{11}B = \left\{\left[\left({}^{11}B/{}^{10}B\right)_{phase} / \left({}^{11}B/{}^{10}B\right)_{NIST\,SRM\,951}\right] - 1\right\} \cdot 1000 \quad (3.3)$$

For the B-isotopic exchange reaction

$$^{10}B_{phase\,1} + {}^{11}B_{phase\,2} \leftrightarrow {}^{11}B_{phase\,1} + {}^{10}B_{phase\,2}, \quad (3.4)$$

the B-isotopic fractionation between two phases is denoted by the isotopic fractionation factor α, an equilibrium coefficient defined as:

$$\alpha = \left({}^{11}B_{phase\,1}/{}^{10}B_{phase2}\right)/\left({}^{11}B_{phase\,1}/{}^{10}B_{phase\,2}\right), \quad (3.5)$$

or as:

$$\Delta^{11}B_{phase\,1-phase\,2} \cong \delta^{11}B_{phase\,1} - \delta^{11}B_{phase\,2}. \quad (3.6)$$

The α factor, rather than $\Delta^{11}B$, is used in thermodynamic modeling of isotopic fractionation because α is equal to the equilibrium constant (K) in the Gibbs free energy equation

$$\Delta G \cong \Delta H - T\Delta S = -RT\ln\alpha \quad (3.7)$$

of the exchange reaction, with ΔG (Gibbs free energy), ΔH (reaction enthalpy), ΔS (entropy), R (universal gas constant), and T (temperature). For ideal gas reactions, ln K follows ~ $1/T$ (in kelvin) for phases containing hydroxyl groups, at low temperatures (below room temperature) and is roughly proportional to $1/T^2$ at high temperatures relevant for geological environments.

The isotopic fractionation factors $\Delta^{11}B$ and α are related by:

$$\Delta^{11}B \cong 1000 \cdot \ln\alpha. \quad (3.8)$$

Only for strong isotopic fractionation and large $\delta^{11}B$-values the approximation (3.8) does not match within the range of experimental uncertainties of $\delta^{11}B$ (see later). Combining Eqs. 3.6 and Eq. (3.8), the isotope fractionation factor α, expressed in terms of $\delta^{11}B$, is defined as:

$$\alpha \cong (1000 + \delta^{11}B_{phase\,1})/(1000 + \delta^{11}B_{phase\,2}). \quad (3.9)$$

3.3 Controls of B-Isotopic Fractionation

Due to the rather low experimentally derived diffusivities of B in melts of haplogranitic compositions (Chakroborty et al. 1993), it can be assumed that kinetic B-isotope exchange is insignificant at melt-crystallization temperatures. This is in strong contrast to the behavior of lithium and its isotopes, for which significant kinetic isotopic fractionation is observed at temperatures above 800 °C (Dohmen et al. 2010). Owing to the limited amount of experimental data on high-T diffusion-induced B-isotope fractionation—which strongly awaits further future high-temperature studies—B-isotope fractionation is regarded here as solely resulting from low-T surface adsorption processes (e.g., smectites) and from equilibrium isotope exchange. Generally, equilibrium isotopic fractionation is driven by differences in the chemical potentials between coexisting phases, which may be represented by the energetic differences in the respective bonding environments. Equilibrium isotope fractionation usually decreases as temperature increases, following the approximation ln α ~ $1/T^2$. For many solid–solid or solid–fluid systems, it has been shown experimentally and by modeling that isotopic fractionation between two phases follows a simple rule: the lighter isotope is preferentially incorporated in the phase where it is at higher coordination. This is accompanied with longer cation-anion bond lengths and weaker bond strength (e.g., Zheng 1993; Schauble 2004; Wunder et al. 2006). Owed to its small ionic radius ($^{[3]}r = 0.01$Å, $^{[4]}r = 0.11$Å after Shannon 1976) B occurs either in trigonal or tetrahedral coordination with oxygen, whereby ^{11}B prefers trigonal coordination and ^{10}B the tetrahedral coordination (Palmer and Swihart 2002). In addition to the concept of coordination-driven isotope fractionation there is also significant isotopic fractionation between phases of the same coordination, e.g., tourmaline—boric acid (Palmer et al. 1992; Meyer et al. 2008) both with trigonal B-coordination (some tourmaline also have $^{[4]}B$, see later). This indicates that beside the often as "first-order criteria" defined coordination rule, also second-order effects, like differences in bond strengths, geometries, energetics, and electronic environments of the local atomic structures of different phases have to be considered. Therefore, structural characterisation of the B-environment in all phases of interest is a prerequisite for the understanding of the principle mechanism and for reliable modeling of B-isotope fractionation in the different systems.

3.4 B-Speciation in Liquids, Vapor, and Melts

3.4.1 B-Speciation in Aqueous Fluids

B in seawater mainly forms two aqueous species which are in a pH-dependent equilibrium (Eq. 3.1): the neutral trigonal boric acid $B(OH)_3$ and the larger tetrahedral borate anion $[B(OH)_4]^-$ (Hershey et al. 1986). At atmospheric conditions, the predominant B species is $B(OH)_3$ at low pH and $[B(OH)_4]^-$ at high pH (Vengosh et al. 1991; Sanchez-Valle et al. 2005). At high aqueous B-concentrations (>0.5 mol/L) other minor B species may be present, e.g., polyborates and borates $NaB(OH)_4$, $[MgB(OH)_4]^+$, $[CaB(OH)_4]^+$ (Christ and Harder 1978), and $[B_3O_3(OH)_4]^-$, $[B_3O_3(OH)_5]^{2-}$ (Kakihana et al. 1977). Hershey et al. (1986) determined the fraction of $B(OH)_3$ and $[B(OH)_4]^-$ species present at ambient conditions as a function of pH. For determination of aqueous B-species in crustal and mantle wedge fluids, Schmidt et al. (2005) studied B-bearing fluids of various compositions in situ at temperatures between 22 and 600 °C and pressures from 0.1 MPa to about 2 GPa using Raman spectroscopy and an externally heated hydrothermal diamond anvil cell (Fig. 3.1a). In acidic fluids, they found $B(OH)_3$ as the main species with traces of HBO_2, whose amount weakly increases with temperature and decreases with the addition of NaCl (Fig. 3.1b, e). In alkaline solutions, ring-like polyboric ions $[B_3O_3(OH)_4]^-$, $[B_4O_5(OH)_4]^{2-}$, $[B_5O_6(OH)_4]^-$, which occur together with $B(OH)_3$ and $[B(OH)_4]^-$ at low pressures and temperatures, completely depolymerise with inceasing temperature, and the amount of $B(OH)_3$ increases relative to $[B(OH)_4]^-$ (Fig. 3.1c,d). In basic fluids, the $[B(OH)_4]^-$ to $B(OH)_3$ ratio increases with increasing pressure.

In summary, Schmidt et al. (2005) showed that $B(OH)_3$ is the predominant aqueous B-species over a wide range of pressure-temperature-pH conditions in slab-related fluids and that the aqueous $[B(OH)_4]^-$ species may become more important in near-neutral to basic fluids at greater depth. Full quantification of the relative proportions of the different B-species from the Raman spectra (Schmidt et al. 2005) would require scattering coefficients of individual species, which are not readily available. Additionally, the aqueous B-species are completely unknown for temperatures above 600 °C and pressures above 2.0 GPa. These are shortcomings, when modeling high pressure and temperature B-isotope fractionation (see later). It should also be kept in mind that due to the high fluid solubilities of alkalis, Si, Al and particularly light elements like Li and Be at high pressures and temperatures (Manning 2004), natural fluids might exhibit other B-species associated with the fluid-compatible elements, more complex than boric acid and the anionic tetrahedral borate.

3.4.2 Boron Speciation in Vapor

To the best of our knowledge, no direct measurements of B-species in vapor are available. Speculations about B speciation in vapor are based on experimental data on B-isotopic vapor - liquid fractionation determined in the *T*-range 400–450 °C for seawater-like H_2O–NaCl model

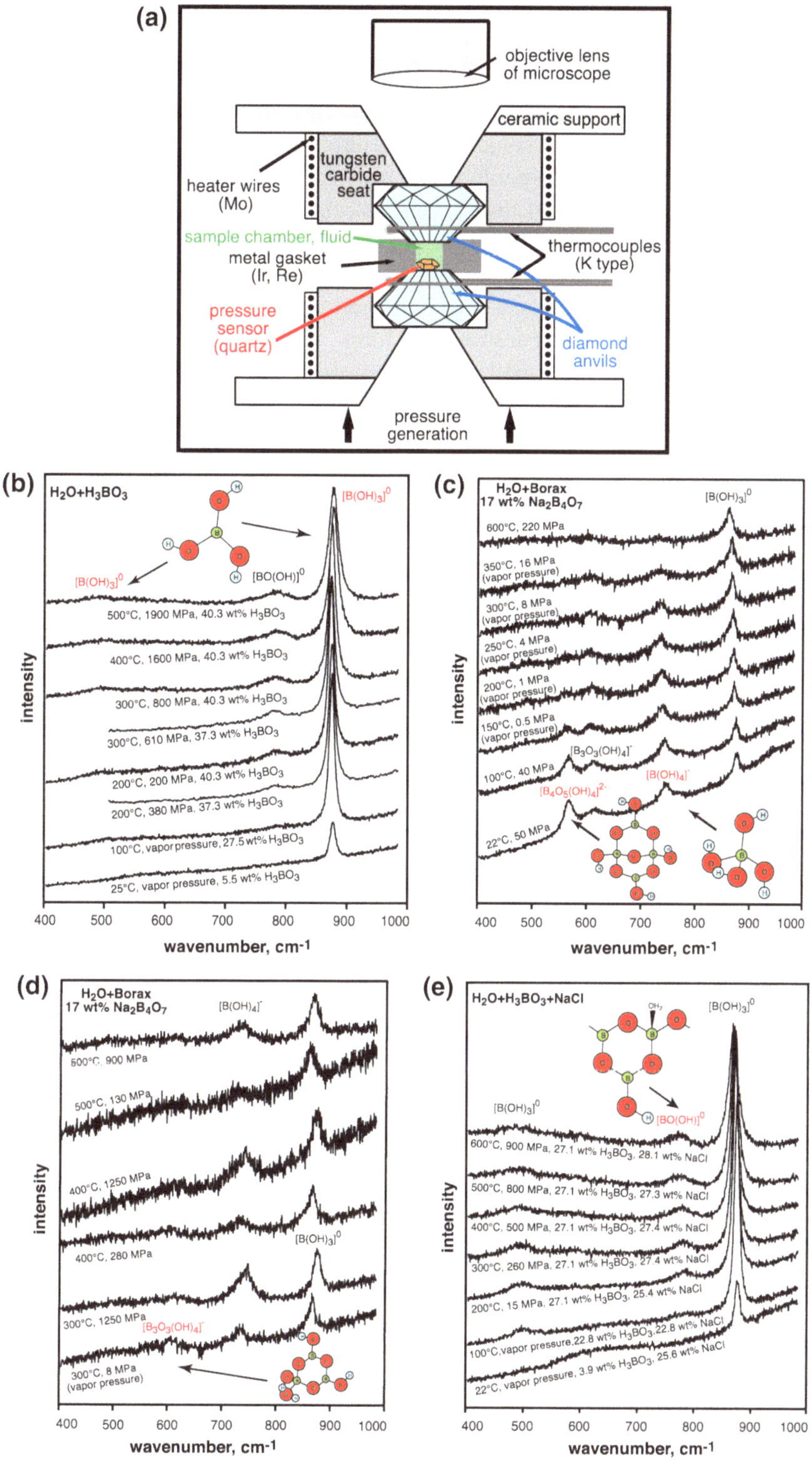

Fig. 3.1 **a** Schematic diagram of the central portion (not to scale) of an externally heated hydrothermal diamond anvil cell (courtesy of Christian Schmidt, GFZ Potsdam). In situ Raman spectra of fluids for three different systems: **b** $H_2O + H_3BO_3$; **c**, **d** $H_2O + Na_2B_4O_7$; **e** $H_2O + H_3BO_3 + NaCl$ (modified after Schmidt et al. 2005). Additionally, the structures of the various B-species dissolved in fluids are shown (*green* boron; *red* oxygen; *blue* hydrogen) according to Liu and Tossel (2005)

systems at near-neutral pH conditions (see below, Spivack et al. 1990; Liebscher et al. 2005). The $\Delta^{11}B_{vapor\text{-}liquid}$-values are small; hence, due to the coordination rule, this suggests similar $B(OH)_3$-speciation in the conjugated aqueous fluid and vapor pairs at the experimental conditions. It is furthermore assumed that B in vapor exhibits the same pH-dependent characteristics as in aqueous fluids: $B(OH)_3$ is predominant at acidic to neutral conditions, while $[B(OH)_4]^-$ is only significantly abundant at strongly basic pH. In the system H_2O–NaCl, the hydrolysis reaction $H_2O + NaCl = HCl + NaOH$ exerts control on the pH. At ambient conditions hydrolysis is small and H_2O–NaCl solutions have near neutral pH. At higher temperatures the concentrations of HCl and NaCl increase. In liquid–vapor systems HCl fractionates into the vapor and the conjugated liquid becomes slightly enriched in NaOH (Shumolvich et al. 2002). Consequently, according to Vakulenko et al. (1989) at 390–470 °C, and 21.7–35.5 MPa, the vapor becomes more acidic (quench pH = 5–6) and the liquid more basic (quench pH = 8–9). This process may influence the boron speciation of the liquid to contain slightly more $[B(OH)_4]^-$ than the coexisting vapor (see later Chap. 5.1).

3.4.3 Boron Speciation in Melt

Boron is only a minor component in natural magmas, but can be significantly enriched by several weight percent of B_2O_3 in S-type granites evolved from evaporitic metasediments (Pichavant et al. 1988), or during late magmatic or pegmatitic stages of granitic intrusions (e.g., London 1997; Thomas et al. 2003). These stages are known to be associated with highly complex internal processes, which may involve more than one phase, i.e., by separation of immiscible silica- and (hydrous) borate-rich liquids (Thomas et al. 2000; Veksler et al. 2002; see also Trumbull and Slack 2017, this book). Boron is the smallest of the network-forming cations in glasses and melts with a high ionic charge. It strongly affects the physical and chemical properties in melts (e.g., glass transition temperature *Tg*, density, thermal expansion coefficient) and therefore has also broad applications in the glass industry (e.g., Scholze 1988). Information on B-speciation in melts and glasses come from neutron diffraction, infrared (IR), Raman, and nuclear magnetic resonance (NMR) spectroscopy. In borosilicate melts and glasses two distinct structural units of boron exist: quasi-planar $[BO_3]^{3-}$ triangles and $[BO_4]^{5-}$ tetrahedra. Both species tend to polymerise and form numerous combinations with network-forming components, i.e., $[SiO_4]^{4-}$, $[AlO_4]^{5-}$. The proportions of 3- and 4-fold coordinations in melts and glasses depend on several factors including composition (e.g., Geisinger et al. 1988; Zhong and Bray 1989; Schmidt et al. 2004; Wu and Stebbins 2013), temperature (e.g., Smirnov et al. 2005; Manara et al. 2009; Wu and Stebbins 2010, 2013), and pressure (e.g., Schmidt et al. 2004; Edwards et al. 2014). Furthermore, the fraction of tetrahedral B significantly increases below the glass transition temperature (Michel et al. 2013). The structural modification during quenching of a B-rich silicate melt changes B from a network-modifier to a network-forming cation, and might explain the low viscosity of B-bearing melts. As the glass transition temperature depends on the cooling rate, the latter also affects the B-coordination. In binary alkali borate melts the addition of up to 35 mol% alkali oxide to a B_2O_3-glass results in a decrease of the $^{[3]}B/^{[4]}B$-ratio (Bray and O'Keefe 1963; Zhong and Bray 1989). At higher alkali contents tetrahedral diborate-complexes depolymerise again and form trigonal B, which is coordinated by one nonbridging oxygen (NBO) with an associated alkali cation. For glasses along the albite (Ab, $NaAlSi_3O_8$)—reedmergnerite (Rd, $NaBSi_3O_8$) join, the $[BO_3]/[BO_4]$-ratio increases from about 0.3 for an Al-free Rd-glass to 0.6 for a glass of $Ab_{80}Rd_{20}$-composition (Geisinger et al. 1988). This indicates that the presence of Al destabilizes the tetrahedral B units. This trend of an increasing $^{[3]}B/^{[4]}B$-ratio was also determined for anhydrous melts of increasing A/CNK-value ($A = Al_2O_3$; CNK = sum of oxides of charge-balancing cations like CaO, Na_2O, K_2O; all in mole) varying from peralkaline (A/CNK < 1) to peraluminous (A/CNK > 1) compositions. Additionally, increasing temperature shifts the reaction $[BO_4] \leftrightarrow [BO_3] + NBO$ to the right (Wu and Stebbins 2010, 2013). Anhydrous boroaluminosilicate glasses synthesized at ambient pressure contain about 2% of tetrahedral species, the amount of which increases to approximately 6% upon hydration with 4.4 wt% water (Schmidt et al. 2004). In the same study, anhydrous glasses synthesized at 1.0 and 2.0 GPa have increased fractions of tetrahedral $[BO_4]$ to 3 and 7, respectively. Spectroscopic measurements of B speciation in three natural rhyolite glasses by Tonarini et al. (2003) revealed 74–92% of B in trigonal sites. Using a highly sophisticated in situ method of high-pressure ^{11}B NMR spectroscopy, Edwards et al. (2014) studied the trigonal $[BO_3]$ to tetrahedral $[BO_4]$ conversion in B-bearing glasses with increasing pressure, and found that the coordination change proceeds via a trigonal bipyramid as precursor phase.

In summary, in natural melts, the amount of tetrahedral boron units is generally small and only slightly increases with higher water content, increasing amounts of alkalis, or at high pressure and low temperature, which might be relevant for the melts of deeply subducted young slabs.

3.5 Experimental Determination of B-Isotope Fractionation

Here we present methods and some examples on the experimental determination of B-isotope fractionation between different phases. Firstly, we résumé experimental studies on vapor–liquid–melt fractionation at medium to high temperatures. A clear distinction needs to be made between diffusive fractionation at the crystal surface, which often represents disequilibrium, and equilibrium exchange due to bulk processes. We, thus, divide the second section of mineral–fluid fractionation into two different parts: (1) solid–aqueous fluid fractionation related to surface B adsorption at low temperatures, and (2) equilibrium solid–fluid fractionation at medium to high temperatures.

3.5.1 Vapor–Liquid–Melt Boron Isotope Fractionation

3.5.1.1 Vapor–Liquid Boron Isotope Fractionation

The B concentration and B-isotopic composition of natural hydrothermal fluids may be altered by various processes, including vapor–liquid separation. Numerous studies on fluid inclusions (e.g., Roedder 1979; Reynolds and Beane 1985; Hedenquist et al. 1998) have confirmed that liquid brines and low-density vapor can coexist across large sections through the altered oceanic crust and also around shallow crustal plutons. The distribution of B between coexisting liquid and vapor has been described for several natural geothermal systems (Smith et al. 1987; Chiodini et al. 1988; Glover 1988; Leeman et al. 1992). At low temperatures B fractionates preferentially into the liquid. $D_B^{liquid/vapor}$ and C_B^{liquid}/C_B^{vapor} decreases with temperature and varies from ~100 to ~10 between 100 and 300 °C (Leeman et al. 1992), and approaches unity at around 450 °C, 34–38 MPa (Spivack et al. 1990). At $T > 500$ °C B fractionates into the vapor phase with $D_B^{liquid/vapor} = 0.84$ at 528 °C, 57 MPa (Audétat et al. 1998) and 0.2 at 800 °C, 100 MPa (Schatz et al. 2004).

Investigating the dependence of $D_B^{liquid/vapor}$ on pressure and salinity and for determination of B-isotope fractionation between liquid and vapor at subcritical conditions, Liebscher et al. (2005) conducted experiments in a large-volume Ti autoclave (Fig. 3.2a), which enabled quasi-isobaric recovery of coexisting liquid and vapor. Liquid and vapor pairs were analyzed separately by ICP-AES for Na and B and by positive thermal ionization mass spectroscopy (p-TIMS) for their B isotopic composition. Experiments were performed in the system H_2O–$NaCl$–B_2O_3 along the 400 and 450 °C isotherms, ranging from 23 to 28 MPa and from 38 to 42 MPa, respectively. The starting fluids had a

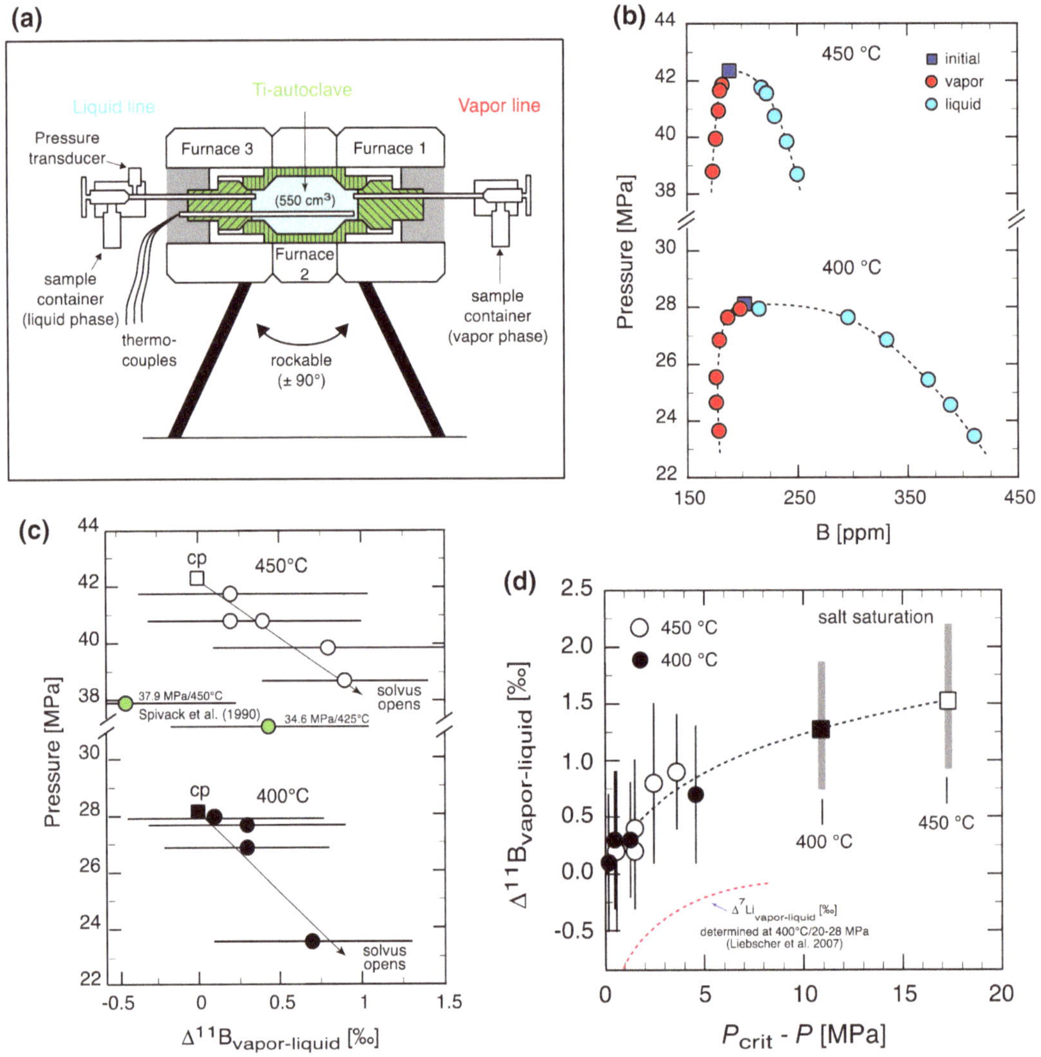

Fig. 3.2 Experimental data on vapor–liquid fractionation of B and its isotopes in the system H_2O–NaCl–B_2O_3 at high-salinity conditions according to Liebscher et al. (2005). **a** Schematic drawing of a large-volume Ti-autoclave to extract immiscible fluids; **b** B concentration in coexisting vapor–liquid pairs along the 450 and 400 °C isotherms; **c** B isotope fraction between coexisting vapor and liquid along the 450 and 400 °C isotherms, with additional low-salinity data (*green* symbols) of Spivack et al. (1990); **d** extrapolation of derived $\Delta^{11}B_{\text{vapor-liquid}}$ at 450 and 400 °C to salt-saturated conditions as a function of the distance to the critical pressure. For comparison, vapor–liquid Li-isotopic fractionation data are also shown (Liebscher et al. 2007). All figures are modified after Liebscher et al. (2005)

neutral pH (at room temperature), contained 200 μg/g B of natural seawater isotopic composition (~39.5‰), and NaCl concentrations that correspond to the composition at the critical point of the respective isotherms. Boron liquid–vapor fractionation is generally small and increases from 1.0 at the critical points with increasing opening of the H_2O–NaCl solvus (Fig. 3.2b). The maximum values at extrapolated salt saturated conditions are $D_B^{\text{liquid/vapor}} = 2.7$ (at 400 °C) and 1.8 (at 450 °C).

B-isotope compositions of coexisting vapor and liquid from natural geothermal systems induced by phase separation at 170–260 °C were studied by Leeman et al. (1992). They determined $\Delta^{11}B_{vapor\text{-}liquid}$ in the range of about 1–3‰, with more pronounced ^{11}B-enrichment in the vapor phase at low temperatures. In experiments within systems of high salinity described above (Liebscher et al. 2005), B-isotope fractionation between vapor and liquid always yielded fractionation of ^{11}B into the vapor phase, analogous to data from the natural systems (Leeman et al. 1992). However, the experimentally derived $\Delta^{11}B_{vapor\text{-}liquid}$ values are smaller and range from +0.2(± 0.7) to +0.9(± 0.5)‰ at 450 °C and from +0.1(± 0.6) to +0.7(± 0.6)‰ at 400 °C (Fig. 3.2c), without any obvious *T*-effect on $\Delta^{11}B_{vapor\text{-}liquid}$. At extrapolated salt-saturated conditions, calculations resulted in $\Delta^{11}B_{vapor\text{-}liquid}$ values of +1.5(± 0.7)‰ at 450 °C and +1.3(± 0.6)‰ at 400 °C (Fig. 3.2d). The experimental data perhaps indicate an increase of vapor - liquid B-isotope fractionation with decreasing pressure during increasing opening of the solvus (Fig. 3.2c). The same trend of an increasingly positive isotope fractionation with opening of the solvus was also observed for vapor–fluid (Fig. 3.2d) from experiments in the system LiCl–H_2O at 400 °C/20–28 MPa (Liebscher et al. 2007). The weak B-isotopic fractionation suggests a similar trigonal speciation of B in vapor and the liquid. Increasing fractionation might be explained by a slight increase of the $[B(OH)_4]^-$ species in the liquid with decreasing pressure (Liebscher et al. 2005) as the pH of the liquid relative to that of the conjugated vapor increases (Vakulenko et al. 1989, see above Chap. 4.1). From these experimental results it was concluded that fluid phase separation is able to significantly alter the B isotope signature of hydrous fluids in open hydrothermal systems. However, due to the low value of $\Delta^{11}B_{vapor\text{-}liquid}$, processes other than vapor–liquid separation might dominate the B isotope geochemistry in oceanic hydrothermal fluids (Liebscher et al. 2005). The B-isotope vapor–liquid fractionation data for high-salinity brines (Liebscher et al. 2005) slightly differ from $\Delta^{11}B_{vapor\text{-}liquid}$ values determined for low-salinity vapor–fluid systems near the critical point (Fig. 3.2c) of seawater in an earlier experimental study (Spivack et al. 1990). They found that $\Delta^{11}B_{vapor\text{-}liquid}$ = + 0.4(± 0.7)‰ (at 34.6 MPa, 425 °C) and −0.4(± 0.7)‰ at 37.9 MPa, 450 °C, which suggests that within the analytical and experimental uncertainties, the fractionation is negligible (Spivack et al. 1990). Thus Spivack et al. (1990) concluded that phase separation in high-temperature geothermal systems is unlikely to have a significant effect on the B-isotopic composition of liquid and vapor.

3.5.1.2 Aqueous Fluid–Melt B-Isotope Fractionation

Hervig et al. (2002) have experimentally investigated B-isotope fractionation between hydrous fluids and silicate melt of basaltic and rhyolitic composition over a pressure–temperature range of 110–500 MPa, 750–1100 °C, which are conditions above the critical point; thus the fluid is defined here as including gas and liquid. The experiments were performed in a piston-cylinder press and an internally heated rapid-quench hydrothermal apparatus, using B-bearing glass chips (containing approximately 2000 µg/g B) and water in a 1:1 ratio as starting materials, sealed in Au or AuPd capsules. The B concentration and B-isotopic composition of the starting and run-product glasses were analyzed by secondary ion mass spectrometry (SIMS); B contents and $\delta^{11}B_{fluid}$ of coexisting fluids were calculated by mass balance considerations.

The partitioning of B between coexisting fluid and melt ($D_B^{fluid/melt}$) is 0.33–0.54 at 950–1100 °C, 110–170 MPa for the system fluid–basaltic melt and 1.19–1.20 at 750–850 °C, 500 MPa for fluid–rhyolitic melt. All experiments on B-isotopic fractionation between melt and aqueous fluid ($\Delta^{11}B_{melt\text{-}fluid}$) showed significant preferential partitioning of ^{11}B into the hydrous fluid over the silicate melts. For the basaltic system, $\Delta^{11}B_{melt\text{-}fluid}$ values were determined to be: −5.2 (± 1.6)‰ (at 950 °C, 140 MPa), −3.6(± 2.0)‰ (at 1080 °C, 110 MPa), −1.3(± 3.4)‰ (at 1100 °C, 170 MPa); and for the rhyolitic system: −7.1(± 2.0)‰ (at 750 °C, 500 MPa) and −4.4 (± 2.0)‰ (at 850 °C, 500 MPa). Extrapolating these $\Delta^{11}B_{melt\text{-}fluid}$ values to low temperatures, the extent of melt–fluid B-isotope fractionation corresponds to that known for the B-isotope fractionation between the 4- and 3-fold coordinated phases illite–fluid (Williams et al. 2001) and mica–fluid, respectively (Wunder et al. 2005, see also later Fig. 3.10). Therefore, Hervig et al. (2002) concluded that the large $\Delta^{11}B_{melt\text{-}fluid}$ results from significant differences in B coordination between the two coexisting phases, i.e., that B mostly exists as tetrahedral $[BO_4]$ in the melts and as trigonal $B(OH)_3$ species in the fluids. However, this is in contradiction to the accepted understanding described above (Sect. 3.4.3), in which $[BO_3]$ is the predominant B species in melts, particularly in peraluminous rhyolitic compositions, which have at most 26% of the B at tetrahedral sites, as determined for rhyolitic glasses (Tonarini et al. 2003); corresponding rhyolitic melts are expected to have an even lower proportion of $[BO_4]$. Therefore, the rational of the strong B-isotopic melt–fluid fractionation at magmatic temperatures observed in the experimental study of Hervig et al. (2002) is not fully understood. Future experimental studies should be completed, for example, to investigate possible kinetically driven B-isotopic fractionation processes, which are known to be significant for lithium and its isotopes (Tomascak et al. 2016).

3.5.2 Solid–Fluid B-Isotope Fractionation

3.5.2.1 Low-Temperature Solid–Aqueous Fluid B-Isotope Fractionation

Diffusive reequilibration of solids with an isotopically distinct B-bearing aqueous fluid is the usual approach to study low-temperature solid–fluid B-isotope fractionation. In most of these exchange experiments, B is added to the solid in a combination of (1) weak adsorption to the mineral's surface and to interlayer positions (in clay minerals) and/or (2) boron incorporation at distinct structural positions by coupled substitutions. The adsorption of weakly bonded B species strongly depends on fluid-pH and charge of the surface. It is therefore essential to characterize the reaction mechanism during adsorption of dissolved boron, for understanding B-isotope fractionation in low-*T* regimes.

Carbonate–Fluid B-Isotope Fractionation

In many studies, the low-temperature B-isotope fractionation in the system carbonate–fluid has been attributed to the exclusive adsorption of the charged borate ion $[B(OH)_4]^-$ (e.g., Hemming et al. 1995; Sanyal et al. 2000). Some more recent results suggest more complex schemes of adsorption (Tossel 2006; Klochko et al. 2009): after surface adsorption of $B(OH)_3$ and/or $[B(OH)_4]^-$ (stage 1), reaction of the two B species with HCO_3^- proceeds in an intermediate stage forming an amorphous carbonate surface layer (stage 2); structural incorporation of trigonal and/or tetrahedral species is completed in the final stage 3.

Boron is reported as a typical trace element in natural carbonates, with concentrations in the order of ~1 to ~100 μg/g (e.g., Furst et al. 1976; Vengosh et al. 1989; Hemming and Hanson 1992; Branson 2017; Marschall 2017, this book). Low-temperature experiments indicate that at ambient conditions and near neutral pH, B uptake by aragonite is greater by a factor of 3–5 compared to calcite; additionally, the presence of Mg in the parent solution results in an increased B incorporation into calcite compared to Mg-free solutions (Hemming et al. 1995). Experiments by Sanyal et al. (2000) yielded an increased incorporation of B into calcite with increasing pH, which was explained by greater $[B(OH)_4]^-$ uptake, due to the increased $[B(OH)_4]^-/B(OH)^{3-}$ratio in the more basic fluids. ^{11}B MAS NMR investigations of Sen et al. (1994) show that in synthetic calcite boron occurs principally in trigonal coordination, whereas it is tetrahedrally coordinated in synthetic aragonite. For calcite it was assumed that $[BO_3]^{3-}$ replaces $[CO_3]^{2-}$ in a coupled substitution; on the other hand, the structural role of $[BO_4]^{5-}$ in aragonite was ambiguous (Sen et al. 1994). These results are contrasted by NMR-investigations of Klochko et al. (2009), who observed mixtures of trigonally and tetrahedrally coordinated B in almost equal abundances in both natural biogenic calcite and aragonite. In a calcite-aragonite co-precipitation study under controlled pH (7.4–9.5), temperature (5–25 °C) and precipitation rate, Mavromatis et al. (2015) showed with ^{11}B MAS NMR that B in aragonite is mostly ($\geq 85\%$) tetrahedrally coordinated. In contrast, proportions of trigonally and tetrahedrally coordinated B are highly variable in calcite, with the amount of $^{[4]}B$ increasing with precipitation rate. These data indicate that adsorbed surface $[B(OH)_4]^-$ mostly preserves its coordination upon structural incorporation into aragonite, but can undergo a structural coordination change during incorporation into calcite. Recent first-principles calculations (Belan et al. 2016) indicate substitution of $[CO_3]^{2-}$ anions by $[BO_2(OH)]^{2-}$ and $[B(OH)_4]^-$ groups in calcite, and mainly by $[B(OH)_4]^-$ species in aragonite. Synchrotron X-ray spectroscopic investigations clearly show that B is hosted solely as trigonal BO_3 in foraminiferal calcite (*Amphistegina lessonii*) and that B-concentration exhibits banding on the micrometer length scale (Branson et al. 2015). Using Atomic Force Microscopy (AFM), Ruiz-Agudo et al. (2012) showed that calcite growth rates are site-specific and strongly depend on pH and B supersaturation within the fluid. For further details see the book chapters by Branson (2017), Rae (2017), and McCulloch (2017).

Hemming et al. (1995) were the first to experimentally investigate B-isotopic fractionation between carbonates and fluids during growth of synthetic carbonates (calcite, aragonite, (Ca,Mg)-carbonate) from solutions at ambient pressure–temperature conditions and a constant pH of approximately 8. The yielded B-isotopic carbonate–fluid fractionation ($\Delta^{11}B_{carbonate\text{-}fluid} = -16.5$ ($\pm$ 0.7)‰) was indistinguishable among the three different carbonates, indicating no dependence of B-isotope fractionation on carbonate structure and chemistry. The determined $\Delta^{11}B_{carbonate\text{-}fluid}$ is in good agreement with B-isotope fractionation observed between modern marine carbonates and seawater with an average pH of 8.2 (Hemming and Hanson 1992). The B-isotopic composition of carbonate corresponded to that of the calculated B-isotopic composition of charged $[B(OH)_4]^-$ species in the experimental fluid. Hence, the experimental data suggest adsorption of $[B(OH)_4]^-$ to the crystal surface combined with preferential ^{10}B fractionation as the dominant influence on B uptake in carbonates. Sanyal et al. (2000) experimentally investigated the dependence of the B-isotopic composition of calcite on the pH of seawater at ambient pressure and temperature conditions. For the pH-range of 7.9–8.6, which approximately covers modern and ancient seawater, B incorporation into calcite strongly increased with increasing pH (pH = 7.9, ~34 µg/g; pH = 8.3, ~48 µg/g; pH = 8.6, ~84 µg/g). The experimental results show a clear relationship between $\delta^{11}B$ of precipitated calcite and the pH of seawater (pH = 7.9, $\delta^{11}B = -20.3$ ($\pm$ 0.3)‰; pH = 8.3, $\delta^{11}B = -17.9(\pm 0.4)$‰; pH = 8.6, $\delta^{11}B = -14.8(\pm 0.6)$‰). These data are consistent with former conclusions (Hemming et al. 1995), determining $[B(OH)_4]^-$ as the dominant species adsorbed on the carbonate surface in contact with seawater. Compared to results from pH-dependent B-isotope exchange experiments by cultivation of foraminifera (*Orbulina universa*) in contact with seawater (Sanyal et al. 1996), the B-isotopic fractionation between inorganic calcite and seawater (Sanyal et al. 2000) is lower by approximately 1‰, which indicates the presence of a small biogenic effect on B-isotope carbonate–seawater fractionation. The extent of such a biogenic effect might vary for different biogenic carbonate species (Sanyal et al. 1996).

In summary, small changes in the acidity of the oceans as a result of changing atmospheric CO_2 levels will significantly change the pH with respect to the proportions of $B(OH)_3$ and $[B(OH)_4]^-$ species dissolved in seawater. From the experimental studies (Hemming et al. 1995; Sanyal et al. 1996, 2000), it is obvious that the B-isotopic composition of carbonates in marine sediments is a powerful tool to reconstruct paleo-ocean pH and ultimately past concentrations of atmospheric CO_2. However, data on B coordination in aragonite and calcite are partly conflicting, which in turn affects the interpretation of measured $\delta^{11}B$ for natural carbonates. This calls for further detailed studies on both structural B incorporation and surface-related mechanisms of B adsorption to different carbonates (calcite, aragonite, carbonates of various solid solutions).

Clay Mineral–Fluid B-Isotope Fractionation

Palmer et al. (1987) proposed that incorporation of B in clay minerals (e.g., illite) is initialized by adsorption of trigonal $B(OH)_3$ to its surface. In a second step, which includes B-isotope fractionation, the coordination of the B species changes from trigonal to pseudo-tetrahedral at the clay's surface. During ongoing exchange, the adsorbed tetrahedral species (enriched in ^{10}B) is then incorporated into the Al-silicate structure, without further isotope fractionation (Palmer et al. 1987). This concept of adsorption mechanism has been modified because more recent studies (e.g., Singh and Mattigod 1992; Goldberg et al. 1996; Williams et al. 2001; Środoň and McCarty 2008) show that adsorption is a function of the charge density in non-homogenous structures, such as phyllosilicates, and the temperature- and pH-dependent mineral's zero point of charge (charge density on the surface is zero). Expandable clay minerals have different kind of charged surfaces: edge(+), exterior(−), basal(−), to which the weak B-adsorption of different aqueous B-species is strongly dependent on fluid pH. For example, at low pH (below 7), the hydroxyl-group of boric acid may absorb to the H^+ on the negatively charged exterior and basal siloxane surfaces; above pH 7 the anionic species $[B(OH)_4]^-$ should become prominent to adsorb to the positively charged clay's edges.

In clastic sediments, B is hosted predominantly in the clay mineral illite, which can contain orders of magnitude more B (up to several hundred μg/g) than other common diagenetic minerals (e.g., smectites, carbonates). Boron in clays is located at three different structural positions: (1) at low temperatures <120 °C clay minerals adsorb B to their surface by weak bonding (see above); (2) in expanded smectite minerals boron might also be adsorbed at interlayer positions as both trigonal and tetrahedral species, perhaps reflecting the composition of the fluid (Williams et al. 2007); (3) in smectite and illite B substitutes for Si by a coupled substitution at structural tetrahedral positions within the layers of the sheet silicates (Spivack and Edmond 1987).

Schwarcz et al. (1969) were the first to study B-isotope clay–fluid fractionation by B adsorption on clays during interaction in contact with synthetic seawater at 23 °C. Fractionation factors α were calculated to be 0.957–0.963 for reversible and 0.970–0.974 for irreversible adsorption, respectively. Based on the interpretation of Schwarcz et al. (1969) that the real adsorption process is partly reversible, and the observation that adsorbed B on clays is isotopically light, the assumption of a pH control on the isotope composition of adsorbed B arose. The pH (pH-range: 6.65–8.45) and temperature (*T*-range: 5–40 °C) controls over B-isotopic fractionation during B adsorption from seawater on marine clay-bearing sediments were studied experimentally by Palmer et al. (1987). In the investigated *T*-range (at constant pH) the B-isotope fractionation was almost invariant; with increasing pH (at constant temperature) the B-isotope fractionation decreased as $[B(OH)_4]^-$ becomes the dominant aqueous species (see below, Fig. 3.10). For the investigated conditions, the clay's surface-adsorbed B is 23.4–34.3‰ lighter than the boron dissolved in the coexisting seawater ($\delta^{11}B$ = + 39.61 ± 0.04‰; Foster et al. 2010). These values are much larger than the adsorption on carbonates at the same pH and temperature (Hemming et al. 1995).

According to Williams and Hervig (2005), the illitization of smectite represents a dissolution/recrystallization process. The clay mineral $\delta^{11}B$ values, therefore, vary during crystal growth as the fluid B composition or temperature change (Clauer et al. 2014). Williams et al. (2001) investigated the B-isotopic change during illitization of smectite under conditions simulating burial digenesis. Experiments were performed at 300 and 350 °C at 100 MPa with smectite as a solid starting material and B-bearing fluid containing 1000 μg/g B, sealed in Au-tubes in a hydrothermal apparatus for different run-durations of up to 150 days. After the experiments, the solids were washed in water and mannitol solution (0.1 N) to facilitate removal of exterior surface-adsorbed B (Tonarini et al. 1997). Adsorbed interlayer B was removed by cation exchange (Moore and Reynolds 1997). Analyses of the B-isotopic composition of the solids were done by SIMS and negative Thermal Ionization Mass Spectrometry (n-TIMS); the B-isotopic composition of the fluids was not measured, but determined from mass-balance calculations. Equilibrium was determined by monitoring the oxygen isotopes of the mineral and fluid, as the equilibrium oxygen isotope fractionation of illite as a function of temperature is well known (Longstaffe 1987). Since B substitution requires breaking of Si–O–bonds, it follows that B is at equilibrium when O is at equilibrium with the fluid. Williams et al. (2001) interpreted their results on B isotope clay–fluid fractionation as follows: during early stages of smectite recrystallization, aqueous B may be adsorbed onto the clay's exterior surfaces and some may be incorporated into the smecitite's interlayer site where the dominant aqueous B species varies according to the layer charge distribution. As smectite recrystallizes to illite, interlayer collapse occurs during ordering (Altaner and Ylagen 1997). Eventually, B is substituted in tetrahedral sites of illite replacing Si (in competition with Al^{3+}) in a coupled substitution. The B-isotopic illite–fluid fractionation of well-ordered illite resulted in a $\Delta^{11}B_{illite\text{-}fluid} = -13‰$ (at 350 °C) and −16‰ (at 300 °C). The long-duration experiments represent fractionation between the trigonal aqueous B (quench pH of the fluid was 6.0, thus the fluid mainly contained $B(OH)_3$ species) and structural tetrahedral B, and is a linear function of temperature. B-isotopic characterization of natural and synthetic clays with various degrees of smectite–illite ordering revealed that trigonal interlayer boron is isotopically heavier than tetrahedral boron by up to 40‰. Williams et al. (2007) observed that the intracrystalline B-isotope fractionation for clays ($\Delta^{11}B_{tetrahedral\text{-}interlayer} = \delta^{11}B_{tetrahedral} - \delta^{11}B_{interlayer}$) is strongly dependent on the degree of clay ordering and thus on the temperature (or reaction time) of recrystallization. Accordingly, Williams et al. (2007) proposed a single-mineral geothermometer approximated by the equation T (°C) = ($\Delta^{11}B_{tetrahedral\text{-}interlayer}$ + 30)/0.05, which however, only applies to samples in which the interlayer B species is in equilibrium with the silicate (closed system).

Boron-Isotope Fractionation in Other Low-Temperature Solid–Fluid Systems

Isotopic fractionation between the $B(OH)_3$ and $[B(OH)_4]^-$ species has been studied experimentally by Kakihana et al. (1977) at low temperatures by adsorption of B from boric acid solution to an ion-exchange resin. The results were used for the theoretical determination of reduced partition function ratios to calculate B-isotope fractionation between trigonal and tetrahedral sites. The calculated fractionation factor α to for Eq. (3.1) decreases from 1.0206 to 1.0177 for the temperature range 0 to 60 °C, with α (Kakihana et al. 1977). A larger fractionation factor α of 1.028 (± 0.001) at 25 °C was determined experimentally by Byrne et al. (2006) using spectrophotometric absorbance measurements of aqueous solutions, which is in good agreement with theoretical assessments (e.g., Zeebe 2005; Liu and Tossel 2005; Rustad et al. 2010a, see below).

Oi et al. (1997) found isotopic fractionation between trigonal and tetrahedral species to vary between 18 and 22‰ for different kinds of hydrocarbon-based ion-exchange resins. They concluded that smaller fractionation results from the adsorption of some trigonal B-species as well as the dominant tetrahedral species, with different resins adsorbing more or less trigonal boron. All the resins studied were hydrocarbon-based, and adsorption was assumed to occur via $B(OH)_2(OC)_2$ clusters (Oi et al. 1997).

Synthesis experiments on B-bearing minerals at the low temperature of 25 °C from aqueous boric acid–sodium hydroxide solutions at various pH conditions (Oi et al. 1991) resulted in the formation of sassolite ($B(OH_3)$) at low pH (5.4–5.9), and borax ($Na_2(B_4O_5(OH)_4{\cdot}3H_2O)$) at neutral to high pH (7.4–11.5). B-isotope analyses of sassolite and water, both containing B in trigonal coordination, showed B isotopic fractionation ($\Delta^{11}B_{sassolite\text{-}fluid}$ = + 4.3‰ at pH = 5.9 and +0.6‰ at pH = 5.5), with preferential fractionation of ^{11}B into the solid phase. Analyses of borax, which contains B in both trigonal and tetrahedral coordination in equal amounts, showed significant pH-dependent fractionation ($\Delta^{11}B_{borax\text{-}fluid}$ = −12.6 to +13.8‰) in the pH-range 7.0–11.5, with ^{11}B fractionating into the fluid phase at pH $\leq$ 8.7 and ^{11}B fractionating into borax at pH $\geq$ 10.3 (Oi et al. 1991).

Vengosh et al. (1992) studied B-isotope variations in brines and coprecipitated salt minerals (gypsum, halite) and potassium-magnesium sulfate minerals during evaporation of seawater at 25 °C. They found similar B-isotope fractionation data as for smectite–fluid (Palmer et al. 1987), which was interpreted as coprecipitation of $[B(OH)_4]^-$ species with the salts during evaporation of seawater (see below, Table 3.1 and Fig. 3.10).

Strong pH-dependent B-isotope solid–fluid fractionation at 25 °C during surface adsorption was determined by Lemarchand et al. (2007) for goethite (α-FeOOH): $\Delta^{11}B_{goethite\text{-}fluid}$ = −40 to 0‰ in the pH-range ~8–10 and birnessite ($K_{0.1}MnO_{2.2}{\cdot}0.9H_2O$): $\Delta^{11}B_{birnesite\text{-}fluid}$ = −15 to +23‰ in the pH-range ~8–10.8. Maximum boron adsorption on both of the hydroxide/oxide surfaces occurred at 8 < pH < 9. Adsorption and the subsequent B-isotope fractionation were rationalized by the formation of trigonal and tetrahedral boron inner-sphere complexes on the goethite surface, and tetrahedral inner-sphere and trigonal outer and inner-sphere complexes on the birnessite surface (Lemarchand et al. 2007).

The strong influence of charge properties at a mineral's surface on B-isotope solid–fluid fractionation was experimentally shown for various (hydr)oxide–fluid systems by Elmaci et al. (2015). For the manganese oxides birnesite ($K_{0.3}MnO_2{\cdot}1.8H_2O$) and cryptomelane ($K_{0.1}MnO_2{\cdot}0.5H_2O$) both possessing negatively charged surfaces, $[^{10}B(OH)_4]^-$ is preferentially fractionated into the solution (pH = 8.5). For $[Mg_{0.58}Al_{0.42}(OH)_2](NO_3)_{0.42}{\cdot}0.6H_2O$, a Mg–Al–layered double hydroxide, selective $[^{10}B(OH)_4]^-$ adsorption was favored by the positively charged layers of the solid at a fluid-pH of 9.5–10.

3.5.2.2 High-Temperature Solid–Aqueous Fluid B-Isotope Fractionation

Some solid phases, e.g., tourmaline, are known to exhibit very low diffusivities. Due to this refractory nature, low-temperature diffusive reequilibrium experiments are generally precluded. In the high-temperature B-isotope fractionation experiments for mica–fluid and tourmaline–fluid (both discussed below), the solids were synthesized in situ from their components in the presence of a coexisting B-bearing fluid. During this procedure, B is incorporated structurally as a main component at distinct crystallographic positions. In the following, boron adsorbed at the crystal surface is neglected, as its volume concentration is dominating over the amount of boron at the low surface area for solids synthesized at high temperatures ($\geq$ 350 °C).

Mica–Fluid B-Isotope Fractionation

White mica is ubiquitous in metamorphic rocks of both igneous and sedimentary origin, and is stable to depths of 300 km in cold subducting slabs where, with progressive subduction, its modal amount continuously decreases and its composition shifts towards phengitic mica (Schmidt 1996). White mica is an important carrier of fluid-mobile trace elements including B, Li, and large ion lithophile elements. Because of its high modal abundance, white mica dominates the B-budget in most tourmaline-free rocks and its B-isotopic composition can be considered as representative of the bulk rock (Peacock and Hervig 1999). Measuring the B-concentration in subduction zone minerals (excluding tourmaline and other boron minerals), white mica contains the highest amount, with up to 269 µg/g, followed by lawsonite, biotite, chlorite, and amphibole all having B concentrations below 100 µg/g (Domanik et al. 1993; Marschall et al. 2006). The maximum B-incorporation capacity in white mica is given for boromuscovite, with

Table 3.1 Compilation of selected experimentally determined vapor, melt, solid–fluid B-isotope fractionation

system	$\Delta^{11}B$ (‰)	T(°C)/P(MPa)/pH	References	#[a]
$^{[3]}$vapor–$^{[3]}$fluid	+0.1 ± 0.6 to +0.7 ± 0.6	400/23–28/~ 7	Liebscher et al. (2005)	1
	+0.2 ± 0.7 to +0.9 ± 0.5	450/38–42/~7		
	+0.4 ± 0.7	425/34.6/~7	Spivack et al. (1990)	2
	−0.4 ± 0.7	450/37.9/~7		
$^{d[3,4?]}$melt–$^{[3]}$fluid	−5.2 ± 1.6	950/140/~7	Hervig et al. (2002)	3
d	−3.6 ± 2.0	1080/110/~7		
d	−1.3 ± 3.4	1100/170/~7		
e	−7.1 ± 2.0	750/500/~7		
e	−4.4 ± 2.0	850/500/~7		
$^{[4,3?]}$carb^{b}–$^{[3]}$fluid	−16.5 ± 0.7	23/0.1/~8	Hemming et al. (1995)	4
	−20.3 ± 0.3 to −14.8 ± 0.6	23/0.1/7.9–8.6	Sanyal et al. (2000)	5
$^{[3,4]}$smectite–$^{[3,4]}$fluid	−34.3 ± 0.2 to −23.4 ± 0.2	5–40/0.1/6.65–8.45	Palmer et al. (1987)	6[f]
$^{[4]}$illite–$^{[3]}$fluid	−16 ± 1.5	300/100/~6	Williams et al. (2001)	7
	−13 ± 1.5	350/100/~6		
$^{[3]}$sassolite–$^{[3]}$fluid	+0.6 ± 0.4 to +4.3 ± 0.6	25/0.1/5.5–5.9	Oi et al. (1991)	8
$^{[3,4]}$borax–$^{[3,4]}$fluid	−12.6 ± 0.4 to +13.8 ± 0.8	25/0.1/7.0–11.5	Oi et al. (1991)	9
$^{[4]}$salt^{c}–$^{[3]}$fluid	−18.7 ± 0.6 to −32.7 ± 0.6	25/0.1/~7	Vengosh et al. (1992)	10
$^{[4]}$mica–$^{[3]}$fluid	−10.9 ± 1.3	500/3000/acidic	Wunder et al. (2005)	11
	−6.5 ± 0.4	700/3000/acidic		
$^{[4]}$mica–$^{[4]}$fluid	−7.4 ± 1.0	400/400/basic	Wunder et al. (2005)	12
	−4.8 ± 1.0	500/400/basic		
$^{[3]}$tour–$^{[3]}$fluid	−8.35 ± 0.4 to −5.88 ± 0.4	350–450/50/acidic	Palmer et al. (1992)	13
	−7.76 ± 0.4 to −2.34 ± 0.4	350–750/100/acidic		
	−6.50 ± 0.4 to −2.24 ± 0.4	350–750/200/acidic		
	−2.7 ± 0.5 to −0.8 ± 0.5	400–700/200/acidic	Meyer et al. (2008)	14
	−2.0 ± 0.6	500/500/acidic		

Abbreviations carb = carbonate; tour = tourmaline
[a]Corresponding numbers in Fig. 3.10; [b]calcite, aragonite, Mg-calcite; [c]gypsum, halite, K–$MgSO_4$; [d]basaltic melt; [e]rhyolitic melt; [f]B-isotope fractionation due to surface adsorption

end member composition $KAl_2[BSi_3O_{10}](OH)_2$, containing approximately 9 wt% B_2O_3. Boromuscovite was chosen as representative of B-bearing white mica in the experiments on B isotope mica–fluid fractionation performed by Wunder et al. (2005), which was determined using three different approaches: (a) synthesis of boromuscovite from gels with boric acid solutions (weakly acidic conditions with $B(OH)_3$ as the major B-bearing species in the fluid); (b) synthesis of boromuscovite from an oxide mixture and a 7.6 M KOH-solution (strongly basic conditions with $[B(OH)_4]^-$species in the fluid); (c) time-dependent exchange experiments lasting up to 28 days with $B(OH)_3$-bearing aqueous solutions, whose initial isotopic composition was far off the expected equilibrium value using the previously synthesized boromuscovite with known isotopic composition (Fig. 3.3a, b). Experiments using approaches (a) and (c) were performed in a piston cylinder press at 3.0 GPa, 500 °C and at 3.0 GPa,

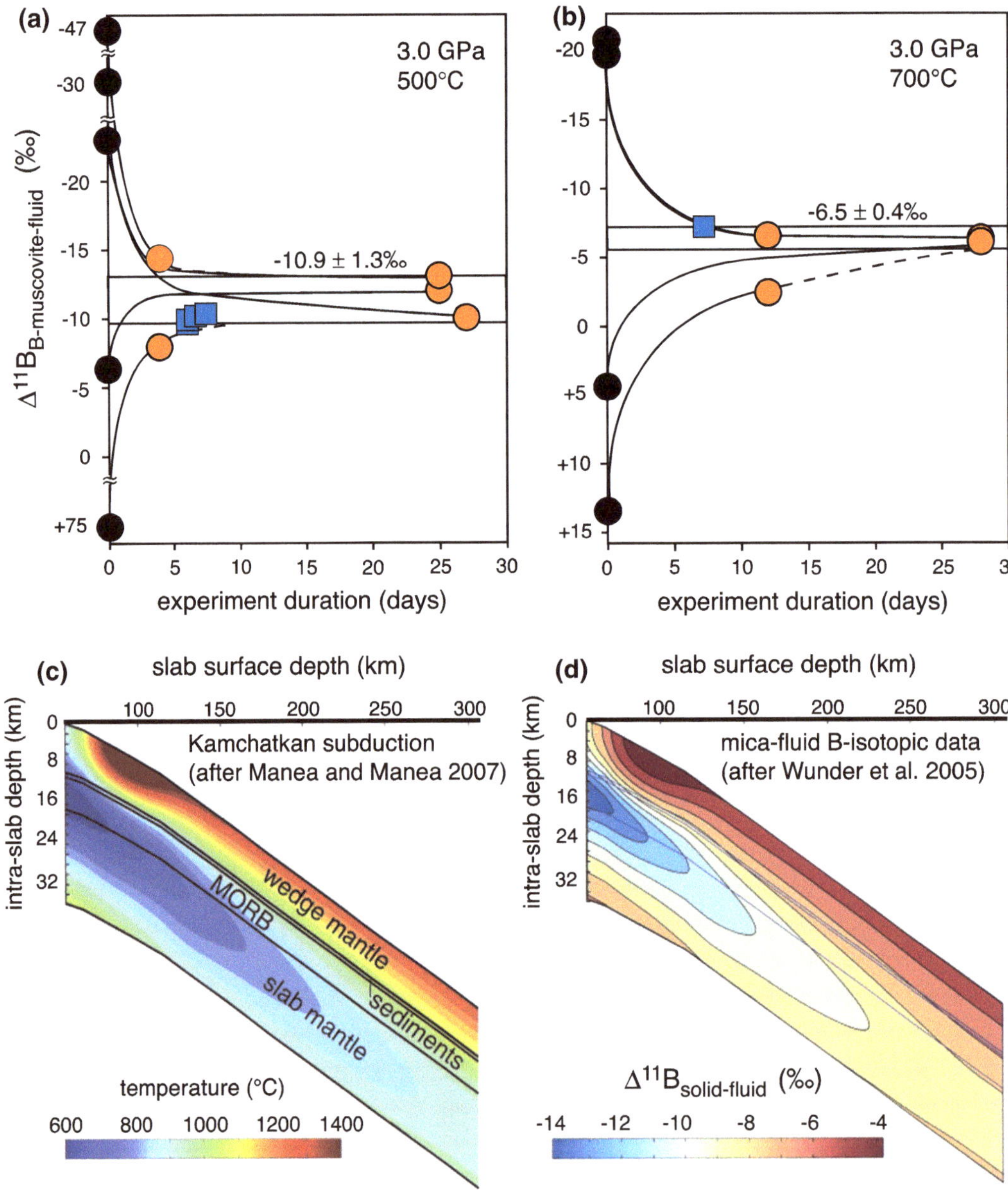

Fig. 3.3 **a**, **b** Experimental data on mica—fluid B-isotope fractionation slightly modified after Wunder et al. (2005) for experiments performed at 500 and 700 °C at weakly acidic conditions. *Black circles* show isotopic composition of exchange experiments at their starting conditions; *orange circles* denote the measured $\Delta^{11}B_{mica\text{--}fluid}$ for time-dependent experiments. Results from synthesis experiments (in *dark blue*) fit within error; **c** the thermal structure of the Kamchatka subduction zone used in the thermodynamic model; **d** the distribution of the B-isotopic fractionation values in the subducting plate (modified after Konrad-Schmolke and Halama 2014) using fractionation data after Wunder et al. (2005)

700 °C; experiments using approach (b) were done in a hydrothermal apparatus at 0.4 GPa, 400 °C and at 0.4 GPa, 500 °C. The solids were characterized by powder X-ray diffraction followed by Rietveld refinement and by transmission electron microscopy (TEM) using electron energy-loss spectroscopy (EELS) for B/K-ratio determination. Both the solid run

products and coexisting fluids were characterized by p-TIMS for their B-isotopic composition. All experiments produced only boromuscovite with traces of quartz in some of the experiments. The boromuscovite consisted mainly of the $2M_1$ polytype (approximately 90%) and approximately 10% of the 1 M polytype (see Chap. 6.3.2 below, for the small effect that different mica-polytypes have on B-isotope fractionation). The synthesis approach (a) and the time-dependent experiments (c) resulted in approximately the same fractionations (Fig. 3.3a, b): $\Delta^{11}B_{mica\text{-}fluid} = -10.9 \pm 1.3‰$ (at 3.0 GPa, 500 °C) and $-6.5 \pm 0.4‰$ (at 3.0 GPa, 700 °C). Experiments at 0.4 GPa that used the strongly basic fluid produced significantly less fractionation $\Delta^{11}B_{mica\text{-}fluid} = -7.4 \pm 1.0‰$ (at 400 °C) and $-4.8 \pm 1.0‰$ (at 500 °C), reflecting the reduced fractionation effect, when large amounts of boron in basic fluids are tetrahedrally coordinated. From these results, Wunder et al. (2005) proposed that the B-isotopic signature in volcanic arcs might result from continuous dehydration of micas, acting as important B carrier during subduction.

Marschall et al. (2007) modeled the release of B in a progressively dehydrating altered oceanic crust during subduction. The derived information on the B-concentration of rocks and fluids at different stages of the *P-T*-path of the subducting slab was combined with *T*-dependent B-isotope fractionation data (Wunder et al. 2005). Assuming low-pH fluids (i.e., trigonally coordinated B in the fluid), this resulted in a model of the B-isotopic evolution of subducting rocks and the released fluids from greenschist to eclogite conditions (Marschall et al. 2007). Konrad-Schmolke and Halama (2014) extended this model by combining thermodynamic modeling of phase relations at subduction conditions for the Kamchatkan subduction zone (Fig. 3.3c), with *T*-dependent mica–fluid B-isotope fractionation (Wunder et al. 2005), in order to quantify the B concentration and B-isotopic composition of coexisting fluid and solid phases during devolatilization, fluid migration and fluid-rock interaction in a subducting slab (Fig. 3.3d). Their derived data provides insights into element cycles, and their model can be used to constrain metamorphic processes and determine thermo-mechanical parameters for subduction zone models (Konrad-Schmolke and Halama 2014). For more details on this topic see De Hoog and Savov (2017).

Tourmaline—Fluid B-Isotope Fractionation
Tourmaline forms in various geochemical environments that have undergone diagenetic, metamorphic, metasomatic, or magmatic processes over a wide range of bulk compositions and has a large *P–T*-stability field ranging from near-surface to mantle conditions (e.g., Dutrow and Henry 2011; van Hinsberg et al. 2011). Negligible diffusion of elements through tourmaline's structure results in the preservation of growth and sector zoning (van Hinsberg and Schumacher 2007). Its refractory behavior has allowed tourmaline to be successfully used for geothermometry (e.g., van Hinsberg and Schumacher 2007), geobarometry (Berryman et al. 2015), for provenance studies (e.g., Dutrow and Henry 2011), as a monitor for fluid-rock interactions in metamorphic rocks (e.g., von Goerne et al. 2001), as well as in tourmaline-bearing fluid–melt–rock systems (e.g., Trumbull and Chaussidon 1999). Tourmaline's B-isotope composition, which can also be affected by growth zoning (van Hinsberg and Marschall 2007), extends its applicability to include its use as a geochemical tracer for geological mass transfer, fluid origin and evolution, and even for evaluating the source and genesis of ore deposits (Marschall and Jiang 2011; Slack and Trumbull 2011).

Tourmaline's compositional variability is due to its complex crystal chemical structure $^{[9]}X^{[6]}Y_3^{[6]}Z_6(^{[4]}T_6O_{18})(^{[3]}BO_3)_3(O,OH)_3(O,OH,F)$. B in tourmaline is situated at two different structural positions: trigonal-planar $^{[3]}B$ and tetrahedral $^{[4]}T$-site. At the latter, B is incorporated via exchange with Si by various coupled substitutions (Kutzschbach et al. 2016a). In most tourmalines, B is exclusively three-fold coordinated; however, some natural $^{[4]}B$-bearing tourmalines with up to 1.23 $^{[4]}B$ per formula unit (pfu) have also been described (e.g., Kalt et al.

2001; Hughes et al. 2004; Marschall et al. 2004; Ertl et al. 2007), mostly from high-pressure environments and for Al-rich stoichiometries. According to Ertl et al. (2008), it is possible to have a maximum of 3 $^{[4]}$B pfu at the T-site of tourmaline; they also concluded that the amount of $^{[4]}$B increases with decreasing temperature and increasing pressure. Kutzschbach et al. (2016a) achieved the incorporation of up to 2.53 ± 0.25 pfu $^{[4]}$B into olenitic tourmaline synthesized at 4.0 GPa, 700 °C, which is the highest B concentration ever detected in either synthetic or natural tourmaline.

Two experimental studies on B-isotope fractionation between tourmaline and aqueous fluid exist (Palmer et al. 1992; Meyer et al. 2008). Both report an enrichment of ^{11}B in the fluid phase under all conditions. Palmer et al. (1992) determined the fractionation of B isotopes between dravitic tourmaline and fluid in hydrothermal experiments at 350–750 °C and 50–200 MPa. The resulting $\Delta^{11}B_{\text{tour-fluid}}$, [−5.88 to −8.35‰ at 50 MPa, 450–350 °C; −2.34 to −7.76‰ at 100 MPa, 750–350 °C; −2.24 to −6.50‰ at 200 MPa, 750–350 °C] display a decrease in fractionation with increasing temperature and pressure (Fig. 3.4a). Meyer et al. (2008) determined the fractionation of B isotopes between dravitic tourmaline and fluid in hydrothermal experiments between 400 and 700 °C at 200 MPa and at 500 °C, 500 MPa. Tourmaline was crystallized from a mix of oxides and boric acid in the presence of water that contained B in excess. In one series of experiments, the fluid to tourmaline B-concentration ratio $C_B^{\text{fluid}}/C_B^{\text{tour}}$ was 9 after the run; in another series it was 0.1 to demonstrate the effect that fractional crystallization has on the B isotope tourmaline–fluid distribution (see below). All experiments produced tourmaline as the only B-bearing solid, along with traces of quartz and talc whose B contents are negligible. Solid samples were characterized by electron microprobe (EMP) and by X-ray diffraction with Rietveld refinement; the isotopic composition of both solids and fluids, were determined by p-TIMS. Rietveld refinements on tourmaline structure revealed no significant amounts of tetrahedrally coordinated B (Meyer et al. 2008). For experiments where $C_B^{\text{fluid}}/C_B^{\text{tour}}$ was 9, a consistent *T*-dependent B-isotope fractionation, valid from 400 to 700 °C, is approximated by $\Delta^{11}B_{\text{tour-fluid}} = -4.20 \cdot (1000/T[\text{K}]) + 3.52$; $R^2 = 0.77$ (Fig. 3.4a). In contradiction to Palmer et al. (1992), no pressure dependence was observed and the $\Delta^{11}B_{\text{tour-fluid}}$ values of Meyer et al. (2008) [−2.7 ± 0.5‰ at 400 °C; −0.8 ± 0.5‰ at 700 °C] are lower by approximately 2 to 4‰ [−4.1‰ at 450 °C, 200 MPa; −2.6 ± 0.3‰ at 750 °C, 200 MPa, see Fig. 3.4 a]. Experiments where $C_B^{\text{fluid}}/C_B^{\text{tour}}$ was 0.1 showed a significantly larger apparent fractionation of up to −4.7‰ at 600 °C (Fig. 3.4a). In one of these runs, the isotopic composition of hand-picked tourmaline crystals of different sizes varied by 1.3‰. This was interpreted as a result of fractional crystallization of B isotopes during tourmaline growth due to the small B reservoir of the fluid relative to tourmaline. The effect is eliminated or minimized in experiments with high amounts of excess B in the fluid. It is, therefore, suggested that values given by the above relationship using a high-B-excess reservoir in the fluid represent true equilibrium fractionation. Marschall et al. (2009) confirmed this interpretation by B-isotope SIMS-measurements of the same synthesized tourmalines. Tourmaline from an experiment with low excess B had significant B-isotope zonation with $^{11}B/^{10}B$ increasing in the growth direction of the crystals (Fig. 3.4b, c). Trend, magnitude, and absolute values strongly support results from the high-B-excess isotope fractionation experiments. Furthermore, the closed system B-isotopic evolution of the experimental fluid, modeled by Rayleigh fractionation, is in excellent agreement with the measured B-isotope composition of the run-product fluid (Fig. 3.4d). In the experiments of Palmer et al. (1992) only small amounts of B excess ($\sim$10%) were present in the fluid; thus, one may hypothesize that fractional crystallization also occurred in their experiments, creating stronger fractionation just as in the low-excess B experiments of Meyer et al. (2008). On the basis of the adsorption mechanism presented for clays (Palmer et al. 1987, see above), Palmer et al.

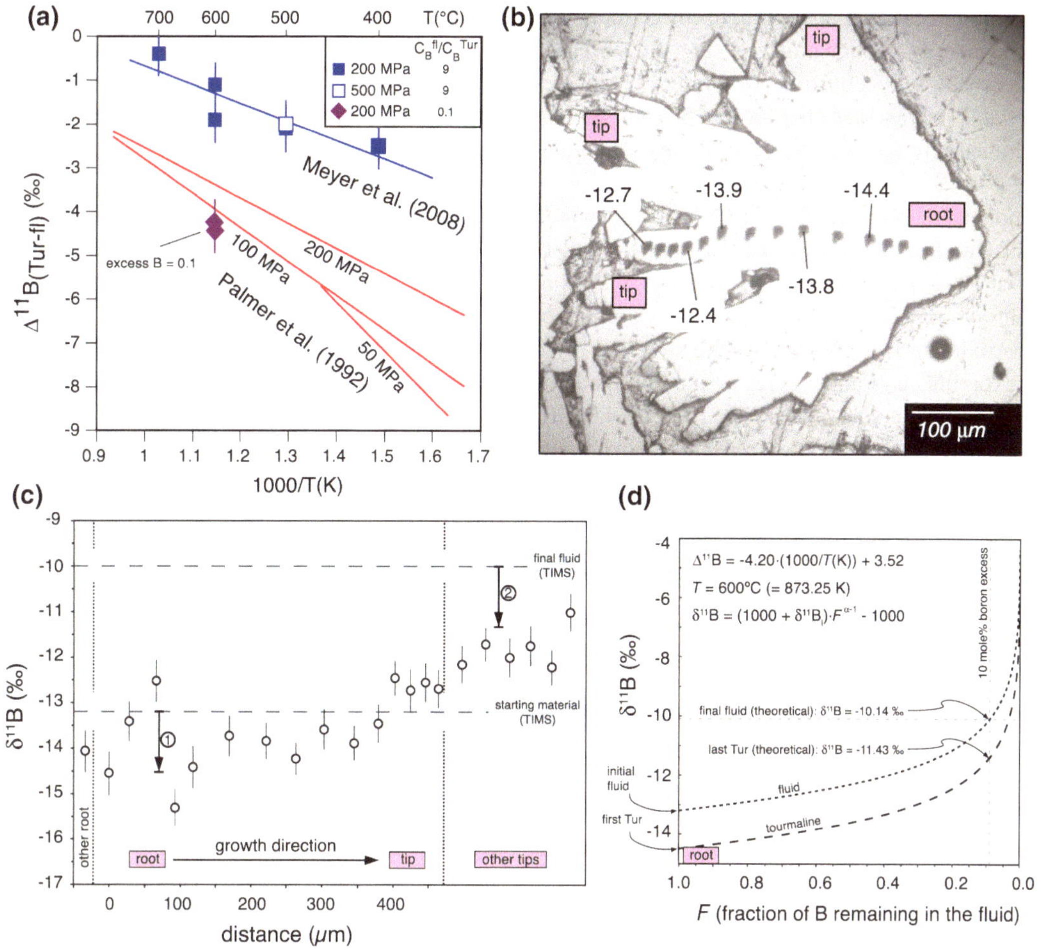

Fig. 3.4 **a** Experimental data on tourmaline–fluid B-isotope fractionation after Palmer et al. (1992) at various pressures (*red*) and after Meyer et al. (2008) for two different fluid/tourmaline B-concentration ratios (C_B^{fluid}/C_B^{tour} = 9, in *blue*; C_B^{fluid}/C_B^{tour} = 0.1, in *violet*). Note that the apparent fractionation from the small B-excess experiments are in the range of those proposed by Palmer et al. (1992) or even larger. **b** Reflected light microscope image of tourmaline from a C_B^{fluid}/C_B^{tour} = 0.1 experiment (#CM2, 600 °C) of Meyer et al. (2008). SIMS spots are visible; numbers refer to measured $\delta^{11}B$ values. **c** B-isotope profile along the growth direction (root to tip) determined from SIMS analyses of various tourmaline crystals of #CM2. *Arrows* show fractionation (according to Meyer et al. 2008) between the starting material and early-stage tourmaline in the root (No. 1), and final fluid and late-stage grown tourmaline at the tip (No. 2), respectively. **d** Evolution of B-isotopic composition of fluid and tourmaline modeled by Rayleigh fractionation for conditions of experiment #CM2. Note that the theoretical $\delta^{11}B$ value of −10.14‰ calculated from the *T*-dependent fractionation value of Meyer et al. (2008) is in very good agreement with the value of −10.0 ± 0.3‰ measured after the experimental run. **b**–**d** are slightly modified after Marschall et al. (2009)

(1992) suggested a modified version aimed at explaining the slight B-isotope fractionation and its *P* dependence for the system tourmaline–aqueous fluid, where both phases have exclusively trigonal $^{[3]}B$. In contrast to clay minerals, in this scheme no coordination-change from trigonal to tetrahedral species occurs after surface adsorption of the trigonal-B species. Instead, the coordination changes to a pseudo-trigonal coordinated adsorption species at tourmaline's surface; this mechanism should be favored at higher pressure. Palmer et al. (1992) also envisaged kinetic effects as a possible reason for the P dependence of fractionation, leaving the observed effect not fully constrained. Moreover, Palmer et al. (1992) detected a small amount of tetrahedrally-coordinated B in their tourmalines by MAS-NMR. However, only significant amounts of tetrahedral B in tourmaline would cause a clear increase in $\Delta^{11}B_{tour\text{-}fluid}$ (Kowalski et al. 2013) and the small observed $^{[4]}B$ concentrations cannot account for the variation of approximately 1–2‰ between low and high P experiments (Palmer et al. 1992). For synthetic olenitic tourmaline formed at 700 °C, 4.0 GPa with significant amounts of $^{[4]}B$, preliminary data by Kutzschbach et al. (2016b) reveal an intracrystalline B-isotope fractionation with $\Delta^{11}B_{(^{[3]}B-^{[4]}B)} = \delta^{11}B^{[3]} - \delta^{11}B^{[4]}$ of +8.4‰, which is in agreement with computationally determined values of B-isotope fractionation for $^{[4]}B$-bearing tourmaline–fluid (Kowalski et al. 2013; see below).

Furthermore, according to Kowalski et al. (2013), tourmaline–fluid B-isotope fractionation might vary for different tourmaline members due to differences in the B–O distances of the trigonal planar groups of different tourmaline (see later Fig. 3.7a).

3.6 Ab Intio Prediction of B-Isotope Fractionation

With the tremendous increase in the availability of computational power that rises by at least a factor of 100 per decade (www.top500.org), the ab initio methods of quantum chemistry, namely density functional theory (DFT), become very popular investigation tools in various research fields, including isotope geochemistry (e.g., Schauble 2004; Kowalski et al. 2013; Jahn and Kowalski 2014). Therefore, reliable ab initio computational methods have been established recently and were successfully applied to predict the isotope fractionation factors among various crystalline solids (Driesner 1997; Yamaji et al. 2001; Schauble 2004; Domagal-Goldman and Kubicki 2008; Hill and Schauble 2008; Méheut et al. 2007, 2009; Schauble et al. 2009; Zeebe 2009, 2010; Hill et al. 2010; Rustad et al. 2010a, b; Kowalski and Jahn 2011; Kowalski et al. 2013; Feng et al. 2014; Méheut and Schauble 2014) and recently also between solids and aqueous solutions (e.g., Zeebe 2005; Otake et al. 2008; Domagal-Goldman and Kubicki 2008; Hill and Schauble 2008; Rustad et al. 2008; Zeebe 2009, 2010; Hill et al. 2010; Rustad et al. 2010a, b; Kowalski and Jahn 2011; Markland and Berne 2012; Kowalski et al. 2013; Fujii et al. 2015; Pinilla et al. 2015; Dupuis et al. 2015). Many of these studies have shown that such calculations are essential for our understanding of the geochemical mechanisms responsible for emerged isotope fractionation and signatures from experiments and nature.

The majority of the ab initio studies focus on the computation of stable isotope fractionation between simple crystalline phases and aqueous solution represented by ion-aqua complexes, whereas calculations of continuous fluids are less abundant. The aqueous solution species are usually represented by an isolated cluster containing the fractionating species and a hydration shell (e.g., Zeebe 2005; Domagal-Goldman and Kubicki 2008; Hill and Schauble 2008; Zeebe 2009, 2010; Hill et al. 2010; Rustad et al. 2010a; b). However, it is known that the distribution of cation coordination and cation–oxygen bond lengths can significantly influence the fractionation of isotopes (Bigeleisen and Mayer 1947). These structural properties of materials are determined by the dynamics of the considered system or its change under compression (Jahn and Wunder 2009; Wunder et al. 2011; Kowalski and Jahn 2011) and can have strong impact on isotope exchange. For instance, considering fractionation of Li isotopes, Kowalski and Jahn (2011) have demonstrated that a cation-hydration shell cluster is not an appropriate representation of aqueous solutions at high pressure ($P > 2$ GPa). Only a few published *ab initio* studies account for the dynamical effects on the isotope fractionation in fluids (Rustad and Bylaska 2007; Kowalski and Jahn 2011; Kowalski et al. 2013). Rustad and Bylaska (2007) considered boron equilibrium isotope fractionation between $B(OH)_3$ and $[B(OH)_4]^-$ species in aqueous solution. They performed ab initio molecular dynamic simulations of these fluids and calculated the $^{11}B/^{10}B$ isotope fractionation coefficient from the vibrational density of states obtained through the Fourier transform of the velocity auto-correlation function. Although the obtained fractionation factor of 0.86 was substantially different from the experimental value of 1.028, the experiment and theory were reconciled by computation of the harmonic frequencies on a set of the quenched ionic configurations extracted from the molecular dynamics trajectory. Using a similar approach together with the "*single atom approximation*" of Bigeleisen and Mayer (1947), in which the isotope fractionation factor was derived from the force constants acting on Li cation, Kowalski and Jahn (2011) computed the high temperature Li isotope fractionation factors between complex minerals (spodumene, staurolite and mica) and aqueous fluid. It is worth noticing that because these simulations were performed using periodic boundary conditions, the computed solids and fluids were treated as continuous media, which allowed for correct accounting for the pressure effects.

With the steadily increasing access to the computing power it becomes possible to apply other methods of quantum chemistry to derive the isotope fractionation factors. For instance, Marland and Berne (2012), Pinila et al. (2014) and Dupuis et al. (2015) used path integral molecular dynamics technique to investigate the fractionation of H and O isotopes in water. We thus expect that further similar studies will be performed in the next years.

3.6.1 Computational Approach

3.6.1.1 The "Single Atom Approximation" Following Bigeleisen and Mayer (1947)

Following the pioneering study of Urey (1947) it is well accepted that the mass-dependent equilibrium fractionation of isotopes is driven by the change in molecular and crystalline vibration frequencies resulting from the different masses of isotopes. The principal parameter which describes the equilibrium isotope fractionation is the so-called isotopic reduced partition function ratio, known also as β factor, which describes the fractionation of isotopes between given species and an ideal monoatomic gas. In the harmonic approximation it is given by the following formula (Bigeleisen and Mayer 1947; Urey 1947; Chacko et al. 2001):

$$\beta = \prod_{i=1}^{N_{\mathrm{dof}}} \frac{u_i^*}{u_i} \exp\left[\frac{u_i - u_i^*}{2} \frac{1 - \exp(-u_i)}{1 - \exp(-u_i^*)}\right] \qquad (3.10)$$

where $u_i = h\nu_i/k_B T$, h is the Planck constant, k_B is the Boltzmann constant, ν_i is the vibrational frequency of the ith degree of freedom, N_{dof} is the number of degrees of freedom and the asterisk marks the heavier isotope. For N being the number of atoms in the considered system N_{dof} is equal to 3 N-5 for diatomic molecule, 3 N-6 for multiatomic molecules and 3 N for crystals. The fractionation factor between two substances A and B, $\alpha_{A\text{-}B}$ is computed as the ratio of the relevant β factors, which is well approximated by the differences in the β factors:

$$\Delta_{A-B} = 1000 \cdot \ln \beta_A - 1000 \cdot \ln \beta_B \qquad (3.11)$$

The calculation of the β factor requires thus only knowledge of the vibrational frequencies that can be measured or computed ab initio (Jahn and Kowalski 2014). The only problem on the computational site is that the calculation of the whole vibrational spectra of complex minerals or fluids requires substantial computational resources and, therefore, is currently limited to systems containing no more than a few dozens of atoms. Many of the current studies thus focus on computation of simple crystalline solids and ion-aqua complexes, accounting for the closest hydration shells only.

Kowalski and Jahn (2011) noticed that by computing only the force constants acting upon the fractionating element, by using for instance the "*single atom approximation*" of Bigeleisen and Mayer (1947), instead of the full phonon spectrum required by Eq. (3.10), one can dramatically reduce the computational demand by a factor of $\sim N$, where N is the number of the atoms constituting the computed system (supercell). Kowalski and Jahn (2011) proposed such a method for an efficient computing of the high temperature isotope fractionation factors between complex materials such as fluids and crystalline solids. The β factor (Eq. 3.12) can be then approximated as:

$$\beta \cong 1 + \sum_{i=1}^{N_{\mathrm{dof}}} \frac{u_i^2 - u_i^{*2}}{24} = 1 + \frac{\Delta m}{mm^*} \frac{\hbar^2}{24 k_B^2 T^2} \sum_{i=1}^{3} A_i, \tag{3.12}$$

where A_i are the three force constants acting on the isotopic atom and $\Delta m = m^* - m$, where m and m^* are the masses of the lighter and heavier isotopes. This approach refers to the "*single atom approximation*" (Bigeleisen and Mayer 1947), because the computation of the β factors using Eq. (3.12) only requires the knowledge of properties of the fractionating element. However, the validity criteria restrict the usage of Eq. (3.12) to frequencies ν [cm^{-1}] smaller than 1.39 T [K] (Kowalski et al. 2013). Because of this difficulty, Kowalski et al. (2013) introduced a modification of the "*single atom approximation*" showing that the error of the computed fractionation factors can be substantially reduced if one uses the three frequencies ν_i derived from the force constants acting on the fractionating element $\nu_i = A_i/4\pi^2 m$ —called *pseudofrequencies*—and combined this with Eq. (3.10) for the calculations. Kowalski et al. (2013) provided a theoretical justification for this method, and tested and applied it successfully to B isotope systems, including cases of tourmaline (dravite, olenite), mica, grandidierite, serendibite, and aqueous solutions, which will be discussed in the next sections.

3.6.1.2 Computational Technique

The calculations reported here were performed by Kowalski et al. (2013), MacGregor et al. (2013) and Belley et al. (2014). These studies are complemented by the discussion of the ab initio data of Zeebe (2005) and Rustadt and Bylaska (2007). The calculations of *pseudofrequencies* and β factors for solids and aqueous solutions were performed following the procedure outlined in Kowalski et al. (2013) by applying density functional theory (DFT), which is currently the most efficient method allowing for quantum-mechanical treatment of extended many-particle systems. The calculations were performed with the planewave DFT code CPMD (Marx and Hutter 2000), which is especially suited for ab initio molecular dynamics simulations of fluids, and with the BLYP exchange-correlation functional (Lee et al. 1988). The norm-conserving Goedecker pseudopotentials were applied to mimic the presence of the core electrons (Goedecker et al. 1996). One advantage of using the BLYP functional is that it usually gives harmonic frequencies that most closely resemble the observed frequencies (e.g., Finley and Stephens 1995; Alecu et al. 2010). The energy cut-off for the plane wave basis set was set to 70 Ryd for geometry relaxations and molecular dynamics simulations and to 140 Ryd for static computation of the force constants and the vibrational *pseudofrequencies*. The periodic boundary conditions were applied for crystalline solids as well as aqueous solutions in order to preserve the continuity of the media. The force constants and resulting *pseudofrequencies* were computed using the finite displacement scheme. In case of the crystal structures these calculations were performed with all atomic positions relaxed to the equilibrium positions. The calculations are restricted to a single phonon wave vector using large supercells, thus avoiding an expensive calculation of full phonon dispersion. Due to the restrictions of CPMD code and the applied "*single atom approximation*", the calculations were performed at Γ-point. Blanchard et al. (2009) have shown that selection of a representative Γ-point, such as the mean value Γ-point (Baldereschi 1973), could improve the accuracy of the calculations. This procedure has been validated at high temperatures (> 400 °C) for Mg-bearing materials (Schauble 2011) and Fe-bearing minerals (Blanchard et al. 2009).

The computation of the force constants and frequencies in case of the aqueous species were performed after relaxing the positions of all the atoms constituting the B-carrying molecule ($B(OH)_3$ or $[B(OH)_4]^-$) to the equilibrium positions, but keeping all other atomic positions fixed. The 10 ps long molecular dynamics trajectories of systems consisting of 64 H_2O molecules and one $B(OH)_3$ or $[B(OH)_4]^-$ molecule were obtained for different temperatures of 1000, 800 and 600 K and a pressure of 0.5 GPa. These parameters closely resemble the experimental conditions of Wunder et al. (2005) and Meyer et al. (2008). The ab initio molecular dynamics simulations were performed for fixed temperature and volume using the Car–Parrinello scheme (Car and Parrinello 1985) and the pressure of aqueous solution was calculated according to the equation of state of Wagner and Pruss (2002). The temperature during each run was controlled by a Nosé–Hoover chain thermostat (Nosé and Klein 1983; Hoover 1985). The β factors were computed on the snapshots extracted uniformly along the molecular dynamics trajectories using an interval of 0.1 ps.

3.6.1.3 Error Estimation Technique

An important aspect of reliable predictions is an estimation of error of the computed fractionation factors, which unfortunately is often omitted in many modeling studies. As shown by Kowalski and Jahn (2011) and Kowalski et al. (2013) the errors in the computed value of the β and Δ fractionation factors can be estimated from an average error of vibrational frequencies computed using the chosen DFT method. For instance, Finley and Stephens (1995), Menconi and Tozer (2002) and Alecu et al. (2010) estimated the errors originating from using different DFT functionals in the calculations of vibrational frequencies of small molecules. According to these studies, the BLYP functional systematically overestimates the harmonic frequencies by 3.5% with a deviation from the mean offset of 1%. Therefore, we expect that using BLYP functional the β and Δ values are systematically overestimated by 7% and that in addition there is a 2% error in the derived β factors. All the values presented here are thus provided together with errors estimated in this way. Similar errors result from using other functionals or even more sophisticated and computationally demanding post-Hartree–Fock methods such as MP2 (Finley and Stephens 1995; Alecu et al. 2010).

3.6.2 Computation of Fractionation Factors

3.6.2.1 Vapor

Kowalski et al. (2013) compared their results for the gas phase $B(OH)_3$ and $[B(OH)_4]^-$ molecules with the published data of Zeebe (2005), where the same DFT BLYP functional was used. The computed vibrational frequencies that are affected by the different B isotope substitutions are in good agreement with earlier theoretical predictions and show similar agreement with the experimental measurements. Figure 3.5a reports the resulting β factors, together with the results of Zeebe (2005). The β factors are smaller by approximately 1‰ at 600–1000 K than the values of Zeebe (2005), although the difference between β factors of $B(OH)_3$ and $[B(OH)_4]^-$ remains nearly identical in both sets of calculations, which points to a systematic error. The discrepancy between the two sets of calculations is substantially reduced when the results obtained with more complete basis set 6-311 + G(d,p) are used (Zeebe 2005). This shows that the here utilized plane-wave based DFT approach provides adequate vibrational frequencies and the resulting isotope fractionation factors are reliable.

3.6.2.2 Aqueous Fluid

Computed β factors for $B(OH)_3$ and $[B(OH)_4]^-$ in aqueous solutions according to Kowalski et al. (2013) are shown in Fig. 3.5c. The presented functions, $\beta = 1 + 2.416\ 10^4/T^2 - 5.823\ 10^8/T^4$ for $B(OH)_3$ and $\beta = 1 + 1.772\ 10^4/T^2 - 4.234\ 10^8/T^4$ for $[B(OH)_4]^-$ were fitted to the computed ab initio values. According to the assessment of uncertainties of the computational method (such as DFT exchange-correlation functional) and the "*single atom approximation*" (Kowalski et al. 2013), the errors of the computed values are T dependent with larger uncertainties at lower temperatures: 27% (at $T = 300$ K), 16% (at

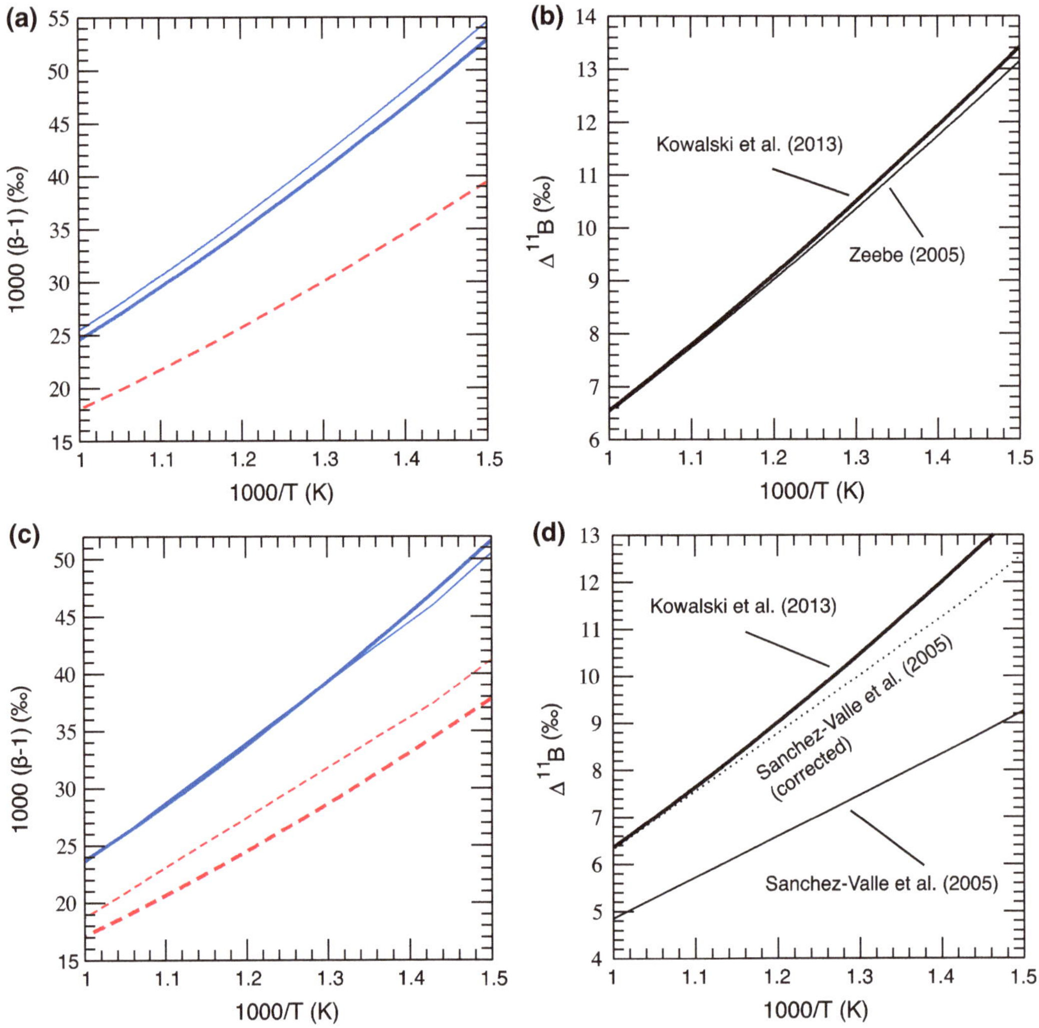

Fig. 3.5 **a** Gas phase β factors for $B(OH)_3$ (*solid blue line*) and $[B(OH)_4]^-$ (*dashed red lines*); **b** gas phase $B(OH)_3$–$[B(OH)_4]^-$ fractionation factors. The *thick lines* represent calculations of Kowalski et al. (2013), and the *thin lines* represent the results derived from data computed by Zeebe (2005); **c** computed aqueous solution β factors for $B(OH)_3$ (*solid blue lines*) and $[B(OH)_4]^-$ (*dashed red lines*); **d** aqueous solution $B(OH)_3$–$[B(OH)_4]^-$ fractionation factors. The *thick lines* represent the results computed by Kowalski et al. (2013). The *thin lines* show the results of Sanchez-Valle et al. (2005) obtained using harmonic frequencies. The *dotted line* represents the corrected Sanchez-Valle et al. (2005) results. The correction is made according to Rustad et al. (2010a). Figure modified after Kowalski et al. (2013)

$T = 600$ K), 11% (at $T = 1000$ K). The high uncertainties of computed β factors at ambient conditions are also shown by Rustad et al. (2010a, their Fig. 3.2) from the spread of fractionation factors calculated using various computational methods.

The β factor of $B(OH)_3$-bearing fluid is substantially larger than the one for the $[B(OH)_4]^-$ and not significantly different from the results obtained for the isolated molecule (Fig. 3.5a). This reflects the substantial difference in the B–O bond lengths exhibited by the two considered B species. For the isolated molecules Kowalski et al. (2013) determined the B–O bond length of 1.40 Å and 1.51 Å for $B(OH)_3$ and $[B(OH)_4]^-$, respectively. Compared to experimental data by Sanchez-Valle et al. (2005), for $B(OH)_3$ both studies give nearly identical β factor; however, for $[B(OH)_4]^-$ Kowalski et al. (2013) predicted a lower value. This discrepancy is caused by the incorrect assignment of a major vibrational mode of $[B(OH)_4]^-$ in the force field by Sanchez-Valle et al. (2005) as discussed by Rustad and Bylaska (2007) and shown by Kowalski et al. (2013). The accordingly corrected results of Sanchez-Valle et al. (2005), which are in good agreement with the prediction of Kowalski et al. (2013), are plotted in Fig. 3.5d.

Liebscher et al. (2005) measured the fractionation of boron isotopes between liquid and vapor systems and found a positive $\Delta^{11}B_{liquid\text{-}vapour}$ of +0.1 to +1.3‰ at 400 °C and of +0.2 to +1.5‰ at 450 °C (see Fig. 3.2). The calculations reported in Fig. 3.5 indicate $\Delta^{11}B_{liquid\text{-}vapour} = +1.0 \pm 1.9$‰ and $+1.5 \pm 2.2$‰ for temperature of 400 and 450 °C, respectively. These values are thus well consistent with the experimental data.

3.6.2.3 Pressure Dependence of the Fractionation Factor in Aqueous Fluids

Kowalski and Jahn (2011) have shown that for $P > 2$ GPa the β factor of Li in aqueous solution increases with increasing pressure. Kowalski et al. (2013) observed the same for $B(OH)_3$ and $[B(OH)_4]^-$ in aqueous fluid, which is consistent with the experimentally observed variation in shift of the vibrational frequencies of different B species with increasing pressure (Sanchez-Valle et al. 2005; Schmidt et al. 2005). The computed β-1 values for $B(OH)_3$ fluid show a linear dependence in pressure, $\beta\text{-}1 = 23.6 + 0.28P$ (P in GPa). Such a relationship is expected as $\beta\text{-}1 \propto v^2 \sim v_0^2 + 2\,_0\Delta v$ (Schauble 2004) and Δv is a linear function of P (Sanchez-Valle et al. 2005; Schmidt et al. 2005). As illustrated in Fig. 3.6c, in case of $[B(OH)_4]^-$, the pressure-dependence is linear up to $P \sim 2$–3 GPa but it becomes less steep at higher pressures. Kowalski et al. (2013) have shown that these predictions are consistent with the measured P-dependent vibrational frequency shifts of Sanchez-Valle et al. (2005). Such a good agreement with the experimental data shows that ab initio calculations can be successfully used in derivation of the pressure dependence of the fractionation factors and the pressure-induced vibrational frequencies shifts. Moreover, first principles calculations can also be useful in extrapolation of the experimental values for β and Δv to more extreme conditions, which are difficult to be reached by experimental techniques.

3.6.3 Computationally Determined Mineral–Fluid B-Isotope Fractionation

Kowalski et al. (2013), MacGregor et al. (2013) and Belley et al. (2014) used the computational approach outlined above to investigate the fractionation between boron-bearing fluids and minerals, such as dravite, olenite, boromuscovite, prismatine, grandidierite and serendibite. The aim of these studies was to investigate the mechanisms driving the fractionation process, including the role of coordination and the B–O bond length to better understand measured B-isotopic signatures. Below we discuss each case separately.

3.6.3.1 Tourmaline–Neutral Fluid B-Isotope Fractionation

Meyer et al. (2008) measured the boron isotope fractionation between tourmaline, represented by dravite, and pH-neutral fluid at $T = 400$–700 °C and $P = 0.2$ GPa (see Fig. 3.4a). The calculated fractionation curve and the experimental data of Meyer et al. (2008) are given in Fig. 3.7a. Calculations reproduce the measurements within the computational accuracy. According to Kowalski et al. (2013) the dravite–fluid fractionation is small because both, dravite and neutral fluid, contain boron in BO_3 units. Kowalski et al.

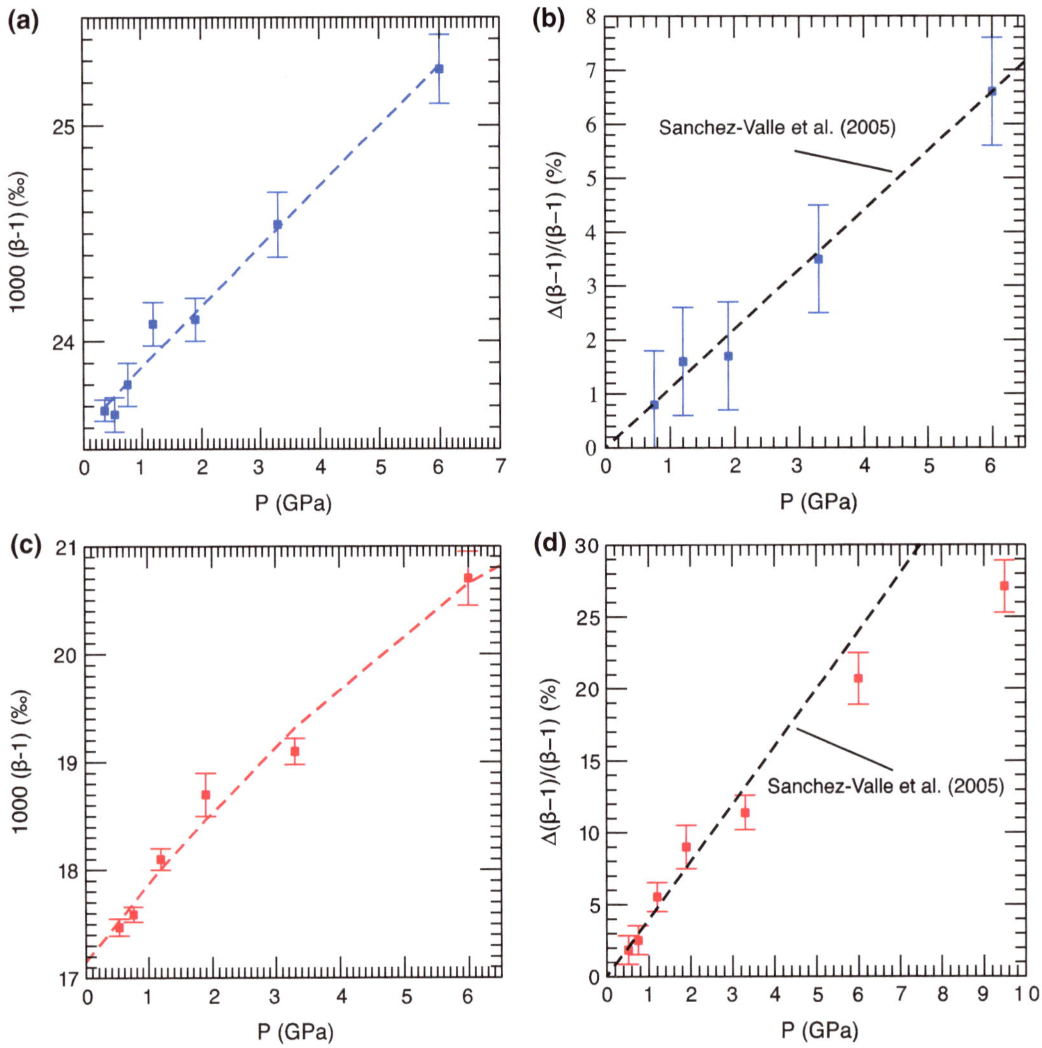

Fig. 3.6 **a** Computed pressure dependent β factor of neutral fluid ($B(OH)_3$) at $T = 1000$ K. *Dashed line* represents the linear regression fit to the calculated values (*squares*); **b** computed change of β factor as a function of pressure (*blue squares*) compared to the results derived from the pressure-induced frequency shifts of the 666 and 1454 cm^{-1} vibrational modes measured by Sanchez-Valle et al. (2005) (*dashed black line*); **c** computed pressure dependent β factor of strongly basic fluid ($[B(OH)_4]^-$) at $T = 1000$ K. *Dashed line* is the regression fit to the calculated values (*squares*); **d** computed change of the β factor with pressure compared to the increase in the β factor derived from the frequency shift of the 975 cm^{-1} line measured by Sanchez-Valle et al. (2005) (*black dashed line*). Figure modified after Kowalski et al. (2013)

(2013) also predicted a small but positive fractionation between olenite and aqueous fluid. It is assumed that the calculated olenite–fluid fractionation differs from that of dravite–fluid due to shorter B–O bond lengths for olenite (1.378 Å) versus 1.397 Å for dravite (Kowalski et al. 2013).

Beside trigonally-coordinated boron, tourmaline can incorporate boron also at the tetrahedral

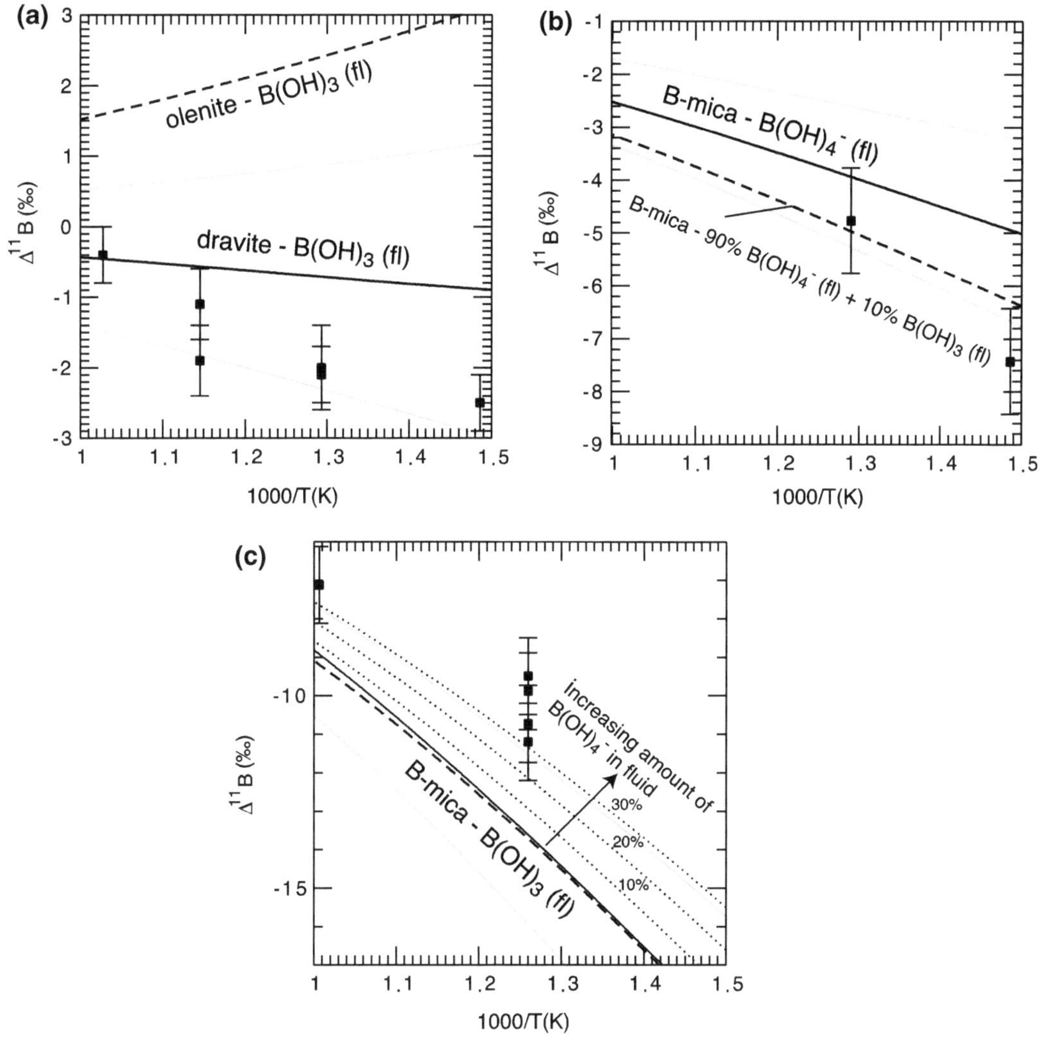

Fig. 3.7 **a** B-isotopic fractionation between tourmaline and aqueous fluid. The *solid line* represents the fractionation between dravite and $B(OH)_3$ neutral fluid computed by Kowalski et al. (2013). The *dashed line* represents the prediction for the fractionation between olenite (without tetrahedral boron) and neutral fluid. The *squares* represent the experimental data measured for the dravite–fluid system by Meyer et al. (2008); **b** B-isotope fractionation factors between boromuscovite and basic aqueous fluid (fluid containing $[B(OH)_4]^-$). The *lines* represent the values calculated by Kowalski et al. (2013) assuming the presence of boron species in the form of $[B(OH)_4]^-$ only (*solid line*) and admixture of 10% of $B(OH)_3$ in the fluid (*dashed line*). The *squares* represent the experimental data (Wunder et al. 2005); **c** B-isotope fractionation factors between boromuscovite and aqueous fluid. The data points are the values measured by Wunder et al. (2005). The *thick lines* represent the fractionation factor computed assuming ambient pressure (*solid line*) and $P = 3$ GPa accounting for compression and thermal expansion (*dashed line*). The *dotted lines* represent the fractionation factors obtained assuming different admixture of $^{[4]}B$ species (represented by $[B(OH)_4]^-$ with abundance indicated in the figure) to the fluid. The *blue regions* represent the computational uncertainty. Figure modified after Kowalski et al. (2013)

site replacing Si in a coupled substitution (e.g., Ertl et al. 1997). Calculations of intra-site B-isotope fractionation $\left(\Delta^{11}B_{([3]B-[4]B)}\right)$ between trigonal and tetrahedral boron sites in dravite resulted in +8.9 ± 1.7‰ (Kowalski et al. 2013) at 600 °C, which is in good agreement to experimental data (Kutzschbach et al. 2016b, see above).

3.6.3.2 Boromuscovite–Strongly Basic Fluid B-Isotope Fractionation

Wunder et al. (2005) measured the isotopic signatures of synthesized boromuscovite that consisted of polytypes, 1 M ($\sim$10%) and $2M_1$ ($\sim$90%). As a comparison with these data, Kowalski et al. (2013) derived β factors for both polytypes and used their weighted average for the analysis. Both polytypes of mica have similar mean B–O bond lengths [1.515 Å (1 M) and 1.525 Å ($2M_1$)], and the calculations show that they are characterized by similar β factors with difference in ($\beta-1$) no larger than 3% (see Fig. 3.9). In order to investigate the impact of the B–O bond length on the fractionation, Kowalski et al. (2013) compared the fractionation between boromusovite, that generally has B in tetrahedral coordination, and a strongly basic fluid containing boron as $[B(OH)_4]^-$. The results of the calculations and the experimental data from the system boromuscovite–strongly basic fluid (Wunder et al. 2005) are shown in Fig. 3.7b. The calculations predict a negative fractionation between mica and the $[B(OH)_4]^-$ fluid, in line with the experimental findings. This result seems reasonable, because the mean B–O bond length of mica is larger than that of aqueous $[B(OH)_4]^-$ (1.513 Å).

Although the data from calculations and experiments agree within errors, the experimental data might indicate a slightly stronger fractionation. Kowalski et al. (2013) explained this discrepancy by admixture of aqueous $^{[3]}B$ species in the experiment. The presence of as little as 10% of $B(OH)_3$ in the measured basic fluid brings predictions and experiments to good agreement (Fig. 3.7b). The experiments were performed with a "*strongly basic*" fluid (Wunder et al. 2005), yet, its exact composition, especially the amount of $^{[3]}B$ species at the conditions of the experiments, was unknown. Nevertheless, good agreement between the prediction and the experimental data exists, which indicates that the strongly basic fluid was dominated by $[B(OH)_4]^-$ species, in line with previous studies (e.g., Zeebe 2005).

3.6.3.3 Boromuscovite–Neutral Fluid B-Isotope Fractionation

The strong B-isotopic fractionation between boromuscovite and neutral fluid results from different coordination of boron in boromuscovite ($^{[4]}B$) and neutral fluid ($^{[3]}B$). Comparison of data from experiments performed at 3.0 GPa (Wunder et al. 2005) with the predicted fractionation shows a difference of $\sim$2.5‰ (Fig. 3.7c). The effect of difference in thermal expansion and compressibility between boromuscovite and fluid was shown to be not responsible for the discrepancy (Kowalski et al. 2013). However, the comparison of calculated results with the experimental data (Kowalski et al. 2013) suggests that the fractionation between boromuscovite and fluid is the same as between $B(OH)_3$ and $[B(OH)_4]^-$ fluids, which is at odds with the non-negligible and negative fractionation between boromuscovite and a strongly basic fluid (Fig. 3.7b). As the β factor derived for boromuscovite seems to be reliable, this points towards another, unaccounted effect that led to the decrease of the B-isotope fractionation between mica and neutral fluid in the high-pressure experiments of Wunder et al. (2005). One possible solution for the discrepancy is a non-negligible amount of $^{[4]}B$ in the neutral solution at 3.0 GPa. This is in line with the studies of Schmidt et al. (2005), who detected admixture of significant amounts of $^{[4]}B$ species in near-neutral, $B(OH)_3$-dominated fluid at high pressures (see Fig. 3.1). Assuming a presence of 15–30% $^{[4]}B$ species in the fluid at 3.0 GPa and a β factor of these species similar to that of $[B(OH)_4]^-$, the fractionation factor between boromuscovite and $B(OH)_3$ in the aqueous fluid decreases, bringing theory and experiment to much better agreement (Fig. 3.7c). If this interpretation is correct, it suggests that boron isotope fractionation could be used to gather information on the B-speciation in aqueous fluids.

3.6.4 B-Isotope Fractionation Among Minerals

The fractionation of B isotopes among solids has received considerable attention (e.g., Hervig et al. 2002; Marschall 2005; Klemme et al. 2011; MacGregor et al. 2013; Belley et al. 2014). From a theoretical point of view, Kowalski et al. (2013) investigated the mica–tourmaline system in which the two coexisting minerals have boron in different coordination, $^{[3]}$B in tourmaline and $^{[4]}$B in mica. This results in large B isotope fractionation between these two minerals, which has been investigated by analyzing natural coexisting pairs of tourmaline and mica (Klemme et al. 2011, and references herein). Comparison of experimental data with measurements from natural samples and the computed *T*-dependent fractionation curve are given in Fig. 3.8. The fractionation factors predicted by Kowalski et al. (2013) are consistent with the natural data (Klemme et al. 2011), which are both larger (taking the absolute values) than the experimental data (Wunder et al. 2005; Meyer et al. 2008). As discussed in Sect. 3.6.3, this discrepancy can be resolved by assuming that in the experiments on mica–fluid B-isotope fractionation (Wunder et al. 2005) the fluid contained a significant admixture of $^{[4]}$B species. The experimental mica–tourmaline fractionation factor corrected for the presence of $^{[4]}$B species in the high-pressure fluids results in a better consistency with the natural data (Fig. 3.8).

The mineral serendibite (ideally $Ca_4(Mg_6Al_6)O_4[Si_6B_3Al_3O_{36}]$) contains boron in tetrahedral coordination. The B-isotope difference of peak metamorphic serendibite and coexisting uvitic tourmaline in the Central Metasedimentary Belt in the Grenville Province of Quebec was determined to be −6.3‰ (Belley et al. 2014), which is

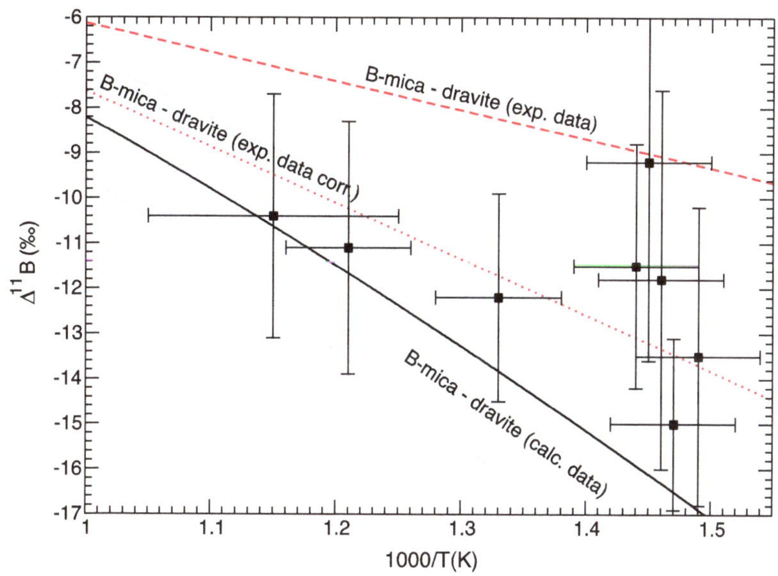

Fig. 3.8 Boron isotope fractionation factors between mica and tourmaline. The *solid line* represents the computed fractionation factor between B-muscovite and dravite. The uncertainties are indicated by the *blue area*. The *dashed red line* is the experimental fractionation factor determined by combining the data of Wunder et al. (2005) and Meyer et al. (2008). The experimental error is ± 2‰. The *dotted red line* is the experimental fractionation factor corrected for the presence of $^{[4]}$B species in the neutral fluid in the experiments of Wunder et al. (2005). The *squares* represent the data on natural samples (Klemme et al. 2011). Figure modified after Kowalski et al. (2013)

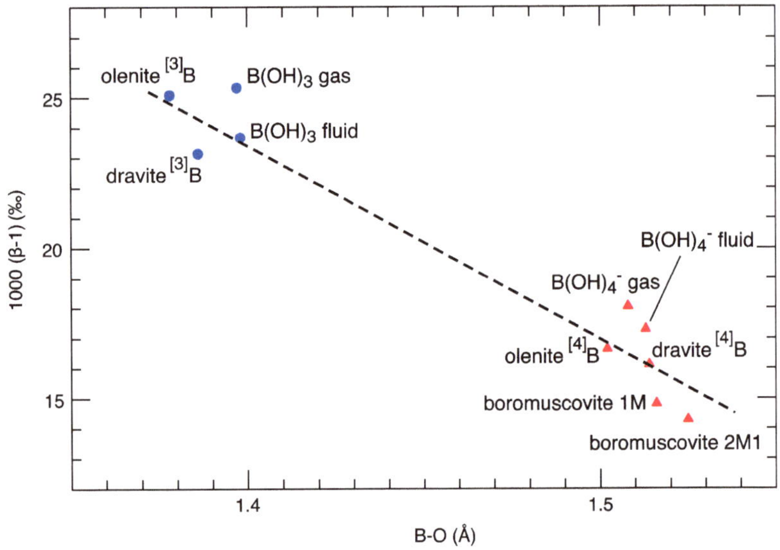

Fig. 3.9 β factor at T = 1000 K for various boron-bearing species as a function of the B–O bond length. *Blue circles* and *red triangles* represent the values obtained for $^{[3]}$B and $^{[4]}$B species, respectively. Figure modified after Kowalski et al. (2013)

consistent with the fractionation factor from ab initio calculations using the method of Kowalski et al. (2013).

Kowalski et al. (2013) have demonstrated that the β factor not only depends on boron coordination but is also strongly correlated with the B–O bond length. Phases with trigonal-coordinated B and short B–O bond lengths have higher β factors than $^{[4]}$B-containing phases and long B–O bond lengths (Fig. 3.9). This observation was used to explain the measured ^{11}B-enrichment of grandidierite (containing structural $^{[3]}$B) and ^{11}B-depletion of prismatine (containing structural $^{[4]}$B) relative to tourmaline (MacGregor et al. 2013). At a temperature of 623 °C, the calculated fractionation factor between grandidierite (B–O bond length of 1.368 Å) and dravitic tourmaline (B–O bond length of 1.385 Å) is +1.8 ± 1.1‰, whereas for the system tourmaline–prismatine the calculations resulted in a B-isotopic fractionation of −6.4 ± 1.3‰. For both systems these data are in good agreement to measured B-isotope mineral–mineral fractionation of +3.3 ± 0.8‰ for coexisting natural grandidierite–tourmaline and −5.0 ± 1.4‰ for tourmaline–prismatine pairs (MacGregor et al. 2013).

3.7 Conclusions

Table 3.1 and Fig. 3.10 summarize the results of experimentally determined B-isotope fractionation data for selected systems presented in this chapter. Due to the discrepancies between assumed B-species in (rhyolitic) melts and the measured large B-isotopic melt–fluid fractionation (Hervig et al. 2002), these data are assumed here to represent disequilibrium fractionation. Therefore, further experimental investigation is greatly needed, which should be complemented by the evaluation of possible kinetic B-isotope fractionation at high temperatures, which might produce considerable effects. Furthermore, inconsistencies in the results on B-incorporation mechanisms for calcite and aragonite and the not-fully understood variations in Δ^{11}B values for different solid–fluid systems (e.g., smectite and carbonates), alert us to the potential role of often poorly known surface-dependent processes in controlling subsequent B-isotope fractionation for different minerals in low-temperature diffusive experiments. Thus, further experimental investigation is needed here too, e.g., by

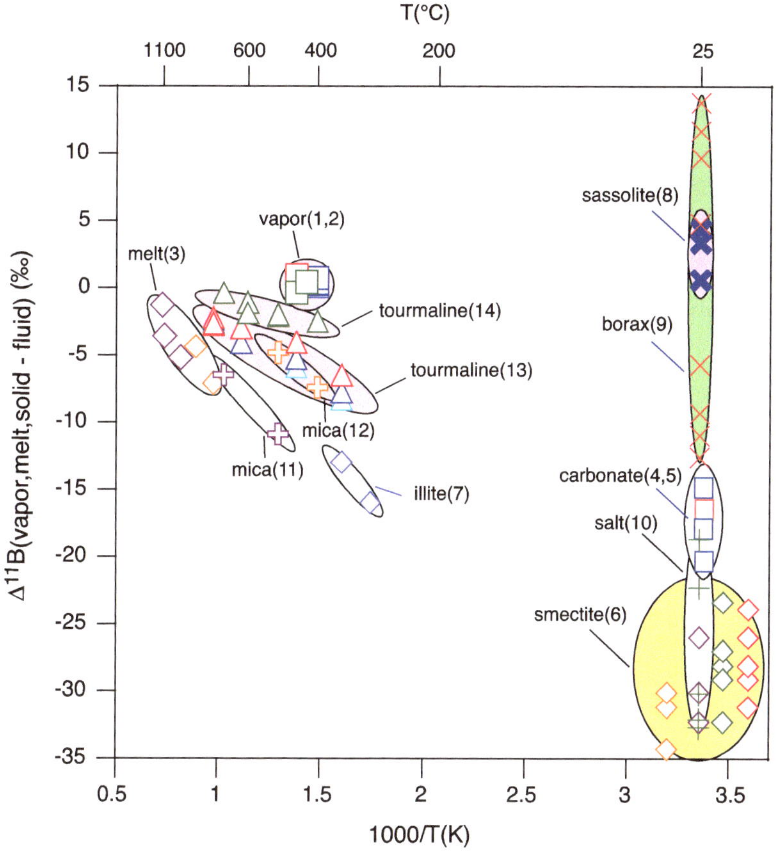

Fig. 3.10 Compilaton of selected experimental data on vapor, melt, solid–fluid B-isotope fractionation. Fractionation data for systems with contrasting B-coordination (i.e., $^{[4]}B$ vs. $^{[3]}B$) are marked by a blue background. Possibly, this does not hold for the melt–fluid (for more details see text). *Pink color* mark fractionation data for systems with uniform B-coordination. Borax (containing mixed $^{[3]}B$ and $^{[4]}B$)–fluid data (in *green*) show a strong variation in fractionation due to variable experimental fluid pH (7, with mainly $^{[3]}B$, to 11.5, with mainly $^{[4]}B$). Smectite–fluid B-isotope fractionation (in *orange*) as result of B-surface adsorption strongly depend on pH. The numbers correspond to the references given in Table 3.1

innovative methods like AFM (Ruiz-Agudo et al. 2012), to develop a robust mechanistic knowledge of B incorporation into minerals during growth. Additionally, it is obvious from Fig. 3.10 that there is a lack of experimental data in the *T*-range 40–350 °C. Further experiments at these temperatures are needed to fill the gap between existing high- and low-*T* data and to better understand the structural controls on B-isotope fractionation during diagenesis, low-*T* metamorphism, low-*T* hydrothermal activity and volcanic exhalation. Generally, it would seem that using experiments with a well-understood B-incorporation mechanism is the favorable approach for acquiring more reliable fractionation data and for avoiding surface-related effects and diffusion problems potentially-occurring in the low-temperature reequilibration experiments.

Recently performed ab initio calculations have the potential to obtain *T*- and *P*-dependent fractionation factors for given materials, which are successfully used not only to complement the experimental data but also to assist their correct interpretation and to derive constraints on unmeasured systems. The computed B-isotope fractionation factors can also help in explaining discrepancies between contrasting measurements and provide valuable insight into the boron speciation in fluids. In consideration of the continuous increase in the efficiency of computing resources and performance of quantum chemistry software, it is concluded that ab initio computational geochemistry will flourish in the coming decades, and that computer-aided first-principles simulations will become standard research tools in isotope geochemistry. They assist in the interpretation of experimental data and the understanding of B behavior in natural settings.

References

Alecu IM, Zheng J, Zhao Y, Truhlar DG (2010) Computational thermochemistry: scale factor databases and scale factors for vibrational frequencies obtained from electronic model chemistries. J Chem Theory Comput 6:2872–2887

Altaner SP, Ylagen RF (1997) Comparison of structural models of mixed-layer illite/smectite and reaction mechanisms of smectite illitization. Clays Clay Miner 45:517–533

Audétat A, Günther D, Heinrich CA (1998) Formation of a magmatic-hydrothermal ore deposit: Insights with LA-ICP-MS analysis of fluid inclusions. Science 279:2091–2094

Balan E, Pietrucci F, Gervais C, Blanchard M, Schott J, Gaillardet J (2016) First-principles study of boron speciation in calcite and aragonite. Geochim Cosmochim Acta. doi:10.1016/j.gca.2016.07.026. (in press)

Baldereschi A (1973) Mean-value point in the Brillouin zone. Phys Rev B 7:5212–5215

Barth S (1993) Boron isotope variations in nature: a synthesis. Geol Rundschau 82:640–651

Belley PM, Grice JD, Fayek M, Kowalski PM, Grew ES (2014) A new occurrence of the borosilicate serendibite in tourmaline-bearing calc-silicate rocks, Portage-du-Fort Marble, Grenville Province, Québec: evolution of boron isotopes and tourmaline compositions in a metamorphic context. Can Mineral 52:595–615

Berryman EJ, Wunder B, Wirth R, Rhede D, Schettler G, Franz G, Heinrich W (2015) An experimental study on K and Na incorporation in dravitic tourmaline and insight into the origin of diamandiferous tourmaline from the Kokchetav Massif Kazakhstan. Contrib Mineral Petrol 129:28

Bigeleisen J, Mayer MG (1947) Calculation of equilibrium constants for isotopic exchange reactions. J Chem Phys 15:262–267

Blanchard M, Poitrasson F, Méheut M, Lazzeri M, Mauri F, Balan E (2009) Iron isotope fractionation between pyrite (FeS_2), hematite (Fe_2O_3) and siderite ($FeCO_3$): a first-principles density functional theory study. Geochim Cosmochim Acta 73:6565–6578

Branson O, Kaczmarek K, Redfern SAT, Misra S, Langer G, Tyliszcczak T, Bijma J, Elderfield H (2015) The coordination and distribution of B in foraminiferal calcite. Earth Planet Sci Lett 416:67–72

Branson O (2017) Boron incorporation into marine $CaCO_3$. In: Marschall HR, Foster GL (eds) Advances in isotope geochemistry. Boron isotopes—the fifth element. Springer, Heidelberg

Bray PJ, O'Keefe JG (1963) Nuclear magnetic resonance investigations of the structure of alkali borate glasses. Phys Chem Glasses 4:37–46

Byrne RH, Yao W, Klochko K, Tossel JA, Kaufman AJ (2006) Experimental evaluation of the isotopic exchange equilibrium $^{10}B(OH)_3 + {}^{11}B(OH)_4^- = {}^{11}B(OH)_3 + {}^{10}B(OH)_4^-$ in aqueous solution. Deep Sea Res I 53:684–688

Car R, Parrinello M (1985) Unified approach for molecular dynamics and density-functional theory. Phys Rev Lett 55:2471–2474

Chacko T, Cole DR, Horita J (2001) Equilibrium oxygen, hydrogen and carbon isotope fractionation factors applicable to geologic systems. Rev Mineral Geochem 43:1–81

Chakraborty S, Dingwell DB, Chaussidon M (1993) Chemical diffusivity of boron in melts of haplogranitic composition. Geochim Cosmochim Acta 57:1741–1751

Chiodini G, Comodi P, Giaquinto S (1988) Ammonia and boric acid in steam and water. Experimental data from geothermal wells in the Phlegrean Fields, Naples, Italy. Geothermics 17:711–718

Christ CL, Harder H (1978) Boron. In: Wedephol KH (ed) Handbook of Geochemistry, vol II/1. Springer, Berlin, pp 5E1–5O3

Clauer N, Williams LB, Fallick AE (2014) Genesis of nanometric illite crystals elucidated by light-element (hydrogen, lithium, boron and oxygen) isotope tracing and K-Ar and Rb-Sr dating. Chem Geol 383:26–50

De Hoog JCM, Savov IP (2017) Boron as tracer of subduction zone processes. In: Marschall HR, Foster GL (eds) Advances in isotope geochemistry. Boron isotopes—the fifth element. Springer, Heidelberg

Dohmen R, Kasemann SA, Coogan L, Chakraborty S (2010) Diffusion of Li in olivine. Part I: experimental observations and multi species diffusion model. Geochim Cosmochim Acta 74:274–292

Domagal-Goldman SD, Kubicki JD (2008) Density functional theory predictions of equilibrium isotope fractionation of iron due to redox changes and organic complexation. Geochim Cosmochim Acta 72:5201–5216

Domanik KJ, Hervig RL, Peacock SM (1993) Beryllium and boron in subduction zone minerals: an ion microprobe study. Geochim Cosmochim Acta 57:4997–5010

Driesner T (1997) The effect of pressure on deuterium–hydrogen fractionation in high-temperature water. Science 277:791–794

Dupuis R, Benoit M, Nardin E, Méheut M (2015) Fractionation of silicon isotopes in liquids: the

importance of configural disorder. Chem Geol 396:239–254
Dutrow BL, Henry DJ (2011) Tourmaline: a geological DVD. Elements 7:301–306
Edwards T, Endo T, Walton JH, Sen S (2014) Observation of the transition state for pressure-induced BO3 → BO4 conversion in glass. Science 345:1027–1029
Elmaci G, Icten O, Nedim Ay A, Zümreoglu-Karan B (2015) Boron isotopic fractionation in aqueous boric acid solution over synthetic minerals: effect of layer and surface charge on fractionation factor. Appl Clay Sci 107:117–121
Ertl A, Pertlik F, Bernhardt H-J (1997) Investigations on olenite with excess boron from the Koralpe, Styria, Austria. Anz Abt I 124:3–10
Ertl A, Hughes JM, Prowatke S, Ludwig T, Brandstätter F, Körner W, Dyar MD (2007) Tetrahedrally coordinated boron in Li-bearing olenite from "mushroom" tourmaline from Momeik, Myanmar. Can Mineral 45:891–899
Ertl A, Tillmanns E, Ntaflos T, Francis C, Giester G, Körner W, Hughes JM, Lengauer C, Prem M (2008) Tetrahedrally coordinated boron in Al-rich tourmaline and its relationship to the pressure-temperature conditions of formation. Eur J Mineral 20:881–888
Feng C, Qin T, Huang S, Wu Z, Huang F (2014) First-principles investigations of equilibrium calcium isotope fractionation between clinopyroxene and Ca-doped orthopyroxene. Geochim Cosmochim Acta 143:132–142
Fijii T, Pringle EA, Chaussidon M, Moynier F (2015) Isotope fractionation of Si in protonation/deprotonation reaction of silicic acid: a new pH proxy. Geochim Cosmochim Acta 168:193–205
Finley JW, Stephens PJ (1995) Density functional theory calculations of molecular structures and harmonic vibrational frequencies using hybrid density functionals. J Mol Struct 357:225–235
Foster GL, Pogge von Strandmann PAE, Rae JWB (2010) Boron and magnesium isotopic composition of seawater. Geochem Geophys Geosys 11:Q08015. doi:10.1029/2010GC003201
Furst M, Lowestam HA, Burnett DS (1976) Radiographic study of the distribution of boron in recent mollusc shells. Geochim Cosmochim Acta 40:1381–1386
Gaillardet J, Allègre CJ (1995) Boron isotopic composition of coral: seawater or diagenesis record? Earth Planet Sci Lett 136:665–676
Geisinger KL, Oestrike R, Navrotsky A, Turner GL, Kirkpatrick RJ (1988) Thermochemistry and structure of glasses along the join $NaAlSi_3O_8$–$NaBSi_3O_8$. Geochim Cosmochim Acta 52:2405–2414
Glover RB (1988) Boron distribution between liquid and vapour in geothermal fluids. In: Proceedings 10th New Zealand Geothermal Workshop, New Zealand
Goedecker S, Teter M, Hutter J (1996) Separable dual-space Gaussian pseudopotentials. Phys Rev B 54:1703–1710
Goldberg S, Forster HS, Lesch SM, Heick EL (1996) Influence of anion competition on boron adsorption by clays and soils. Soil Sci 161:99–103
Hedenquist JW, Arribas A, Reynolds TJ (1998) Evolution of an intrusion-centered hydrothermal system, far southeast-Lepanto porphyry and epithermal Cu–Au deposits, Phillipines. Econ Geol 93:373–404
Hemming NG, Hanson GN (1992) Boron isotopic composition and concentration in modern marine carbonates. Geochim Cosmochim Acta 56:537–543
Hemming NG, Reeder RJ, Hanson GN (1995) Mineral-fluid partitioning and isotopic fractionation of boron in synthetic calcium carbonate. Geochim Cosmochim Acta 59:371–379
Hershey JP, Fernandez M, Milne PJ, Millero FJ (1986) The ionization of boric acid in NaCl, Na–Ca–Cl, Na–Mg–Cl solutions at 25 °C. Geochim Cosmochim Acta 50:143–148
Hervig RL, Moore GM, Williams LB, Peacock SM, Holloway JR, Roggensack K (2002) Isotopic and elemental partitioning of boron between hydrous fluid and silicate melt. Am Mineral 87:769–774
Hill PS, Schauble EA (2008) Modeling the effects of bond environment on equilibrium iron isotope fractionation in ferricaquo–chloro complexes. Geochim Cosmochim Acta 72:1939–1958
Hill PS, Schauble EA, Young ED (2010) Effects of changing solution chemistry on Fe^{3+}/Fe^{2+} isotope fractionation in aqueous FeCl solutions. Geochim Cosmochim Acta 74:6669–6689
Hoover WG (1985) Canonical dynamics equilibrium phase-space distributions. Phys Rev A 31:1695–1697
Hughes JM, Ertl A, Dyar MD, Grew E, Wiedenbeck M, Brandstätter F (2004) Structural and chemical response to varying $^{[4]}$B content in zoned Fe-bearing olenite from Koralpe, Austria. Can Mineral 38:447–454
Ishikawa T, Nakamura E (1994) Origin of the slab component in arc lavas from across-arc variation of B and Pb isotopes. Nature 370:205–208
Jahn S, Kowalski PM (2014) Theoretical approaches to structure and spectroscopy of Earth materials. In: Henderson GS, Neuville DR, Downs RT (eds) Spectroscopic methods, vol 78. Rev Mineral Geochem, chap 17. Mineralogical Society of America, Washington DC, pp 691–743
Jahn S, Wunder B (2009) Lithium speciation in aqueous fluids at high P and T studied by ab initio molecular dynamics and consequences for Li isotope fractionation between minerals and fluids. Geochim Cosmochim Acta 73:5428–5434
Kakihana H, Kotaka M, Satoh S, Nomura M, Okamoto M (1977) Fundamental studies on the ion-exchange separation of boron isotopes. Bull Chem Soc Japan 50:158–163
Kalt A, Schreyer W, Ludwig T, Prowatke S, Bernhardt H-J, Ertl A (2001) Complete solid solution between magnesium schorl and lithium excess-boron olenite in pegmatite from the Koralpe (eastern Alps, Austria). Eur J Mineral 13:1191–1205

Klemme S, Marschall HR, Jacob DE, Prowatke S, Ludwig T (2011) Trace-element partitioning and boron isotope fractionation between white mica and tourmaline. Can Mineral 49:165–176

Klochko K, Cody GD, Tossell JA, Dera P, Kaufman AJ (2009) Re-evaluation boron speciation in biogenic calcite and aragonite using ^{11}B MAS NMR. Geochim Cosmochim Acta 73:1890–1900

Konrad-Schmolke M, Halama R (2014) Combined thermodynamic-geochemical modeling in metamorphic geology: boron as tracer of fluid-rock interaction. Lithos 208:393–414

Kowalski PM, Jahn S (2011) Prediction of equilibrium Li isotope fractionation between minerals and aqueous solutions at high P and T: an efficient ab initio approach. Geochim Cosmochim Acta 75:6112–6123

Kowalski PM, Wunder B, Jahn S (2013) Ab initio prediction of equilibrium boron isotope fractionation between minerals and aqueous fluids at high P and T. Geochim Cosmochim Acta 101:285–301

Kutzschbach M, Wunder B, Rhede D, Koch-Müller M, Ertl A, Giester G, Heinrich W, Franz G (2016a) Tetrahedral boron in natural and synthetic HP/UHP tourmaline: evidence from Raman spectroscopy, EMPA, and single-crystal XRD. Am Mineral 101:93–104

Kutzschbach M, Wunder B, Trumbull RB, Meixner A, Heinrich W, Franz G (2016b) The effect of tetrahedral B on the B isotope fractionation between olenitic tourmaline and fluid. In: Abstract, EMC^{2016}-conference, Rimini, Italy

Lea DW (2003) Elemental and isotopic proxies of past ocean temperatures. In: Elderfield H, Holland HD, Turekian KK (eds) The oceans and marine geochemistry. Elsevier, Oxford, p 365

Lee C, Yang W, Parr RG (1988) Development of the Colle-Salvetti correlation-energy formula into a functional of the electron density. Phys Rev B 37:785–789

Leeman WP, Sisson VB (2002) Geochemistry of boron and its petrological applications. In: Grew ES, Anovitz LM (eds) Boron: mineralogy, petrology and geochemistry, vol 33, 2nd edn. Mineralogical Society of America, Washington DC, pp 645–708

Leeman WP, Vocke RD, McKibben MA (1992) Boron isotopic fractionation between coexisting vapor and liquid in natural geothermal systems. In: Kharaka YF, Maest AS (eds) Proc 7th Int Sympos on water-rock interaction-WRI-7. Balkema, Rotterdam, pp 1007–1010

Lemarchand E, Schott J, Gaillardet J (2007) How surface complexes impact boron isotope fractionation: evidence from Fe and Mn oxides sorption experiments. Earth Planet Sci Lett 260:277–296

Liebscher A, Meixner A, Romer RL, Heinrich W (2005) Liquid-vapor fractionation of boron and boron-isotopes: Experimental calibration at 400 °C/23 MPa to 450 °C/42 MPa. Geochim Cosmochim Acta 69:5693–5704

Liebscher A, Meixner A, Romer RL, Heinrich W (2007) Experimental calibration of the vapor-liquid phase relations and lithium isotope fractionation in the system H_2O–LiCl at 400 °C/20–28 MPa. Geofluids 7:369–375

Liu Y, Tossell JA (2005) Ab initio molecular orbital calculations for boron isotope fractionations on boric acids and borates. Geochim Cosmochim Acta 69:3995–4006

London D (1997) Estimating abundances of volatile and other mobile components in evolved silicate melts through mineral-melt equilibria. J Petrol 38:1691–1706

Longstaffe FJ (1987) Stable isotope studies of diagenetic processes. In: Kyser K (ed) Short course in stable isotope geochemistry of low-temperature fluids. Mineralogical Association of Canada, Saskatoon, pp 187–257

MacGregor JR, Grew ES, De Hoog JCM, Harley SL, Kowalski PM, Yates MG, Carson CJ (2013) Boron isotopic composition of tourmaline, prismatine, and grandidierite from granulite facies paragneisses in the Larsemann Hills, Prydz Bay, East Antarctica: evidence for a non-marine evaporite source. Geochim Cosmochim Acta 123:261–283

Manara D, Grandjean A, Neuville DR (2009) Advances in understanding the structure of borosilicate glasses: a Raman spectroscopy study. Am Mineral 94:777–784

Manning CE (2004) The chemistry of subduction-zone fluids. Earth Planet Sci Lett 223:1–16

Markland TE, Berne BJ (2012) Unraveling quantum mechanical effects in water using isotopic fractionation. Proc Natl Acad Sci USA 109:7988–7991

Marschall HR (2005) Lithium, beryllium, and boron in high-pressure metamorphic rocks from Syros (Greece). PhD thesis, University of Heidelberg, Heidelberg, Germany, 411 pp. www.ub.uni-heidelberg.de/archiv/5634

Marschall HR, Jiang S-Y (2011) Tourmaline isotopes: no element left behind. Elements 7:313–319

Marschall HR, Ertl A, Hughes JM, McCommon C (2004) Metamorphic Na- and OH-rich disordered dravite with tetrahedral boron associated with omphacite, from Syros, Greece: chemistry and structure. Eur J Mineral 16:817–823

Marschall HR, Altherr R, Ludwig T, Kalt A, Gméling K, Kasztovszky Z (2006) Partitioning and budget of Li, Be and B in high-pressure metamorphic rocks. Geochim Cosmochim Acta 70:4750–4769

Marschall HR, Altherr R, Rüpke L (2007) Squeezing out the slab—modeling the release of Li, Be and B during progressive high-pressure metamorphism. Chem Geol 239:323–335

Marschall HR, Meyer C, Wunder B, Ludwig T, Heinrich W (2009) Experimental boron isotope fractionation between tourmaline and fluid: confirmation from in situ analyses by secondary ion mass spectrometry and from Rayleigh fractionation modelling. Contrib Miner Petrol 158:675–681

Marx D, Hutter J (2000) Ab initio molecular dynamics: theory and implementation. In: Grotendorst J (ed) Modern methods and algorithms of quantum chemistry. NIC, FZ Jülich, pp 301–449, CPMD code: J. Hutter et al. see: www.cpmd.org

Mavromatis V, Montouillout V, Noireaux J, Gaillardet J, Schott J (2015) Characterization of boron incorporation and speciation in calcite and aragonite from co-precipitation experiments under controlled pH, temperature and precipitation rate. Geochim Cosmochim Acta 150:299–313

McCulloch MT (2017) Boron isotopic and element constraints on coral calcification. In: Marschall HR, Foster GL (eds) Advances in isotope geochemistry. Boron isotopes—the fifth element. Springer, Heidelberg

Méheut M, Schauble EA (2014) Silicon isotope fractionation in silicate minerals: insights from first-principles models of phyllosilicates, albite and pyrope. Geochim Cosmochim Acta 134:137–154

Méheut M, Lazzeri M, Balan E, Mauri F (2007) Equilibrium isotopic fractionation in the kaolinite, quartz, water system: prediction from first-principles density-functional theory. Geochim Cosmochim Acta 71:3170–3181

Méheut M, Lazzeri M, Balan E, Mauri F (2009) Structural control over equilibrium silicon and oxygen isotopic fractionation; a first principles density functional theory study. Chem Geol 258:28–37

Menconi G, Tozer DJ (2002) Diatomic bond lengths and vibrational frequencies: assessment of recently developed exchange-correlation functionals. Chem Phys Lett 360:38–46

Meyer C, Wunder B, Meixner A, Romer RL, Heinrich W (2008) Boron-isotope fractionation between tourmaline and fluid: an experimental re-investigation. Contrib Miner Petrol 156:259–267

Michel F, Cormier L, Lombard P, Beuneu B, Galoisy L, Calas G (2013) Mechanisms of boron coordination change between borosilicate glasses and melts. J Non-Cryst Solids 379:169–176

Moore DM, Reynolds RC (1997) X-ray diffraction and the identification and analysis of clay minerals, 2nd edn. Oxford University Press, New York, p 378

Mottl MJ, Holland HD (1978) Chemical exchange during hydrothermal alteration of basalt by seawater - I Experimental results for major and minor components of seawater. Geochim Cosmochim Acta 42:1103–1115

Nosé S, Klein ML (1983) Constant pressure molecular dynamics for molecular systems. Mol Phys 50:1055–1076

Oi T, Kato J, Ossaka T, Karihana H (1991) Boron isotope fractionation accompanying boron mineral formation from aqueous boric acid-sodium hydroxide solutions at 25 °C. Geochem J 25:377–385

Oi T, Shimazaki H, Ishii R, Hosoe M (1997) Boron isotope fractionation in liquid chromatography with boron-specific resins as column packing material. Sep Sci Technol 32:1821–1834

Otake T, Lasaga AC, Ohmoto H (2008) Ab initio calculations for equilibrium fractionations in multiple sulfur isotope systems. Chem Geol 249:357–376

Palmer MR, Swihart GH (2002) Boron isotope geochemistry: an overview. In: Grew ES, Anovitz LM (eds) Boron: mineralogy, petrology and geochemistry Vol 33, 2nd edn. Reviews in Mineralogy, Mineralogical Society of America, Washington, DC, pp 709–744

Palmer MR, Spivack AJ, Edmond JM (1987) Temperature and pH controls over isotopic fractionation during adsorption of boron on marine clays. Geochim Cosmochim Acta 51:2319–2323

Palmer MR, London D, Morgan VIGB, Babb HA (1992) Experimental determination of fractionation of $^{11}B/^{10}B$ between tourmaline and aqueous vapor: a temperature- and pressure-dependent isotopic system. Chem Geol 101:123–129

Palmer MR, Pearson PN, Cobb SJ (1998) Reconstructing past ocean pH-depth profiles. Science 282:1468–1471

Peacock SM, Hervig RL (1999) Boron isotopic composition of subduction-zone metamorphic rocks. Chem Geol 160(4):281–290

Pichavant M, Kontak DJ, Briquen L, Herrera JV, Clark AH (1988) The Miocene-Pliocene Macusani volcanics, SE Peru II. Geochemistry and origin of a felsic peraluminous magma. Contrib Miner Petrol 100:325–338

Pinilla C, Blanchard M, Balan E, Ferlat G, Vuilleumier R, Mauri F (2014) Equilibrium fractionation of H and O isotopes in water from path integral molecular dynamics. Geochim Cosmochim Acta 135:203–216

Pinilla C, Blanchard M, Balan E, Natarajan SK, Vuilleumier R, Mauri F (2015) Equilibrium magnesium isotope fractionation between aqueous $Mg2^{+}$ and carbonate minerals: Insights from path integral molecular dynamics. Geochim Cosmochim Acta 163:126–139

Rae JWB (2017) Boron isotopes in foraminifera. In: Marschall HR, Foster GL (eds) Advances in isotope geochemistry. Boron isotopes—the fifth element. Springer, Heidelberg

Reynolds TJ, Beane RE (1985) Evolution of hydrothermal fluid characteristics at the Santa Rita, New Mexico, porphyr copper deposit. Econ Geol 80:1328–1347

Roedder E (1979) Fluid inclusions as samples of ore fluids. In: Barnes HL (ed) Geochemistry of hydrothermal ore deposits, 2nd edn. Wiley, New York, pp 684–737

Rosner M, Erzinger J, Franz G, Trumbull RB (2003) Slab-derived boron isotope signatures in arc volcanic rocks from the Central Andes and evidence for boron isotope fractionation during progressive slab dehydration. Geochem Geophys Geosys 4:1–25

Ruiz-Agudo E, Putnis CV, Kowacz M, Ortega-Huertas M, Putnis A (2012) Boron incorporation into calcite during growth: implications for the use of boron in carbonates as a pH proxy. Earth Planet Sci Lett 345:9–17

Rustad JR, Bylaska EJ (2007) Ab initio calculation of isotopic fractionation in $B(OH)_3$ (aq) and $B(OH)_4^-$ (aq). J Am Chem Soc 129:2222–2223

Rustad JR, Nelmes SL, Jackson VE, Dixon DA (2008) Quantum-chemical calculations of carbon-isotope

fractionation in CO_2(g), aqueous carbonate species, and carbonate minerals. J Phys Chem A 112(3):542–555
Rustad JR, Bylaska EJ, Jackson VE, Dixon DA (2010a) Calculation of boron-isotope fractionation between B$(OH)_3$(aq) and $B(OH)_4^-$(aq). Geochim Cosmochim Acta 74:2843–2850
Rustad JR, Casey WH, Yin QZ, Bylaska EJ, Felmy AR, Bogatko SA, Jackson VE, Dixon DA (2010b) Isotopic fractionation of Mg^{2+}(aq), Ca^{2+}(aq), and Fe^{2+}(aq) with carbonate minerals. Geochim Cosmochim Acta 74:6301–6323
Ryan JG, Langmuir CH (1993) The systematics of boron abundances in young volcanic rocks. Geochim Cosmochim Acta 57:1489–1498
Sanchez-Valle C, Reynard B, Daniel I, Lecuyer C, Martinez I, Chervin J-C (2005) Boron isotopic fractionation between minerals and fluids: new insights from in situ high pressure-high temperature vibrational spectroscopic data. Geochim Cosmochim Acta 69:4301–4313
Sanyal A, Hemming NG, Hanson GN, Broecker WS (1995) Evidence for a higher pH in the glacial ocean from boron isotopes in foraminifera. Nature 373:234–236
Sanyal A, Hemming NG, Broecker WS, Lea DW, Spero HJ, Hanson GN (1996) Oceanic pH control on the boron isotopic composition: evidence from culture experiments. Paleooceanography 11:513–517
Sanyal A, Nugent M, Reeder RJ, Bijma J (2000) Seawater pH control on the boron isotopic composition of calcite: evidence from inorganic calcite precipitation experiments. Geochim Cosmochim Acta 64:1551–1555
Schatz OJ, Dolejš D, Stix J, Williams-Jones AE, Layne GD (2004) Partitioning of boron among melt, brine and vapor in the system haplogranite-H_2O–NaCl at 800 °C and 100 MPa. Chem Geol 210:135–147
Schauble EA (2004) Applying stable isotope fractionation to new systems. In: Johnson CM, Beard BL, Albarede F (eds) Geochemistry of non-traditional stable isotopes, vol 55. Rev Mineral Geochem, chap 3. Mineralogical Society of America, Washington DC, pp 65–111
Schauble EA (2011) First-principles estimates of equilibrium magnesium isotope fractionation in silicate, oxide, carbonate and hexaaquamagnesium(2+) crystals. Geochim Cosmochim Acta 75:844–869
Schauble EA, Méheut M, Hill PS (2009) Combining metal stable isotope fractionation theory with experiments. Elements 5:369–374
Schmidt MW (1996) Experimental constraints on recycling of potassium from subducted oceanic crust. Science 272:1927–1930
Schmidt BC, Zotov N, Dupree R (2004) Structural implications of water and boron dissolution in albite glass. J Non-Cryst Solids 37:207–219
Schmidt C, Thomas R, Heinrich W (2005) Boron speciation in aqueous fluids at 22 to 600 °C and 0.1 MPa to 2 GPa. Geochim Cosmochim Acta 69:275–281
Scholze H (1988) Glas. Springer, Berlin, 339 pp
Schwarcz HP, Agyei EK, McMullen CC (1969) Boron isotopic fractionation during clay adsorption from sea-water. Earth Planet Sci Lett 6:1–5
Sen S, Stebbins JF, Hemming NG, Ghosh B (1994) Coordination environments of B impurities in calcite and aragonite polymorphs: A ^{11}B MAS NMR study. Am Mineral 79:819–825
Shannon RD (1976) Revised effective ionic radii and systematic studies of interatomic distances in halides and chalcogenides. Acta Cryst A 32:751–767
Shmulovich K, Heinrich W, Möller P, Dulski P (2002) Experimental determination of REE fractionation between liquid and vapour in the system NaCl–H_2O and $CaCl_2$–H_2O up to 450 °C. Contrib Miner Petrol 144:257–273
Singh SPN, Mattigod SV (1992) Modeling boron adsorption on kaolinite. Clays Clay Miner 40:192–205
Slack JF, Trumbull RB (2011) Tourmaline as a recorder of ore-forming processes. Elements 7:321–326
Smirnov SZ, Thomas VG, Demin SP, Drebushchak VA (2005) Experimental study of boron solubility and speciation in the Na_2O–B_2O_3–SiO_2–H_2O system. Chem Geol 223:16–34
Smith CL, Ficklin WH, Thompson JM (1987) Concentrations of arsenic, antimony, and boron in steam and steam condensate at the Geyers, California. J Volcanol Geotherm Res 32:329–341
Spivack AJ, Edmond JM (1987) Boron isotope exchange between seawater and the oceanic crust. Geochim Cosmochim Acta 51:1033–1041
Spivack AJ, Berndt ME, Seyfried WE Jr (1990) Boron isotope fractionation during supercritical phase separation. Geochim Cosmochim Acta 54:2337–2339
Spivack AJ, You CF, Smith HJ (1993) Foraminifera boron isotope ratios as a proxy for surface ocean pH over the past 21 Myr. Nature 363:149–151
Środoň J, McCarty DK (2008) Surface area and layer charge of smectite from CEC and EGME/H_2O retention measurements. Clays Clay Miner 56:142–161
Thomas R, Webster JD, Heinrich W (2000) Melt inclusions in pegmatitic glass: complete miscibility between silicate melts and hydrous fluids at low pressure. Contrib Miner Petrol 139:394–401
Thomas R, Förster H-J, Heinrich W (2003) The behaviour of boron in a peraluminious granite-pegmatite system and associated hydrothermal solutions: a melt and fluid-inclusion study. Contrib Miner Petrol 144:457–472
Tomascak PB, Magna T, Dohmen R (2016) Advances in lithium isotope geochemistry, Springer, Berlin, 195 pp
Tonarini S, Pennisi M, Leeman WP (1997) Precise boron isotopic analysis of complex silicate (rock) samples using alkali carbonate fusion and ion-exchange separation. Chem Geol 142:129–137
Tonarini S, Forte C, Petrini R, Ferrara G (2003) Melt/biotite $^{11}B/^{10}B$ isotopic fractionation and the

boron local environment in the structure of volcanic glasses. Geochim Cosmochim Acta 67:1863–1873
Tossel JA (2006) Boric acid adsorption on humic acids: ab initio calculation of structures, stabilities, ^{11}B NMR and ^{11}B, ^{10}B isotopic fractionations of surface complexes. Geochim Cosmochim Acta 70:5089–5103
Trumbull RB, Chaussidon M (1999) Chemical and boron isotopic composition of magmatic and hydrothermal tourmalines from Sinceni granite-pegmatite system in Swaziland. Chem Geol 153:125–137
Trumbull RB, Slack JF (2017) Boron isotopes in the continental crust: granites, pegmatites and related hydrothermal systems. In: Marschall HR, Foster GL (eds) Advances in isotope geochemistry. Boron isotopes—the fifth element. Springer, Heidelberg
Urey HC (1947) The thermodynamic properties of isotopic substances. J Chem Soc 562–581
Vakulenko AG, Alekhin YV, Razina MV (1989) Solubility and thermodynamic properties of alkali chlorides in steam. In: Proceedings II International Symposium “Properties of water and steam”, Prague, pp 395–401
van Hinsberg VJ, Marschall HR (2007) Boron isotope and light element sector zoning in tourmalines: Implications for the formation of B-isotopic signatures. Chem Geol 238:141–148
van Hinsberg VJ, Schumacher JC (2007) Using estimated thermodynamic properties to model accessory phases in the cases of tourmaline. J Metamorph Geol 25:769–779
van Hinsberg VJ, Henry DJ, Dutrow BL (2011) Tourmaline as a petrological forensic mineral. A unique recorder of its geologic past. Elements 7:327–332
Veksler IV, Dorfman AM, Dingwell DB, Zotov N (2002) Element partitioning between immiscible borosilicate liquids: A high-temperature centrifuge study. Geochim Cosmochim Acta 66:2603–2614
Vengosh A, Chivas AR, McCulloch MT (1989) Direct determination of boron and chlorine isotopic compositions in geological materials by negative thermal-ionization mass spectrometry. Chem Geol 79:333–343
Vengosh A, Kolodny Y, Starinsky A, Chivas AR, McCulloch M (1991) Coprecipitation and isotopic fractionation in boron in modern biogenic carbonates. Geochim Cosmochim Acta 55:2901–2910
Vengosh A, Starinsky A, Kolodny Y, Chivas AR, Raab M (1992) Boron isotope variations during fractional evaporation of sea water: new constraints on the marine vesurs nonmarine debate. Geology 20:799–802
von Goerne G, Franz G, Heinrich W (2001) Synthesis of tourmaline solid solutions in the system Na_2O–MgO–Al_2O_3–SiO_2–B_2O_3–H_2O–HCl and the distribution of Na between tourmaline and fluid at 300–700 °C and 200 MPa. Contrib Miner Petrol 141:160–173
Wagner W, Pruss A (2002) The IAPWS formulation 1995 for the thermodynamic properties of ordinary water substance for general and scientific use. J Phys Chem Ref Data 31:387–535
Williams LB, Hervig RL (2005) Lithium and boron isotopes in illite-smectite: the importance of crystal size. Geochim Cosmochim Acta 69:5705–5716
Williams LB, Hervig RL, Holloway JR, Hutcheon I (2001) Boron isotope geochemistry during diagenesis. Part I Experimental determination of fractionation during illitization of smectite. Geochim Cosmochim Acta 65:1769–1782
Williams LB, Turner A, Hervig RL (2007) Intracrystalline boron isotope partitioning in illite-smectite: testing the geothermometer. Am Mineral 92:1958–1965
Wu J, Stebbins JF (2010) Quench rate and temperature effects on boron coordination in aluminoborosilicates. J Non-Cryst Solids 356:2097–2108
Wu J, Stebbins JF (2013) Temperature and modifier cation field strength effects on aluminoborosilicate glass network structure. J Non-Cryst Solids 362:73–81
Wunder B, Meixner A, Romer RL, Wirth R, Heinrich W (2005) The geochemical cycle of boron: constraints from boron isotope partitioning experiments between mica and fluid. Lithos 84:206–216
Wunder B, Meixner A, Romer RL, Heinrich W (2006) Temperature-dependent isotopic fractionation of lithium between clinopyroxene and high-pressure hydrous fluids. Contrib Miner Petrol 151:112–120
Wunder B, Romer RL, Meixner A, Jahn S (2011) Li-isotope silicate fluid fractionation: pressure dependence and influence of the bonding environment. Eur J Miner 23:333–342
Yamaji K, Makita Y, Watanabe H, Sonoda A, Kanoh H, Hirotsu T, Ooi K (2001) Theoretical estimation of lithium isotopic reduced partition function ratio for lithium ions in aqueous solution. J Phys Chem A 105:602–613
Zeebe RE (2005) Stable boron isotope fractionation between dissolved $B(OH)_3$ and $B(OH)_4^-$. Geochim Cosmochim Acta 69:2753–2766
Zeebe RE (2009) Hydration in solution is critical for stable oxygen isotope fractionation between carbonate ion and water. Geochim Cosmochim Acta 73:5283–5291
Zeebe RE (2010) A new value for the stable oxygen isotope fractionation between dissolved sulfate ion and water. Geochim Cosmochim Acta 74:818–828
Zheng Y-F (1993) Oxygen isotope fractionation in SiO_2 and Al_2SiO_5 polymorphs: effect of crystal structure. Eur J Mineral 5:651–658
Zhong J, Bray PJ (1989) Change of coordination in alkali borate glasses, and mixed alkali effects as elucidatd by NMR. J Non-Cryst Solids 111:67–76

Boron Incorporation into Marine $CaCO_3$

4

Oscar Branson

Abstract

The isotopic composition ($\delta^{11}B$) and abundance (B/Ca) of boron in the marine $CaCO_3$ minerals calcite and aragonite are used as paleoceanographic tracers for past oceanic pH and carbon chemistry. These environmental proxies depend upon the ability of $CaCO_3$ minerals to incorporate trace concentrations of B within their structure, and record the state of the pH-dependent equilibrium between $B(OH)_3$ and $B(OH)_4^-$, and the relative abundance of B and C in seawater. To achieve this $CaCO_3$ minerals must either incorporate a single species of aqueous B, or take up a predictable mixture of both species. Initial investigations found evidence to suggest the sole incorporation of aqueous $B(OH)_4^-$ into the anion site of $CaCO_3$ minerals. These observations established the required link between aqueous B chemistry and $CaCO_3^-$ hosted B, and provided the foundation for the development and application of the $\delta^{11}B$ and B/Ca proxies. However, advances in our understanding of aqueous B chemistry, improvements in the accuracy of B isotopic measurements of carbonates, and new data from controlled precipitation experiments have since revealed more complex, structure-dependent mechanisms of B incorporation into $CaCO_3$. Studies of aragonite appear to support a relatively straightforward substitution of $B(OH)_4^-$ into the mineral anion site. Conversely, a growing number of studies of calcite suggest either that

O. Branson (✉)
Department of Earth and Planetary Sciences, University of California, Davis, One Shields Ave, Davis, CA 95616, USA
e-mail: obranson@ucdavis.edu

Present Address:
O. Branson
Research School of Earth Sciences, Australian National University, 142 Mills Rd, Acton, ACT 2601, Australia

H. Marschall and G. Foster (eds.), *Boron Isotopes*, Advances in Isotope Geochemistry,
https://doi.org/10.1007/978-3-319-64666-4_4

both aqueous $B(OH)_3$ and $B(OH)_4^-$ are taken up into the mineral, or that B is subject to a significant isotopic fractionation during incorporation. While a growing body of theoretical and experimental work are moving toward an understanding of B uptake in $CaCO_3$, we currently lack a systematic description of this key process, particularly in calcite. As long as the mechanisms of B incorporation remain unknown, the relationships between $\delta^{11}B$ and B/Ca and ocean chemistry must be treated as empirical, adding uncertainty to the paleoceanographic records derived from them. This chapter will explore our current understanding of B incorporation into marine $CaCO_3$ minerals, in context of their structure and growth mechanisms. We will consider the broad question of 'how does B get from seawater into calcite and aragonite?'

Keywords
Boron · Cacite · Aragonite

4.1 Introduction

It has so far been impossible to directly observe the uptake of B from solution into a carbonate mineral. Our mechanistic understanding of B incorporation is inferred from indirect geochemical and structural observations, in context of our understanding of mineral growth processes. Theories of B incorporation have therefore evolved with advances in our understanding of carbonate mineral growth processes and B solution chemistry, and our ability to measure B geochemistry.

The study of B as an impurity in carbonate minerals began with descriptions of boron concentrations in synthetic and biogenic carbonates. These studies revealed that B is more easily incorporated into aragonite than calcite (Leutwein and Waskowiakr 1962; Furst et al. 1976; Kitano et al. 1978), and that the abundance of B within both minerals is directly proportional to its concentration in the host solution (Kitano et al. 1978). This demonstrated a dependence of B incorporation on both external B concentration and mineral structure. Subsequent study of diverse authigenic and biogenic marine carbonates revealed a broad range of naturally occurring B concentrations (Hemming and Hanson 1992), indicating that natural variability in the conditions and mechanisms of mineral growth are also important factors in B uptake.

Studies of B isotopic content of biogenic (Vengosh et al. 1991; Hemming and Hanson 1992) and synthetic (Hemming et al. 1995) carbonates revealed a systematic variability in B isotopic composition with solution chemistry. Based on the best available measurement of aqueous B fractionation (Kakihana et al. 1977), carbonate $\delta^{11}B$ appeared most similar to the predicted $\delta^{11}B_{B(OH)4}$ of seawater. This early body of work, combined with observations that $B(OH)_4^-$ appeared to interact directly with calcite growth surfaces (Hemming et al. 1998), culminated in the development of a working model of B incorporation in which only the tetrahedral $B(OH)_4^-$ anion is incorporated from solution. In this model $B(OH)_4^-$ is adsorbed onto the growing surface of carbonate minerals, and directly incorporated into the anion site of aragonite, or transformed into trigonal HBO_3^{2-} during incorporation into calcite with little or no isotopic fractionation (Hemming et al. 1995; Hemming and Hanson 1992). This mechanism was supported by all available experimental data, and was based on a fundamental difference between the aqueous B species: the charged $B(OH)_4^-$ ion will interact more readily with a growing mineral surface, compared to uncharged $B(OH)_3$. This model provided the basis for the subsequent development and application of the B carbonate proxies.

The first evidence to suggest that B incorporation may be more complex came from the detailed study of isotope fractionation between aqueous $B(OH)_3$ and $B(OH)_4^-$. Numerous theoretical and experimental studies concluded that B fractionation is significantly different from Kakihana et al. (1977) initial estimate of ~19‰, with estimates ranging between 18–30‰, depending on the specifics of the experimental conditions or theoretical approach used (Klochko et al. 2006; Y. Liu and Tossell 2005; Nir et al. 2015; Oi 2000; Sanchez-Valle et al. 2005; Zeebe 2005). Today, the widely accepted fractionation factor for B in seawater is 27.2 ± 0.6‰ (Klochko et al. 2006). In light of this revised fractionation factor, it became apparent that previous $\delta^{11}B$ measurements of carbonates (Hemming et al. 1995; Hemming and Hanson 1992; Sanyal et al. 1996, 2000, 2001) were significantly different from aqueous $\delta^{11}B_{B(OH)_4^-}$, which undermines the conveniently straightforward B incorporation mechanism proposed by Hemming and Hanson (1992). Subsequent measurements of $\delta^{11}B$ in other synthetic and biomineral carbonates largely reinforce this observation, exhibiting $\delta^{11}B$ values that are often heavier (more positive) or lighter (more negative) than the $\delta^{11}B_{B(OH)_4^-}$ predicted by a ~27.2‰ fractionation (Fig. 4.1; e.g. Sanyal et al. 2001; Foster 2008; Henehan et al. 2013). Some studies report $\delta^{11}B$ that is very similar to aqueous $\delta^{11}B_{B(OH)_4^-}$ (e.g. Rae et al. 2011; Kaczmarek et al. 2016), but given the wide $\delta^{11}B$ variability in the literature, and the complexity and diversity of B data processing methods (e.g. choice of equilibrium coefficients and fractionation factors), these studies must be treated as the exceptions, rather than the rule, when seeking to understand B incorporation mechanisms.

The $\delta^{11}B$ of the majority of naturally and synthetically formed carbonates are significantly offset in both intercept and slope from 'ideal'

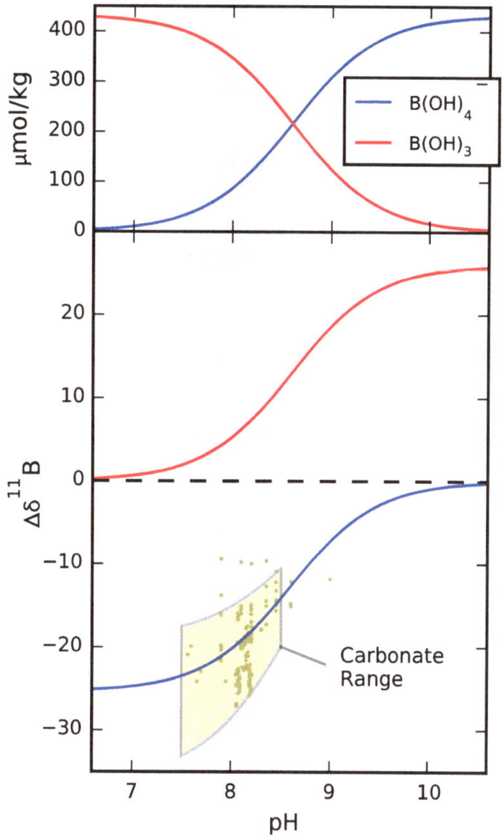

Fig. 4.1 Boron in solution and carbonates. Aqueous B exists in a pH-dependent equilibrium between $B(OH)_3$ and $B(OH)_4^-$ (*top*), and there is an isotope fractionation of ~26‰ (Nir et al. 2015) between these species (*bottom*). The $\delta^{11}B$ of trace B within carbonate minerals varies with pH, although the absolute $\delta^{11}B$ value and slope of the change is highly variable, depending on the provenance of the mineral (e.g. specific organism or precipitation environment). In this plot pK_B = 8.6 (Dickson et al. 2007) and σ_{4-3} = 1.026 (Nir et al. 2015). The 'calcite range' envelope is described by two σ_{4-3} values of 1.019 and 1.037, and encompasses 95% of measurements of synthetic and biomineral calcite and aragonite reported by systematic studies across a broad pH range (Hemming et al. 1995; Sanyal et al. 1995; Sanyal et al. 2000; Sanyal et al. 2001; Hönisch et al. 2003; Hönisch et al. 2004; Foster 2008; Rollion-Bard & Erez 2010; Krief et al. 2010; Trotter et al. 2011; McCulloch et al. 2012; Anagnostou et al. 2012; Henehan et al. 2013; Martinez-Boti et al. 2015; Noireaux et al. 2015; Henehan et al 2016). Data points used to calculate this range are shown as *grey dots*. The variability within this data reflects a combination of both constant offsets and deviations in slope from σ_{4-3} lines

predicted $\delta^{11}B_{B(OH)_4^-}$ values. The offsets from predicted $\delta^{11}B_{B(OH)_4^-}$ may be caused by kinetic isotope fractionation during the incorporation of B, the incorporation of both $B(OH)_4^-$ and $B(OH)_3$ from solution or, in biominerals, the influences of biological controls on solution chemistry and mineral precipitation processes. However, despite these offsets it is notable that carbonates from a single, distinct origin (e.g. precipitates formed under similar conditions, or shells produced by the same organism) consistently show relationships between mineral $\delta^{11}B$ and solution $\delta^{11}B_{B(OH)_4^-}$. Based on these relationships, careful calibrations linking the $\delta^{11}B$ and B/Ca[1] of carbonates to solution chemistry have allowed the development of empirical carbonate chemistry and pH palaeoproxies. However, the unknown factors influencing B uptake into carbonate minerals introduces considerable uncertainties into their interpretation, particularly when seeking to interpret the B chemistry of biominerals beyond the range of extant species where unknown 'vital effects' complicate empirical calibrations. A systematic understanding of B uptake into carbonate minerals would allow us to separate the influences of crystallographic, biological and climatic signals on carbonate $\delta^{11}B$ and B/Ca, and significantly increase our confidence in their application as paleoproxies.

Uncovering the mechanisms of B incorporation in carbonates is inherently complex, because it involves the interaction of two kinetically and thermodynamically distinct aqueous B species with a complex growing crystal surface. There are a lot more 'moving parts' to consider than with the relatively straightforward cationic trace elements (e.g. Mg^{2+} or Sr^{2+}). There have also been historical uncertainties surrounding the equilibrium values of B speciation and isotope fractionation in solution (Pagani et al. 2005; Hemming and Hönisch 2007), and differences in precipitation conditions and choices of data processing methods between studies (e.g. Uchikawa et al. 2015 vs. Kaczmarek et al. 2016), all of which combine to complicate interpretation of published data. Recent inorganic precipitation experiments seek to take these considerations into account, and offer insights into the dynamics of B uptake into calcite and aragonite through systematic, controlled experiments (Farmer et al. 2015; Mavromatis et al. 2015; Noireaux et al. 2015; Uchikawa et al. 2015; Kaczmarek et al. 2016; Holcomb et al. 2016). These studies provide compelling evidence to suggest that the incorporation of B into synthetic calcite and aragonite occurs via distinct mechanisms. Aragonite appears to solely incorporate $B(OH)_4^-$, while calcite may take up a mixture of both $B(OH)_4^-$ and $B(OH)_3$ from solution, possibly with a significant isotopic fractionation associated with B incorporation. This offers confidence in our understanding of B incorporation into aragonite, but highlights a significant shortfall in our understanding of B uptake into calcite.

The lack of a comprehensive model of B uptake mechanism in carbonates has significant implications for our interpretation of carbonate-derived B proxy records, and is sorely needed if we are to confidently interpret these records. To explore the mechanisms of B incorporation in calcite and aragonite, we will first review the structure and growth mechanisms of marine carbonate minerals, and the mechanics of impurity incorporation. We will then review the B geochemistry of synthetic carbonates, and consider how they fit with our current understanding of B solution chemistry and mineral growth processes. Finally, we will consider the complicating factors that may separate carbonate biominerals from these synthetic precipitates.

[1]Strictly, it is more appropriate to consider B/CO_3^{2-} ratios in carbonates because B is an anion. However, B is routinely measured by mass spectrometry relatve to Ca, so is conveniently expressed as B/Ca. Stoichiometrically, $B/CO_3^{2-} = B/Ca / (1 - B/Ca)$, so as long as B is a trace element, the difference between B/Ca and B/CO_3^{2-} will be vanishingly small, and they can be treated as equivalent.

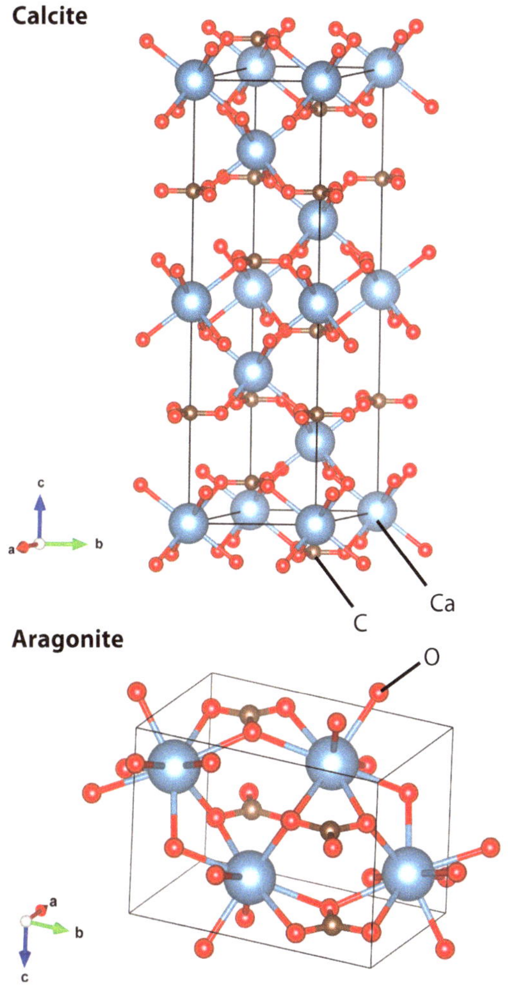

Fig. 4.2 The atomic-scale structure and calcite and aragonite. Calcite is a trigonal hexagonal structure with *R3c* symmetry, while aragonite is an orthorhombic structure with *mmm* symmetry. Calcite ($\Delta G_{formation}$ – 1128.8 kJ mol^{-1}) is more stable than aragonite (−1127.8), with a higher solubility product (−log K_{sp} = 8.30 vs 8.12). Aragonite is denser than calcite (2.95 vs 2.71 g cm^{-3})

4.2 Marine Calcium Carbonate Minerals

Calcium Carbonate ($CaCO_3$) minerals are abundant throughout the ocean, and are a major component of the global carbon cycle. The majority of $CaCO_3$ produced in the modern ocean is in the form of biomineral skeletal components, made by a diverse range of organisms (Milliman 1993; Morse et al. 2007). The two most abundant forms of $CaCO_3$ are calcite and aragonite. These polymorphs are stoichiometrically identical, formed from a lattice of Ca^{2+} and CO_3^{2-} ions, but differences in the arrangement and bonding of these ions give them disparate physical, chemical and structural properties (Fig. 4.2; Reeder 1983).

4.2.1 Carbonate Mineral Formation

While the composition and structure of carbonate minerals are relatively simple, their formation processes can be surprisingly diverse and complex (Fig. 4.3). Calcite and aragonite may form via monomer addition to a growing mineral surface, as described by the Terrace Ledge Kink (TLK) model (Fig. 4.4; Burton et al. 1951; Chernov 1984; de Yoreo and Vekilov 2003), or 'non-classical' processes such as the agglomeration and crystallization of amorphous precursor phases via Ostwald ripening (Bots et al. 2012; Ostwald 1897), or the oriented assembly of nano-crystalline particles (Fig. 4.3; de Yoreo et al. 2015). Numerous extensive reviews offer a more complete discussion of carbonate nucleation and growth (e.g. de Yoreo et al. 2015; de Yoreo and Vekilov 2003; Morse et al. 2007; Weiner 2008), and we will focus here on those aspects of mineral precipitation most relevant to B incorporation in marine carbonates.

Many carbonate biominerals are likely to be formed via non-classical growth pathways, involving amorphous precursor phases or nano-particle assembly (de Yoreo et al. 2015; Weiner and Addadi 2011; Weiner et al. 2002). Therefore, deciphering the uptake of B via non-classical mineralization pathways may ultimately offer the most comprehensive and useful understanding of B incorporation in marine carbonates. However, these growth mechanisms are diverse and complex, and studies seeking to

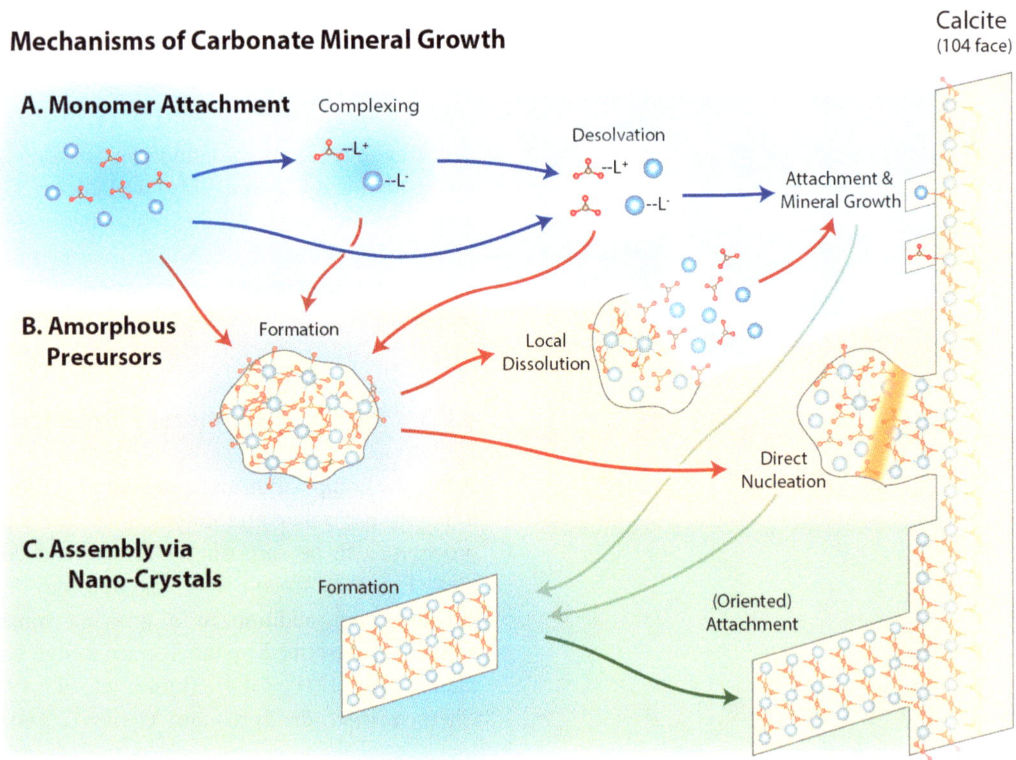

Fig. 4.3 **Carbonate precipitation mechanisms.** Carbonates form via a diverse range of processes, from **a** the addition of monomers from solution onto a mineral growth surface (Fig. 4.4), to **b** the accumulation and transformation of amorphous precursor phases, or **c** the formation and assembly of crystalline nano-particles. These processes are more accurately considered as mechanistic end-members, and may often coincide (e.g. de Yoreo et al. 2015). This schematic illustrates the general steps involved in each end member and they ways they can interact. The mechanisms of mineral growth in carbonates produce a diverse range of both structural and non-structural mechanisms of trace element incorporation (Fig. 4.5)

determine the mechanisms of B uptake into carbonates have thus far focussed on simpler synthetic precipitates, produced in well-constrained laboratory conditions. In these idealized systems, crystal growth tends to proceed via the relatively simple and well-understood 'classical' TLK growth model.

4.2.1.1 Terrace-Ledge-Kink (TLK) Growth

The overall drive of mineral precipitation is thermodynamic. A carbonate mineral may form from a solution when the concentrations of Ca^{2+} and CO_3^{2+} are high enough that the precipitation of a mineral reduces the free energy of the system. At the level of TLK growth, this is described in terms of the kinetics of monomer or molecule attachment and detachment at the growing surface of a crystal (de Yoreo and Vekilov 2003; Zhang and Nancollas 1990). Crystal growth will occur when the rate of ion attachment exceeds the rate of detachment (Fig. 4.4a). These rates can be considered instantaneously in terms of the probability of a given ion overcoming the energy barriers associated with attachment or detachment.

The first step to attachment at a crystal surface is the encounter of a solute ion with the surface. The probability of this encounter is directly proportional to the activity of the ion in solution, which in simple solutions is equivalent to its concentration. Next, the ion must undergo

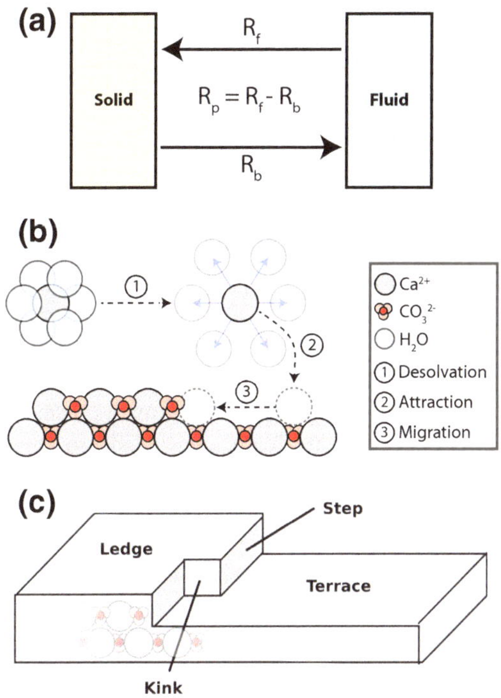

Fig. 4.4 The TLK growth model. **a** Crystal precipitation rate (R_p) may be considered as the difference between the rates of molecule attachment (R_f) and detachment (R_b). **b** Attachment rates at the mineral surface are governed by the energy barriers to desolvation of the solution ion and/or surface, bond making, and migration across the surface to a low-energy binding environment at 'growth step' edges. The energy barrier to bond breaking at the mineral surface determines detachment rates. **c** In three dimensions, this leads to the development of the 'Terrace-Ledge-Kink' growth environment, where monomer-thick layers propagate across the crystal surface, adding new ions at the low energy bonding environments available at 'step' and 'kink' sites

'desolvation', and break free of an ordered shell of water molecules, known as its hydration shell. Once free of its associated water molecules, the ion may form bonds with the mineral surface (Fig. 4.4b). These steps each involve an energy barrier, and the probability that these barriers are overcome depends upon the amount of free energy in the environment (i.e. temperature). The energy barrier to desolvation of the ion in solution or the mineral surface is often a rate limiting step in this processes, and variations in the strength of hydration shells between different ions drives a degree of specificity in the attachment of solute ions to the surface. The energy required for desolvation can be increased or decreased by ion-specific interactions with other solutes, or ions and compounds that are already attached to the crystal surface. This can lead to complex relationships between solute and surface composition, surface structure, and ion uptake into the crystal (e.g. Piana et al. 2008). Once attached to the mineral surface, the ion may detach from the surface, migrate across the surface to adjacent attachment sites, or remain attached and become incorporated into the bulk mineral structure.

The probability of ion detachment is determined by the energy barrier to break the bonds formed with the surface, and is nearly independent of its concentration of solution (DePaolo 2011). The migration of ions on the surface is governed by the principles of surface diffusion, where the energy barrier for an ion to 'jump' between adjacent binding sites tends to be much lower than the energy barrier to detachment. The energy barriers to both surface diffusion and detachment depend on the strength and number of bonds formed with the mineral surface, which is largely dependent on the structure of the bonding environment. During precipitation, ions on the surface are most strongly attached at 'step' edges, where they are bonded on two sides, and 'kink' sites where they bond on three sides (Fig. 4.4c). At near-equilibrium conditions, the majority of ion detachment and attachment occurs at kink sites, while at high supersaturations the attachment of ions to step edges occurs more frequently, forming '1D nuclei' which generate new kink sites (Zhang and Nancollas 1990). Once attached to a step edge or kink site, the energy barriers to ion detachment and diffusion are significantly greater, and the ion is more likely to be incorporated into the bulk mineral. This drives the propagation of kink sites and growth terraces across a mineral face. Because atoms can be unevenly arranged within the crystal lattice, different growth faces of the mineral can present distinct bonding environments, with corresponding differences in binding energies. In the case of a crystal like calcite, differences in the atomic structure of these binding sites leads to strong preferential attachment and growth parallel to the crystallographic 'c axis', which determines the shape of the overall crystal.

4.2.1.2 Non-classical Crystal Growth

Crystals may also grow via more complex processes, involving the formation and transformation of intermediate metastable phases. Mineralization via amorphous calcium carbonate (ACC, Fig. 4.3b, c) involves the concentration of disordered, or poorly ordered Ca^{2+} and CO_3^{2-} at similar density to the mature mineral (Faatz et al. 2004; Navrotsky 2004; Radha et al. 2010), with stabilizing additives (e.g. amino acids, proteins, or inorganic ligands), that prevents the nucleation or growth of a coherent crystalline structure (Aizenberg et al. 2002; Raz et al. 2003; Wang et al. 2009). Note that 'ACC' is an umbrella term describing a diverse range of poorly crystalline precursor phases, with a wide range of stabilization and crystallization mechanisms, and a corresponding range in geochemistry (de Yoreo et al. 2015). On contact with an external nucleation sites, or on removal of the stabilizing element, the nucleation and growth of the mature mineral phase occurs rapidly from ACC, either through a highly localized dissolution-reprecipitation reaction (Bots et al. 2012; Rodriguez-Blanco et al. 2011; Rodriguez-Navarro et al. 2015, 2016), or via a propagating solid phase transformation (Gal et al. 2013). The assembly of ACC nano-particles can follow similar TLK-type patterns, adding ACC particles to the growing surface instead of monomer ions (Rodriguez-Navarro et al. 2016). These ACC 'building blocks' can transform into coherent, single crystal structures, or leave a distinctive, 'nodular' microstructure, depending on the specific stabilizing additive in the ACC phase (Rodriguez-Navarro et al. 2016). The same thermodynamic and kinetic principles governing TLK crystal growth still apply to the transformation of these ACC particles, but the chemical and physical environment of mineral growth is far removed from an 'open' equilibrated system, and precipitation or transformation occurs in a restricted, diffusion-limited environment that is distinct from the parent solution.

Nano-crystal assembly may be involved in the formation of complex biomineral structures (e.g. Oaki et al. 2008; Killian et al. 2009), and is best considered as a mechanism for controlling the overall architecture of a mineral structure, rather than a process of mineral precipitation in itself. Nano-crystals 'building blocks' may form by either classical or non-classical growth processes, most likely in chemically controlled biological environments (de Yoreo et al. 2015).

4.2.2 Impurity Incorporation in Carbonates

Calcite and aragonite are able to accommodate trace impurities within their structure. Impurities can be hosted in a diverse range of atomic environments (Fig. 4.5), from straightforward substitution of individual monomers within the mineral structure (e.g. Branson et al. 2013), to coupled substitutions (e.g. Binder and Troll 1989), concentration along crystal grain boundaries (e.g. Watson 1996), distinct micro-domains within the mineral structure (e.g. Schmalz 1965; Weber and Kaufman 1965), or in association with non-mineral skeletal components in composite biomineral structures (e.g. Branson et al. 2016; Erez 2003). The environment in which an impurity is hosted within the mineral structure depends on the chemical and physical characteristics of the impurity, and the structure and precipitation mechanism of the host mineral.

The dynamics of trace cation incorporation by direct substitution in simple, TLK growth systems have been extensively studied (Davis et al. 2000; DePaolo 2011; Gabitov et al. 2013; Katz 1973; Nielsen et al. 2012; Oomori et al. 1987; Paquette and Reeder 1995), and form the basis of our understanding of trace element incorporation in carbonates as a whole. At the bulk scale, as with mineral precipitation, impurity incorporation can be considered thermodynamically in

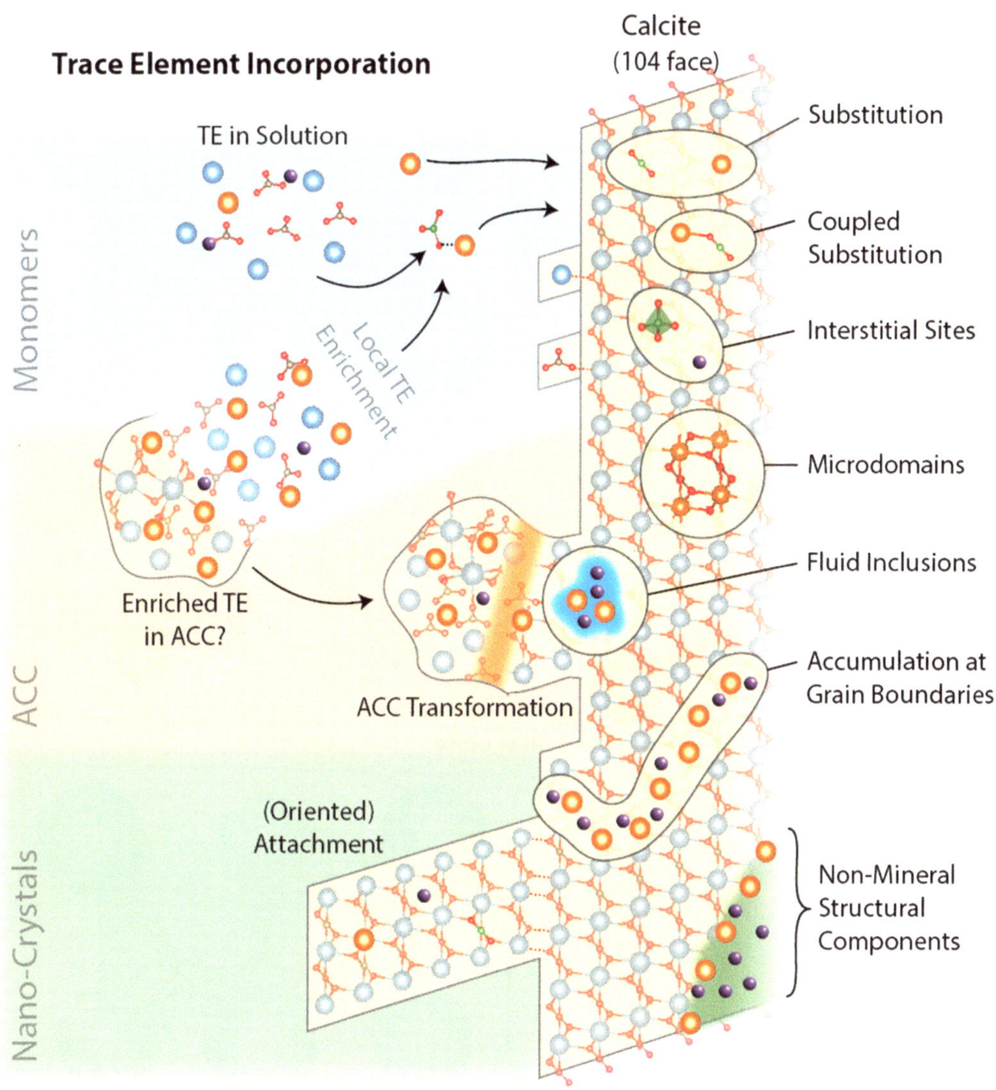

Fig. 4.5 The environments of trace elements in carbonates, arranged in approximate correspondence to the mineral precipitation mechanisms most likely to produce them. Simple monomer-addition growth, as described by the TLK mechanism (Fig. 4.4) is most likely to incorporate ideally substituted impurities, either as individual ions or in coupled ion complexes. If precipitation is rapid, incorporation in interstitial sites is also likely, via entrapment mechanisms. This is particularly true during the rapid dissolution-reprecipitation of ACC phases in diffusion-limited environments. The crystallization of ACC phases brings the possibility of excluding impurities to grain boundaries, or in extreme cases concentrating them into micro-domains of a distinct mineral phase. This may occur in either dissolution-reprecipitation or solid-state transformation, as the crystal formation excludes impurities present in the ACC phase away from the propagating crystallization edge. Finally, the assembly of mineral structures via the agglomeration of crystalline or ACC nano-particles provides opportunity to include fluid inclusions that capture trace elements from solution, or may be built around non-crystalline structural components (e.g. organic templates), which can contribute to the bulk chemical signature of the mineral structure

terms of the free energy contribution of the impurity to the host mineral. Any disparity in size, charge or shape between the impurity and the host ion it is replacing will increase the free energy of the mineral, and decrease the likelihood of its incorporation. For example, Ca^{2+} in

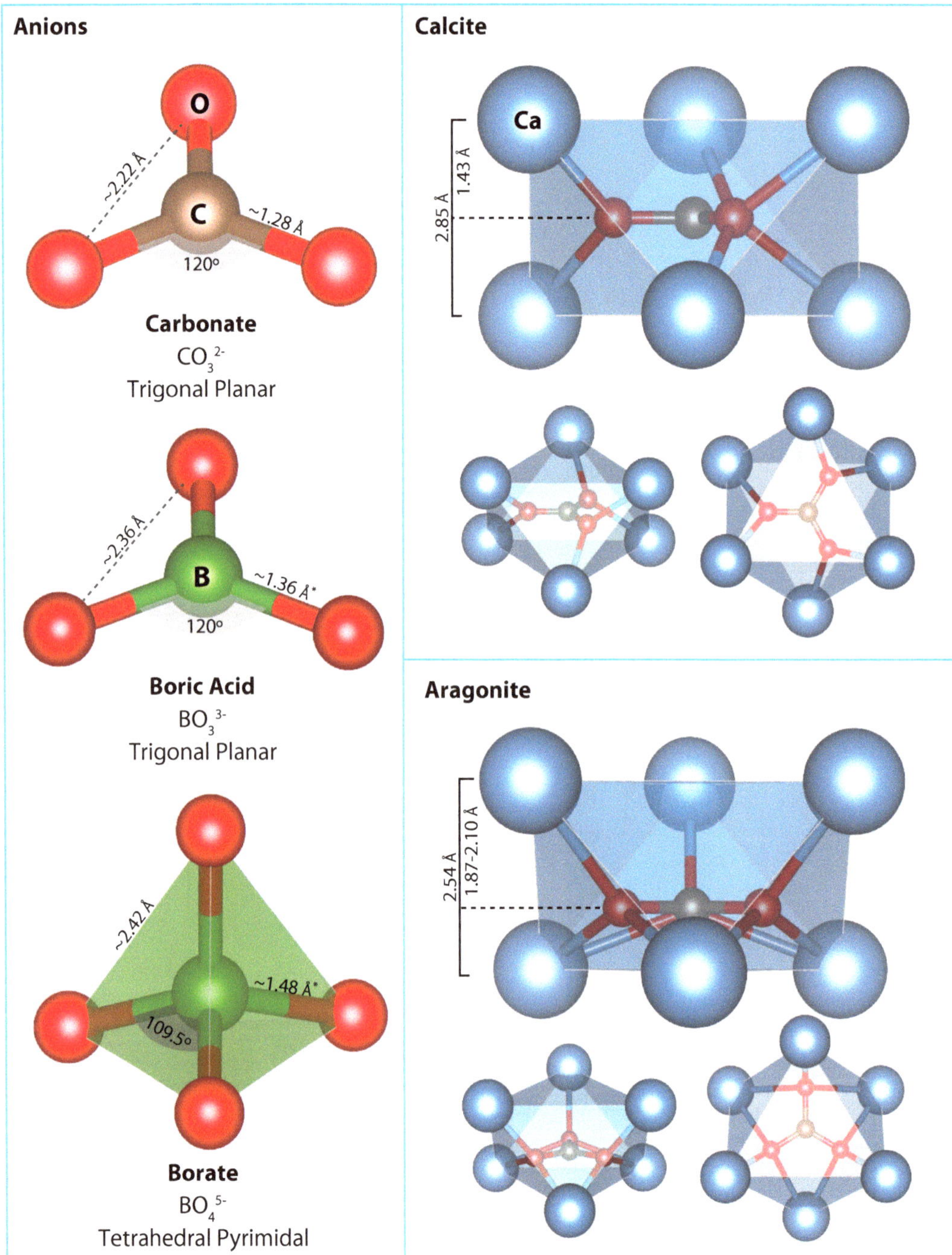

Fig. 4.6 The size and shape of C and B anions, and the structure of the anion site in calcite and aragonite. The anion sites are defined by an octahedral network of nearest-neighbor Ca^{2+} ions. The anion site structure is superficially similar, but differences in the bounding symmetry of Ca^{2+} ions alter the anion bonding environment. Significantly, the CO_3^{2-} ion in aragonite is shifted towards the lower edge (as presented here) of the octahedron, providing more space to incorporate a non-planar impurity ion

calcite is bonded to eight CO_3^{2-} anions in a tight arrangement, while in aragonite it is bonded to nine anions in a larger, looser arrangement (Fig. 4.6) (Reeder 1983). Because of the differences in the size of the cation site, impurity cations smaller than Ca^{2+} (e.g. Mg^{2+}) will substitute more readily into calcite than aragonite, as less distortion is required to accommodate the impurity. Conversely, cations larger than Ca^{2+} (e.g. Sr^{2+}) will substitute more easily into aragonite (Mucci and Morse 1983).

While it is convenient to reduce the substitution of cation impurities to considerations of ionic radii, this is only a good predictor of trace element content when the mineral is growing in equilibrium with the solution. That is, when the speed of reactions with the mineral surface and the diffusion of ions in solution are both faster than the net rate of ion addition to the mineral. When precipitation rate exceeds the speeds of surface reactions or ion diffusion, kinetically controlled non-equilibrium processes become a much stronger control over impurity incorporation, and drive deviations from ideal thermodynamic behavior. A complete kinetic view of mineral growth involves the interaction of every solute with the mineral surface, and with each other. Any solute that encounters the surface may attach to it, everything attached to the surface may become detached, and specific ion-ion interactions both in solution and at the mineral surface can alter these attachment and detachment rates. Given the prevalent super-saturation of seawater with respect to $CaCO_3$ minerals, and the tight controls over biological mineralization, it is likely that the majority of marine carbonate minerals are precipitated under reaction-limited conditions (DePaolo 2011; Nielsen et al. 2012), possibly within diffusion-limited environments (DePaolo 2011; Watson and Liang 1995; Watson 1996), where non-equilibrium kinetics are the main drivers of impurity incorporation.

4.2.2.1 Reaction-Limited Impurity Incorporation

A complete understanding of impurity incorporation should consider the interaction of all solutes with the mineral surface and with each other, but in practice this is prohibitively complex. Instead, ion-by-ion growth models considering subsets of solute-mineral interactions are beginning to uncover some of the kinetic controls over isotope fractionation and trace element incorporation during mineral growth (Surface Kinetic Model, SKM; DePaolo 2011; Nielsen et al. 2012, 2013). In these models impurity incorporation is considered in terms of the attachment and detachment rates of impurities relative to the host constituents of the mineral (Fig. 4.7), and the influence that these impurities exert on the kinetic 'landscape' of the mineral growth surface. The rate of impurity attachment is determined by its activity in solution relative to the host ions, the availability of binding sites on the mineral surface, and the energy barriers to bond-making with the mineral surface.

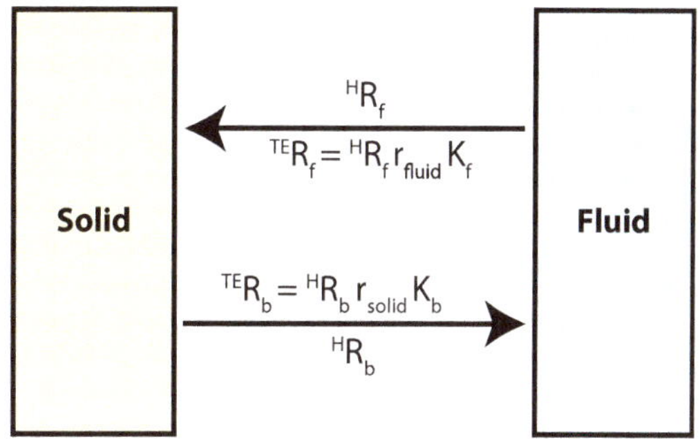

Fig. 4.7 **The incorporation of trace elements (TE)** may be reduced to the consideration of the attachment and detachment of trace impurity ions (^{TE}R), relative to the host ions (^{H}R). This may be expressed in terms of the ratio of the impurity trace element to the host constituent (TE/H) in the fluid or solid, and a partition coefficient (K) (DePaolo 2011)

The availability of bonding sites is driven by the relatively well-defined controls over TLK growth geometry, which link solution chemistry to the density of 'kink' ion binding sites (Zhang and Nancollas 1990; Nielsen et al. 2012). As with the host mineral constituents, the energy barriers to bond-making with the mineral surface are governed by the desolvation of the impurity and the mineral surface, and the electrostatic interaction between the impurity and the surface. Desolvation energies are highly ion-specific, and depend upon the charge density (or ionic radius) of the surface or ion. Electrostatic interactions are determined by the difference in charge between the growth surface and ion.

Most surfaces in water are charged, and this charge is measureable as the 'zeta potential'—the electrostatic potential between the surface and solution under a particular set of conditions. The sign of this charge dictates whether the surface will interact more easily with cations or anions. Numerous processes combine to determine surface charge (e.g. Chap. 4 of Butt et al. 2003), but at a basic level it can be considered as the steady-state balance of ion attachment and detachment rates, and the pH of the solution. At low pH protons can accumulate at the mineral surface, making it positively charged, while at high pH protons move from the mineral surface to the proton-deplete solution, and the surface becomes negatively charged. Alongside the action of pH, the relative attachment and detachment rates and the ionic potential (charge/size) of its main constituent ions will also influence surface charge. For example, in a solution that is Ca^{2+} rich but CO_3^{2-} poor, available Ca^{2+} sites on the surface will be occupied more quickly than CO_3^{2-} sites, leading to a more positive surface charge (Sawada 1997; Chibowski et al. 2003; Durán-Álvarez et al. 2016; Plank and Bassioni 2007; Vdović 2001). In carbonates, the interaction of pH with aqueous carbon speciation, and the roles of both Ca^{2+} and CO_3^{2-} in determining both binding site availability and surface charge can create surprisingly complex relationships between calcite surface charge and solution chemistry.

The zeta potential of calcite has been measured by numerous studies, with a wide range of results (reviewed in Wolthers et al. 2008), although very few studies explicitly consider aragonite (Moulin and Roques 2003). Both the magnitude and sign of the surface charge, and its response to pH variations are highly variable across these experiments, with approximately equal numbers of studies reporting both negative and positive charges for carbonate surfaces (Moulin and Roques 2003; Wolthers et al. 2008). However, the majority of these differences are accounted for by changes in solution chemistry, mineral surface structure, and mixing rates between the experiments. Importantly, studies that investigate the influence of fluid composition on surface charge find that a major control over the surface charge is the relative abundance of divalent cations and carbon anions in solution (Chibowski et al.; Fuerstenau and Herrera-Urbina 1992; Huang et al. 1991; Moulin and Roques 2003). Across a seawater- and biomineral-relevant pH range ($\sim$7 to $\sim$10), most of these studies find that carbonates have a weakly positive surface charge (zeta potential of >4 mV), as long as Ca^{2+} or Mg^{2+} are present in the environment at higher concentrations than CO_3^{2-} or HCO_3^- (Chibowski et al. 2003; Fuerstenau and Herrera-Urbina 1992; Huang et al. 1991). A notable exception to this is the study of Moulin and Roques (2003), who report predominantly negative surface charges in both calcite and aragonite across a range of experimental conditions, with aragonite being $\sim$2–4 mV more negative than calcite. The authors argue that previous reports of positive surface charges are primarily attributable to insufficient equilibration of the mineral surface before observation. However, given that natural carbonate minerals are unlikely to precipitate under surface-equilibrated conditions, and Ca^{2+} is relatively abundant in seawater, this may be taken to suggest that carbonate surfaces are most likely weakly positive in most natural formation environments, with relatively more vacant CO_3^{2-} than Ca^+ binding sites on the surface. This must be accompanied by the major caveat that studies of carbonate zeta potential are all conducted in relatively simple solutions, which do not reflect the ionic complexity of seawater.

Once attached to the surface, the rate of detachment of impurities will depend upon the strength of the bonds it forms with the mineral surface, which is related to how well the impurity fits in the host mineral structure. The probability of an attached ion becoming detached may be considered in terms of the amount of time it remains on the surface, relative to the growth rate of the mineral. The longer the impurity remains on the surface without being structurally incorporated, the higher its chance of detachment.

Attached impurities interact with the growth processes of minerals, and can influence the kinetic 'landscape' of a mineral surface (Davis et al. 2004; Ruiz-Agudo et al. 2012), and therefore their chance of being incorporated (de Yoreo and Vekilov 2003; Dove et al. 2004; Nielsen et al. 2013). Impurities may alter mineral growth dynamics either by being incorporated and increasing the local solubility of the mineral surface, by binding to the surface and blocking the propagation of growth steps or kink sites, or by acting as a surfactant to modify the energetics of the mineral surface (Dove et al. 2004). Impurities can alter the growth surface of the crystal, and ultimately its trace chemistry and morphology.

4.2.2.2 Diffusion-Limited Surface Environments

The incorporation of trace elements may be considered in terms of the attachment and detachment of impurities from a well-mixed, bulk solution. However, marine carbonates may often be precipitated rapidly, in diffusion-limited environments. This can generate boundary layers with distinct composition, and further complicate the incorporation of trace elements (DePaolo 2011; Watson and Liang 1995; Watson 1996).

Disequilibrium boundary layers can develop on both the fluid and mineral sides of a growing crystal face. In the fluid, a boundary layer will form if the net rate of ion attachment to the surface exceeds the diffusivity of the ions in solution (Watson 1996). The precise composition and thickness of this surface layer depends upon the mineral precipitation rate, the ease of incorporation of impurity ions into the solid, and the diffusivity of individual ions in solution. Because the uptake of host constituents into the mineral will always exceed impurity uptake, the boundary layer will be enriched in impurity ions relative to the bulk solution. This surface boundary layer has the potential to separate the chemistry 'seen' by the mineral growth surface from the chemistry of the bulk solution (Watson and Liang 1995; Watson 1996). For synthetic carbonates growing in well-mixed solutions, DePaolo (2011) points out that the effects of this boundary layer will be minimal, given the diffusivities of Ca^{2+} and CO_3^{2-}, even at high precipitation rates. However, fluid boundary layers may be important in natural settings, where minerals often form in confined, poorly-mixed microenvironments.

On the mineral side, the first few layers of attached ions can act as a second, diffusion-limited boundary layer, and provide a route to impurity incorporation via 'growth entrapment' (Watson 2004). This mechanism considers the diffusivity of individual ions within the mineral surface, and their ability to escape a 'partially-crystallized' advancing mineralization front. If the mineral precipitation rate exceeds the rate at which an impurity ion can diffuse out of the mineral surface, the impurity is trapped within the growing mineral. This entrapment mechanism can be considered in terms of the 'critical thickness' of the surface boundary layer, which defines the depth below which an impurity is trapped within the mineral structure (Watson and Liang 1995; Watson 1996). This growth entrapment model can be used to explain precipitation-rate dependent trends in the trace element content of several inorganically precipitated calcites (Watson 2004). However, DePaolo (2011) points out that to explain these trends, the model must employ diffusivities many orders of magnitude faster than those measured in carbonate minerals. It is therefore likely that diffusion-limited boundary layers on the fluid side of the interface have the most potential to influence trace element uptake. In this case, reaction limited impurity incorporation via the Surface Kinetic Model (SKM; Fig. 4.7; DePaolo 2011; Nielsen et al. 2012, 2013) from a modified fluid boundary layer offers the highest potential for deviation from thermodynamic impurity incorporation behavior. Interestingly, while DePaolo's surface-reaction and Watson's entrapment models describe significantly different processes, the two models predict functionally similar trends of impurity incorporation. Both offer a route to impurity uptake in growing minerals that is distinct from equilibrium growth, and depends on the growth rate of the mineral. Both are able to account for patterns of growth-rate dependent trace element incorporation in calcite (DePaolo 2011; Watson 2004), but the Surface Kinetic Model (SKM; DePaolo 2011) can simultaneously explain kinetic isotope fractionation during incorporation.

4.2.2.3 Impurity Incorporation in Non-classical Growth

All mechanisms discussed thus far are derived from the consideration of relatively simple TLK growth processes. Non-classical precipitation mechanisms can significantly alter impurity incorporation processes and alter the bulk geochemistry of a mineral (Fig. 4.5). In particular, the involvement of ACC presents a variety of mechanisms to incorporate impurities into both structural and non-structural sites within the mineral. The composition of ACC can be distinct from the solution in which it formed, and is controlled by a variety of factors (e.g. Wang et al. 2009). The transformation of ACC to crystalline carbonate may proceed via dissolution-reprecipitation (Bots et al. 2012; Rodriguez-Blanco et al. 2011; Rodriguez-Navarro et al. 2015, 2016) or solid phase transformation (Gal et al. 2013). Both transformation mechanisms involve a rapid precipitation of a mineral in a diffusion-limited poorly-crystalline environment, where the Growth Entrapment Model (Watson 2004) may become an important mechanism of trace element uptake. With limited room for escape, trace elements included in the ACC phase that are incompatible with the crystalline phase may either be incorporated in the mineral via growth entrapment, excluded to ACC grain boundaries, or concentrated in micro-domains of a distinct mineral phase. The agglomeration of ACC or crystalline nano-particle assembly also provides the opportunity for incorporating fluid inclusions, which could capture the chemistry of the surface boundary layer. Finally, the inclusion of non-mineral structural components in biominerals may significantly alter the geochemistry of the bulk structure (e.g. Branson et al. 2016). While impurities in these latter mechanisms are not strictly 'incorporated' into the mineral structure, they will impact the geochemistry of the bulk mineral. The involvement of non-classical mineralization processes significantly complicates the incorporation of impurities within a mineral, and it is unclear how well our current understanding of impurity incorporation derived from classical growth systems translates to these more complex processes.

4.3 Boron Incorporation in Calcium Carbonates

The incorporation of B into carbonate minerals is governed by the same fundamental mineral growth and impurity incorporation processes as any other element, and should be predictable following the same principles. However, B incorporation has several extra layers of complexity, and there are many more 'moving parts' to consider than in direct cation-cation substitutions. The first layer of complexity comes from the aqueous chemistry of B.

4.3.1 Aqueous Boron Chemistry

Boron in solution exists in a pH-dependent equilibrium between two aqueous species, trigonal $B(OH)_3$ and tetrahedral $B(OH)_4^-$. The pK^*_B of this equilibrium is sensitive to temperature and solution composition (Pagani et al. 2005), and is ~8.6 in typical Holocene seawater conditions (DOE 1994). The stronger B–O bonds in $B(OH)_3$ lead to an accumulation of heavy ^{11}B in this species (Hemming and Hanson 1992; Urey 1947). In seawater, the isotope fractionation between the species is somewhere between 26–27‰ (Klochko et al. 2006; Nir et al. 2015). Most applications of the B palaeoproxies use Klochko et al. (2006) seawater value of 27.2 ± 0.6‰, but it is worth noting that the fractionation of B is not set in stone. Nir et al. recently reported value of 26‰ had a measurement error of ± 1‰, and while Klochko et al. 27.2‰ value has a smaller measurement error, they report a wide range of fractionations (~24 to ~30‰) depending on solution chemistry (pure water, KCl and seawater). This may be significant, given that organisms are known to modify the composition of the fluids in their calcifying environments.

The equilibration and isotopic fractionation between the aqueous B species is rapid (~95 and ~123 μs), several orders of magnitude faster than the incorporation of ions during biomineral carbonate mineral growth, as estimated from gross biomineral extension rates (Zeebe et al. 2001). The relative speed of isotope equilibration should preclude any isotope fractionation during B incorporation into carbonates, if the surface remains in equilibrium with the solution. However, this reasoning only holds true if crystal precipitation rate is directly related gross extension rate. In TLK growth this is likely the case, but if precipitation proceeds via a more complex mechanism, it may be more accurately considered as a series of episodic events with a fast instantaneous precipitation rate (e.g. crystallization of ACC granules). If this instantaneous precipitation rate approaches the equilibration time of the B species, kinetic fractionation effects may become important.

The size, shape and charge of the major aqueous B species make them most suited to substitution in the anion (CO_3^{2-}) site in $CaCO_3$ minerals (Fig. 4.6), particularly when considered in their dehydrated form (BO_3^{3-} and BO_4^{4-}). Trigonal BO_3^{3-} is most similar to the CO_3^{2-} ion, with a similar size (B–O and C–O bonds of 1.28 vs 1.36 Å) and trigonal planar shape (Fig. 4.6). If, as suggested by the literature, carbonate growth surfaces in marine environments carry a weak positive charge (Sect. 4.2.2.1), the negatively charged $B(OH)_4^-$ ion will interact more readily with the carbonate growth surface than $B(OH)_3$. However, the tetrahedral pyramidal structure of BO_4^{5-}, combined with its greater charge make it a poor fit the CO_3^{2-} ion site (Balan et al. 2016). Based purely on size and charge considerations, the most straightforward substitution for the CO_3^{2-} ion in carbonate would be a trigonal B species HBO_3^{2-} (Balan et al. 2016; Hemming et al. 1995; Hemming and Hanson 1992).

4.3.2 Boron in Synthetic Carbonates

Data from systematic inorganic precipitation experiments allow the examination of B incorporation mechanisms in a relatively well-constrained environment. The growth mechanism of the synthetic carbonates discussed here was not experimentally determined, but the simple, open-solution chemistry of the experiments, combined with the euhedral structure of the resulting precipitates strongly suggests that mineral growth occurred via TLK mechanisms (Mavromatis et al. 2015; Uchikawa et al. 2015).

When considering the mechanisms of trace element uptake into minerals grown under different conditions, it is necessary to normalize measurements to the same scale. In simple ionic substitutions this is straightforward, and the ‘partitioning coefficient’ (K) of a trace element can be defined relative to solution chemistry. In the case of Mg:

$$K = \frac{Mg/Ca_{solid}}{Mg/Ca_{fluid}}$$

Because B exists in two aqueous species, and most likely interacts with the anion site of carbonates, its partitioning can be defined in a variety of ways relative to the aqueous B and C species. In the literature partition coefficients have been calculated by normalizing carbonate B/Ca (which is Stoichiometrically equivalent to B/C) to solution $B(OH)_4^-/HCO_3^-$ (Hemming et al. 1995), B/DIC (Uchikawa et al. 2015), $B(OH)_4^-/CO_3^{2-}$ (Mavromatis et al. 2015), $B(OH)_4^-/(CO_3^{2-})^{0.5}$, and the Nernst partition coefficient (mass% B_{solid}/mass% $B_{solution}$, Holcomb et al. 2016). The diversity of partition coefficient definitions complicates the direct comparison of results from different studies. To allow direct comparison, data presented here are re-calculated to the λ notation of Uchikawa et al. (2015), where $\lambda = 1000 \times (B/Ca)_{solid}/[B/DIC]_{solution}$.

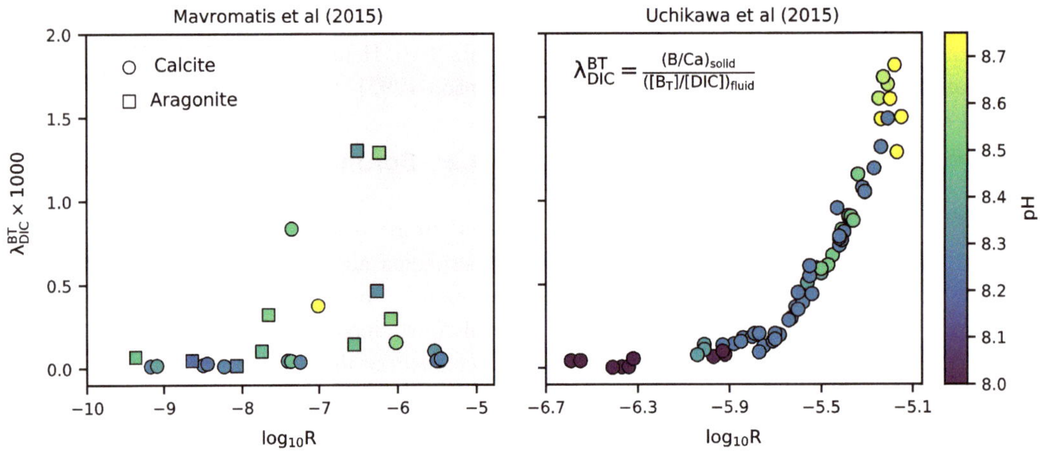

Fig. 4.8 The partitioning of B into calcite and aragonite from two recent inorganic studies, presented following the λ notation of Uchikawa et al. (2015). Both studies note a relationship between B incorporation and mineral growth rate, although the relationship is not apparent in the data of Mavromatis et al. (2015) when the data are converted to the λ scale. Mavromatis et al. (2015) also show that aragonite incorporates significantly more B than calcite under the same conditions (Fig. 4.9). A secondary dependence on solution pH is evident in the data, although Uchikawa et al. (2015) specifically decoupled pH and precipitation rate, revealing that this is less significant than precipitation rate. The shaded area in the left panel highlights the log(R) range shown in the right panel

4.3.2.1 Synthetic Aragonite

Boron is incorporated more easily into aragonite than calcite (Figs. 4.8 and 4.9; Kitano et al. 1978, 1979; Hemming et al. 1995; Mavromatis et al. 2015). The $\delta^{11}B$ of synthetic aragonite shows clear correspondence to $\delta^{11}B_{B(OH)_4^-}$ of the parent solution (Fig. 4.10; Noireaux et al. 2015), except in an early study (Hemming et al. 1995) where analytical offsets account for the discrepancy (Foster et al. 2013). The partitioning of B into aragonite increases with mineral precipitation rate, but is relatively insensitive to temperature (Fig. 4.9; Mavromatis et al. 2015; Holcomb et al. 2016). Of the various studies investigating B uptake into aragonite, only the two most recent characterized the aqueous carbonate system (Holcomb et al. 2016; Mavromatis et al. 2015). These studies show that B uptake is best predicted by either the boron to dissolved inorganic carbon ratio ([B]/[DIC]; Holcomb et al. 2016; Mavromatis et al. 2016), or $[B(OH)_4^-]/[CO_3^{2-}]^{0.5}$ of solution (Holcomb et al. 2016).

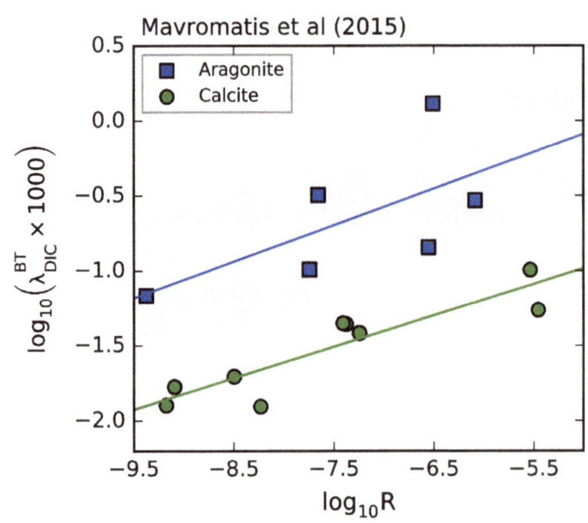

Fig. 4.9 Growth dependence of B incorporation. At similar temperature and pH the partitioning of B increases exponentially with growth rate in both calcite and aragonite, although aragonite consistently incorporates ~1 order of magnitude more B than calcite. Data presented here are from precipitates grown at 25°C, and 8.26<pH<8.38 or 8.42<pH<8.54 for calcite and aragonite, respectively (following Mavromatis et al. 2015). Solid lines show least squares linear regressions through the data, with a shaded 95% confidence envelope

Structural studies of the coordination state of B in synthetic and biomineral aragonite (Klochko et al. 2009; Mavromatis et al. 2015; Rollion-Bard et al. 2011; Sen et al. 1994) show that B is predominantly hosted in a tetrahedral ($H_nBO_4^{(4-n)-}$) coordination within the mineral structure, although variable proportions of trigonal B can also be present ($H_nBO_3^{(3-n)-}$). These data are in line with Density Functional Theory (DFT) calculations of B coordination within aragonite, which demonstrate that the direct substitution of $B(OH)_4^-$ for CO_3^{2-} within the mineral structure is energetically favourable in aragonite (Balan et al. 2016). These calculations also show that the variable amounts of trigonal B observed in aragonite may be explained by the less energetically-favorable entrapped or substituted $B(OH)_3$ groups, particularly in biogenic samples. Together, the correspondence between aragonite $\delta^{11}B$ and aqueous $\delta^{11}B_{B(OH)_4^-}$, the predominant tetrahedral coordination of B within aragonite, and the relatively weak partitioning against B incorporation during mineral growth offer strong evidence to support the direct incorporation of aqueous $B(OH)_4^-$ into aragonite.

4.3.2.2 Synthetic Calcite

Calcite takes up 2–10 times less B than aragonite (Fig. 4.9; Hemming et al. 1995; Kitano et al. 1978, 1979; Mavromatis et al. 2015), and the partitioning of B into synthetic calcite is strongly dependent on the rate (Gabitov et al. 2014; Mavromatis et al. 2015; Uchikawa et al. 2015; Kaczmarek et al. 2016) and mechanism (Hobbs and Reardon 1999) of mineral precipitation (Figs. 4.8 and 4.9). Of these studies, only Uchikawa et al. (2015) and Kaczmarek et al. (2016) explore different B partition coefficient definitions, and only Uchikawa et al's. experiments offer the comprehensive experimental matrix necessary to discriminate between them. Uchikawa et al. (2015) demonstrate that total solution [B]/[DIC] is the best predictor of B partitioning in their data, particularly at high precipitation rates (Fig. 4.8).

Kaczmarek et al. (2016) report a strong negative correlation between B incorporation and temperature, while Mavromatis et al. (2015) find no such effect. Kaczmarek et al ascribe this discrepancy to a lack of directly comparable data in Mavromatis et al. study. An alternate possibility is that the temperature effect observed by Kaczmarek et al is at least partly attributable to large changes in Ca^{2+} concentrations used to control precipitation rate between their temperature treatments. There is some evidence to suggest that Ca^{2+} concentration can influence B uptake (Uchikawa et al. 2015), possibly as a result of changes in surface binding site availability driven by solution Ca^{2+}:CO_3^{2-} ratio (Nielsen et al. 2012). Further study is required to explore the influence of temperature on B uptake into calcite.

Structural studies of calcite growth surfaces by atomic force microscopy reveal a specific preference of B for a particular type of kink site during TLK growth (Hemming et al. 1998; Ruiz-Agudo et al. 2012). The symmetry of the calcite crystal structure gives the terraces and ledges in TLK growth a distinct rhombohedral structure, with both obtuse- and acute-angle kink sites. These studies noted that under high pH conditions, when the majority of B in solution exists as $B(OH)_4^-$, B interacted specifically with the obtuse angle kink sites, inhibiting their progression along the step edges, and disrupting the rhombohedral terrace structure of pure calcite. Combined with the structural argument that larger $B(OH)_4^-$ ions will prefer obtuse-angle kink sites simply because they are larger, this observation offers compelling evidence for the specific interaction of $B(OH)_4^-$ ions with the growing crystal.

Measurements of B coordination in calcite reveal a wide range of trigonal and tetrahedral coordination environments, from entirely trigonal (Branson et al. 2015; Sen et al. 1994), to entirely tetrahedral (Mavromatis et al. 2015) and everything in between (Klochko et al. 2006; Mavromatis et al. 2015; Rollion-Bard et al. 2011; Sen et al. 1994). First-principles DFT simulations of B in calcite reveal that the most stable coordination is a partially deprotonated $BO_2(OH)^{2-}$

hosted in the CO_3^{2-} site, along with a lesser fraction of $B(OH)_4^-$ groups. This supports the majority of observational data, with the exception of the few observations of purely tetrahedral B by Mavromatis et al. (2015), but does not offer a mechanism to explain variations in the tetrahedral B content of calcite. In this regard, Mavromatis et al. (2015) observed that the proportion of trigonal B tends to decrease at faster precipitation rates. This suggests that precipitation rate may be a controlling factor in the incorporation mechanism of B, and presents the intriguing possibility that the $B(OH)_4^-$ ion is somehow incorporated more easily at high precipitation rates, possibly via a growth-entrapment mechanism. This is counter-intuitive, because the $B(OH)_4^-$ actively inhibits kink advancement (Ruiz-Agudo et al. 2012), and is at odds with Uchikawa et al. (2015) suggestion that more $B(OH)_3$ is incorporated at faster precipitation rates. A secondary possibility is that the tetrahedral B observed in NMR studies of natural samples could come from intra-skeletal organic components, which were not removed by pre-analysis cleaning procedures, which is consistent with the complete absence of tetrahedral B in microscopic studies of B coordination (Branson et al. 2015).

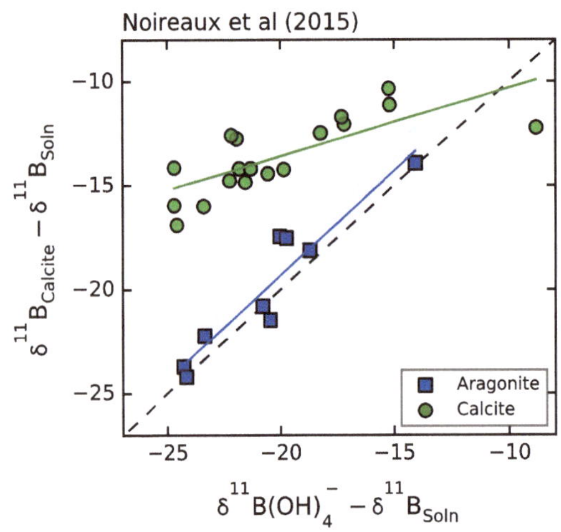

Fig. 4.10 **The offset of carbonate δ^{11}B from solution $\delta^{11}B_{B(OH)_4^-}$** as calculated by Noireaux et al. (2015; α_{4-3} = 1.026). Aragonite exhibits no systematic offsets from the 1:1 line (dashed), while calcite is significantly enriched in δ^{11}B. The offset of calcite δ^{11}B from solution $\delta^{11}B_{B(OH)_4^-}$ decreases with increasing pH (Noireaux et al. 2015). Solid lines show least squares linear regressions through the data, with a shaded 95% confidence envelope

Recent B isotope studies report that synthetic calcites tend to be significantly enriched in ^{11}B, relative to aqueous $\delta^{11}B_{B(OH)_4^-}$ (Fig. 4.10; Farmer et al. 2015; Noireaux et al. 2015). This is also true in older studies, despite analytical offsets (Hemming et al. 1995; Sanyal et al. 2000). The degree of δ^{11}B enrichment in calcite has also been shown to vary systematically with either pH of solution (Noireaux et al. 2015) or mineral growth rate (Farmer et al. 2015) in different studies. The absolute magnitude of calcite offsets from $\delta^{11}B_{B(OH)_4^-}$ varies widely between studies, and the exact cause of these differences is unclear, although large differences in major ion chemistry, ionic strength and B concentrations between the experiments may account for some of the variability. For example, the B concentrations used for the majority of Mavromatis et al. (2015) and Noireaux et al. (2015)'s precipitates were around an order of magnitude higher than that used by Uchikawa et al. (2015). These large differences in solution chemistry may drive differences in the kinetics of mineral precipitation and the incorporation of B into the mineral. Finally, a recent precipitation study reported negligible offsets from solution $\delta^{11}B_{B(OH)_4^-}$, and a ~1‰ δ^{11}B increase on doubling crystal growth rate (Kaczmarek et al. 2016). At face value, these data significantly expand the range of B isotope behavior observed in calcites. However, the results of Kaczmarek et al. study are difficult to interpret because they employ a highly unusual fractionation factor of 25‰ derived from KCl solutions (Klochko et al. 2006), but applied to precipitates formed in a NaCl medium. It is unclear whether the lack of offset from $\delta^{11}B_{B(OH)_4^-}$ in this study is real, or attributable to the use of an inappropriate fractionation factor. The use of an NaCl-relevant fractionation factor would introduce a ~2 ‰

offset from $\delta^{11}B_{B(OH)_4^-}$ similar to that seen in other studies.

Together, the wide variety of $\delta^{11}B$ offsets between calcite and aqueous $B(OH)_4^-$, the dependence of B partitioning on total B in solution, rather than $B(OH)_4^-$ concentration, and the inconsistent B coordination environment within synthetic calcite combine to provide compelling evidence for a complex B incorporation mechanism, where B in calcite is not solely related to $B(OH)_4^-$ in solution.

4.3.2.3 Differences Between Calcite and Aragonite

The combination of data from synthetic calcite and aragonite provide evidence for two, distinct B incorporation mechanisms. In synthetic aragonite, $B(OH)_4^-$ appears to be incorporated directly from solution and substituted into the carbonate anion site (Balan et al. 2016; Holcomb et al. 2016; Mavromatis et al. 2015; Noireaux et al. 2015). In synthetic calcite, most recent studies show that B incorporation is not solely related to $B(OH)_4^-$ in solution, exhibits a dependence upon crystal growth rate and solution chemistry, and requires some degree of B re-coordination during incorporation (Balan et al. 2016; Mavromatis et al. 2015; Noireaux et al. 2015; Uchikawa et al. 2015). Given that the precipitation of both polymorphs is governed by the same kinetic and thermodynamic principles, and that they are grown under similar major-ion solution chemistries, these distinct B incorporation pathways must be attributable to differences in the interaction of trigonal $B(OH)_3$ and tetrahedral $B(OH)_4^-$ with the unique structures of calcite and aragonite.

The importance of mineral structure on B incorporation is demonstrated by both first principles calculations (Balan et al. 2016), and measurements of B coordination within calcite and aragonite (Sen et al. 1994; Klochko et al. 2009; Rollion-Bard et al. 2011; Branson et al. 2015; Mavromatis et al. 2015). In combination, these models and data demonstrate that tetrahedrally coordinated B can be easily accommodated within the aragonite structure, and that while calcite can accommodate both trigonal and tetrahedral B, the trigonal form is more stable. A key experimental observation in support of this is the transformation of tetrahedral to trigonal boron associated with an aragonite-calcite phase transformation (Sen et al. 1994), which demonstrates the link between mineral structure and B coordination.

Hemming et al. (1995) discuss B coordination in terms of the relative size of the anion sites in calcite and aragonite. The authors found the preference of aragonite for incorporating the larger tetrahedral B ion surprising, given that the anion site in aragonite (37.8 $Å^3$) is smaller than calcite (40.9 $Å^3$, sizes calculated based on the position and geometry of nearest-neighbor Ca^{2+} ions). While this volumetric consideration is important, it ignores differences in both the elasticity (Liu et al. 2005) and bonding structure of the anion sites between calcite and aragonite, which can account for this coordination difference (Balan et al. 2016). The anion site in both calcite and aragonite is enclosed by an octahedral framework of Ca^{2+} ions (Fig. 4.6). In calcite, this framework is symmetric, with the CO_3^{2-} ion uniformly bonded in the center of the Ca^{2+} octahedron. In aragonite the octahedron is slightly asymmetric, and the CO_3^{2-} ion forms twice as many bonds on one face of the octahedra than the other, resulting in the CO_3^{2-} ion being closer to that side (Fig. 4.6). In combination with the greater elasticity of aragonite (Blundy and Wood 1994; Liu et al. 2005), the asymmetry of the anion position in aragonite may facilitate the incorporation of tetrahedral impurity ions with relatively little lattice distortion, compared to the symmetric, less elastic site in calcite. Thus, while the anion site in aragonite is slightly smaller than in calcite, it is better suited to accommodating tetrahedral B (Balan et al. 2016).

In context of TLK mineral growth, a major factor governing impurity incorporation by direct substitution is availability of low-energy binding sites on the growth surface, which reduce the probability of impurity detachment from the surface. The relative stability of tetrahedral B in the aragonite anion site suggests that $B(OH)_4^-$

can attain a lower energy state on the surface of aragonite than calcite, be less likely to detached, and therefore be more easily incorporated into the mineral. This is evident in the relatively low partitioning against B incorporation in aragonite, relative to calcite (Fig. 4.9; Kitano et al. 1978; Mavromatis et al. 2015), and in the agreement between aragonite $\delta^{11}B$ and solution $\delta^{11}B_{B(OH)_4^-}$ (Fig. 4.10; Noireaux et al. 2015). The $\delta^{11}B$ of aragonite further reveals that this incorporation proceeds without any significant isotope fractionation, as predicted by the rapid equilibration of B isotopes relative to the speed of carbonate precipitation (Zeebe et al. 2001). In contrast, $B(OH)_4^-$ does not fit as easily into the anion site of calcite, and requires a tetrahedral-trigonal transformation to be incorporated into the mineral (Balan et al. 2016). Thus, attached $B(OH)_4^-$ will remain on the calcite surface until it is either transformed to a trigonal species, incorporated as relatively-incompatible $B(OH)_4^-$, or detaches from the surface. The difficulty of $B(OH)_4^-$ incorporation is evident in its 'kink blocking' influence on calcite (Ruiz-Agudo et al. 2012), where attachment of $B(OH)_4^-$ to kink sites prevents the subsequent attachment of Ca^{2+}, and inhibits kink propagation. Because of this incompatibility, attached $B(OH)_4^-$ spends longer on the mineral surface, greatly increasing its probability of detachment, driving the relatively strong partitioning against the direct incorporation of $B(OH)_4^-$ in calcite.

These structural differences can account for the stronger partitioning against B in calcite, and offer an explanation for why B uptake in aragonite is directly related to $B(OH)_4^-$ in solution, despite having a smaller anion site. However, understanding the complex B/Ca and $\delta^{11}B$ chemistry of calcite requires a more detailed consideration of the kinetics of B uptake, and the mechanisms that may drive systematic offsets from solution $\delta^{11}B_{B(OH)_4^-}$.

4.3.2.4 The Variability of B in Calcite: Possible Causes

The deviations of calcite $\delta^{11}B$ from solution $\delta^{11}B_{B(OH)_4^-}$ (Hemming et al. 1995; Noireaux et al. 2015; Sanyal et al. 2000) suggest either a significant isotopic fractionation during $B(OH)_4^-$ incorporation, the uptake of B complexes into the mineral, the incorporation of both $B(OH)_4^-$ and $B(OH)_3$ from solution, or a combination of all three processes. Crucially, any B incorporation mechanism must be able to simultaneously explain both the partitioning and isotopic content of B in calcite.

Isotopic fractionation during $B(OH)_4^-$ incorporation has previously been discounted, because of the speed of isotopic equilibration between aqueous B species relative to ion attachment rates in calcification (Zeebe et al. 2001). However, this argument did not consider possible fractionation associated with a tetrahedral-trigonal transformation of ions attached to the mineral surface, which appears to be a significant step in B incorporation (Balan et al. 2016). The tetrahedral-trigonal transformation requires the breaking of a single B–O bond of an attached $B(OH)_4^-$. Because the ^{10}B–O bond is slightly weaker than the ^{11}B–O bond, it follows that $^{10}B(OH)_4^-$ will transform more easily than its heavier counterpart. The probability of ion detachment from the surface is proportional to the amount of time it spends 'exposed' on the mineral surface. Thus, if the tetrahedral-trigonal B transformation is the rate-limited step in $B(OH)_4^-$ incorporation, the detachment probability will be greater for $^{11}B(OH)_4^-$ than $^{10}B(OH)_4^-$, driving an isotope fractionation that enriches calcite in ^{10}B (drives $\delta^{11}B$ negative). This is the opposite direction to the observed offsets between calcite $\delta^{11}B$ and solution $\delta^{11}B_{B(OH)_4^-}$. Therefore, while fractionation associated with a tetrahedral-trigonal transformation at the mineral surface is conceivable, it is unable to account for the discrepancies between calcite $\delta^{11}B$ and solution $\delta^{11}B_{B(OH)_4^-}$. Calcite $\delta^{11}B$ offsets must therefore be driven by either the incorporation of B complexes from solution, or the simultaneous uptake of both $B(OH)_3$ and $B(OH)_4^-$.

Aqueous B interacts with other solutes to form both large, poly-atomic complexes (e.g. $B(OH)_2CO_3^-$, McElligott and Byrne 1997; and

B-[humic acid], Tossell 2006), and simpler ion-pairs (e.g. $NaB(OH)_4$; Bassett 1980; Hershey et al. 1986; Simonson et al. 1987). The poly-atomic ions are known to exhibit distinct isotopic fractionations (Lemarchand et al. 2005; Tossell 2006), while the isotopic influence of B ion pairing has not been explored. These complexes are a relatively minor component of total aqueous B in seawater-like conditions, but have the potential to have a significant influence on the partitioning and isotopic content of B, either through preferential interaction with mineral growth surfaces or by modifying the activity (and therefore reactivity) of the free B species.

The direct uptake of complexes or ion pairs is difficult to exclude as a mechanism of B uptake, given the relatively simple solutions employed by precipitation studies thus far. However, the lack of aragonite $\delta^{11}B$ offsets from aqueous $\delta^{11}B_{B(OH)_4^-}$ either suggest that complexes are an insignificant component of B incorporation, or that the B complexes incorporated are isotopically indistinguishable from $B(OH)_4^-$. If non-fractionated complexes were involved in B uptake in aragonite, this could lead to a decoupling between B partitioning and $[B(OH)_4^-]$ in solution, while the isotopic content remains the same as $\delta^{11}B_{B(OH)_4^-}$. While data in the literature are limited, the results of Mavromatis et al. (2015) and Noireaux et al. (2015) do not support this. Rather, the majority of data in the literature supports a straightforward $B(OH)_4^-$ substitution in aragonite, and it is seems unlikely that the direct incorporation of complexes is a significant B uptake pathway. In calcite, where there are large discrepancies in both B partitioning and $\delta^{11}B$, isotopically distinct B complexes may play a role. However, the relatively low abundance (McElligott and Byrne 1997), large size and complex structure (Tossell 2006) of the more poly-atomic B complexes, makes them unlikely candidates for preferential incorporation into calcite, compared to the simpler, more abundant B species.

A much more likely influence of B complexing is through the alteration of $B(OH)_3$ and $B(OH)_4^-$ activities by ion pairing in solution (Bassett 1980; Hershey et al. 1986; Simonson et al. 1987). This would be evident in variations in B partitioning, and possibly isotope fractionation in both calcite and aragonite as a function of the ionic strength and major ion composition of the precipitation environment. Unfortunately, this is hard to evaluate with currently available data, as experiments that vary either ionic strength or major ion chemistry simultaneously change other parameters (Mavromatis et al. 2015; Kaczmarek et al. 2016). However, because ion-pairing effects should influence both calcite and aragonite, the apparent straightforward connection between solution $B(OH)_4$ and B incorporation in aragonite again suggests that ion pairing may not be a major factor in controlling B uptake. While it is impossible to conclusively evaluate the contribution of B complexes and their influence on aqueous B activity to B incorporation, they should not be invoked as a primary influence on B uptake before exhausting simpler, kinetically driven incorporation processes, which may explain both the complex B incorporation patterns in calcite, as well as the differences between calcite and aragonite.

4.3.2.5 The Variability of B in Calcite: A Surface Kinetic Explanation?

As with any trace element, the incorporation of B can be considered in terms of its interactions with the growing mineral surface. Studies of B uptake have had some success explaining B partitioning using Watson's (2004) Growth Entrapment Model (Gabitov et al. 2014; Kaczmarek et al. 2016), which describes impurity 'entrapment' in the mineral when precipitation rate exceeds the impurity's ability to diffuse out of a boundary layer. However, significant weaknesses of this model, as applied in the literature, are that it relies on the poorly-constrained diffusivity of B within the mineral surface, and on the development of significant diffusion-limited boundary layers, which are unlikely in a well-mixed open precipitation environment (DePaolo 2011). Most importantly, applications of the model only consider total B concentration in solution. While this is appropriate for simple cation substitutions, this cannot be extended to consider B incorporation because it neglects the disparate physical and chemical properties of $B(OH)_3$ and $B(OH)_4^-$ and the different stabilities (and diffusivities) of each species within the mineral, and is unable to explain the isotopic content of B in the mineral. While the GEM could be expanded to consider each species independently, a more intuitive solution is offered by DePaolo's (2011) Surface Kinetic Model (Fig. 4.7; SKM), which explicitly considers the attachment and detachment fluxes of impurities at the mineral surface. The SKM, and micro-scale extensions of it (Nielsen et al. 2012, 2013), are functionally similar to Watson's (2004) GEM model, but can simultaneously predict both stable isotope fractionation and trace element partitioning into inorganic precipitates.

A complete kinetic consideration of B uptake into carbonates should consider the attachment and detachment fluxes of both $B(OH)_3$ and $B(OH)_4^-$, the feedbacks between precipitation rate and the kink blocking effect of $B(OH)_4^-$, and the energy barriers to B re-coordination on the mineral surface. This is considerably more complex than straightforward cation-cation substitutions, and is most suited to a micro-kinetic approach that explicitly considers all these processes (e.g. Nielsen et al. 2012, 2013). However, the micro-kinetic parameterization required to fully describe B uptake becomes complex to the point where it hinders the intuitive understanding of the underlying processes, so we will focus here on a first-order 'thought experiment' considering the kinetic controls on B uptake, using DePaolo's (2011) simpler formulation.

DePaolo's (2011) SKM model describes the precipitation rate of a mineral (R_p) as the balance between the forward (R_f) and backward (R_b) reaction rates of its main constituents at the growth surface:

$$ {}^{C}R_p = {}^{C}R_f - {}^{C}R_b \tag{4.1} $$

where C represents Ca^{2+} or CO_3^{2-}, the monomer components of $CaCO_3$ (Fig. 4.4a). The rate of attachment of a trace element (${}^{TE}R_p$) is then defined relative to the fluxes of the host constituent it is replacing (Fig. 4.7):

$$ \begin{aligned} {}^{TE}R_p &= {}^{TE}R_f - {}^{TE}R_b \\ &= {}^{TE}K_f^{TE}\, r_L^C R_f - {}^{TE}K_b^{TE}\, r_S^C R_b \end{aligned} \tag{4.2} $$

where K represents a reaction-specific fractionation factor of the trace element (${}^{TE}K_{(b\ or\ f)}$), and r is the ratio of the trace element to the constituent it is replacing in either the liquid (L) or solid (S). Following DePaolo (2011), during steady state precipitation the composition (TE/C) of the mineral surface does not change with time (i.e. is the same as the bulk mineral):

$$ \begin{aligned} \frac{\delta\,{}^{TE}r_S}{\delta t} = 0 &= \frac{1}{N_C}\left(\frac{\delta N_{TE}}{\delta t} - {}^{TE}r_S \frac{\delta N_C}{\delta t}\right) \\ &= \frac{1}{N_C}\left({}^{TE}R_P - {}^{TE}r_S\,{}^{C}R_P\right) \end{aligned} \tag{4.3} $$

where N is the number of TE or C atoms in the structure. This may then be combined with Eqs. 4.1 and 4.2 to solve the steady-state ${}^{TE}r_S$ of the mineral:

$$ {}^{TE}r_S = \frac{{}^{TE}K_f^{TE} r_L\left({}^{C}R_p + {}^{C}R_b\right)}{\left({}^{TE}K_b^{C}R_b + {}^{C}R_p\right)} \tag{4.4} $$

In this formulation, mineral composition depends solely upon precipitation rate (${}^{C}R_p$), the backward detachment rate of host constituents (${}^{C}R_b$), the forward and backward reaction fractionation factors (${}^{TE}K_f$ and ${}^{TE}K_b$), and the ratio of TE/C in solution. Both the precipitation (${}^{c}R_p$) and solution TE/C (${}^{TE}r_L$) must be experimentally determined, if the model is to be applied to understand the chemistry of the resulting precipitates. The detachment rate of ions from calcite (${}^{C}R_b$) has been estimated by various methods to be around 6×10^{-7} mol/m^{2}s^{-1} across a seawater-relevant pH range (Chou et al. 1989; DePaolo 2011). The detachment rate of ions from aragonite is less well constrained, though likely higher than in calcite because of its greater solubility. The reaction fractionation factors can loosely be equated to the combination of the energy barriers associated with the attachment or detachment of an ion. In attachment, this is primarily desolvation and bond-making with the surface, in detachment it is dominated by

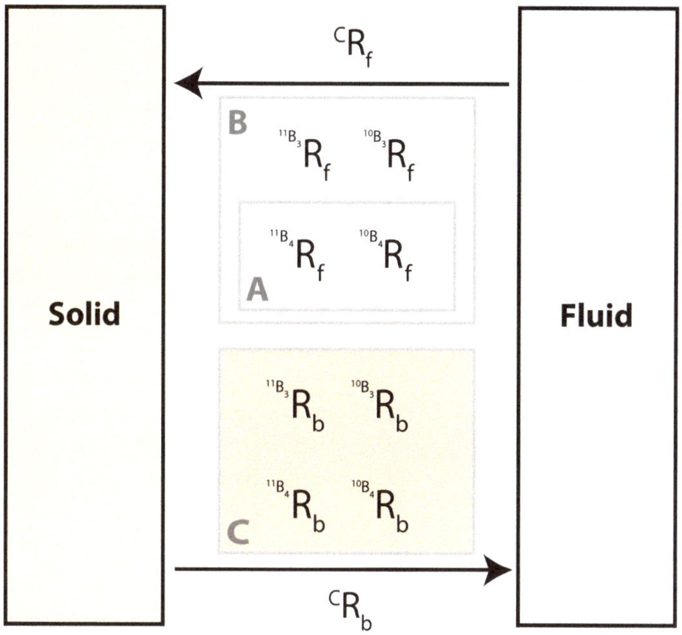

Fig. 4.11 A kinetic description of B incorporation. The uptake of B may be considered in terms of the rates of attachment (R_f) and detachment (R_b) of $B(OH)_3$ (B_3) and $B(OH)_4^-$ (B_4), relative to the attachment rate of C (^{C}R) to the crystal surface, as outlined by the Surface Kinetic Model (DePaolo 2011). For completeness, each B term should be further divided into separate terms for ^{11}B and ^{10}B. The attachment and incorporation of $B(OH)_4^-$, as proposed by Hemming et al. (1992, 1995), only explicitly considers the attachment of $B(OH)_4^-$ (box A), although the backward $B(OH)_4^-$ flux is implicit in their mechanism. From inorganic precipitation experiments, it appears that the fluxes of both $B(OH)_3$ and $B(OH)_4^-$ must be considered in calcite. The attachment rates of each aqueous species (box B) may be considered in terms of their abundance in solution relative to C, and a species-specific partition coefficient (K), based on how easily $B(OH)_3$ and $B(OH)_4^-$ are incorporated into the mineral. The relative uptake of each isotope may be considered in terms of a species-specific fractionation factor ($R(^{11}B)$ / $R(^{10}B)$), which is likely to be close to 1 (Zeebe et al. 2001). The detachment rates of each species (box C) are impossible to constrain, given the unknown dynamics of re-coordination during uptake

bond-breaking, combined with the amount of time it spends on the surface without being incorporated—i.e. structurally incompatible ions are more likely to detach. Because of the complexity of ion-surface interactions, these factors are best determined by fitting experimental data, although some constraint may be placed on their values by measuring the 'equilibrium partition coefficient' ($^{TE}K_{eq}$)—the trace element ratio of a precipitate formed at slow precipitation rates, when the mineral is close to thermodynamic equilibrium with the solution, and is related to the forward and backward attachment rates ($^{TE}K_{eq} = {}^{TE}K_f / {}^{TE}K_b$). Backward reaction coefficients tend to be significantly higher than forward reaction coefficients, because of the incompatibility of the impurity with the solid—i.e. the energy barriers to attach to a mineral surface are much higher than the energy barriers to detach.

Ignoring the recoordination of B on the surface and the kink blocking effects of $B(OH)_4^-$, this simple model can be extended to consider the incorporation of B as the combination of two, independent $B(OH)_3$ and $B(OH)_4^-$ steady states (Fig. 4.11):

$$^{B}r_S = {}^{B3}r_S + {}^{B4}r_S = \frac{^{B3}K_f^{B3}\, r_L\left(^{C}R_p + {}^{C}R_b\right)}{\left(^{B3}K_b^{C}\, R_b + {}^{C}R_p\right)} + \frac{^{B4}K_f^{B4}\, r_L\left(^{C}R_p + {}^{C}R_b\right)}{\left(^{B4}K_b^{C}\, R_b + {}^{C}R_p\right)} \quad (4.5)$$

where the denominator in the $^{x}r_L$ term may be either $[CO_3^{2-}]$, $[HCO_3^-]$ or [DIC], each of which implies that B competes with different carbon species for binding sites on the mineral surface. Finally, the B isotopic content of the resulting mineral can be calculated by a linear mixing between the incorporated $B(OH)_3$ and $B(OH)_4^-$:

$$^{B}A_{11} = \frac{^{B3}A_{11(L)}\,{}^{B3}r_S + {}^{B4}A_{11(L)}\,{}^{B4}r_S}{^{B}r_S} \quad (4.6)$$

where A_{11} is the fractional abundance of ^{11}B, and is related to δ^{11}B via:

$$A_{11} = \frac{^{11}B}{^{10}B + {}^{11}B}$$
$$^{11}R = \frac{^{11}B}{^{10}B} = \frac{A_{11}}{1 - A_{11}}$$
$$\delta^{11}B = \left(\frac{^{11}R_{sample}}{^{11}R_{standard}} - 1\right) \times 1000 \quad (4.7)$$
$$= \left(\frac{\left(\frac{A_{11}}{1-A_{11}}\right)_{sample}}{^{11}R_{standard}} - 1\right) \times 1000$$

The fractional abundance is used in place of the more familiar δ^{11}B notation because 'δ' values do not mix linearly (e.g. Zeebe and Wolf-Gladrow 2001).

While ignoring B re-coordination and kink-blocking effects over-simplifies the possible kinetics of B incorporation, this model provides a starting point from which to explore potential kinetic controls on B uptake into calcite and aragonite (Fig. 4.12). Distinct B uptake regimes can be considered by employing different relative magnitudes of reaction partition coefficients for $B(OH)_3$ and $B(OH)_4^-$. For example, in the case of aragonite, it is apparent from both the partitioning and isotopic content of B in inorganic precipitates (e.g. Mavromatis et al. 2015; Noireaux et al. 2015) that $B(OH)_4^-$ substitutes relatively easily into aragonite, and that $B(OH)_4^-$ is relatively stable in the anion site (Balan et al. 2016). This could be parameterized in the model by setting $^{B4}K_f > {}^{B3}K_f$ and $^{B4}K_b < {}^{B3}K_b$ (Fig. 4.12). Calcite, on the other hand, may incorporate a combination of both $B(OH)_3$ and $B(OH)_4^-$ (Mavromatis et al. 2015; Noireaux et al. 2015; Uchikawa et al. 2015), and trigonal $BO(OH)_2^-$ appears to be most stable in the anion site (Balan et al. 2016). If the calcite surface carries a net is positive charge, $B(OH)_4^-$ would be attracted to the mineral surface more than $B(OH)_3$, so $^{B4}K_f > {}^{B3}K_f$, but owing to the incompatibility of tetrahedral B with the calcite structure (and, implicitly, the energy barrier to phase transformation) the detachment rate of $B(OH)_4^-$ would be much greater than for the more compatible trigonal species, and $^{B4}K_b > {}^{B3}K_b$. Conversely, if the calcite surface was negatively charged, repulsive forces between the surface and $B(OH)_4^-$ would drive $^{B4}K_f < {}^{B3}K_f$, while detachment rates remain similar (Fig. 4.12).

Model predictions based on these scenarios (Fig. 4.12) predict patterns of B partitioning and isotopic offsets from $\delta^{11}B_{B(OH)_4^-}$ which are sensitive to both growth rate and pH, and offer insights into the possible mechanisms of B incorporation. Models were calculated using [DIC] in the denominator for all B/C_{Liquid} ($^{x}r_L$) terms. Changing the carbon species in the denominator altered the magnitude, and to a lesser extent shape of the predicted patterns, but produced qualitatively similar results. The $CaCO_3$ backward reaction rate ($^{C}R_b$) was 6×10^{-7} mol m^2 s^{-1}.

In the case of aragonite, the model predicts increased B partitioning with both growth rate and pH, and a negligible isotopic offset from solution $\delta^{11}B_{B(OH)_4^-}$. This is in agreement with the majority of inorganic data, and is driven by the relative ease of both $B(OH)_4^-$ attachment and incorporation, relative to $B(OH)_3$. For calcite with a positively charged surface that interacts with both aqueous B species, partitioning is more dependent upon precipitation rate than pH, and the combination between the incorporation of both B species and their changing relative abundance with pH drives a pH-dependent offset from $\delta^{11}B_{B(OH)_4^-}$, with more $B(OH)_3$ incorporated at lower pH. This matches patterns in the partitioning data of Uchikawa et al. (2015; Fig. 4.13) and the $\delta^{11}B_{B(OH)_4^-}$ data of Noireaux et al. (2015; Fig. 4.14) well, but does not agree

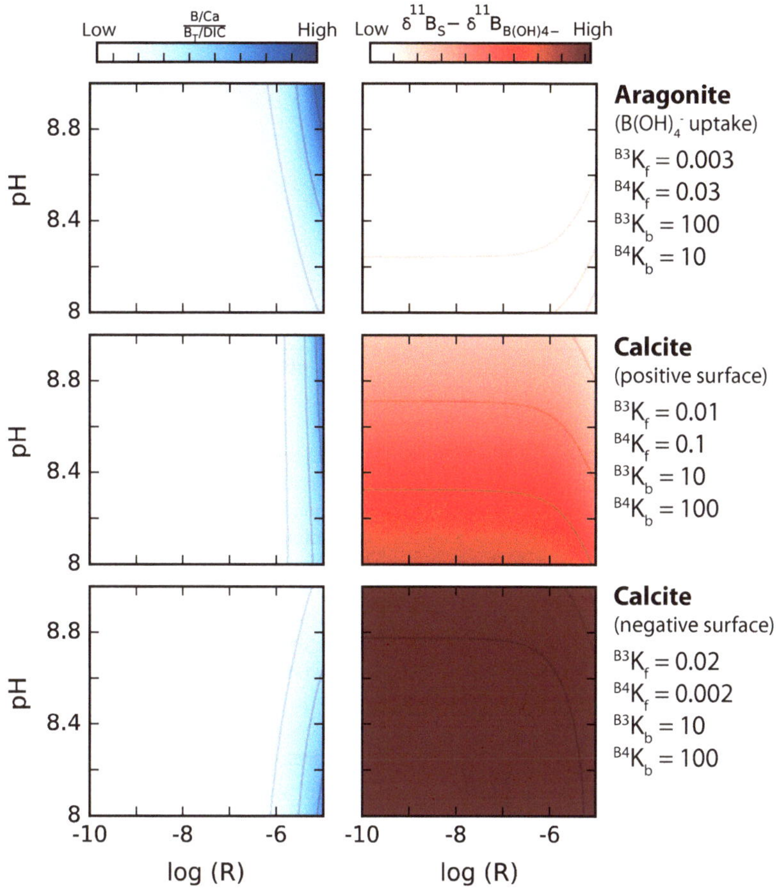

Fig. 4.12 Patterns of B partitioning and isotopic content in calcite and aragonite as a function of pH and growth rate, based on a Surface Kinetic Model (SKM). Changes in $B(OH)_3$ and $B(OH)_4^-$ abundance driven by pH interact with species-specific precipitation rate effects predicted by the SKM to produce distinct patterns in both B partitioning and offset from $\delta^{11}B_{B(OH)_4^-}$ in calcite and aragonite, under distinct incorporation regimes. Three contours on each plot highlight the shape and location of maximal gradients on the plot. Note that, all else being constant, precipitation rate and pH are inherently linked, as exemplified by the data of Uchikawa et al. (2015) in Fig. 4.13. This means that within a single set of precipitation conditions the B partitioning and $\delta^{11}B$ offsets will move from the lower left, to the upper right of this parameter space as pH and growth rate increase, and the exact slope and form of this relationship depends upon solution chemistry and the mineral polymorph. At all stages in the model [DIC] was used in the denominator for ${}^{x}rL$ calculations. Using $[CO_3^{2-}]$ or $[HCO_3^-]$ slightly changes the shape and magnitude of the patterns, but produces qualitatively similar results. The reaction partition coefficients chosen are purely illustrative, and slight differences in magnitude were required to plot all panels on the same scale. The backward reaction coefficients are orders of magnitude higher than the forward coefficients, to reflect the strong equilibrium partitioning (K_{eq} = K_f / K_b) against B incorporation in carbonates. The rate of backward reaction (${}^{c}R_b$) was $6x10^{-7}$

with the patterns in Mavromatis et al. (2015; Fig. 4.13). This is surprising, as Noireaux et al and Mavromatis et al analyzed the same sample material. A possible source of the discrepancy in Mavromatis et al. data may derive from the broad range of major ion chemistry and ionic strength in their study, combined with the use of the MINTEQA2 database within PHREEQC (Parkhurst and Appelo 1999) to calculate their B and C solution chemistries. Dissociation constants calculated by PHREEQC are known to diverge from empirical constants across large shifts in solution chemistry (Hain et al. 2015), which could lead to inaccuracies in calculated solution compositions, and the resulting calculated B partition coefficients into carbonates. If

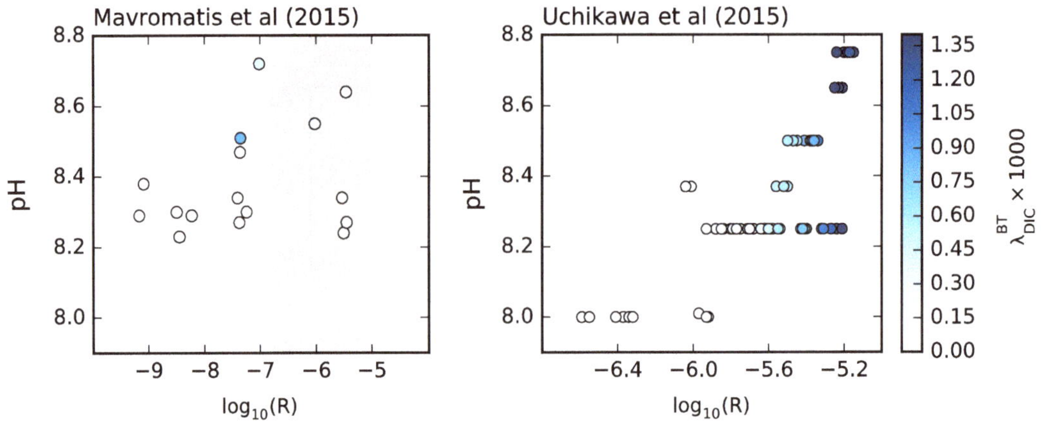

Fig. 4.13 **B partitioning data from recent B precipitation studies of calcite**, plotted as a function of precipitation rate and pH. The data from Uchikawa et al. (2015) reproduce the pattern predicted by the Surface Kinetic Model well (Fig. 4.12), with the highest B distribution coefficients in the high growth rate, high-pH quadrant of the plot. Data from Mavromatis et al. (2015) do not conform to the model, possibly because their DIC concentrations are less well constrained (see main text). The lack of agreement with the model predictions, and the lower overall partitioning of B in the Mavromatis et al. (2015) data may be driven by around 80% of their precipitation solutions having B concentrations at least an order of magnitude higher than Uchikawa et al. (2015). At these high concentrations the inhibitory effect of B on calcite growth (Ruiz-Agudo et al. 2012) may reduce B partitioning from solution, and alter its uptake dynamics. The grey bar in the left panel denotes the log(R) range shown in the right panel

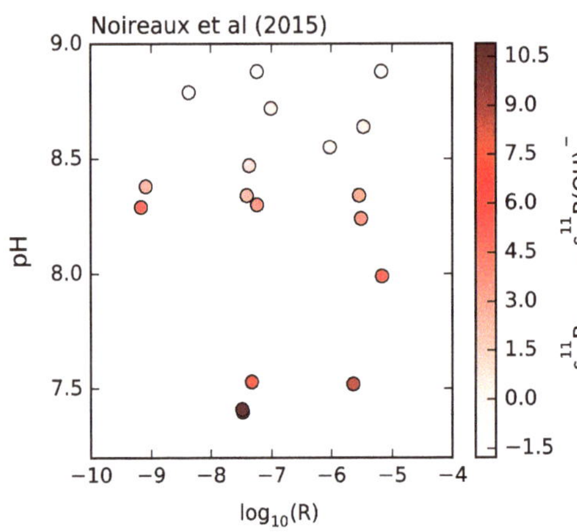

Fig. 4.14 **Calcite δ^{11}B offsets from solution $\delta^{11}B_{B(OH)_4^-}$** as a function of precipitation rate and pH (α = 1.026; Nir et al. 2015). The pattern of offset between calcite δ^{11}B and aqueous $\delta^{11}B_{B(OH)_4^-}$ is similar to that predicted by the Surface Kinetic Model for calcite with a positive surface charge incorporating both $B(OH)_3$ and $B(OH)_4^-$ (Fig. 4.12)

this were the case, it might also alter the B isotope offsets observed by Noireaux et al. (2015), making this pair of studies particularly difficult to interpret. Finally, calcite with a negative surface charge, which rejects $B(OH)_4^-$ from the surface more strongly than neutral $B(OH)_3$ while maintaining similar relative detachment rates, exhibits an increase in B partitioning with growth rate, but a decrease with pH, owing to the scarcity of $B(OH)_3$ at higher pH. Given the relationship between pH and growth rate, this would manifest in a negligible increase in B partitioning with both pH and growth rate (i.e. moving across the plot panel from lower left to upper right). Alongside this, the B isotopes are relatively uniform and offset from $\delta^{11}B_{B(OH)_4^-}$, consistent with the predominant incorporation of $B(OH)_3$ into the mineral. These patterns do not fit with any synthetic calcite precipitation data, and suggest that $B(OH)_4^-$ must account for a significant proportion of B incorporated into calcite.

The qualitative similarity between patterns predicted by the kinetic model and significant trends in both B partitioning (Uchikawa et al. 2015; Fig. 4.13) and isotope fractionation (Noireaux et al. 2015; Fig. 4.14) offer compelling evidence to suggest that surface kinetics may exert a significant influence on B incorporation in carbonates. However, the simplistic implementation presented here neglects several key processes, which could play an important role in B uptake:

- **B re-coordination at the surface.** Coordination (e.g. Branson et al. 2013; Sen et al. 1994) and computational (Balan et al. 2016) studies suggest that a tetrahedral-trigonal transformation is an important step in B incorporation into calcite. The energy barrier to re-coordination will prevent the incorporation of $B(OH)_4^-$, and leave it 'exposed' on the mineral surface for longer, and more likely to become detached. The influence of this process on B partitioning can be implicitly accommodated in the kinetic model by increasing the detachment fractionation coefficient of $B(OH)_4^-$ (${}^{B4}K_b$). The effect on isotopes, however, is more complex. When a $B(OH)_4^-$ undergoes re-coordination on the surface it alters the isotopic composition of the pool of trigonally coordinated B within the solid, changing the ${}^{B3}A_{11}$ of detached trigonal B, and invalidating Eq. 4.6. If the rate of tetrahedral-trigonal transformation were constant this would manifest as a constant, slightly negative offset from $\delta^{11}B_{B(OH)_4^-}$. However, if the transformation rate changed with growth conditions (e.g. rate), this could manifest in a dynamic offset from $\delta^{11}B_{B(OH)_4^-}$.
- **Kink-blocking by $B(OH)_4^-$.** Atomic force microscopy has demonstrated a specific interaction between $B(OH)_4^-$ and obtuse-angle kink sites (Ruiz-Agudo et al. 2012). At low precipitation rates, the extra time spent on the surface would lead to an increased probability of $B(OH)_4^-$ detachment (${}^{B4}K_b$). However, at faster precipitation rates, where the rate of 1D nucleation on step edges increases, it is possible that the propagation of new kink sites could collide with the kink-blocking $B(OH)_4^-$, stabilizing it in the structure and allowing its direct incorporation. This kink-collision entrapment mechanism could provide a route to incorporate tetrahedral B within the calcite structure. If this is the case, then the often large proportions of tetrahedral B observed in coordination studies (e.g. Mavromatis et al. 2015) might suggest that this is an important B incorporation mechanism. This could lead to a growth-rate dependent increase in $B(OH)_4^-$ incorporation in calcite through a reduction in ${}^{B4}K_b$, driving an increase in B partitioning and a negative shift in mineral $\delta^{11}B$ with precipitation rate. A precipitation-rate dependent shift in $\delta^{11}B$ of this type was observed by Kaczmarek et al. (2016) and Farmer et al. (2015), although not by Noireaux et al. (2015).
- **Alteration of $B(OH)_4^-$ activity by ion-pairing interactions.** The pairing of aqueous B with free ions in solution has the potential to alter the activity of B in solution, particularly in the case of the charged $B(OH)_4^-$ species (Bassett 1980; Hershey et al. 1986; Kitano et al. 1978; Simonson et al. 1987). This would effectively alter the B/C activity ratio in solution (r_L), as a function of solution chemistry, ionic strength and pH.

Each of these processes has the potential to influence both B partitioning and isotope fractionation in distinct ways, which will not be apparent when considering patterns of partitioning or isotopic content in isolation. The relative importance of these effects can only be constrained by simultaneously fitting a model to paired analyses of B partitioning and isotopic offset across a range of precipitation conditions and solution chemistries. The only such paired data published today are the combined studies of Mavromatis et al. (2015) and Noireaux et al. (2015), but the the lack of clear trends in their B partitioning data preclude model comparison (Fig. 4.13). This complexity may either be the result of some aspect of their experimental design or analytical approach, or could be taken as evidence that surface kinetic processes are insufficient to explain B uptake in calcite. However, the agreement between patterns predicted by the kinetic model, Noireaux et al. (2015)

isotopic analyses, and Uchikawa et al. (2015) partitioning data, highlight the possible utility of this approach. Future studies that simultaneously consider both the partitioning and isotopic composition of B across a range of precipitation conditions will offer a powerful system for testing surface reactive models that explicitly consider the independent reaction of both B species with the growing mineral surface, and may yield the next significant advance in our understanding of B incorporation in carbonates. To that end, more sophisticated parameterizations of the surface kinetic model, taking into account the atom-scale processes that govern B re-coordination and kink blocking (e.g. Nielsen et al. 2013, 2012), placed in context of ion-pairing reactions in solution, may be developed and applied to new and existing data.

The data and models considered here demonstrate that, while we are approaching a general understanding of the relationship between solution chemistry, mineral precipitation rate, and calcite B geochemistry, we currently lack a quantitative mechanistic description of B incorporation in calcite. New experimental studies that carefully control growth rates and solution chemistry are beginning to offer significant new insights into these processes, and the combination of these systematic studies with quantitative models of trace element incorporation into calcite offer the potential for great advances in the near future. However, there are clear, unresolved differences between inorganic precipitation studies that lead to significant variability in precipitate B geochemistry. Owing to the complexity of the aqueous B and C system, it is often unclear whether these differences stem from differences in B incorporation mechanism, choices in data processing or solution speciation calculations, or some overlooked aspect of experimental design. These discrepancies must be understood and explained in any general model of B incorporation that is proposed. However, even when these differences are resolved, it is vital to recognize that our understanding of B incorporation will be based on synthetic precipitates, which grow via relatively straightforward mechanisms from well-mixed solutions; far removed from the formation environments of naturally occurring marine carbonates. The complex chemistry of natural precipitation environments and the involvement of non-classical crystal growth processes both have the potential to dramatically alter the incorporation mechanisms of B. For example, although B incorporation into inorganic aragonite appears to be relatively straightforward, the biomineralisation processes of corals introduce significant, complex offsets from $\delta^{11}B$ and B/Ca values predicted by seawater chemistry (e.g. Anagnostou et al. 2012; see Chap. 6). Conversely, the $\delta^{11}B$ of calcite produced by epifaunal benthic foraminifera is indistinguishable from seawater $\delta^{11}B_{B(OH)_4^-}$, despite the complexities of B uptake into synthetic calcite (Rae et al. 2011; see Chap. 5). Thus, the consideration of the B geochemistry of natural carbonates must always proceed carefully, on a case-by-case basis.

4.4 Boron in Carbonate Biominerals

The involvement of biology brings significant additional complexity to mineralization and trace element incorporation, which can distance biomineral B geochemistry from seawater chemistry. The primary mechanisms drive deviations from 'ideal' B incorporation are the biological control over the chemistry of the calcification environment, and the mode of mineral precipitation employed by the organism.

The mineral precipitation mechanisms of biomineralising organisms are poorly characterized. It is likely that numerous organisms precipitate skeletal minerals via amorphous precursor phases (de Yoreo et al. 2015; Weiner 2008; Weiner and Addadi 2011; Weiner and Dove 2003), which are distinct from the classical TLK growth of inorganic precipitates that form the basis of our understanding of B incorporation. The involvement of amorphous precursors could drive isotope fractionation during B incorporation, or increase the non-specific incorporation of B from solution via growth entrapment mechanisms. Owing to the unknown variety of crystallization processes employed by biomineralization, the magnitude, direction and importance of the effects of precipitation mechanism on B incorporation remain unclear. Once the incorporation mechanisms of B into simple, synthetic carbonates is understood, study of the incorporation of B via non-classical precipitation mechanisms in in vitro systems would provide a useful evaluation of the importance of growth mechanism on B incorporation.

Whichever crystal growth mechanism is employed by biominerals, the chemistry of the calcification environment has the potential to significantly alter the B uptake. Organisms exhibit varying degrees of control over the chemistry of calcification, ranging from external, biologically facilitated mineralisation by relatively simple bacteria, to the controlled production of specific mineral structures in tightly controlled, internal environments by complex metazoans (Dove 2010; Knoll 2003). To produce a mineral structure, organisms must either elevate the saturation state of seawater by increasing the concentration of Ca^{2+} or CO_3^{2-}, or removing inhibitory Mg^{2+} (Berner 1975) and organic components (Walter and E.A. Burton 1986) from seawater. The pH of the calcification environment may either be actively controlled (Bentov et al. 2009; de Nooijer et al. 2009; Anagnostou et al. 2012; Venn et al. 2013), or be passively altered by the proton flux associated with $CaCO_3$ precipitation (Glas et al. 2012). The manipulation of solution chemistry occurs in a calcification environment that may be completely 'closed', encapsulated in layers of membranes or complex tissues (e.g. coccolithophorids; Young and Henriksen 2003; Young et al. 1999), or more 'open' to seawater (e.g. corals, Gagnon et al. 2012; foraminifera, Bentov et al. 2009). The size and geometry of calcification environments varies widely between organisms, from picoliter-sized vacuoles in coccoliths, to much larger reservoirs in complex bivalves and gastropods.

At present, no comprehensive measurements of the chemistry of the calcification environment are available. However, the concentration and isotopic composition of B in the environment may be considered in terms of the 'openness' of the system with respect to seawater, and the pH of the environment. Given the rapid equilibration time of aqueous B (Zeebe et al. 2001), the speciation and fractionation of B within the calcification environment will be predominantly determined by solution pH and ionic composition, both of which may alter the equilibrium dissociation constant (pK^*_B) and isotope fractionation of B (DOE 1994; Klochko et al. 2006). The total B concentration and $\delta^{11}B$ of the fluids surrounding calcification will be determined by the balance of B transport in and out of the environment. In relatively open calcification environments with high seawater throughput, the [B] and $\delta^{11}B$ if the fluid will be closely linked to seawater. In more closed environments, additional B fluxes from diffusive and active transport, as well as B removal by precipitation may become more important. Neutral $B(OH)_3$ is able to diffuse relatively freely across biological membranes, compared the membrane-impermeable charged $B(OH)_4^-$ ion (Dordas and Brown 2001, 2000; Dordas et al. 2000; Hu and Brown 1997). Passive trans-membrane $B(OH)_3$ diffusion will therefore provide a constant, high-$\delta^{11}B$ flux into the calcification environment, particularly in calcification sites with elevated pH, which will reduce internal $[B(OH)_3]$ and create a strong inward concentration gradient. In contrast, $B(OH)_4^-$ may enter the calcification environment either through specific transport channels (Dordas et al. 2000), or leakage through transporters targeting similar ions, like HCO_3^-. The active transport of $B(OH)_4^-$ to the calcification site is unlikely, as it would involve an expenditure of energy for no mineralization benefit, and the possibility of leakage through bicarbonate transporters is entirely theoretical, and has not been observed in nature (Rae et al. 2011). The diffusive transport of $B(OH)_3$ alone has been shown to be sufficient to drive B/Ca in coccolith calcite produced in a closed biomineralisation environment (Stoll et al. 2012), demonstrating the feasibility of these processes, but their importance have yet to be explored in other organisms.

Despite the complications introduced by the unknown chemical environment and mineral precipitation mechanisms of biominerals, numerous measurements of B geochemistry of both calcite and aragonite consistently reveal sensitivity to external pH and B chemistry (Chaps. 5 and 6). Intriguingly, measurements of $\delta^{11}B$ from biomineral calcite are often closer to seawater $\delta^{11}B_{B(OH)_4^-}$ (Foster 2008; Rae et al. 2011; Sanyal et al. 1996) than the $\delta^{11}B$ of biomineral aragonite (Allison and Finch 2010; Anagnostou et al. 2012; Blamart et al. 2007; Hönisch et al. 2004; Krief et al. 2010; McCulloch et al. 2012; Trotter et al. 2011)—opposite to the patterns observed in inorganic precipitate studies (Noireaux et al. 2015). Physiological controls over calcification chemistry (Anagnostou et al. 2012; Bentov et al. 2009; Rollion-Bard and Erez 2010), and the mineral precipitation mechanism (Bentov and Erez 2005; Hobbs and Reardon 1999) of biominerals may each influence B incorporation into biominerals, and drive the differences between biomineral and synthetic carbonates. In coral aragonite, several studies suggest a direct relationship between the pH of the calcification site and mineral $\delta^{11}B$ (Chap. 6; e.g. Anagnostou et al. 2012; Martin et al. 2015; McCulloch et al. 2012) based on elevated aragonite $\delta^{11}B$. While this is likely related to the pH of the calcification environment, studies making this inference do not consider the influence of calcification chemistry on the dissociation and isotopic fractionation of B at the site of calcification. For example, if $B(OH_3)$ diffusion were an important factor in corals, it would drive similar shifts in $\delta^{11}B$ to those currently attributed to pH elevation. However, aside from possible solution chemistry complications, these studies do not offer any evidence to suggest that B incorporation into coral aragonite is significantly more complex than in inorganic systems. Deconvolving the relative importance of biological controls over calcification chemistry and mineral precipitation processes in calcite biominerals, on the other hand, is significantly more complicated because of our lack of understanding of B incorporation, in synthetic calcite.

Within our current understanding of B incorporation into carbonate minerals, the relationships between carbonate B geochemistry and environmental conditions, and therefore the B/Ca and $\delta^{11}B$ paleoceanographic proxies, must remain empirical. While there is undoubtedly a connection to seawater pH, B and C chemistry, the specifics of this relationship currently evade quantitative, mechanistic description. Developments in our understanding of B incorporation into synthetic calcite, and the application of systematic modeling to both B incorporation and ion transport in biomineralisation have the potential to significantly reduce the necessary empirical elements in the B paleoproxies.

Acknowledgements I would like to thank Alex Gagnon for useful boron discussions throughout, and Jesse Farmer for his invaluable thoughts on B incorporation, and open discussion of unpublished data. I would also like to thank Howard Spero, Alex Gagnon and Steve Eggins for their generous support of all my academic activities. Finally, I would like to thank the editors, Gavin Foster and Horst Marschall, for their patience and good humor during my lengthly writing process, and Michael Henehan and Joji Uchikawa for their constructive reviews of this chapter.

References

Aizenberg J, Lambert G, Weiner S, Addadi L (2002) Factors involved in the formation of amorphous and crystalline calcium carbonate: a study of an ascidian skeleton. J Am Chem Soc 124:32–39. doi:10.1021/ja0169901

Allison N, Finch AA (2010) $\delta^{11}B$, Sr, Mg and B in a modern Porites coral: the relationship between calcification site pH and skeletal chemistry. Geochim Cosmochim Ac 74:1790–1800. doi:10.1016/j.gca.2009.12.030

Anagnostou E, Huang KF, You CF, Sikes EL, Sherrell RM (2012) Evaluation of boron isotope ratio as a pH proxy in the deep sea coral Desmophyllum dianthus: Evidence of physiological pH adjustment. EPSL 349–350:251–260. doi:10.1016/j.epsl.2012.07.006

Balan E, Pietrucci F, Gervais C, Blanchard M, Schott J, Gaillardet J (2016) First-principles study of boron speciation in calcite and aragonite. Geochim Cosmochim Acta 193:119–131. doi:10.1016/j.gca.2016.07.026

Bassett RL (1980) A critical evaluation of the thermodynamic data for boron ions, ion pairs, complexes, and polyanions in aqueous solution at 298.15 K and 1 bar.

Geochim Cosmochim Acta 44:1151–1160. doi:10.1016/0016-7037(80)90069-1

Bentov S, Erez J (2005) Novel observations on biomineralization processes in foraminifera and implications for Mg/Ca ratio in the shells. Geology 33(11):841–844. doi:10.1130/G21800.1

Bentov S, Brownlee C, Erez J (2009) The role of seawater endocytosis in the biomineralization process in calcareous foraminifera. Proc Natl Acad Sci USA 106 (51):21500–21504. doi:10.1073/pnas.0906636106

Berner RA (1975) The role of magnesium in the crystal growth of calcite and aragonite from sea water. Geochim Cosmochim Ac 39:489–504. doi:10.1016/0016-7037(75)90102-7

Binder G, Troll G (1989) Coupled anion substitution in natural carbon-bearing apatites. Contr Miner Pet 101:394–401. doi:10.1007/BF00372213

Blamart D, Rollion-Bard C, Meibom A, Cuif JP, Juillet Leclerc A, Dauphin Y (2007) Correlation of boron isotopic composition with ultrastructure in the deep-sea coral Lophelia pertusa: implications for biomineralization and paleo-pH. Geochem Geophys Geosyst 8:Q12001. doi:10.1029/2007GC001686

Blundy J, Wood B (1994) Prediction of crystal-melt partition coefficients from elastic moduli. Nature 372:452–454. doi:10.1038/372452a0

Bots P, Benning LG, Rodriguez-Blanco JD, Roncal-Herrero T, Shaw S (2012) Mechanistic insights into the crystallization of amorphous calcium carbonate (ACC). Cryst Growth Design 12:3806–3814. doi:10.1021/cg300676b

Branson O, Redfern SAT, Tyliszczak T, Sadekov AY, Langer G, Kimoto K, Elderfield H (2013) The coordination of Mg in foraminiferal calcite. EPSL 383:134–141. doi:10.1016/j.epsl.2013.09.037

Branson O, Kaczmarek K, Redfern SAT, Misra S, Langer G, Tyliszczak T, Bijma J, Elderfield H (2015) The coordination and distribution of B in foraminiferal calcite. EPSL 416:67–72. doi:10.1016/j.epsl.2015.02.006

Branson O, Bonnin EA, Perea DE, Spero HJ, Zhu Z, Winters M, Honisch B, Russell AD, Fehrenbacher JS, Gagnon AC (2016) Nanometer-scale chemistry of a calcite biomineralization template: implications for skeletal composition and nucleation. Proc Natl Acad Sci USA 113(46):12934–12939. doi:10.1073/pnas.1522864113

Burton WK, Cabrera N, Frank FC (1951) The Growth of Crystals and the Equilibrium Structure of their Surfaces. Philos Trans Royal Society A Math Phys Eng Sci 243:299–358. doi:10.1098/rsta.1951.0006

Chernov AA, (1984) Modern crystallography III - crystal growth. Springer,Berlin, ISBN 978-3-642-81835-6

Chibowski E, Hotysz L, Szcześ A (2003) Time dependent changes in zeta potential of freshly precipitated calcium carbonate. Colloids Surf A 222(1–3):41–54. doi:10.1016/S0927-7757(03)00232-2

Davis KJ, Dove PM, de Yoreo JJ (2000) The Role of Mg^{2+} as an impurity in calcite growth. Science 290:1134–1137. doi:10.1126/science.290.5494.1134

de Nooijer LJ, Toyofuku T, Kitazato H (2009) Foraminifera promote calcification by elevating their intracellular pH. Proc Natl Acad Sci USA 106:15374–15378. doi:10.1073/pnas.0904306106

de Yoreo JJ, Vekilov PG (2003) Principles of crystal nucleation and growth. Rev Min Geochem. doi:10.2113/0540057

de Yoreo JJ, Gilbert PUPA, Sommerdijk NAJM, Penn RL, Whitelam S, Joester D, Zhang H, Rimer JD, Navrotsky A, Banfield JF, Wallace AF, Michel FM, Meldrum FC, Cölfen H, Dove PM, (2015) Crystallization by particle attachment in synthetic, biogenic, and geologic environments. Science 349: aaa6760. doi:10.1126/science.aaa6760

DePaolo DJ (2011) Surface kinetic model for isotopic and trace element fractionation during precipitation of calcite from aqueous solutions. Geochim Cosmochim Ac 75:1039–1056. doi:10.1016/j.gca.2010.11.020

Dickson AG, Sabine CL, Christian JR, (2007) Guide to best practices for ocean CO2 measurements,PICES Special Publication 3. doi:10.1159/000331784

DOE (1994) Handbook of methods for the analysis of the various parameters of the carbon dioxide system in sea water; version 2. ORNL/CDIAC-74

Dordas C, Brown PH (2000) Permeability of boric acid across lipid bilayers and factors affecting it. J Memb Biol 175:95–105

Dordas C, Brown PH (2001) Permeability and the mechanism of transport of boric acid across the plasma membrane of Xenopus laevis oocytes. Biol Trace Elem Res 81:127–139

Dordas C, Chrispeels MJ, Brown PH (2000) Permeability and channel-mediated transport of boric acid across membrane vesicles isolated from squash roots. Plant Physiol 124:1349–1361

Dove PM (2010) The Rise of Skeletal Biominerals. Elements 6:37–42. doi:10.2113/gselements.6.1.37

Dove PM, de Yoreo JJ, Davis KJ, (2004). Inhibition of $CaCO_3$ crystallization by small molecules: the magnesium example. In: Liu XY, de Yoreo JJ (eds) Nanoscale structure and assembly at solid-fluid interfaces volume II: assembly in hybrid and biological systems. Kluwer Academic Publishers, Dordrecht, pp 56–82

Durán-Álvarez A, Maldonado-Domínguez M, González-Antonio O, Durán-Valencia C, Romero-Ávila M, Barragán-Aroche F, López-Ramírez S (2016) Experimental-theoretical approach to the adsorption mechanisms for anionic, cationic, and zwitterionic surfactants at the calcite-water interface. Langmuir 32 (11):2608–2616. doi:10.1021/acs.langmuir.5b04151

Erez J, (2003) The source of ions for biomineralization in foraminifera and their implications for paleoceanographic proxies. Rev Min Geochem 54:115–149. doi:10.2113/0540115

Faatz M, Gröhn F, Wegner G (2004) Amorphous calcium carbonate: synthesis and potential intermediate in biomineralization. Adv Mat 16:996–1000. doi:10.1002/adma.200306565

Farmer JR, Uchikawa J, Penman DE, Hönisch B, Zeebe RE, (2015) Outside the pH box: boron isotopes in synthetic calcite precipitated under varying solution chemistry. AGU Fall Meet Abstr

Foster GL (2008) Seawater pH, pCO_2 and $[CaCO_3]$ variations in the Caribbean Sea over the last 130 kyr: A boron isotope and B/Ca study of planktic foraminifera. EPSL 271:254–266

Foster GL, Hönisch B, Paris G, Dwyer GS, Rae JWB, Elliott T, Gaillardet J, Hemming NG, Louvat P, Vengosh A (2013) Interlaboratory comparison of boron isotope analyses of boric acid, seawater and marine $CaCO_3$ by MC-ICPMS and NTIMS. Chem Geol 358:1–14. doi:10.1016/j.chemgeo.2013.08.027

Fuerstenau DW, Herrera-Urbina R (1992) The surface chemistry of bastnaesite, barite and calcite in aqueous carbonate solutions. Colloids Surf 68(1–2):95–102

Furst M, Lowenstam HA, Burnett DS (1976) Radiographic study of the distribution of boron in recent mollusc shells. Geochim Cosmochim Ac 40:1381–1386. doi:10.1016/0016-7037(76)90128-9

Gabitov RI, Gagnon AC, Guan Y, Eiler JM, Adkins JF (2013) Accurate Mg/Ca, Sr/Ca, and Ba/Ca ratio measurements in carbonates by SIMS and NanoSIMS and an assessment of heterogeneity in common calcium carbonate standards. Chem Geol 356:94–108. doi:10.1016/j.chemgeo.2013.07.019

Gabitov RI, Rollion-Bard C, Tripati A, Sadekov AY (2014) In situ study of boron partitioning between calcite and fluid at different crystal growth rates. Geochim Cosmochim Acta 137:81–92. doi:10.1016/j.gca.2014.04.014

Gagnon A, Adkins JF, Erez J (2012) Seawater transport during coral biomineralisation. Earth Planet Sci Lett 329–330:150–161. doi:10.1016/j.epsl.2012.03.005

Gal A, Habraken W, Gur D, Fratzl P, Weiner S, Addadi L (2013) Calcite crystal growth by a solid-state transformation of stabilized amorphous calcium carbonate nanospheres in a hydrogel. Angew Chem Int Ed 52:4867–4870. doi:10.1002/anie.201210329

Glas MS, Langer G, Keul N (2012) Calcification acidifies the microenvironment of a benthic foraminifer (Ammonia sp.). J Exp Mar Biol Ecol 424–425:53–58. doi:10.1016/j.jembe.2012.05.006

Hain MP, Sigman DM, Higgins JA, Haug GH (2015) The effects of secular calcium and magnesium concentration changes on the thermodynamics of seawater acid/base chemistry: Implications for Eocene and Cretaceous ocean carbon chemistry and buffering mathis. Global Biogeochem Cycles 29:517–533. doi:10.1002/2014GB004986

Hemming NG, Hanson GN (1992) Boron isotopic composition and concentration in modern marine carbonates. Geochim Cosmochim Ac 56:537–543. doi:10.1016/0016-7037(92)90151-8

Hemming NG, Hönisch B (2007) Chapter seventeen boron isotopes in marine carbonate sediments and the ph of the ocean. Dev Mar Geol 1:717–734

Hemming NG, Reeder RJ, Hanson GN (1995) Mineral-fluid partitioning and isotopic fractionation of boron in synthetic calcium carbonate. Geochim Cosmochim Ac 59:371–379. doi:10.1016/0016-7037(95)00288-B

Hemming NG, Reeder RJ, Hart SR (1998) Growth-step-selective incorporation of boron on the calcite surface. Geochim Cosmochim Ac 62:2915–2922. doi:10.1016/S0016-7037(98)00214-2

Henehan MJ, Rae JWB, Foster GL, Erez J, Prentice KC, Kucera M, Bostock HC, Martínez-Botí MA, Milton JA, Wilson PA, Marshall BJ, Elliott T (2013) Calibration of the boron isotope proxy in the planktonic foraminifera Globigerinoides ruber for use in palaeo-CO2 reconstruction. EPSL 364:111–122. doi:10.1016/j.epsl.2012.12.029

Henehan MJ, Foster GL, Bostock HC, Greenop R, Marshall BJ, Wilson PA, Southampton OC, Way E, Parade EB, Zealand N, Building I, Sciences O (2016) A new boron isotope-pH calibration for Orbulina universa , with implications for understanding and accounting for " vital effects." Earth Planet. Sci. Lett. 454:282–292. doi:10.1016/j.epsl.2016.09.024

Hershey JP, Fernandez M, Milne PJ, Millero FJ (1986) The ionization of boric acid in NaCl, Na–Ca–Cl and Na–Mg–Cl solutions at 25°C. Geochim Cosmochim Ac 50:143–148. doi:10.1016/0016-7037(86)90059-1

Hobbs MY, Reardon EJ (1999) Effect of pH on boron coprecipitation by calcite: further evidence for nonequilibrium partitioning of trace elements. Geochim Cosmochim Acta 63:1013–1021. doi:10.1016/S0016-7037(98)00311-1

Holcomb M, DeCarlo TM, Gaetani GA, McCulloch M (2016) Factors affecting B/Ca ratios in synthetic aragonite. Chem Geol Accepted. doi:10.1016/j.chemgeo.2016.05.007

Hönisch B, Bijma J, Russell AD, Spero HJ, Palmer MR, Zeebe RE, Eisenhauer A (2003) The influence of symbiont photosynthesis on the boron isotopic composition of foraminifera shells. Mar. Micropaleontol. 49:87–96. doi:10.1016/S0377-8398(03)00030-6

Hönisch B, Hemming NG, Grottoli AG, Amat A (2004) Assessing scleractinian corals as recorders for paleo-pH: empirical calibration and vital effects. Geochim Cosmochim Acta 68:3675–3685

Hu H, Brown PH (1997) Absorption of boron by plant roots. Plant Soil 193:49–58

Huang YC, Fowkes FM, Lloyd TB, Sanders ND (1991) Adsorption of calcium ions from calcium chloride solutions onto calcium carbonate particles. Langmuir 7(8):1742–1748. doi:10.1021/la00056a028

Kaczmarek K, Nehrke G, Misra S, Bijma J, Elderfield H (2016) Investigating the effects of growth rate and temperature on the B/Ca ratio and $\delta^{11}B$ during inorganic calcite formation. Chem Geol 421:81–92. doi:10.1016/j.chemgeo.2015.12.002

Kakihana H, Kotaka M, Satoh S, Nomura M, Okamoto M (1977) Fundamental studies on the ion-exchange separation of boron isotopes. B Chem Soc Jpn 50:158–163. doi:10.1246/bcsj.50.158

Katz A (1973) The interaction of magnesium with calcite during crystal growth at 25–90°C and one atmosphere.

Geochim Cosmochim Ac 37:1563–1586. doi:10.1016/0016-7037(73)90091-4
Killian CE, Metzler RA, Gong YUT, Olson IC, Aizenberg J, Politi Y, Wilt FH, Scholl A, Young A, Doran A, Kunz M, Tamura N, Coppersmith SN, Gilbert PUPA (2009) Mechanism of calcite co-orientation in the sea urchin tooth. J Am Chem Soc 131(51):18404–18409. doi:10.1021/ja907063z
Kitano Y, Okumura M, Idogaki M (1978) Coprecipitation of borate-boron with calcium carbonate. Geochem J 12:183–189. doi:10.2343/geochemj.12.183
Kitano Y, Okumura M, Idogaki M (1979) Influence of borate-boron on crystal form of calcium carbonate. Geochem J 13:223–224
Klochko K, Kaufman AJ, Yao W, Byrne RH, Tossell JA (2006) Experimental measurement of boron isotope fractionation in seawater. EPSL 248:276–285. doi:10.1016/j.epsl.2006.05.034
Klochko K, Cody G, Tossell JA, Dera P (2009) Re-evaluating boron speciation in biogenic calcite and aragonite using ^{11}B MAS NMR. Geochim Cosmochim Ac. 73:1890–1900
Knoll A, (2003) Biomineralization and evolutionary history. Rev Min Geochem 54(1):329–356
Krief S, Hendy EJ, Fine M, Yam R, Meibom A, Foster GL, Shemesh A (2010) Physiological and isotopic responses of scleractinian corals to ocean acidification. Geochim Cosmochim Ac 74:4988–5001. doi:10.1016/j.gca.2010.05.023
Lemarchand E, Schott J, Gaillardet J (2005) Boron isotopic fractionation related to boron sorption on humic acid and the structure of surface complex formed. Geochim Cosmochim Acta 69:3519–3533. doi:10.1016/j.gca.2005.02.024
Leutwein F, Waskowiak R (1962) Geochemische Untersuchungen an reszenten mariner Molluskenschalen. N Jahrb Miner Abh 99:45–78
Liu Y, Tossell JA (2005) Ab initio molecular orbital calculations for boron isotope fractionations on boric acids and borates. Geochim Cosmochim Acta 69:3995–4006. doi:10.1016/j.gca.2005.04.009
Liu L-G, Chen C-C, Lin C-C, Yang Y-J (2005) Elasticity of single-crystal aragonite by Brillouin spectroscopy. Phys Chem Miner 32:97–102. doi:10.1007/s00269-005-0454-y
Martin P, Goodkin NF, Stewart JA, Foster GL, Sikes EL, White HK, Hennige S, Roberts JM (2015) Deep-sea coral δ13C: a tool to reconstruct the difference between seawater pH and δ11B-derived calcifying fluid pH. Geophys Res Lett 43:299–308. doi:10.1002/2015GL066494
Martinez-Boti MA, Marino G, Foster GL, Ziveri P, Henehan MJ, Rae JWB, Mortyn PG, Vance D (2015) Boron isotope evidence for oceanic carbon dioxide leakage during the last deglaciation. Nature 518:219–222. doi:10.1038/nature14155
Mavromatis V, Montouillout V, Noireaux J, Gaillardet J, Schott J (2015) Characterization of boron incorporation and speciation in calcite and aragonite from co-precipitation experiments under controlled pH, temperature and precipitation rate. Geochim Cosmochim Ac 150:299–313. doi:10.1016/j.gca.2014.10.024
McCulloch M, Trotter J, Montagna P, Falter J, Dunbar R, Freiwald A, Försterra G, López Correa M, Maier C, Rüggeberg A, Taviani M (2012) Resilience of cold-water scleractinian corals to ocean acidification: boron isotopic systematics of pH and saturation state up-regulation. Geochim Cosmochim Acta 87:21–34. doi:10.1016/j.gca.2012.03.027
McElligott S, Byrne RH (1997) Interaction of $B(OH)_3$ 0 and HCO - in seawater: formation of $B(OH)_2CO_3$. Aquat Geochem 3:345–356. doi:10.1023/A:1009633804274
Milliman JD (1993) Production and accumulation of calcium carbonate in the ocean: budget of a nonsteady state. Glo Biogeochem Cyc 7:927–957
Morse JW, Arvidson RS, Luttge A (2007) Calcium carbonate formation and dissolution. Chem Rev 107:342–381. doi:10.1021/cr050358j
Moulin P, Roques H (2003) Zeta potential measurement of calcium carbonate. J Colloid Interface Sci 261(1):115–126. doi:10.1016/S0021-9797(03)00057-2
Mucci A, Morse W (1983) The incorporation of Mg^{2+} and Sr^{2+} into calcite overgrowths: influences of growth rate and solution composition. Geochim Cosmochim Acta 47:217–223. doi:10.1029/93GB02524
Navrotsky A (2004) Energetic clues to pathways to biomineralization: precursors, clusters, and nanoparticles. Proc Natl Acad Sci USA 101:12096–12101
Nielsen LC, Depaolo DJ, de Yoreo JJ (2012) Self-consistent ion-by-ion growth model for kinetic isotopic fractionation during calcite precipitation. Geochim Cosmochim Acta. doi:10.1016/j.gca.2012.02.009
Nielsen LC, de Yoreo JJ, DePaolo DJ (2013) General model for calcite growth kinetics in the presence of impurity ions. Geochim Cosmochim Acta 115:100–114. doi:10.1016/j.gca.2013.04.001
Nir O, Vengosh A, Harkness JS, Dwyer GS, Lahav O (2015) Direct measurement of the boron isotope fractionation factor: reducing the uncertainty in reconstructing ocean paleo-pH. EPSL 414:1–5. doi:10.1016/j.epsl.2015.01.006
Noireaux J, Mavromatis V, Gaillardet J, Schott J, Montouillout V, Louvat P, Rollion-Bard C, Neuville DR (2015) Crystallographic control on the boron isotope paleo-pH proxy. EPSL 430:398–407. doi:10.1016/j.epsl.2015.07.063
Oaki Y, Kajiyama S, Nishimura T, Imai H, Kato T (2008) Nanosegregated amorphous composites of calcium carbonate and an organic polymer. Adv Mater 20(19):3633–3637. doi:10.1002/adma.200800773
Oi T (2000) Calculations of reduced partition function ratios of monomeric and dimeric boric acids and borates by the ab initio molecular orbital theory. J Nucl Sci Technol 37:166–172. doi:10.1080/18811248.2000.9714880

Oomori T, Kaneshima H, Maezato Y, Kitano Y (1987) Distribution coefficient of Mg^{2+} ions between calcite and solution at 10–50°C. Mar Chem 20:327–336

Ostwald W (1897) Studien über die Bildung und Umwandlung fester Körper. Z Phys Chem 22:289–330

Pagani M, Lemarchand D, Spivack A, Gaillardet J (2005) A critical evaluation of the boron isotope-pH proxy: the accuracy of ancient ocean pH estimates. Geochim Cosmochim Acta 69:953–961. doi:10.1016/j.gca.2004.07.029

Paquette J, Reeder RJ (1995) Relationship between surface structure, growth mechanism, and trace element incorporation in calcite. Geochim Cosmochim Acta 59:735–749. doi:10.1016/0016-7037(95)00004-J

Parkhurst DL, Appelo CAJ (1999) User's guide to PHREEQC (Version 2): a computer program for speciation, batch-reaction, one-dimensional transport, and inverse geochemical calculations. Water-resources investigations report 99-4259. USGS, Denver CO, USA

Piana S, Jones F, Taylor Z, Raiteri P, Gale JD (2008) Exploring the role of ions and amino acids in directing the growth of minerals from solution. Mineral Mag 72 (1):273–276. doi:10.1180/minmag.2008.072.1.27

Plank J, Bassioni G (2007) Adsorption of Carboxylate Anions on a CaCO 3 Surface. Zeitschrift Fur Naturforschung Section B-a. J Chem Sci 62(10):1277–1284. doi:10.1515/znb-2007-1008

Radha A, Forbes T, Killian C, Gilbert P, Navrotsky A (2010) Transformation and crystallization energetics of synthetic and biogenic amorphous calcium carbonate. Proc Natl Acad Sci USA 107:16438–16443

Rae JWB, Foster GL, Schmidt DN, Elliott T (2011) Boron isotopes and B/Ca in benthic foraminifera: proxies for the deep ocean carbonate system. EPSL 302:403–413. doi:10.1016/j.epsl.2010.12.034

Raz S, Hamilton PC, Wilt FH, Weiner S, Addadi L (2003) The transient phase of amorphous calcium carbonate in sea urchin larval spicules: the involvement of proteins and magnesium ions in its formation and stabilization. Adv Func Mat 13:480–486. doi:10.1002/adfm.200304285

Reeder RJ (ed) (1983) Carbonates: mineralogy and chemistry. Mineralogical Society of America, Washington DC

Rodriguez-Blanco JD, Shaw S, Benning LG (2011) The kinetics and mechanisms of amorphous calcium carbonate (ACC) crystallization to calcite, viavaterite. Nanoscale 3:265–271. doi:10.1039/C0NR00589D

Rodriguez-Navarro C, Kudtacz K, Cizer Ö, Ruiz-agudo E (2015) Formation of amorphous calcium carbonate and its transformation into mesostructured calcite. CrystEngComm 17:58–72. doi:10.1039/C4CE01562B

Rodriguez-Navarro C, Burgos Cara A, Elert K, Putnis CV, Ruiz-Agudo E (2016) Direct nanoscale imaging reveals the growth of calcite crystals via amorphous nanoparticles. Cryst Growth Des 16(4):1850–1860. doi:10.1021/acs.cgd.5b01180

Rollion-Bard C, Erez J (2010) Intra-shell boron isotope ratios in the symbiont-bearing benthic foraminiferan Amphistegina lobifera: implications for $\delta^{11}B$ vital effects and paleo-pH reconstructions. Geochim Cosmochim Acta 74:1530–1536. doi:10.1016/j.gca.2009.11.017

Rollion-Bard C, Blamart D, Trebosc J, Tricot G, Mussi A, Cuif J-P (2011) Boron isotopes as pH proxy: a new look at boron speciation in deep-sea corals using ^{11}B MAS NMR and EELS. Geochim Cosmochim Acta 75:1003–1012. doi:10.1016/j.gca.2010.11.023

Ruiz-Agudo E, Putnis CV, Kowacz M, Ortega-Huertas M, Putnis A (2012) Boron incorporation into calcite during growth: implications for the use of boron in carbonates as a pH proxy. EPSL 345–348:9–17. doi:10.1016/j.epsl.2012.06.032

Sanchez-Valle C, Reynard B, Daniel I, Lécuyer C, Martinez I, Chervin J-C (2005) Boron isotopic fractionation between minerals and fluids: new insights from in situ high pressure-high temperature vibrational spectroscopic data. Geochim Cosmochim Acta 69:4301–4313. doi:10.1016/j.gca.2005.03.054

Sanyal A, Hemming N, Hanson G, Broecker W (1995) Evidence for a higher pH in the glacial ocean from boron isotopes in foraminifera. Nature 373:234–236

Sanyal A, Hemming NG, Broecker WS, Lea DW, Spero HJ, Hanson GN (1996) Oceanic pH control on the boron isotopic composition of foraminifera: evidence from culture experiments. Paleoceanography 11:513–517. doi:10.1029/96PA01858

Sanyal A, Nugent M, Reeder RJ, Bijma J (2000) Seawater pH control on the boron isotopic composition of calcite: Evidence from inorganic calcite precipitation experiments. Geochim Cosmochim Acta 64:1551–1555

Sanyal A, Bijma J, Spero HJ, Lea DW (2001) Empirical relationship between pH and the boron isotopic composition of Globigerinoides sacculifer: implications for the boron isotope paleo-pH proxy. Paleoceanography 16:515–519

Sawada K (1997) The mechanisms of crystallization and transformation of calcium carbonates. Pure Appl Chem 69(5):921–928. doi:10.1351/pac199769050921

Schmalz RF (1965) Brucite in carbonate secreted by the red alga Goniolithon sp. Science 149:993–996

Sen S, Stebbins JF, Hemming NG, Ghosh B (1994) Coordination enyironments of B impurities in calcite and aragonite polymorphs: a ^{11}B MAS NMR study. Am Min 79:819–825

Simonson JM, Roy RN, Roy LN, Johnson DA (1987) The thermodynamics of aqueous borate solutions I. Mixtures of boric acid with sodium or potassium borate and chloride. J Solution Chem 16(10):791–803. doi:10.1007/BF00650749

Stoll H, Langer G, Shimizu N, Kanamaru K (2012) B/Ca in coccoliths and relationship to calcification vesicle pH and dissolved inorganic carbon concentrations. Geochim Cosmochim Acta 80:143–157. doi:10.1016/j.gca.2011.12.003

Tossell JA (2006) Boric acid adsorption on humic acids: ab initio calculation of structures, stabilities, ^{11}B NMR and ^{11}B, ^{10}B isotopic fractionations of surface complexes. Geochim Cosmochim Acta 70:5089–5103

Trotter J, Montagna P, McCulloch M, Silenzi S, Reynaud S, Mortimer G, Martin S, Ferrier-Pagès C, Gattuso J-P, Rodolfo-Metalpa R (2011) Quantifying the pH "vital effect" in the temperate zooxanthellate coral Cladocora caespitosa: validation of the boron seawater pH proxy. EPSL 303:163–173. doi:10.1016/j.epsl.2011.01.030

Uchikawa J, Penman DE, Zachos JC, Zeebe RE (2015) Experimental evidence for kinetic effects on B/Ca in synthetic calcite: Implications for potential $B(OH)_4^-$ and $B(OH)_3$ incorporation. Geochim Cosmochim Acta 150:171–191. doi:10.1016/j.gca.2014.11.022

Urey HC (1947) The thermodynamic properties of isotopic substances. J Chem Soc 562–581

Vdović N (2001) Electrokinetic behaviour of calcite—the relation with other calcite properties. Chem Geol 177 (3–4):241–248. doi:10.1016/S0009-2541(00)00397-1

Vengosh A, Kolodny Y, Starinsky A, Chivas AR, McCulloch MT (1991) Coprecipitation and isotopic fractionation of boron in modern biogenic carbonates. Geochim Cosmochim Acta 55:2901–2910. doi:10.1016/0016-7037(91)90455-E

Venn AV, Tambutté E, Holcomb M, Laurent J, Allemand D, Tambutté S (2013) Impact of seawater acidification on pH at the tissue-skeleton interface and calcification in reef corals. Proc Nat Acad Sci USA 110:1634–1639. doi:10.1073/pnas.1216153110

Walter LM, Burton EA (1986) The effect of orthophosphate on carbonate mineral dissolution rates in seawater. Chem Geol 56:313–323

Wang D, Wallace AF, de Yoreo JJ, Dove PM (2009) Carboxylated molecules regulate magnesium content of amorphous calcium carbonates during calcification. Proc Natl Acad Sci USA 106:21511–21516. doi:10.1073/pnas.0906741106

Watson EB (1996) Surface enrichment and trace-element uptake during crystal growth. Geochim Cosmochim Acta 60:5013–5020. doi:10.1016/S0016-7037(96)00299-2

Watson E (2004) A conceptual model for near-surface kinetic controls on the trace-element and stable isotope composition of abiogenic calcite crystals. Geochim Cosmochim Ac 68:1473–1488. doi:10.1016/j.gca.2003.10.003

Watson E, Liang Y (1995) A simple model for sector zoning in slowly grown crystals: implications for growth rate and lattice diffusion, with emphasis on accessory minerals in crustal rocks. Am Min 80:1179–1187

Weber JN, Kaufman JW (1965) Brucite in the caleareous alga Goniolithon. Science 149:996–997

Weiner S (2008) Biomineralization: a structural perspective. J Struct Biol 163:229–234

Weiner S, Addadi L (2011) Crystallization pathways in biomineralization. Annu Rev Mater Res 41:21–40. doi:10.1146/annurev-matsci-062910-095803

Weiner S, Dove P (2003) An overview of biomineralization processes and the problem of the vital effect. Rev Min Geochem 54:1

Weiner S, Levi-Kalisman Y, Raz S, Weiss I, Addadi L (2002) A basic strategy for biomineralization: taking advantage of disorder. Microsc Microanal 8:164–165

Wolthers M, Charlet L, van Cappellen P (2008) The surface chemistry of divalent metal carbonate minerals; a critical assessment of surface charge and potential data using the charge distribution milti-site ion complex model. Am J Sci 308:905–941. doi:10.2475/08.2008.02

Young J, Henriksen K (2003) Biomineralization within vesicles: the calcite of coccoliths. Rev Min Geochem 54

Young JR, Davis S, Bown P, Mann S (1999) Coccolith ultrastructure and biomineralisation. J Struct Biol 126:195–215

Zeebe RE (2005) Stable boron isotope fractionation between dissolved $B(OH)_3$ and $B(OH)_4^-$. Geochim Cosmochim Ac 69:2753–2766

Zeebe RE, Wolf-Gladrow D (2001) CO_2 in Seawater: equilibrium. Elsevier Science, Kinetics, Isotopes. doi:10.1016/S0422-9894(01)80002-7

Zeebe RE, Sanyal A, Ortiz JD, Wolf-Gladrow DA (2001) A theoretical study of the kinetics of the boric acid–borate equilibrium in seawater. Mar Chem 73:113–124

Zhang J, Nancollas GH (1990) Kink densities along a crystal surface step at low temperatures and under nonequilibrium conditions. J Cryst Growth 106(2–3):181–190. doi:10.1016/0022-0248(90)90062-P

5 Boron Isotopes in Foraminifera: Systematics, Biomineralisation, and CO_2 Reconstruction

James W.B. Rae

Abstract

The boron isotope composition of foraminifera provides a powerful tracer for CO_2 change over geological time. This proxy is based on the equilibrium of boron and its isotopes in seawater, which is a function of pH. However while the chemical principles underlying this proxy are well understood, its reliability has previously been questioned, due to the difficulty of boron isotope ($\delta^{11}B$) analysis on foraminferal samples and questions regarding calibrations between $\delta^{11}B$ and pH. This chapter reviews the current state of the $\delta^{11}B$-pH proxy in foraminfera, including the pioneering studies that established this proxy's potential, and the recent work that has improved understanding of boron isotope systematics in foraminifera and applied this tracer to the geological record. The theoretical background of the $\delta^{11}B$-pH proxy is introduced, including an accurate formulation of the boron isotope mass balance equations. Sample preparation and analysis procedures are then reviewed, with discussion of sample cleaning, the potential influence of diagenesis, and the strengths and weaknesses of boron purification by column chromatography versus microsublimation, and analysis by NTIMS versus MC-ICPMS. The systematics of boron isotopes in foraminifera are discussed in detail, including results from benthic and planktic taxa, and models of boron incorporation, fractionation, and biomineralisation. Benthic taxa from the deep ocean have $\delta^{11}B$ within error of borate ion at seawater pH. This is most easily explained by simple incorporation of borate ion at the pH of seawater. Planktic foraminifera have $\delta^{11}B$ close to borate ion, but with minor offsets. These may be driven by physiological influences on the

Electronic supplementary material The online version of this chapter (doi:10.1007/978-3-319-64666-4_5) contains supplementary material, which is available to authorized users.

J.W.B. Rae (✉)
School of Earth and Environmental Sciences,
University of St Andrews, St Andrews, Scotland, UK
e-mail: jwbr@st-andrews.ac.uk

H. Marschall and G. Foster (eds.), *Boron Isotopes*, Advances in Isotope Geochemistry,
https://doi.org/10.1007/978-3-319-64666-4_5

foraminiferal microenvironment; a novel explanation is also suggested for the reduced $\delta^{11}B$-pH sensitivities observed in culture, based on variable calcification rates. Biomineralisation influences on boron isotopes are then explored, addressing the apparently contradictory observations that foraminifera manipulate pH during chamber formation yet their $\delta^{11}B$ appears to record the pH of ambient seawater. Potential solutions include the influences of magnesium-removal and carbon concentration, and the possibility that pH elevation is most pronounced during initial chamber formation under favourable environmental conditions. The steps required to reconstruct pH and pCO_2 from $\delta^{11}B$ are then reviewed, including the influence of seawater chemistry on boron equilibrium, the evolution of seawater $\delta^{11}B$, and the influence of second carbonate system parameters on $\delta^{11}B$-based reconstructions of pCO_2. Applications of foraminiferal $\delta^{11}B$ to the geological record are highlighted, including studies that trace CO_2 storage and release during recent ice ages, and reconstructions of pCO_2 over the Cenozoic. Relevant computer codes and data associated with this article are made available online.

Keywords
Boron isotopes · Foraminifera · CO_2 · Biomineralisation · Proxy

5.1 Introduction

The boron isotope composition of marine carbonates has been widely shown to reflect ocean pH (Vengosh et al. 1991; Hemming and Hanson 1992; Sanyal et al. 1996; Foster 2008; Rae et al. 2011; Henehan et al. 2013). Boron isotope measurements on fossil carbonates thus have the potential to constrain changes in pH and CO_2 over geological time (Pearson and Palmer 2000; Hönisch et al. 2009; Rae et al. 2014; Penman et al. 2014; Greenop et al. 2014; Martínez-Botí et al. 2015a). The shells, or "tests", of foraminifera, which are single-celled protists, provide a convenient host for the $\delta^{11}B$-pH proxy, being widely distributed throughout the world's oceans, relatively well-preserved in marine sediment cores, and having chemistry that closely reflects surrounding seawater conditions (Wefer et al. 1999; Erez 2003). Foraminifera have thus become key agents in geochemical reconstructions of past climates, and boron isotope analyses on foraminifera may be readily compared to tracers of various other conditions determined on the same samples (Katz et al. 2010).

5.1.1 Aqueous Boron Isotope Systematics

The ability of boron isotopes in foraminifera to record seawater pH stems from the acid-base equilibrium between boric acid ($B(OH)_3$) and borate ion $\left(B(OH)_4^-\right)$:

$$B(OH)_3 + H_2O \rightleftharpoons B(OH)_4^- + H^+. \quad (5.1)$$

The equilibrium constant of this reaction (K_B) is $10^{-8.6}$ (Dickson 1990), giving pK_B of 8.6, close enough to typical ocean pH (~ 8) to give large changes in the relative abundance of these molecules as a function of seawater pH (Fig. 5.1a):

$$K_B = \frac{[B(OH)_4^-][H^+]}{[B(OH)_3]}; \quad pK_B \approx 8.6. \quad (5.2)$$

Boron has two stable isotopes, ^{11}B (~80%) and ^{10}B (~20%), and an equilibrium fractionation of 27.2‰ (Klochko et al. 2006) exists between boric acid and borate ion, due to the stronger bonding environment of boron in the trigonal boric acid molecule:

$$\alpha_B = \frac{{}^{11}_{10}R_{B(OH)_3}}{{}^{11}_{10}R_{B(OH)_4^-}} = 1.0272. \quad (5.3)$$

The boron isotope composition of seawater ($\delta^{11}B_{SW}$) is 39.61‰ (Foster et al. 2010). As boron in seawater is overwhelmingly made up of boric acid and borate,[1] the combined boron nuclides within these molecules must give $^{11}B/^{10}B$ equal to that in seawater. Therefore as the abundance of these molecules vary, so too must their isotopic composition (Fig. 5.1b). At low pH, where seawater boron is dominantly present as boric acid, this molecule must have $\delta^{11}B = \delta^{11}B_{SW}$, with an infinitesimal quantity of borate ion offset below this by a function of the fractionation factor:

$$\delta^{11}B_{B(OH)_3} = \delta^{11}B_{B(OH)_4^-} \cdot \alpha_B + \varepsilon_B, \quad (5.4)$$

where $\varepsilon_B = 1000 \cdot (\alpha_B - 1)$. The opposite is true at high pH, while at intermediate pH the isotopic composition of these molecules will vary as a predictable function of pH. This is commonly expressed as the mass balance:

$$\begin{aligned} {}^{11}_{10}R_{SW} \times [B]_{SW} =& {}^{11}_{10}R_{B(OH)_3} \times [B(OH)_3] \\ &+ {}^{11}_{10}R_{B(OH)_4^-} \times [B(OH)_4^-], \end{aligned} \quad (5.5)$$

from which, rearranging and substituting Eq. (5.4), the following expression for borate isotopic composition as a function of pH can be derived:

$$\delta^{11}B_{B(OH)_4^-} = \frac{\delta^{11}B_{(sw)}[B]_{sw} - \varepsilon_B[B(OH)_3]}{[B(OH)_4^-] + \alpha_B[B(OH)_3]}. \quad (5.6)$$

[1]Minor polynuclear species of aqueous boron exist, such as $B_3O_3(OH)_4^-$, but are only present in negligible quantities at boron concentrations <25 mmol/kg (Su and Suarez 1995), and can thus safely be ignored at seawater concentrations of ~0.4 mmol/kg.

However note that this is an approximation, as mass balance cannot accurately be formulated in terms of isotope ratios. More strictly, we must write the mass balance for each nuclide and then combine these equations to obtain the isotope ratio mass balance. This gives the formula:

$$\begin{aligned} &\frac{{}^{11}_{10}R_{SW} \cdot ([B(OH)_4^-] + [B(OH)_3])}{{}^{11}_{10}R_{SW} + 1} \\ &= \frac{{}^{11}_{10}R_{B(OH)_4^-} \cdot \alpha_B \cdot [B(OH)_3]}{{}^{11}_{10}R_{B(OH)_4^-} \cdot \alpha_B + 1} \\ &+ \frac{{}^{11}_{10}R_{B(OH)_4^-} \cdot [B(OH)_4^-]}{{}^{11}_{10}R_{B(OH)_4^-} + 1}. \end{aligned} \quad (5.7)$$

which can be rearranged and solved (best done by computer; see supplementary code) for the isotopic composition of borate. This is given by the root:

$$\begin{aligned} R_{B(OH)_4^-} = &((H^2 \cdot R_{SW}^2 + 2H^2 \cdot R_{SW} \cdot \alpha_B \\ &+ H^2 \cdot \alpha_B^2 + 2H \cdot K_B \cdot R_{SW}^2 \cdot \alpha_B \\ &- 2H \cdot K_B \cdot R_{SW} \cdot \alpha_B^2 \\ &+ 8 \cdot H \cdot Kb \cdot R_{SW} \cdot \alpha_B \\ &- 2 \cdot H \cdot K_B \cdot R_{SW} \\ &+ 2H \cdot K_B \cdot \alpha_B + K_B^2 \cdot R_{SW}^2 \cdot \alpha_B^2 \\ &+ 2K_B^2 \cdot R_{SW} \cdot \alpha_B + K_B^2)^{(1/2)} \\ &- H \cdot \alpha_B - K_B + H \cdot R_{SW} \\ &+ K_B \cdot R_{SW} \cdot \alpha_B)/(2 \cdot \alpha_B \cdot (H + K_B)); \end{aligned} \quad (5.8)$$

where R is the $^{11}B/^{10}B$ ratio and H is the hydrogen ion concentration. Equation (5.4) can then be used to calculate the composition of boric acid. The error in calculated boron isotopic compositions resulting from the use of the approximation in Eq. (5.6) compared to the accurate formula in Eq. (5.8) is illustrated in Fig. 5.1c, and may be up to 0.15‰. Matlab codes and an Excel spreadsheet that carry out these calculations are provided in the supplementary online materials.

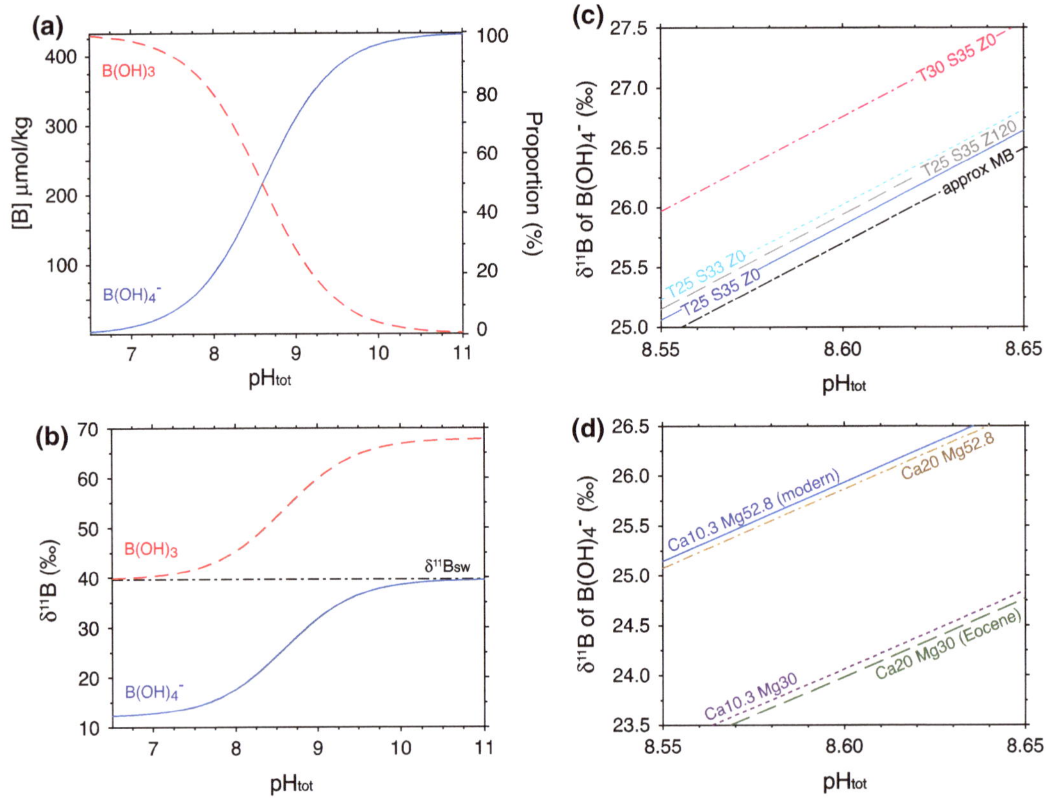

Fig. 5.1 Concentration (**a**) and isotopic composition (**b**) of boric acid and borate ion as a function of seawater pH, at 25 °C, 35 psu, and 0 m depth (Dickson 1990), with modern [Mg] and [Ca], $\delta^{11}B_{SW}$ = 39.61‰ (Foster et al. 2010), and α_B = 27.2‰ (Klochko et al. 2006), calculated using the accurate form of the boron isotope mass balance equation (Eqs. 5.5 and 5.6). **c** Influence of possible changes in temperature, salinity, and pressure at a given site on $\delta^{11}B$ of borate. Note that this plot is focused on the region of pH values close to pK_B, where these influences have greatest sensitivity. Also shown is $\delta^{11}B$ of borate at 25 °C, 35 psu, 0 m depth calculated using the approximate boron isotope mass balance equation (Eqs. 5.5 and 5.6; labeled "approx. MB"). **d** Influence of changes in [Mg] and [Ca] to Eocene values on $\delta^{11}B$ of borate, using the MyAMI model (Hain et al. 2015), are shown individually and together. Code used to make this figure is available in the online materials accompanying this article

5.1.2 The Carbonate $\delta^{11}B$-pH Proxy

Marine carbonates have boron isotopic compositions of ~15–25‰, substantially lower than that of seawater (39.61‰). This is explained by the dominant incorporation of borate ion into growing carbonate (Vengosh et al. 1991; Hemming and Hanson 1992). As the boron isotopic composition of borate varies with pH, so too should $\delta^{11}B$ of marine carbonate.

Early calibrations of boron isotopes in foraminifera showed strong control by pH (Sanyal et al. 1996, 2001), leading to several applications to reconstruct past CO_2 system change and its causes (Sanyal et al. 1995; Pearson and Palmer 2000; Palmer and Pearson 2003). However the reliability of this proxy has been challenged, notably by Pagani et al. (2005a). Concerns with the $\delta^{11}B$-pH proxy have centered on: (1) boron isotope fractionation; (2) mechanisms of boron incorporation into carbonate; (3) vital effects in foraminifera; (4) analytical uncertainties; (5) the evolution of $\delta^{11}B_{SW}$. While several of these topics remain the subject of considerable ongoing research, significant progress has been made on each of these fronts.

Box 1: A quick primer on the ocean carbonate system

The behaviour of boron isotopes in foraminifera discussed in this chapter is intimately linked to the chemistry of CO_2 in seawater, so a brief review of this topic may be helpful. For more detailed treatment see Zeebe and Wolf-Gladow (2001) and Sarmiento and Gruber (2006).

CO_2 reacts with water to form bicarbonate (HCO_3^-) and carbonate ion (CO_3^{2-}):

$$CO_2 + H_2O \overset{K_1}{\Leftrightarrow} H^+ + HCO_3^{2-} \overset{K_2}{\Leftrightarrow} 2H^+ + CO_3^{2-}$$

where K_1 and K_2 are the first and second dissociation constants for carbonic acid, and have values of $\sim 10^{-6}$ and $\sim 10^{-9}$ respectively (varying with temperature, salinity, pressure, and major ion chemistry).

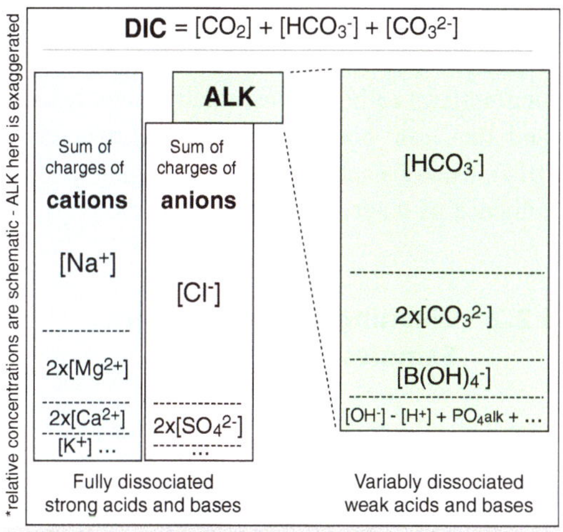

pH (and thus $\delta^{11}B$ of borate) is a sensitive tracer of the equilibrium state of this acid-base system. However this state is actually controlled by the balance between the master variables alkalinity (ALK) and dissolved inorganic carbon (DIC). DIC is simply the total concentration of inorganic carbon molecules. ALK is analytically defined as the number of moles of strong acid required to drive seawater to the equivalence point of bicarbonate. More intuitively, ALK is equivalent to the charge imbalance between strong bases and strong acids in seawater (see inset figure based on Broecker 2005). The requirement for charge balance is taken up largely by variable dissociation of weak acids and bases, predominantly those of CO_2 and boron.

Therefore if alkalinity is large and the pool of DIC to balance it is relatively small, the carbonate system must be pulled toward doubly charged CO_3^{2-}; vice versa, if a large pool of DIC exists but alkalinity is small, much of the DIC must be present as uncharged CO_2. pCO_2, $[CO_3^{2-}]$, pH (and $[B(OH)_4^-]$) are thus primarily controlled by the balance of ALK and DIC, and as a result are tightly coupled and closely traced by $\delta^{11}B$ of borate. This is well illustrated by a ALK-DIC contour plot (inset figure, plotted for 25 °C, 35 psu, and 0 m water depth).

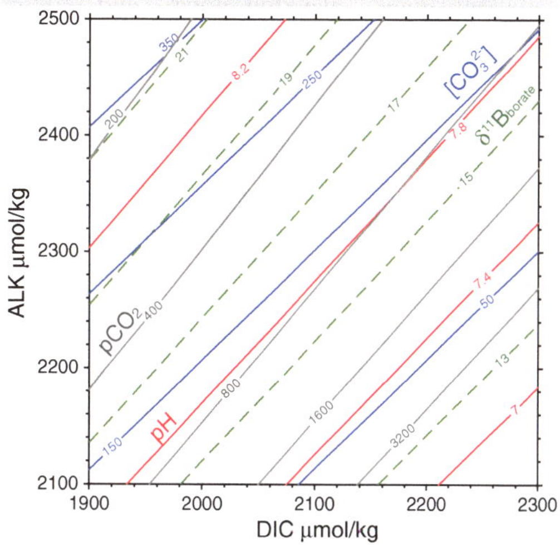

A final important concept is calcium carbonate mineral saturation (Ω):

$$\Omega = \frac{[Ca^{2+}] \times [CO_3^{-}]}{K_{sp}}$$

where K_{sp} is the equilibrium solubility product, $\sim 10^{-6.4}$ for calcite and $\sim 10^{-6.2}$ for aragonite in the surface ocean. Undersaturated conditions correspond to $\Omega < 1$, supersaturated conditions to $\Omega > 1$. $\Omega_{calcite}$ ranges from $\sim$4–6 in most of the modern surface ocean, dominantly as a function of $[CO_3^{2-}]$, and decreases to <1 at depth.

5.2 Methods of Boron Isotope Analysis in Foraminifera

The analysis of boron isotopes is discussed in detail by Foster et al. (2013) for carbonates, and in Chap. 2 of this volume for a broad range of natural samples. Below I provide a brief overview of the specifics of $\delta^{11}B$ analysis of foraminifera, which is challenging due to the small sample sizes and complex matrix involved, and has been a topic of considerable discussion (Ni et al. 2010; Foster et al. 2013; Farmer et al. 2016).

5.2.1 Samples

5.2.1.1 Sample Size and Preparation

Foraminiferal samples are taken from ocean sediment cores, plankton tows, and culture experiments. For solution analyses between ~2 and 20 ng boron are typically used. Boron concentrations and shell masses vary considerably between different species of foraminifera (Yu and Elderfield 2007; Henehan et al. 2016), dictating the number of foraminifera required: for low mass (5–20 µg) low B/Ca (40–60 µmol/mol) planktic species such as *G. bulloides* and *N. pachyderma*, ~400 specimens are typically required; for bulkier (10–30 µg) or higher B/Ca (80–120 µmol/mol) planktic species such as *T. sacculifer* and *G. ruber*, ~100 specimens are required; for higher B/Ca epifaunal benthic species such as *C. wuellerstorfi* and *C. mundulus* (~30–100 µg; 100–250 µmol/mol) it is possible to analyze ~2–20 specimens. Note that analytical uncertainty increases as sample sizes are reduced (Rae et al. 2011).

For sediment core samples, standard 10 cm^3 volumes of pelagic sediment are often sufficient, though this may vary considerably depending on setting. As with any foraminifera-based work, sediment disaggregation and sieving is required; detergents and disaggregation agents (Feldmeijer et al. 2013) may have high levels of boron and should be avoided.

5.2.1.2 Preservation and Diagenesis

Preservation state of foraminifera is not currently thought to exert major influence on boron isotope ratios. For instance, no change in $\delta^{11}B$ with partial dissolution is seen in recent *G ruber* (Ni et al. 2007; Seki et al. 2010; Henehan et al. 2013) or *C. wuellerstorfi* (Rae et al. 2011). Although changes in $\delta^{11}B$ have been noted with partial dissolution in recent *T. sacculifer*, this is thought to be due to the presence of dissolution resistant and anomalously low-$\delta^{11}B$ gametogenic calcite in this species (Hönisch and Hemming 2004; Ni et al. 2007; Seki et al. 2010).

Edgar et al. (2015) show similar $\delta^{11}B$ between well-preserved "glassy" and partially recrystallized "frosty" planktic foraminifera of Eocene age, despite lower B/Ca in the recrystallized samples. This may result from only partial exchange with pore waters during foraminiferal recrystalisation and from low partitioning of boron into inorganically (re)precipitated calcite (Uchikawa et al. 2015) compared to the original foraminiferal carbonate, especially under low-pH (and thus low borate) conditions (Edgar et al. 2015). Mass balance may thus help minimize the influence of diagensis on foraminiferal $\delta^{11}B$.

5.2.2 Cleaning of Foraminiferal Samples

Samples of foraminifera for $\delta^{11}B$ analysis are typically subjected to physical and chemical cleaning, of the type developed for trace element determinations (Boyle 1981; Barker et al. 2003; Ni et al. 2007). This involves gentle crushing to open chambers, repeat ultrasonication in deionized water to remove clays, and oxidation of organic matter in bleach or buffered hydrogen peroxide. Extensive tests of the influence of cleaning on B/Ca ratios suggest that reductive cleaning, commonly employed for Cd/Ca, has little effect on B/Ca, but can result in substantial loss of shell material (Yu et al. 2007). This step is thus typically omitted when cleaning foraminifera for $\delta^{11}B$ (though may still be used if certain other trace element analyses on the same sample are required). Note that given the high contents of organic matter in cultured and towed foraminifera, these samples are typically subjected to a more thorough oxidative cleaning (Russell et al. 2004; Henehan et al. 2013). To check for sample cleanliness and obtain complementary proxy data, several laboratories analyse trace element ratios (e.g. Al/Ca) on aliquots of the same samples prior to boron isotope analyses. No strict cut-off value for contamination exists, as the boron to aluminium ratio and the boron isotope composition of the leachate will vary between samples, but samples with Al/Ca of a few hundred μmol/mol or more may be treated with suspicion (see also Deyhle and Kopf 2004).

5.2.3 Chemical Purification

Chemical purification of boron from the sample matrix is required prior to analysis by Multi-Collector Inductively-Coupled-Plasma Mass Spectrometry (MC-ICPMS; see Wei et al. 2014). Some Negative Thermal Ionisation Mass Spectrometry (NTIMS) methods also employ chemical purification, whereas others load samples in their dissolved carbonate matrix (compare Duke and Lamont methods in Foster et al. (2013) and see discussion in Farmer et al. 2016).

5.2.3.1 Column Chromatography

Boron purification is typically carried out using column chromatography, taking advantage of the Amberlite 743 boron-specific resin (Kiss 1988; Lemarchand et al. 2002). Dissolved samples are buffered to elevate pH, which allows borate to complex with the resin's N-methyl-glucamine groups (Yoshimura et al. 1998). $CaCO_3$ matrix is then removed using deionized water, and boron is eluted using dilute acid. Some laboratories use a cation column to reduce major ion concentrations prior to Amberlite purification (Paris et al. 2010a; Trotter et al. 2011); and see Paris and Duke methods in Foster et al. 2013). As boron is volatile in acidic conditions (Gaillardet et al. 2001) it is not possible to pre-concentrate samples by evaporation, so column volumes must be minimized to concentrate boron for analysis. The column chromatography approach is common in isotope geochemistry and can provide data with reproducibility of ~0.2‰ (2SD; e.g. Rae et al. 2011). However the relatively long exposure time of samples purified by this technique requires laboratory contamination ("blank") to be carefully controlled. This likely accounts for poor inter-laboratory reproducibility between MC-ICPMS techniques at small sample sizes (see discussion in Foster et al. 2013). Control of laboratory blank is a unique challenge for boron, due to the common use of boro-silicate fibers in HEPA filters, lab insulation, and fire-retardants (Rosner et al. 2005). However as boron contamination is also a problem for the semi-conductor industry, boron-free alternatives are available, allowing total procedural blanks to be kept to <100 pg under carefully controlled conditions (Wang et al. 2010; Rae et al. 2011).

5.2.3.2 Microsubliation

Boron purification has also been carried out by "micro-sublimation" (Gaillardet et al. 2001; Wang et al. 2010; Misra et al. 2014). This takes advantage of the volatility of boron to evaporate it from the sample matrix. The dissolved sample is loaded onto the upturned lid of a Teflon beaker, which is then closed and placed on a hotplate, allowing boron to evaporate from the matrix and condense along with water in the beaker's upturned base. This method has the advantage of being totally enclosed, minimizing contamination. Microsublimation has yielded promising results for solution standards, seawater, and multiple aliquots of dissolved carbonate samples (Gaillardet et al. 2001; Wang et al. 2010; Misra et al. 2014). However results on foraminifera are more limited and, to date, show poorer reproducibility; for instance Misra et al. (2014) show that multiple solid splits of a foraminiferal sample have variability of ~1.7‰ (2SD). This is significantly larger than replicate aliquots of a single solution by this method (~0.5‰, 2SD), and cannot be attributed to sample heterogeneity, as the same levels of scatter are seen in subsamples of 25 and 50 foraminiferal tests (Misra et al. 2014). This may suggest variable mass fractionation during boron evaporation as a function of subtle differences in sample matrix.

5.2.4 Mass Spectrometry

5.2.4.1 NTIMS

The pioneering early work on boron isotopes in foraminifera was carried out by Negative Thermal Ionisation Mass Spectrometry (Hemming and Hanson 1994; Sanyal et al. 1995; Pearson and Palmer 2000). By switching polarity from typical positive TIMS and analyzing boron as BO_2^- beams instead of $Cs_2B_2^+$, sample sizes were reduced to levels permitting analysis of small foraminiferal samples (Hemming and Hanson 1994). As described in greater detail in Chap. 2, the carbonate matrix promotes ionization of boron, so dissolved foraminifera are typically loaded directly onto filaments with no further purification (Kasemann et al. 2009; Farmer et al. 2016). However this sensitivity to matrix has the potential to introduce inaccuracies, with NTIMS analyses deviating from MC-ICPMS values in samples with low boron/matrix ratios (see inter-laboratory comparison by Foster et al. (2013) and intra-laboratory comparison by Farmer et al. 2016), though the exact cause of these offsets remains unknown.

5.2.4.2 MC-ICPMS

MC-ICPMS is increasingly used for analysis of boron isotopes on marine carbonates. As described above and in Chap. 2, this technique requires careful separation of boron from the carbonate matrix, and has further challenges including memory effects in the sample introduction system due to boron's volatility (Al-Ammar et al. 2000), large mass bias ($\sim$15% per amu) requiring stable-enough conditions for accurate correction by sample-standard bracketing (Foster 2008), and relatively low sensitivity ($\sim$10 mV/ppb using a 50 μl/min nebuliser). Recent inter-laboratory comparison shows good agreement ($\sim$0.5‰ 2SD) between MC-ICPMS analyses for relatively large samples (Foster et al. 2013; Gutjahr et al. 2014). However reproducibility is notably poorer for smaller samples, of the size typically used in foraminiferal analyses (Foster et al. 2013), highlighting the need for careful control of laboratory blank.

5.2.4.3 In Situ Analysis

Laser-Ablation MC-ICPMS (LA-MC-ICPMS) and Secondary Ionisation Mass Spectrometry (SIMS) can be used to make in situ analyses of $\delta^{11}B$ in foraminifera (Rollion-Bard and Erez 2010; Fietzke et al. 2010; Kaczmarek et al. 2015). Until recently, the precision of these methods has been relatively low, due to low count rates. However recent technical developments, along with pooling of multiple spots have considerably improved LA-MC-ICPMS precision (Fietzke et al. 2015; Sadekov et al. 2016). Accurate correction for mass bias in situ analyses requires homogenous and carefully characterized standards, and matrix matching is important for SIMS (Kasemann et al. 2009), though less so for LA-MC-ICPMS (Fietzke et al. 2015). While in situ analysis is unlikely to replace bulk/solution techniques for generating foraminiferal $\delta^{11}B$ records, it may provide valuable insights into the distribution of boron isotopes within foraminiferal carbonate, and the statistical spread of $\delta^{11}B$ values of individuals within a sample population (Rollion-Bard and Erez 2010; Kaczmarek et al. 2015; Sadekov et al. 2016).

5.3 Boron Isotope Systematics in Modern Foraminifera

A particular attraction of the $\delta^{11}B$-pH proxy is its underlying chemical framework: pH controls $\delta^{11}B$ of borate, and as borate is thought to be incorporated in foraminifera, foraminiferal $\delta^{11}B$ should track pH. However it is possible that equilibrium or kinetic isotopic fractionations occur on boron incorporation into carbonate, and/or that foraminiferal physiology may impart further offsets from inorganic systematics (known as "vital effects"). Thus it is crucial to test the relationship between $\delta^{11}B$ of borate, calculated from measured seawater pH, and foraminiferal δ^{11}B. This has been achieved with measurements of natural modern samples taken from sediment core-tops, plankton tow nets and sediment traps, and using culturing experiments, where foraminifera are taken from the wild and grown in controlled laboratory conditions over a range of pH.

As δ^{11}B proxy calibration studies aim to test the relationship between δ^{11}B in foraminifera and borate—on which this proxy is based—presenting calibration data on cross-plots of foram δ^{11}B versus δ^{11}B of borate is recommended. This also has the advantage over presentation on δ^{11}B-pH figures that different environmental conditions may be readily plotted together, whereas δ^{11}B-pH figures may only be constructed for a single set of environmental conditions (T, S, P, and resulting K_B). However given that several previous studies have shown calibration data on δ^{11}B-pH figures, these are also shown here for reference, with data plotted at the pH of the original growth conditions and δ^{11}B re-calculated for standard surface water conditions of 25 °C and 35 psu. This is achieved by applying the offset between foraminiferal and borate δ^{11}B found at the original growth conditions to the δ^{11}B of borate calculated with K_B for 25 °C, 35 psu, and 0 m depth, at the original growth pH, i.e.

1. calculate δ^{11}B offset at in situ conditions: $\Delta\delta^{11}B = \delta^{11}B_{foram} - \delta^{11}B_{borate\text{-}in\ situ}$;
2. calculate K_B at the T, S, P conditions of the desired pH-δ^{11}B plot, using Eq. (5.11) (and 5.12 and 5.13 if plotting at non-zero water depths);
3. calculate δ^{11}B of borate for the desired plot conditions using Eq. (5.8), with K_B from step 2 above, and pH from the original measurement at in situ conditions;
4. apply the δ^{11}B offset from step 1 to δ^{11}B of borate calculated in step 3, to give the adjusted foram δ^{11}B: $\delta^{11}B_{foram\text{-}adjusted} = \delta^{11}B_{borate\text{-}plot} + \Delta\delta^{11}B$.

Other approaches to this adjustment are possible (for instance using pH rather than δ^{11}B offsets), but the assumptions and steps in the approach above are the most straightforward.

Boron isotope data on modern foraminifera are discussed below, from deep sea calcitic benthic, planktic, and other benthic foraminifera in turn. The implications of these data for δ^{11}B-pH systematics are then synthesized, including modes of boron incorporation into foraminiferal carbonate, isotopic fractionation and physiological influences, and constraints on pH during biomineralisation.

5.3.1 Results of Boron Isotope Calibration Studies on Modern Foraminifera

5.3.1.1 Deep Sea Benthic Foraminifera —A Model System?

The low-Mg calcite hyaline benthic foraminifera found in deep ocean environments provide a useful starting point for assessing foraminiferal boron isotope systematics. This grouping includes a wide range of genera, lacks symbionts, is thought to have relatively low metabolic rate, and inhabits stable deep-sea environments that span a wide range of conditions (pH, T, S, P, ΔCO_3^{2-} etc.).

Deep sea benthic foraminifera appear to record the δ^{11}B of borate ion—and thus pH—of the surrounding water with little or no modification. This is most readily demonstrated for the epifaunal taxa *Cibicides* and *Planulina* (Rae et al. 2011; Yu et al. 2010), which inhabit the sediment surface and thus have the closest association with bottom water conditions (Figs. 5.2 and 5.3). In a set of 45 foraminiferal δ^{11}B samples spanning 17 different sites in the Atlantic, *C. wuellerstorfi, C. mundulus, C. lobatus, C. ungerianus, and P. ariminensis* all have δ^{11}B that closely matches δ^{11}B of borate at bottom water pH (Rae et al. 2011). There are no systematic differences in δ^{11}B values with different foraminiferal test size nor between species (Rae et al. 2011). *C. wuellerstorfi* appears to track bottom water pH conditions with least scatter, as previously observed for other geochemical tracers (δ^{13}C, δ^{18}O, element/calcium ratios), likely due to its

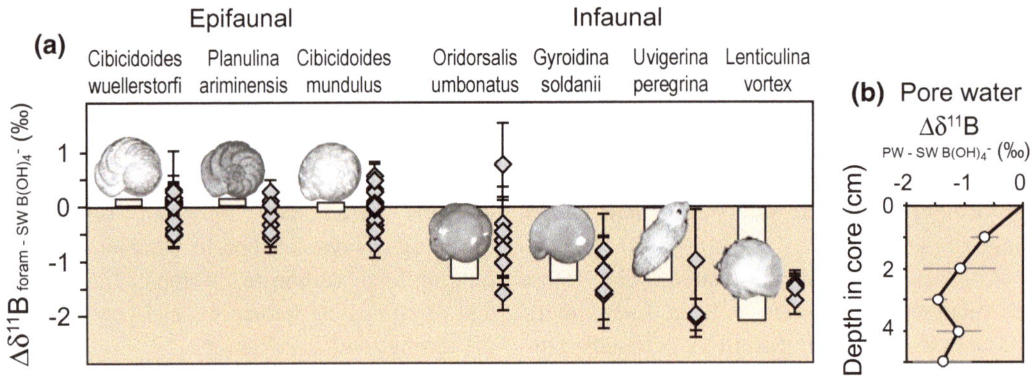

Fig. 5.2 **a** Offsets between δ^{11}B measured in core-top benthic foraminifera from deep ocean sediment cores and δ^{11}B of borate ion calculated for overlying seawater (Rae et al. 2011). Uncertainties are equivalent to 2SD on repeat analyses. Calcite epifaunal foraminifera have δ^{11}B within error of seawater borate, while infaunal foraminifera are offset below bottom water by up to 2‰. **b** Offset between δ^{11}B of borate in bottom water and pore waters, averaged from 3 sediment cores from the Namibian margin. Pore water δ^{11}B of borate values are calculated from measured pH and δ^{11}B of pore water (Rae et al. 2011); the mean and 2SE of δ^{11}B of borate in each of the top 5 cm is shown for three nearby cores. These offsets are similar to those seen between bottom water and infaunal benthic foraminifera (some of which were taken from these cores), suggesting that these species record δ^{11}B of borate in their ambient pore water conditions

elevated epifaunal habitat (Lutze and Thiel 1987).

Infaunal benthic foraminifera also appear to record δ^{11}B of borate, but at conditions reflecting their pore water environment (Fig. 5.2). Infaunal species analysed to date include *C. robertsonianus, Oridorsalis umbonatus, Gyroidina soldanii, Uvigerina peregrine, Melonis zaandamae, Ammonia beccarii,* and *Lenticulina vortex,* inhabiting the top few cm of deep sea sediment. Infaunal foraminifera have δ^{11}B up to 2‰ lower than δ^{11}B of borate in bottom waters. However pH may be lowered in pore waters by oxidation of organic matter, and δ^{11}B of pore water itself may be lowered by interaction with light boron from clays, carbonate, opal, and organic matter (Rae et al. 2011). These factors combine to decrease δ^{11}B of borate by around −0.5 to −2.0‰ in the top 5 cm of pore waters compared to overlying bottom waters (Fig. 5.2b), similar to the δ^{11}B offsets seen in infaunal benthic foraminifera (Fig. 5.2a). More detailed comparison of pore-water chemistry and infaunal δ^{11}B is limited by knowledge of sedimentary habitat depths; nonetheless, the most simple explanation of δ^{11}B in infaunal foraminifera is that it follows the same systematics as in epifaunal species, incorporating borate ion but at the conditions of the surrounding pore water (Corliss 1985).

The close match between δ^{11}B of benthic foraminifera and δ^{11}B of borate at the pH of the surrounding water, across a wide range of environmental conditions and different taxa, provides support for simple models of δ^{11}B-pH proxy systematics (e.g. Hemming and Hanson 1992 and see Sect. 3.2.1).

5.3.1.2 Planktic Foraminifera—Key Proxy Carriers

Planktic foraminifera have been the focus of much of the development work on foraminiferal boron isotopes, due to the importance of surface ocean δ^{11}B-derived pH records in studies of past atmospheric CO_2. Modern planktic foraminifera evolved from benthic species (Hemleben et al. 2012), and have relatively simple low-Mg calcite tests compared to the variety found in benthic taxa. However despite their low taxonomic diversity, planktic foraminifera span a wide range of ecological niches and lifestyles, ranging from shallow-dwelling taxa with photosynthetic symbionts, to deep-dwelling taxa lacking symbionts.

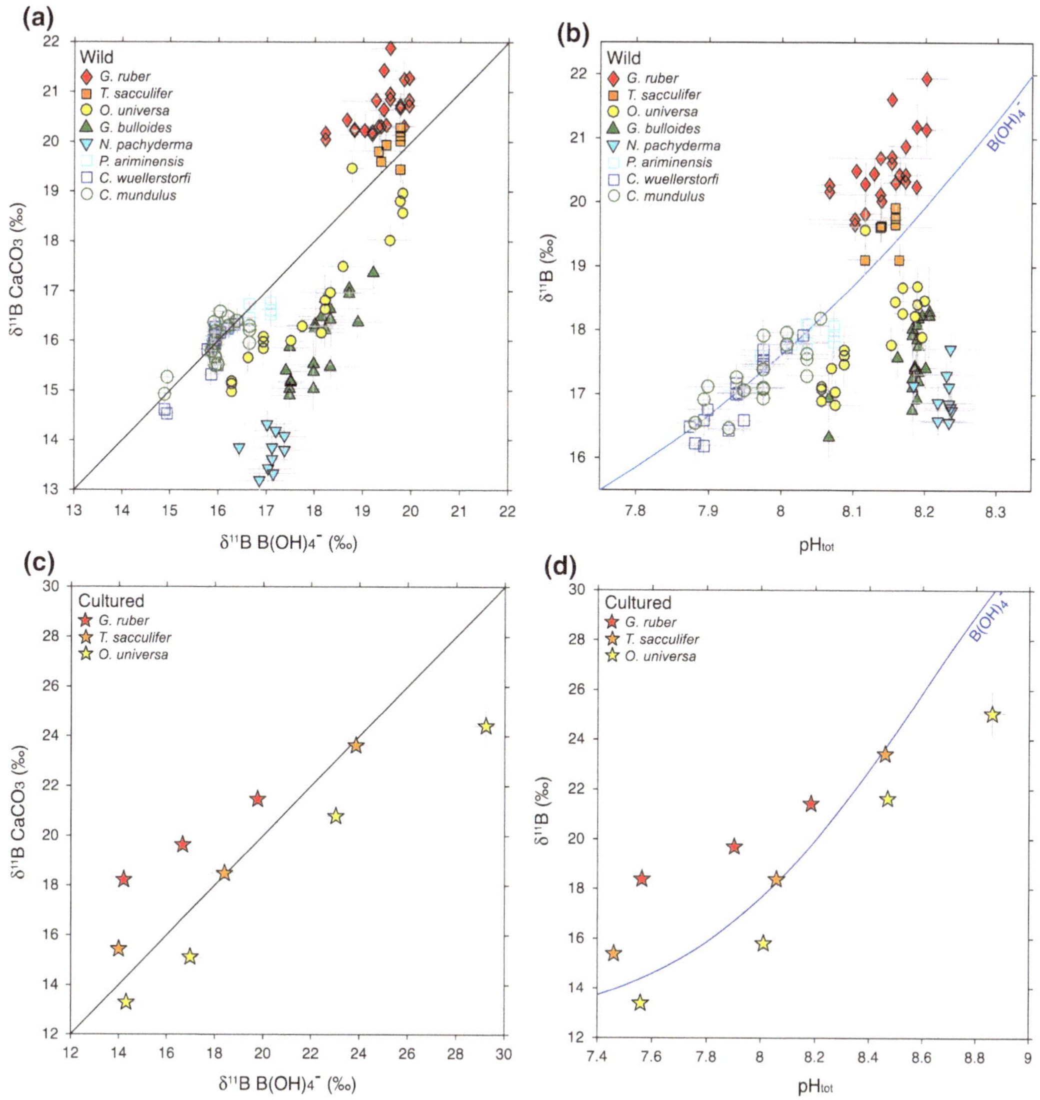

Fig. 5.3 **a** Boron isotope data from modern planktic and benthic foraminifera from core-tops and plankton tows, compared to δ^{11}B of borate calculated from carbonate system data in ambient seawater. **b** Boron isotope data from panel (**a**) re-plotted at 25 °C, 35 psu, and 0 m water depth, compared to pH from carbonate system data in ambient seawater. To re-plot boron isotope data, the δ^{11}B offsets between foraminifera and borate ion found at the original growth conditions are added to the δ^{11}B of borate calculated with pKb for 25 °C, 35 psu, and 0 m depth at the original growth pH (see main text Sect. 5.3). **c**, **d** are equivalent to (**a**) and (**b**) but for cultured planktic foraminifera. Wild planktics: *G. ruber* data are from Henehan et al. (2013) and Foster (2008), *T. sacculifer* (formerly known as *G. sacculifer*) data are from Foster (2008) and Martínez-Botí et al. (2015b), *O. universa* data are from Henehan et al. (2016), *G. bulloides* data are from Martínez-Botí et al. (2015b), *N. pachyderma* data are from Yu et al. (2013). Wild benthics: *P. ariminensis*, *C. wuellerstorfi*, and *C. mundulus* data are from Rae et al. (2011). Cultured planktics: *G. ruber* data are from Henehan et al. (2013), *O. universa* data are from Sanyal et al. (1996), with a −3.3‰ offset to account for interlaboratory bias (Henehan et al. 2016), and *T. sacculifer* data are from Sanyal et al. (2001), with a −5.3‰ offset to account for interlaboratory bias (Foster et al. 2012). Uncertainties are 2σ and taken from original publications. Code and data used to make this figure are available in the online materials accompanying this article

The life span of most planktic species is relatively short (weeks to months), so metabolic rates are high: compared to benthic foraminifera, planktics tend to live fast and die young. Some species have preferences for certain seasons, and some may migrate through different water depths during their life cycle. However individual species tend to occupy a relatively limited set of environmental (e.g. pH and temperature) conditions (Bijma et al. 1990). These factors, along with limited depth and seasonally-resolved water-column pH data, may complicate proxy calibration in planktic foraminifera from the wild (e.g. core-tops, tows, sediment traps). Culturing experiments have thus been employed in $\delta^{11}B$ calibrations to provide an expanded pH range with controlled environmental conditions, but do not necessarily provide analogous conditions to the open ocean.

Boron isotope data from all planktic foraminifera analysed to date lie within a few permil of $\delta^{11}B$ of borate and show sensitivity to seawater pH (Fig. 5.3). However while $\delta^{11}B$ values lie close to $\delta^{11}B$ of borate (compared to $\delta^{11}B$ of seawater or boric acid), there are notable offsets. Species lacking photosynthetic symbionts (including *G. bulloides* and *N. pachyderma*) have $\delta^{11}B$ below that of borate ion (i.e. an offset to lower pH), while symbiont-bearing species (e.g. *G. ruber* and *T. sacculifer*) tend to have $\delta^{11}B$ above borate at seawater pH (i.e. an offset to higher pH; Hönisch et al. 2003).

These offsets may be explained by the physiological influences of photosynthesis, respiration, and calcification on the foraminiferal microenvironment (Zeebe et al. 2003; Fig. 5.4a). Photosynthesis draws down CO_2, raising microenvironmental pH, while respiration releases CO_2, lowering pH (Wolf-Gladrow et al. 1999). Calcification draws down two moles of alkalinity (via Ca^{2+}) per one mole of dissolved inorganic carbon (either as HCO_3^- or CO_3^{2-}), resulting in a net lowering of pH (see Box 1 for a quick review of carbonate chemistry). The influence of these processes on the microenvironment of planktic foraminifera has been demonstrated using microelectrodes (Rink et al. 1998; Köhler-Rink and Kühl 2000), which reveal pronounced gradients in pH and O_2 in the diffusive boundary layer surrounding planktic foraminifera. Culturing studies demonstrate that these processes are reflected in boron isotope composition (Hönisch et al. 2003), with an increase in $\delta^{11}B$ of *O. universa* grown at increasing light levels, due to enhanced photosymbiont activity. These offsets have also been reproduced in a diffusion-reaction model of the foraminiferal microenvironment (Zeebe et al. 2003; and see further discussion below). Physiological alteration of foraminiferal microenvironment may also account for the influence of size on $\delta^{11}B$ in symbiont-bearing planktic foraminifera (Zeebe et al. 2003). Increases in $\delta^{11}B$ with shell size are seen in *G. ruber* and *T. sacculifer* (Henehan et al. 2013; Ni et al. 2007; Hönisch and Hemming 2004), which may result from documented increases in photosynthetic rates relative to respiration with increasing test size (Lombard et al. 2010; Hemleben and Bijma 1994).

Given the influence of physiological processes on planktic foraminiferal $\delta^{11}B$, it is crucial for proxy calibration to test how these are realized in modern ocean settings. This is well-illustrated by recent open ocean $\delta^{11}B$ data from *O. universa* (Henehan et al. 2016), which are offset below $\delta^{11}B$ of borate at seawater pH, despite the presence of dinoflagellate symbionts. This may be explained by the fact that *O. universa* tends to inhabit water depths where light attenuation may limit photosynthetic activity (Morard et al. 2009; Jorgensen et al. 1985). Respiration and calcification thus exert a greater net influence than photosynthesis at the habitat depth of *O. universa*, leading to an offset to lower $\delta^{11}B$ and pH. Indeed Hönisch et al. (2003) note that their culture experiments that used low light levels lie closest to their *O. universa* data from plankton tows (though not core-tops).

Early culture work (Sanyal et al. 1996) suggested that $\delta^{11}B$ in *O. universa* showed a lower sensitivity to pH than $\delta^{11}B$ of borate (Fig. 5.3c, d). However this is not seen in the wild data of Henehan et al. (2016), which closely follow $\delta^{11}B$ of borate with a small constant offset (see further discussion in Sect. 3.2.2). Indeed despite the influence of physiology on planktic foraminiferal $\delta^{11}B$, Henehan et al. (2016)'s open ocean *O. universa* data—from plankton tows and core tops—demonstrate that these vital effects may be remarkably stable, or closely coupled, over a wide range of environmental conditions.

5.3.1.3 Other Benthic Foraminifera—Enigmatic Signals in High-Mg Calcite and Aragonite

Large, high-Mg calcite, reef-dwelling benthic foraminifera of the genus *Amphistegina* are commonly used in culture studies of biomineralisation, being relatively robust to environmental change, easy to collect, and large enough to image effectively. However boron isotope data on *Amphistegina* from different studies are currently somewhat conflicting. While Rollion-Bard and Erez (2010)'s SIMS data, when averaged, give bulk $\delta^{11}B$ values lying ~5–12‰ above $\delta^{11}B$ borate, Kaczmarek et al. (2015)'s recent laser ablation data range from values ~2‰ above borate to ~7‰ below. Kaczmarek et al. (2015) also find a range of ~6‰ between different specimens cultured under the same environmental conditions, with the lowest $\delta^{11}B$ values in these specimens lying below the lower limit for $\delta^{11}B$ of borate ion.

The cause of these offsets between specimens and studies is currently unknown. However the range of $\delta^{11}B$ in specimens taken from the same environmental conditions suggests strong but variable internal control of $\delta^{11}B$ by individual foraminifera of this species under these conditions. Internal control of $\delta^{11}B$ in Amphistegina was also suggested by Rollion-Bard and Erez (2010), who found a notable change in the spread of $\delta^{11}B$ values from individual SIMS spots in a given specimen as a function of pH (see Sect. 3.2.3 on Biomineralisation).

Enigmatic boron isotope results have also been obtained from the only common modern aragonitic foraminifera, *Hoeglundina elegans*. Indeed the life style of this species is in itself an enigma, being widely distributed in the deep ocean at depths and conditions where aragonite is highly undersaturated. Boron isotope values from this epifaunal species lie below ambient borate by ~0.5–3.5‰ (Rae et al. 2011), notably different to low-Mg epifaunal benthic foraminifera. Indeed the lowest $\delta^{11}B$ values from this species (11.97‰; Rae et al. 2011) lie below the lower limit of seawater borate ion, suggesting further isotopic fractionation during biomineralisation in this species.

5.3.2 Discussion of Boron Isotope Calibration on Modern Foraminifera

The synthesis of boron isotope data on modern foraminifera presented above provides a basis to re-evaluate the assumptions underlying the $\delta^{11}B$-pH proxy, and provides powerful constraints on models of calcification.

5.3.2.1 Boron Incorporation in Foraminifera

A key assumption in the use of boron isotopes in marine carbonates to reconstruct pH is that only borate ion is incorporated into carbonate (for more detailed discussion see Chap. 3 by Branson and recent paper by Balan et al. 2016). Given the physiological influences on planktic foraminiferal $\delta^{11}B$, the deep sea benthic foraminifera provide a useful model system to test this assumption. The close match between epifaunal benthic $\delta^{11}B$ and $\delta^{11}B$ of borate (Figs. 5.2 and 5.3) strongly supports the exclusive incorporation of borate ion from seawater into foraminiferal carbonate: incorporation of just 4% boric acid would offset benthic $\delta^{11}B$ from borate ion by more than 1‰, in contrast with the close match observed. Indeed across a wide range of life styles and environments, and despite physiological influences, almost all foraminiferal $\delta^{11}B$ data lie relatively close to $\delta^{11}B$ of borate, and are substantially below $\delta^{11}B$ of seawater or boric acid.

Although NMR data have shown the presence of both tetrahedrally and trigonally coordinated boron in biogenic carbonates (Sen et al. 1994; Klochko et al. 2009; Rollion-Bard et al. 2011; Noireaux et al. 2015; Mavromatis et al. 2015), this does not necessarily reflect the coordination of the molecule incorporated from solution

(Klochko et al. 2009; Balan et al. 2016). Indeed some re-coordination of tetrahedral borate is to be expected upon incorporation into trigonal calcite (Hemming et al. 1995; Klochko et al. 2009), with recent quantum mechanical modelling showing that boron is structurally substituted for carbonate groups in the calcite lattice, being found as partially deprotonated trigonal $BO_2(OH)^{2-}$ and minor $B(OH)_4^-$ (Balan et al. 2016). Although under equilibrium conditions isotopic fractionation may be expected between borate and $BO_2(OH)^{2-}$, if this re-coordination occurs during incorporation without further exchange with seawater, it may not be expressed. Indeed recent synchrotron work (Branson et al. 2015) suggests that boron may be exclusively present in trigonal form in calcitic benthic foraminifera. If this trigonal boron were to reflect boric acid incorporation, foraminiferal $\delta^{11}B$ values should lie ~28‰ above borate. However $\delta^{11}B$ data on these samples lie within a few ‰ of $\delta^{11}B$ of borate ion (Kaczmarek et al. 2015), supporting exclusive addition of borate ion to growing calcite, with re-coordination to trigonal form upon incorporation into foraminiferal calcite with no further isotopic exchange (Hemming and Hanson 1992).

Although some recent experimental work has suggested that boric acid may be incorporated into inorganically precipitated calcites, results from different studies are conflicting. Based on boron concentration data, Uchikawa et al. (2015) suggest that partitioning of boron into calcite is best described in terms using total boron in solution rather than borate ion, especially at high precipitation rates. However preliminary $\delta^{11}B$ data on these same precipitates lie close to borate ion (Farmer et al. 2015), inconsistent with substantial incorporation of boric acid. Noireaux et al. (2015) provide $\delta^{11}B$ data on inorganic aragonite and calcite precipitates (as used in Mavromatis et al. 2015). While Noireaux et al. (2015)'s aragonites have $\delta^{11}B$ within error of borate ion, the calcites have $\delta^{11}B$ up to ~10‰ higher than borate, and these authors suggest incorporation of boric acid may account for this offset. However in a separate study, Kaczmarek et al. (2016) give $\delta^{11}B$ values for inorganic calcite precipitates within 1.5‰ of borate ion, supporting the predominant incorporation of borate. Further work is required to reconcile these conflicting results and this will doubtless lead to improved understanding of the fundamental mechanisms of boron incorporation into carbonate (Balan et al. 2016). However it should be noted that inorganic experiments do not provide a perfect analogue to precipitation of calcite by foraminifera, and that foraminiferal $\delta^{11}B$ data are currently most easily explained by exclusive incorporation of borate ion at pH close to that in surrounding seawater.

5.3.2.2 Boron Isotope Fractionation in Foraminifera

The boron isotope fractionation factor between boric acid and borate ion was, for a long time, poorly constrained (see further discussion in Chap. 8). Early studies frequently used a value of 1.0194, based on calculations by Kakihana and Kotaka (1977) using vibrational frequency data from Kotaka and Kakihana (1977). However it was later shown that a major vibrational frequency mode had been improperly assigned in these calculations (Rustad and Bylaska 2007), and recalculation using various methods yields higher values, though with considerable range (~1.020–1.050; Oi 2000; Liu and Tossell 2005; Zeebe 2005). Experimental determination of the fractionation factor was thus crucial. Klochko et al. (2006) achieved this with an elegant experiment that takes advantage of the fact that pure ^{10}B boric acid and pure ^{11}B boric acid have subtly different dissociation constants, which can be linked to the fractionation factor. This yields an isotopic fractionation of 1.0272 ± 0.0006 (95% confidence) between borate and boric acid in seawater, which has been widely adopted. This has recently been corroborated using an independent experimental set-up, based on separation of boric acid from borate ion (and associated ion pairs) using reverse osmosis, and isotopic measurement of these solutions (Nir et al. 2015). This yields a value of 1.0260 ± 0.001, within error of the Klochko et al. (2006) value.

Although the aqueous fractionation factor is now well constrained, there has been continued

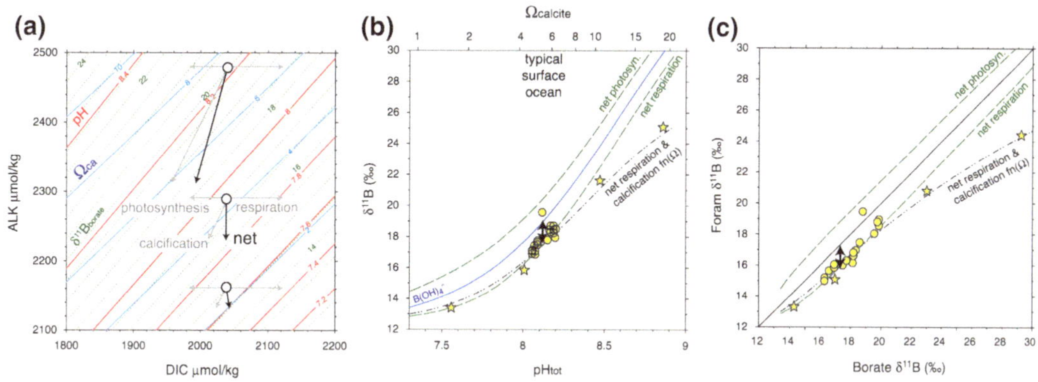

Fig. 5.4 **a** Schematic of the influence of foraminiferal physiology on the carbonate chemistry of the microenvironment. *Circles* indicate water conditions; *grey arrows* indicate DIC drawdown via photosynthesis, DIC release via respiration, and alkalinity and DIC drawdown (in a 2:1 ratio) via calcification; *black arrows* show the net influence of these processes. For simplicity, the minor alkalinity changes due to nitrate uptake and release have been ignored. While respiration and photosynthesis stay constant with changing carbonate chemistry, calcification is likely to vary as a function of saturation state, driving variable offsets in δ^{11}B of borate between the foraminiferal microenvironment and the surrounding seawater. **b**, **c** An equilibrium calculation of δ^{11}B offsets in the foraminiferal microenvironment. Note that this is a major simplification compared to the Wolf-Gladrow et al. (1999) and Zeebe et al. (2003) approach, but useful for building intuition. Bulk solution conditions have DIC of 1800 µmol/kg and alkalinity of 1800–3000 µmol/kg, giving a range of pH and δ^{11}B of borate values. The *green dashed lines* show the influence of 75 µmol/kg DIC drawdown via photosynthesis and 75 µmol/kg DIC release via respiration on δ^{11}B in the foraminiferal microenvironment. The *gray dot-dash line* shows the combined influence of calcification, which draws down alkalinity and DIC in a 2:1 ratio as a function of 2 $(\Omega - 1)^{1.9}$, and 50 µmol/kg net DIC release via respiration. Also plotted are the wild *O. universa* data of Henehan et al. (2016) and the cultured *O. universa* data of Sanyal et al. (1996), plotted as in Fig. 5.3. The *black arrow* indicates the magnitude of δ^{11}B offset in *O. universa* observed between *light* and *dark* culture conditions (Hönisch et al. 2003), and the *gray* band indicates $\Omega_{calcite}$ found in typical modern surface ocean conditions (i.e. covering the full habitat range of *O. universa* and excluding only low Ω waters from the high latitude Southern Ocean and North Pacific). Changes in calcification as a function of carbonate ion saturation may drive variable δ^{11}B offsets in the extremely wide range of saturation states created under culture conditions, whereas offsets remain constant for surface ocean values of $\Omega_{calcite}$. All plots are given for 25 °C, 35 psu and 0 m water depth, with calculations carried out in CO2SYS.m (van Heuven et al. 2009)

discussion about fractionation of boron into carbonates (for instance see supplementary information of Hönisch et al. 2009). While inorganic precipitate data might be hoped to provide constraints on inorganic equilibrium and/or kinetic fractionation, the unexplained discrepancies between different studies (discussed in Sect. 3.2.1) currently make it difficult to derive unambiguous fractionation factors from this work. Much of the discussion of boron fractionation during carbonate formation has thus been based on δ^{11}B in biogenic carbonates.

As discussed above, benthic foraminifera appear to follow δ^{11}B of borate with no significant offset (Figs. 5.2 and 5.3). However planktic foraminifera show a range of offsets (Fig. 5.3), some of which vary as a function of pH (i.e. the δ^{11}B data have a different "slope" or sensitivity to pH compared to borate). In corals, large offsets above δ^{11}B of borate are commonly attributed to internal pH elevation (Hönisch et al. 2004; Krief et al. 2010; McCulloch et al. 2012b; Allison and Finch 2010; Allison et al. 2014 and see Chap. 5 by McCulloch), but this appears not to exert major influence in most foraminifera (see further discussion below). In foraminifera, the physiological processes of photosynthesis, respiration, and calcification are well-documented to cause pH modification in the microenvironment (Rink et al. 1998; Köhler-Rink and Kühl 2000). Zeebe et al. (2003)'s reaction-diffusion model suggested that boron isotope offsets due to these processes may stay constant across a range of pH, making the variable offsets hard to explain. However variable offsets may occur if the nature of carbonate precipitation changes over the range of culture conditions.

Variable δ^{11}B offsets may be expected to arise due to changes in the buffering capacity of the carbonate system at different pH. At low pH, a given change in DIC (for instance via photosynthesis) will drive a larger change in pH than at higher absolute pH values (Fig. 5.4a; and see Allison and Finch 2010). This appears at first to provide a possible explanation for the larger offsets from δ^{11}B of borate observed in foraminifera cultured at low pH. However the effect of this on the boron isotope system is countered by the reduced sensitivity of δ^{11}B of borate at lower pH (the flattening of the δ^{11}B of borate curve in Fig. 5.2b). Thus although the size of the pH offset resulting from a constant photosynthetic DIC-drawdown may change at different bulk solution pH, the offset in δ^{11}B remains relatively constant. This is illustrated by the schematic in Fig. 5.4a: the photosynthesis DIC-drawdown vectors cross 0.1 pH units at high pH and 0.2 at low pH, but in both cases cross $\sim$2 permil in δ^{11}B of borate. This is also shown with simple steady state calculations in Fig. 5.4b, c: the offset between δ^{11}B of borate in the bulk solution and the green dashed lines (depicting a constant DIC-drawdown or addition) stays relatively constant with solution pH, as shown more comprehensively in Zeebe et al. (2003)'s diffusion-reaction model.

Variable δ^{11}B offsets may instead be explained by changes in physiology as a function of pH. Photosynthesis and respiration rates are not thought to be substantially impacted by changes in the carbonate system of this magnitude (Glas et al. 2012a). (Though a potential exception to this is noted by Henehan et al. (2013), who observe that cultured *G. ruber* holds its symbionts closer to its test under low pH conditions, which may increase the intensity of photosynthetic elevation of pH and δ^{11}B at low pH.) Calcification rate, on the other hand, is expected to vary substantially as a function of changes in carbonate saturation (Ω) over the range of pH values used in culture studies (e.g. Bijma et al. 2002). For instance calcite saturation state varies from $\sim$1.5 to 20 over the pH_{tot} range of 7.6–8.9 used in Sanyal et al. (1996); a generalized calcite precipitation rate law of $R = 2(\Omega - 1)^{1.9}$, gives a factor of 1000 change in precipitation rate over this range. Carbonate precipitation draws down local pH by removing 2 mol of ALK for every 1 of DIC (Fig. 5.4a). Rapid $CaCO_3$ precipitation at high saturation states may thus drive large decreases in microenvironment pH (Glas et al. 2012b; Toyofuku et al. 2017), compared to slower $CaCO_3$ precipitation at lower saturation states giving smaller local pH decrease (Fig. 5.4a). This may act to decrease δ^{11}B values in foraminiferal carbonate in cultures at high pH and saturation state, and thus give δ^{11}B in cultured foraminifera a relatively "shallow slope" or reduced sensitivity to δ^{11}B of borate as a function of pH.

The potential influence of variable calcification rates on foraminiferal microenvironment is illustrated with the simple steady state calculation in Fig. 5.4b, c. This calculation alters the carbonate system for a modelled microenvironment compared to the bulk solution by ALK and DIC drawdown in a 2:1 ratio as a function of saturation state, along with a constant net DIC-drawdown due to net respiration. This equilibrium treatment is a (gross!) simplification of variable rates of carbonate precipitation, and it should also be noted that foraminifera may not follow this precipitation rate law (see Allen et al. 2016, though note that crystal growth rate may vary substantially from mass added in culture). However this calculation does illustrate the potential for the extremely large range of saturation states experienced in culture to cause reduced sensitivity of foraminiferal δ^{11}B to pH ("shallow slopes"), as observed in culture (Fig. 5.4b, c).

The reduced sensitivity of planktic foraminiferal δ^{11}B to pH displayed in culture is not readily observed in wild samples of planktic or benthic foraminifera (Figs. 5.3 and 5.4b, c), although it should be noted that few wild calibration studies span a large-enough pH range to precisely constrain this sensitivity. Nonetheless, the recent wild *O. universa* data of Henehan et al. (2016) follow the pH sensitivity of $\delta^{11}B$ of borate more closely than previous *O. universa* cultures (Sanyal et al. 1996). This might be explained by the different character of carbonate system changes under culture versus natural conditions. While many culture experiments make large alkalinity changes at constant DIC, resulting in an extremely large range of carbonate saturation states ($\Omega_{calcite}$ of ~1.5–20), in natural conditions alkalinity and DIC changes are often coupled, and changes in saturation state are muted by carbonate compensation. Furthermore on long timescales, changes in ocean carbonate saturation are buffered compared to changes in pH (Tyrrell and Zeebe 2004; Hönisch et al. 2012). Therefore if the large range of carbonate saturation is the chief cause of reduced sensitivity of cultured $\delta^{11}B$ to pH, then calibration data obtained from wild samples may be more applicable to the geological record. Further work is required to test this idea and to further improve calibrations of foraminiferal $\delta^{11}B$ to pH under both culture and wild conditions.

5.3.2.3 Boron Isotope Constraints on Biomineralisation

The sensitivity of boron isotopes to pH makes them a powerful constraint on mechanisms of calcification in foraminifera. Here I briefly discuss the influence of some potential calcification processes on boron isotopes in foraminifera, and speculate on models that may match the constraints from $\delta^{11}B$. Note that as the timescale of boron isotope equilibration is extremely fast (~125 μs; Zeebe et al. 2001), kinetic processes are unlikely to exert a primary influence, simplifying interpretations compared to $\delta^{18}O$ and $\delta^{13}C$ (Zeebe and Wolf-Gladow 2001).

A common feature of many models of calcification is the elevation of pH in a somewhat isolated volume of seawater (de Nooijer et al. 2014). pH elevation of ~0.5–1 units above ambient seawater has indeed been observed during calcification in foraminifera (de Nooijer et al. 2009; Bentov et al. 2009) and corals (Venn et al. 2011) using fluorescent dyes. However while high $\delta^{11}B$ in corals appears to reflect this pH elevation (see Chap. 6 and McCulloch et al. 2012a; Krief et al. 2010; Anagnostou et al. 2012; Allison et al. 2014), $\delta^{11}B$ in most foraminifera does not: pH elevation of 1 pH unit should give $\delta^{11}B$ ~7–14‰ above seawater borate (Fig. 5.5), but most foraminifera lie within a few permil of borate ion at seawater pH (Figs. 5.3 and 5.5).

The impact of pH elevation on foraminiferal $\delta^{11}B$ may be even greater if boric acid is able to diffuse from seawater to the site of pH elevation, where it would re-equilibrate and elevate the total (or "seawater") $\delta^{11}B$ of the calcifying solution above 39.61‰. Rayleigh fractionation also has the potential to elevate $\delta^{11}B$ in an isolated volume of seawater, due to progressive drawdown of borate into carbonate, but the low partitioning of boron into calcite negates this effect (Rae et al. 2011; Stewart et al. 2016).

To date, substantial influence of pH elevation on foraminiferal $\delta^{11}B$ has only been suggested from one study, based on SIMS analyses of cultured *Amphistegina lobifera* (Rollion-Bard and Erez 2010). In these specimens the lowest $\delta^{11}B$ values from individual SIMS spots fall close to $\delta^{11}B$ of borate at ambient culture water pH, while the highest $\delta^{11}B$ values reach an upper limit of ~30‰, equivalent to a pH of ~8.9. Notably, this is similar to pK_2, the second equivalence point of carbonic acid, above which the overwhelming majority of DIC is present as CO_3^{2-}. pH elevation above this point would thus give no further increase in carbonate saturation and may also impact carbon concentration mechanisms. Whatever the causal mechanism, these data thus appear to suggest well-regulated pH elevation up to a chemically-determined limit (see also Gagnon et al. 2013 and Toyofuku et al. 2017), with calcification at pHs between this limit and ambient seawater (Rollion-Bard and Erez 2010; and see green lines in Fig. 5.5a). However in a separate study using LA-MC-ICPMS, Kaczmarek et al. (2015) find large negative $\delta^{11}B$ offsets in cultured *Amphistegina*, which are hard to reconcile with pH elevation (though one possible explanation is discussed below). The results from these high Mg foraminifera thus remain somewhat enigmatic.

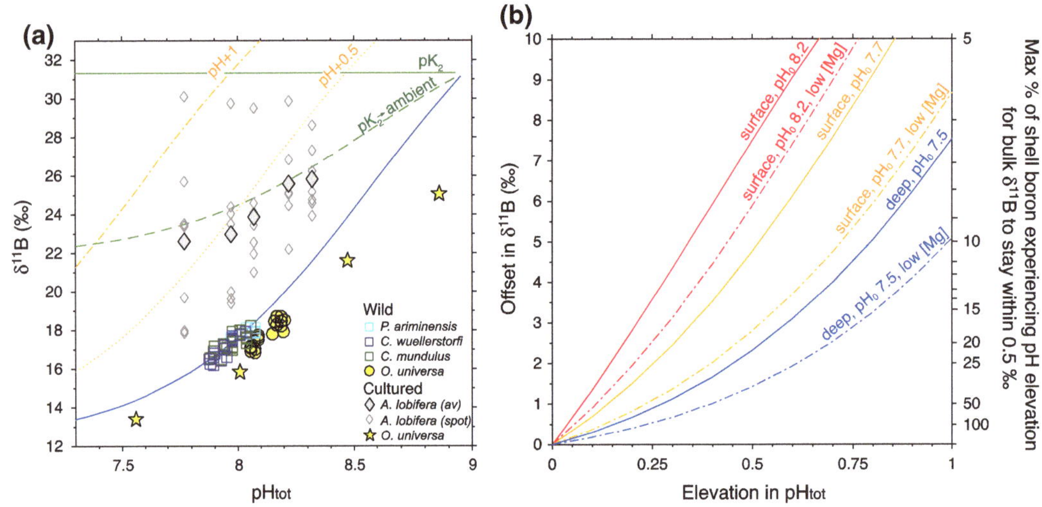

Fig. 5.5 **a** Boron isotope composition of borate ion under different pH elevation scenarios. The *blue curve* shows ambient water conditions. The *orange dotted* and *dot-dash lines* show pH elevation of 0.5 and 1.0. The *green solid line* shows pH elevation to the equivalence point of HCO_3^- and CO_3^{2-} (pK_2), and the *green dashed line* shows the mean of conditions at pK_2 and in ambient water. This illustrates a scenario where pH elevation initiates calcification, and pH then falls as calcification takes place, giving $\delta^{11}B$ offsets that fall between elevated and ambient conditions. Foraminiferal boron isotope data are also shown, including wild epifaunal benthic species (Rae et al. 2011; *open squares*), wild *O. universa* (Henehan et al. 2016; *yellow circles*), cultured *O. universa* (Sanyal et al. 1996), and cultured *A. lobifera* (Rollion-Bard and Erez 2010; open diamonds showing individual SIMS spots on a single foraminifera and filled diamonds their mean). All scenarios are calculated for surface water conditions at 25 °C, 35 psu and 0 m water depth, and foraminiferal data have been adjusted to these conditions (as described in Fig. 5.5 3 and Sect. 5.3). **b** Offset between boron isotope composition of borate under different pH elevation scenarios compared to ambient seawater conditions. *Red* and *Orange lines* show surface conditions (25 °C, 35 psu and 0 m water depth), with initial ambient water pH (pH_0) of 8.2 and 7.7. *Blue lines* show deep water conditions (0 °C, 35 psu and 2000 m water depth) with pH_0 of 7.5. *Dot-dash lines* show conditions with magnesium lowered to 3 mmol/kg (Hain et al. 2015). This is similar to the magnesium concentration predicted for the site of calcification in low-Mg foraminifera using inorganic partition coefficients. (In contrast, calcium concentration has relatively little influence on these relationships; Fig. 5.5 1d). Right hand axis shows the maximum proportion of boron that could have been incorporated into a foraminiferal shell under elevated pH, while still allowing bulk shell $\delta^{11}B$ to fall within 0.5‰ of borate at ambient seawater pH, as observed in symbiont-barren epifaunal benthic taxa

As the low-Mg benthic and planktic foraminifera have $\delta^{11}B$ values close to $\delta^{11}B$ of borate at seawater pH, the majority of the boron incorporated into these genera seems unlikely to have been derived directly from a seawater pool that experienced large pH elevation (Fig. 5.5). However this must be reconciled with the observations that at least some seawater *is* present during chamber formation (de Nooijer et al. 2009; Bentov et al. 2009; Nehrke et al. 2013) and that pH elevation *does* occur in a range of genera (de Nooijer et al. 2009; Bentov et al. 2009). Possible mechanisms that may satisfy these apparently contradictory observations include:

(1) borate that reaches the calcifying space being (selectively) removed from seawater at seawater pH;
(2) notable pH elevation occurring only during initial chamber formation, while the bulk of the shell is precipitated at a pH close to seawater, perhaps aided by Mg-removal;
(3) initial pH elevation being countered by carbon concentration and/or carbonate precipitation;
(4) pH elevation being less pronounced in the wild than during culture observations.

If borate ion is selectively transported from seawater to the calcifying space, without input of bulk seawater or boric acid, then pH could vary without impacting shell $\delta^{11}B$, provided that all of the borate in this pool is incorporated (Rae et al. 2011). Selective transport could potentially be achieved by trans-membrane ion channels, which have been proposed by some authors to deliver significant portions of calcium (Nehrke et al. 2013; de Nooijer et al. 2009; Toyofuku et al. 2008), and potentially also DIC (de Nooijer et al. 2014), to the calcifying space. It has been suggested that Ca^{2+} channels might be "leaky" to Mg^{2+}, given its similar size and charge (Raitzsch et al. 2010; de Nooijer et al. 2009); likewise it is possible that HCO_3^- transporters could be leaky to $B(OH)_4^-$. However the presence of membrane-impermeable dyes and beads at the site of calcification (Bentov and Erez 2006; Nehrke et al. 2013; de Nooijer et al. 2009) and in calcite itself (Bentov and Erez 2006) shows that bulk seawater—and not just channel-derived ions—must reach the site of calcification. Given the high concentration of boron in seawater this would likely dominate the total boron in the calcification space, so pH elevation would elevate foraminiferal $\delta^{11}B$. Selective transport of borate is thus unlikely to offer a silver bullet to explain the match between seawater borate and foraminiferal $\delta^{11}B$. However if borate transport does occur, it could potentially help counter elevation of $\delta^{11}B$ due to pH elevation, by reducing $\delta^{11}B$ of bulk boron in the calcification space.

Alternatively the lack of an obvious signal of pH elevation in low-Mg foraminifera could be explained if pH elevation only occurs during initial chamber formation, with the majority of calcification taking place at pH closer to seawater. Bulk shell $\delta^{11}B$ could be kept within 0.5‰ of seawater borate if pH elevation of 0.5 units is only experienced by ~5–25% of the boron incorporated into foraminiferal calcite (depending on environmental conditions; Fig. 5.5b). The rest of the boron could be sourced from borate ion adsorbed and incorporated during thickening of calcite onto existing chambers, if this process takes place at pH close to ambient seawater. The challenge to such a scenario is finding mechanisms to drive calcification under these conditions.

Calcification at pH close to ambient may perhaps be facilitated by active magnesium removal (Bentov and Erez 2006). Mg-removal acts to increase the activity of carbonate ions and reduce Mg "poisoning" of calcite growth (Zeebe and Sanyal 2002), and has been suggested to occur via mitochondrial activity in pseudopods surrounding the shell (Spero et al. 2015). Mg-removal has particular relevance to boron isotopes because reduced Mg concentrations notably increase K_B (Fig. 5.1d), reducing the sensitivity of $\delta^{11}B$ to changes in pH (Fig. 5.5b). For instance reducing Mg concentrations at the site of calcification to ~3 mol/kg (as predicted using inorganic partition coefficients and Mg concentrations in low-Mg foraminifera) may reduce the impact of 0.5 units pH elevation on $\delta^{11}B$ from ~5 to ~3‰ depending on initial conditions (Fig. 5.5b). Mg-removal would also reduce the impact of local pH decrease during calcification (see below). Different degrees of Mg removal from the calcifying space may also partially explain the contrasting $\delta^{11}B$ offsets between taxa with different Mg/Ca (Fig. 5.5): extensive Mg-removal in low-Mg benthic and planktic species would minimise the impact of local pH changes on $\delta^{11}B$; in contrast minimal Mg-removal in high-Mg taxa such as *Amphistegina* renders their boron isotope composition highly sensitive to pH changes during biomineralisation.

The influence of pH elevation during initial chamber formation on $\delta^{11}B$ may also be countered by local reductions in pH due to DIC concentration or during calcification. Bentov et al. (2009) demonstrate the presence of low-pH vacuoles during calcification, which they suggest may drive carbon concentration via CO_2 diffusion to low-CO_2/high-pH seawater vacuoles (see Allison et al. 2014 for analogous process in corals). Indeed were CO_2 diffusion infinitely efficient, DIC addition would keep pace with pH elevation, resulting in minimal net elevation in pH or $\delta^{11}B$. The fact that high pH vacuoles are observed demonstrates that DIC addition is not as efficient as in this end member scenario.

Nonetheless, concentration of CO_2 seems likely to play a role in minimizing offsets in $\delta^{11}B$.

Acidification of the calcifying microenvironment may also occur due to net alkalinity drawdown during calcification (Fig. 5.4a), and indeed could act as a negative feedback on calcification. To combat this, foraminifera have been shown to actively extrude H + from the calcifying space into surrounding seawater (Glas et al. 2012b), using a V-ATPase "proton-pump" (Toyofuku et al. 2017). Nonetheless, depending on the efficiency of this pump, some degree of acidification at the site of calcification may still occur. Acidification of the seawater surrounding the foraminifera is also likely to drive further CO_2 diffusion to the site of calcification, increasing DIC and countering some of the initial pH elevation (Toyofuku et al. 2017). Furthermore, if this low-pH water is vacuolised and transported to the site of calcification, it could not only counter the influence of initial pH elevation, but may even produce negative offsets between foraminiferal $\delta^{11}B$ and seawater borate (Fig. 5.4). The magnitude of these influences will be highly dependent on calcification rate and diffusion in the shell's microenvironment, which may differ between settings, and could perhaps provide a mechanism to explain differences in *Amphistegina* $\delta^{11}B$ between studies (Rollion-Bard and Erez 2010; Kaczmarek et al. 2015).

Finally it is possible that active pH manipulation may be most pronounced under the warm and energy-rich conditions in lab cultures, and less pronounced in the wild, in particular in cold deep-sea environments with low food. The reduced sensitivity of $\delta^{11}B$ to pH in cold, low-pH environments (Fig. 5.5b) may also contribute to the absence of a pH elevation signal in wild *Cibicidoides* from the deep ocean, despite the fact that pH elevation has been observed in this genera under culture conditions (de Nooijer et al. 2009). Higher turbulence in the wild may also help minimize pH gradients in the microenvironments surrounding calcifying foraminifera (Glas et al. 2012b; Toyofuku et al. 2017).

In summary, the close correspondence between $\delta^{11}B$ of foraminifera and seawater borate provides useful constraints on pH during calcification. Trans-membrane transport of borate ion is unlikely to explain this signal (in contrast to the suggestion of Rae et al. 2011), as it will be overwhelmed by boron from the seawater that is present at the site of calcification. DIC concentration and acidification during calcification could in part counter pH elevation; however at least some of this acidification is a result of actively extruded protons, which will maintain elevated pH at the site of chamber formation. Furthermore although scenarios that balance extremely high and low pH influences are possible, it seems unlikely that they could consistently produce small $\delta^{11}B$ offsets if they account for the majority of the shell's boron. One possible scenario is that the majority of the shell's boron is incorporated at pH close to seawater during chamber thickening, with pH elevation only employed during initial chamber formation and partially offset by concentration of DIC. Magnesium removal may help drive this secondary thickening, and would also reduce the sensitivity of $\delta^{11}B$ to further change in pH. Indeed variable Mg-removal, along with environmental influences on pH-$\delta^{11}B$ sensitivity, foraminiferal energy budgets, and variable efficiency of DIC concentration, may contribute to variable $\delta^{11}B$ offsets between taxa. Further observations are required to test this model, including better estimates of the mass balance of calcite and boron from primary chamber formation versus secondary thickening (Allen et al. 2011; Raitzsch et al. 2011). Nonetheless, it is interesting to consider that pH elevation may only be used to promote rapid initial chamber formation, with foraminifera expending energy to quickly overcome the thermodynamic barriers and physiological vulnerabilities of this process, whereas the bulk of the shell—and its boron—may be precipitated onto an existing calcite template at pH closer to seawater.

5.4 $\delta^{11}B$-Derived pH and CO_2

Interest in the boron isotope composition of foraminifera stems largely from its application as a tracer of the CO_2 system in the geological past (Fig. 5.6). Below I describe the processes and assumptions that allow foraminiferal $\delta^{11}B$ to be converted to pH and CO_2. However it should be noted that even without further conversion, $\delta^{11}B$ of borate, as tracked by foraminiferal $\delta^{11}B$, provides a powerful tracer of carbonate system changes in its own right. Indeed, over much of the ALK-DIC space in the modern ocean, $\delta^{11}B$ of borate is a more linear tracer of ALK/DIC ratio—and $[CO_3^{2-}]$ and pCO_2—than is pH (see inset figure in Box 1). Thus provided timescales are short enough that $\delta^{11}B_{sw}$ can be assumed to have remained relatively constant (<10 Myr; see below), foraminiferal $\delta^{11}B$ alone provides a valuable tracer of relative changes in past CO_2 system conditions, and plotting raw $\delta^{11}B$ data is a valuable first step in any record. Indeed this is somewhat analogous to foraminiferal $\delta^{18}O$, which reflects both temperature and $\delta^{18}O_{sw}$, but is typically provided with no further conversion, and provides a valuable tracer of past oceanographic conditions in this raw form. Furthermore, it should be noted that despite uncertainties in absolute pH and CO_2 values arising from estimates of $\delta^{11}B_{sw}$ and a second carbonate system parameter, relative changes in pH and CO_2 may still be well constrained. Indeed Foster and Rae (2016) demonstrate that estimates of climate sensitivity in the past, based on comparison of $\delta^{11}B$-derived CO_2 to sea surface temperature, are relatively insensitive to uncertainties in $\delta^{11}B_{sw}$ and estimates of alkalinity.

5.4.1 pH from $\delta^{11}B$

5.4.1.1 $\delta^{11}B$ of Borate and pH

The first step in obtaining pH from foraminiferal $\delta^{11}B$ is conversion to $\delta^{11}B$ of borate. This is achieved using the calibrations described above and shown in Fig. 5.3 (see also Foster and Rae 2016). Ideally the calibration used would be based on foraminifera collected under wild conditions, as these should most accurately reflect the combination of influences felt by open ocean foraminifera taken from samples "down-core". However in many cases these wild calibrations do not span a wide range of pH and so culture calibrations are also commonly applied. As discussed above, some culture calibrations have lower sensitivity to $\delta^{11}B$ borate changes than wild calibrations, and more work is required to determine which is most appropriate for the geological record. Careful micropalaeontoloty is also required to assess the most appropriate calibrations to use for extinct species (Anagnostou et al. 2016).

With $\delta^{11}B$ of borate determined, conversion to pH is typically achieved using Eq. (5.9), which is based on the approximation provided by Eq. (5.6).

$$pH = pK_B - \log\left(-\frac{\delta^{11}B_{SW} - \delta^{11}B_{B(OH)_4^-}}{\delta^{11}B_{SW} - \alpha_B.\delta^{11}B_{B(OH)_4^-} - \varepsilon_B}\right) \quad (5.9)$$

More accurately, using Eqs. (5.2) and (5.7), we obtain the expression in Eq. (5.10) below. Conversion of $\delta^{11}B$ borate to pH thus requires knowledge of K_B and $\delta^{11}B_{sw}$.

$$pH = -\log_{10}\frac{K_B\cdot{}^{\frac{11}{10}}R_{B(OH)_4^-} - K_B\cdot{}^{\frac{11}{10}}R_{SW} + \alpha_B\cdot K_B\cdot{}^{\frac{11}{10}}R^2_{B(OH)_4^-} - \alpha_B\cdot K_B\cdot{}^{\frac{11}{10}}R_{SW}\cdot{}^{\frac{11}{10}}R_{B(OH)_4^-}}{{}^{\frac{11}{10}}R_{SW} + {}^{\frac{11}{10}}R_{SW}\cdot{}^{\frac{11}{10}}R_{B(OH)_4^-} - \alpha_B\cdot{}^{\frac{11}{10}}R_{B(OH)_4^-} - \alpha_B\cdot{}^{\frac{11}{10}}R_{B(OH)_4^-}} \quad (5.10)$$

5.4.1.2 K_B

The equilibrium constant of boric acid, K_B, is a well-defined function of temperature, salinity, and pressure, given by the expression (Dickson 1990):

$$\begin{aligned} K_B = \exp[((&-8966.9 - 2890.53 * S^{1/2} \\ &- 77.942 * S + 1.728 * S^{3/2} - 0.0996 * S^2))/Tk \\ &+ 148.0248 + 137.1942 * S^{1/2} + 1.62142 * S \\ &+ (-24.4344 - 25.085 * S^{1/2} - 0.2474 * S) * \ln(Tk) \\ &+ 0.053105 * S^{1/2} * Tk] \end{aligned} \tag{5.11}$$

with the formula below giving the correction for different pressures:

$$\begin{aligned} P_{factor} = &(29.48 - 0.1622 * Tc \\ &+ 0.002608 * Tc^2)/(R^*Tk) * P\ldots \\ &+ 0.5 * (-0.00284)/(R^*Tk) * P^2 \end{aligned} \tag{5.12}$$

$$K_B = \exp\,(\ln K_B + P_{factor}) \tag{5.13}$$

where S is salinity, Tk is temperature in Kelvin, Tc is temperature in Celcius, P is pressure in bars (equivalent to depth in meters divided by 10), and R is the gas constant (83.14472 cm^3 bar K^{-1} mol^{-1}; Dickson et al. 2007). Note that the influence of these environmental parameters on the conversion of $\delta^{11}B$ of borate to pH is relatively minor over the conditions likely to be found at a given site: changes of ±5 °C, ±1 psu, and ±100 m result in differences in $\delta^{11}B$ borate of ~0.35‰, 0.05‰, and 0.03‰ respectively, for $\delta^{11}B$ borate of ~15‰ at 25 °C, 35 psu and 100 m water depth (Fig. 5.1c). Matlab codes and an Excel spreadsheet that carry out these calculations are provided in the supplementary online materials.

The composition of seawater can also influence K_B, in particular due to ion pairing of Ca and Mg ions with borate (Hershey et al. 1986; Nir et al. 2015; Hain et al. 2015). Hain et al. (2015)'s MyAMI ion pairing model shows that pK_B is relatively insensitive to the changes in [Ca] seen over the Cenozoic, though more sensitive to [Mg] (Fig. 5.1d). For instance $\delta^{11}B$ of borate of 16‰, at 25 °C, 35 psu, 0 m water depth, and with modern $\delta^{11}B_{sw}$, yields pH of 7.82 under modern [Ca] of 10 M and [Mg] of 52 M, 7.84 under Miocene [Ca] of 14 M and [Mg] of 48 M, and 7.94 under Eocene [Ca] of 20 M and [Mg] of 30 M (Horita et al. 2002; Hain et al. 2015). Note that equivalent calculations using the Pitzer ion pairing model available with PHREEQ give slightly larger changes in pK_B (Nir et al. 2015).

5.4.1.3 $\delta^{11}B$ of Seawater

The boron isotope composition of modern seawater is 39.61 ± 0.04‰ (Foster et al. 2010). Boron has a long residence time in seawater of ~10 Myr, and modelling suggests that typical rates of change for $\delta^{11}B_{sw}$ over the last ~100 Myr are of the order 0.1‰/Myr (Lemarchand et al. 2000). $\delta^{11}B_{sw}$ may thus be assumed to have remained within analytical error of modern values for the last ~3 Myr, and $\delta^{11}B$ records spanning a few million years should be dominantly a function of change in pH rather than $\delta^{11}B_{sw}$. However for accurate reconstruction of absolute pH values beyond ~3 Ma and for interpretation of $\delta^{11}B$ records spanning multiple millions of years, constraints on past $\delta^{11}B_{sw}$ are required (see Figs. 5.6 and 5.8).

Three general approaches have been used to reconstruct $\delta^{11}B_{sw}$:

(1) box modelling, using estimates of oceanic input and output fluxes over time;
(2) measurement of substances, such as pore waters and halites, with $\delta^{11}B$ close to that of ancient seawater;
(3) measurement of carbonate $\delta^{11}B$ where independent constraints can be placed on pH.

Box models of $\delta^{11}B_{sw}$ must take account of source terms, including weathering flux in rivers (Lemarchand et al. 2000), hydrothermal venting (Smith et al. 1995), and fluids from accretionary prisms (You et al. 1993), and sink terms, including oceanic crust alteration (Spivack and Edmond 1987), adsorption onto clays (Spivack et al. 1987), and incorporation in carbonates (Hemming and Hanson 1992). These terms are not well

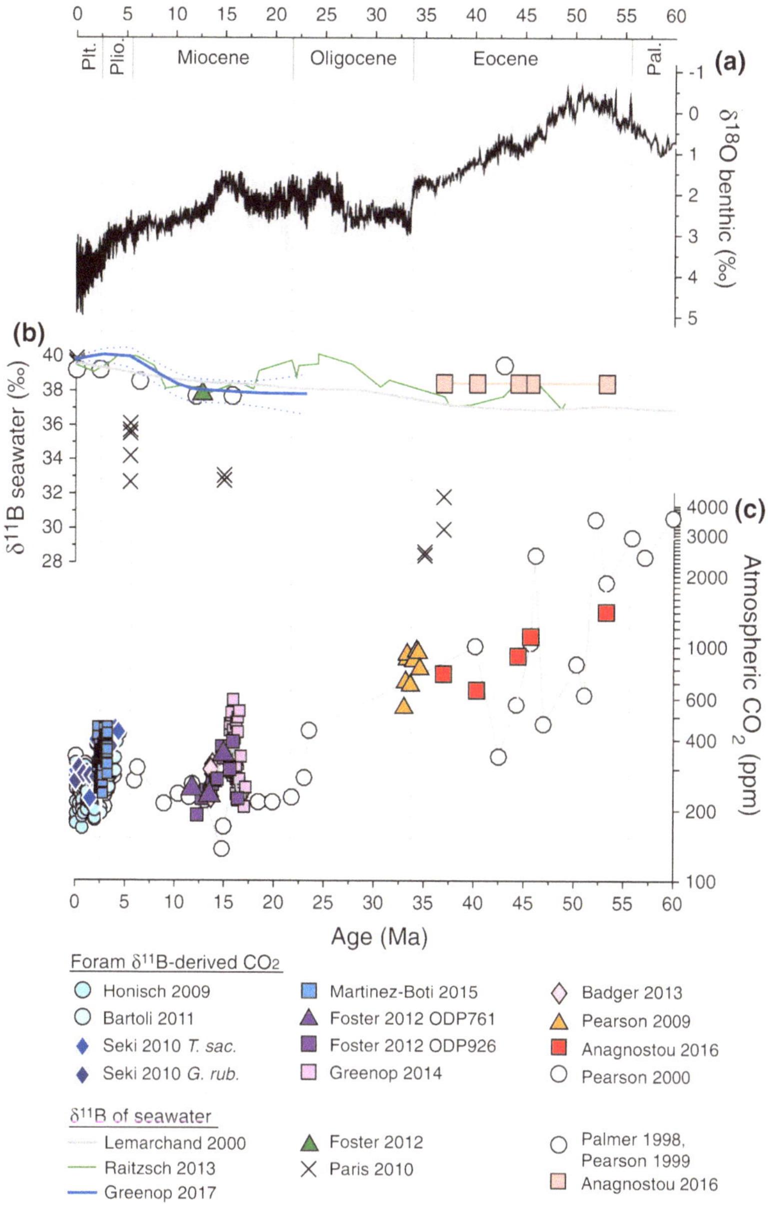

Fig. 5.6 Compilation of δ^{11}B-based CO_2 reconstructions and $\delta^{11}B_{SW}$ estimates for the Cenozoic. **a** Benthic δ^{18}O compilation (Zachos et al. 2001, 2008; *grey line*) with a 5 point moving average (*black line*). Heavier values indicate cooler temperatures and greater continental ice volume (note reverse scale). **b** Estimates of δ^{11}B of seawater. Halite data lighter than 28‰ are not plotted; note that Paris et al. (2010b) also suggest that several of the lightest halite values in the intervals shown may be excluded. The Lemarchand et al. (2000) model output is shown for the scenario of constant riverine input. The Greenop et al. (2017) curve is shown with its 2σ error envelope. The Raitzsch and Hönisch (2013) estimate is their scenario using alpha from Klochko et al. (2006) and assuming a linear change in pH over the Cenozoic. **c** pCO_2 reconstructions based on foraminiferal δ^{11}B. Where different estimates of $\delta^{11}B_{SW}$ or second carbonate system parameters were given, the assumptions favored in the original publication are shown. Note the long term decrease in atmospheric CO_2 as climate cools over the Cenozoic and the correspondence between more rapid atmospheric CO_2 change and major climate transitions

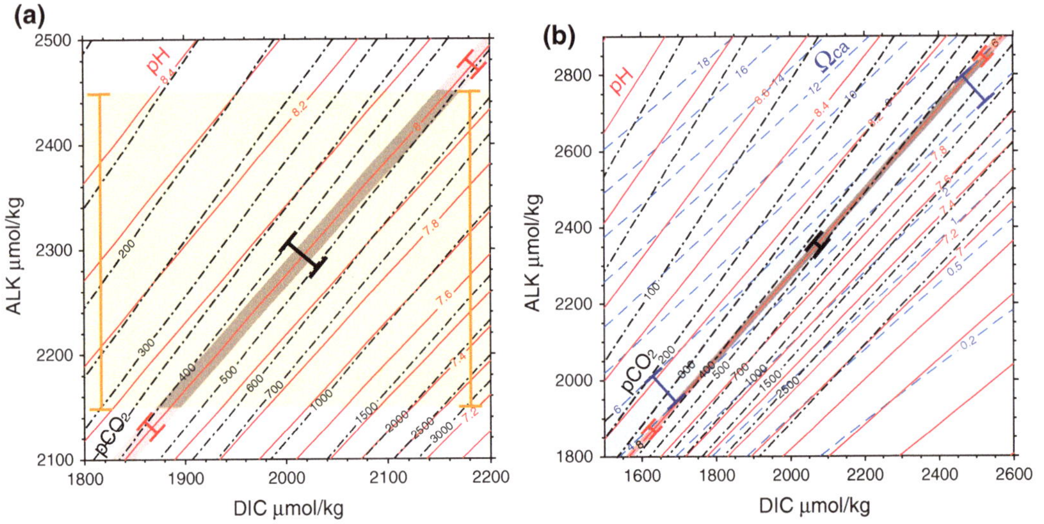

Fig. 5.7 Determination of pCO_2 from $\delta^{11}B$-derived pH, using an assumption of surface ocean alkalinity (**a**) or calcite saturation (**b**). The *red shaded* area indicates $\delta^{11}B$-derived pH of 8.0 ± 0.02; the *yellow area* in (**a**) shows alkalinity of 2300 ± 150 μmol/kg (close to the whole range in the modern surface ocean); the *light blue shaded* area in (**b**) shows $\Omega_{calcite}$ of 4–6 (thought to reflect the long-term range in the surface ocean). Overlap between the pH and second carbonate system parameter fields (*dark gray area*) constrains the range of pCO_2 values (*black error bar*). Note the close relationship between pH and pCO_2, which allows a precise determination of pH, combined with conservative estimates of alkalinity or $\Omega_{calcite}$, to still give relatively tight constraints on pCO_2. As contours of $\Omega_{calcite}$ are relatively closely aligned with pH, the use of $\Omega_{calcite}$-derived $[CO_3^{2-}]$ (**b**) provides a less precise constraint on pCO_2 than does alkalinity (**a**). All plots are given for 25 °C, 35 psu and 0 m water depth, with calculations carried out in CO2SYS.m (van Heuven et al. 2009). The scale of (**b**) is expanded compared to (**a**)

constrained over geological time, but may be proxied by more general geological processes, such as seafloor spreading rates (Simon et al. 2006) and total carbonate deposition (Lemarchand et al. 2000). Despite the uncertainties in how each of these fluxes has changed, this approach may be valuable when paired with constraints on the secular evolution of seawater chemistry from other isotopic systems (e.g. $^{87}Sr/^{86}Sr$, δ^7Li).

Measurement of archives with $\delta^{11}B$ close to ancient seawater is appealing, but it is hard to find settings where these signatures are pristine. Paris et al. (2010b) use $\delta^{11}B$ measurements on halites in evaporite deposits to reconstruct $\delta^{11}B_{sw}$ at 3 intervals over the last 40 Myr (Fig. 5.6b). This approach takes advantage of the fact that halites are one of the only solid materials with $\delta^{11}B$ similar to seawater rather than borate ion, likely due to the trapping of brine inclusions of fluid inclusions during halite growth. However $\delta^{11}B_{sw}$ determined by this method lies ~3–6‰ below other estimates, and it is possible that evaporite $\delta^{11}B$ may be influenced by fractionation during evaporation or incorporation into halite (Vengosh et al. 1992) or by riverine input in the partially isolated conditions of evaporite formation. Spivack et al (1993) used pore fluid $\delta^{11}B$ data from a carbonate-rich ODP core to estimate changes in $\delta^{11}B_{sw}$. However given the notable changes in $\delta^{11}B_{sw}$ seen in even the top few cm of pore fluids (Rae et al. 2011), use of this approach requires careful constraints on boron desorption from clays and dissolution of carbonates. Some deep pore fluids have also been shown to have quite exotic boron isotope compositions (Brumsack and Zuleger 1992), which further complicates this approach.

If pH can be independently constrained, then measurements of $\delta^{11}B$ in carbonates can be used to estimate $\delta^{11}B_{sw}$. Raitzsch and Hönisch (2013) assume a linear increase in deep ocean pH over the last 50 Myr, and combined this with $\delta^{11}B$ measurements of benthic foraminifera to estimate $\delta^{11}B_{sw}$ (Fig. 5.6). However, it is unlikely that pH change has been linear through time, as evidenced by the unrealistically rapid changes in $\delta^{11}B_{sw}$; in this record, given boron's ~10 Myr residence time. Furthermore, use of this $\delta^{11}B_{sw}$ record with benthic $\delta^{11}B$ data would only ever reproduce the original assumption of linear change in pH.

An alternative approach instead uses measurements of carbonate $\delta^{11}B$ across an assumed or estimated pH gradient. This takes advantage of the curved relationship between $\delta^{11}B$ of borate and pH, which means that for a given change in pH, a measured $\delta^{11}B$ difference will be a function of $\delta^{11}B_{sw}$. This approach was first proposed by Palmer et al. (1998), who assumed that the pH gradient between surface and thermocline-dwelling planktic foraminifera has remained constant, and made $\delta^{11}B$ measurements on a suite of these species over the Cenozoic. Pagani et al. (2005a) point out the influence of uncertainty in habitat depth on this calculation. However Anagnostou et al. (2016) have recently shown the power of this method to give bounds on possible changes in $\delta^{11}B_{sw}$, when supported by measurements of $\delta^{18}O$ and $\delta^{13}C$, and underpinned by conservative assumptions with fully propagated uncertainties. Foster et al. (2012) and Greenop et al. (2017) have also proposed a variant of this approach based on comparison of planktic and benthic $\delta^{11}B$ measurements, with the pH gradient between them constrained by measurements of $\delta^{13}C$.

Thus while accurate determination of $\delta^{11}B_{sw}$ remains a challenge for $\delta^{11}B$-based reconstructions of the carbonate system, progress is being made—and should continue—on a number of fronts.

5.4.2 CO_2 from pH

While records of pH may be used to study the history and impact of ocean acidification, much of the interest in foraminiferal boron isotope records stems from the ability of a paleo-pH meter to reconstruct CO_2. pH and aqueous CO_2 concentrations are closely coupled in seawater, both being governed by the primary carbonate system variables alkalinity and dissolved inorganic carbon (Box 1 and Fig. 5.7). From knowledge of aqueous [CO_2], along with estimates of temperature and salinity, Henry's law allows calculation of the partial pressure of atmospheric CO_2 in equilibrium with this water. Measurement of $\delta^{11}B$ in planktic foraminifera may thus be used to reconstruct atmospheric CO_2 in regions such as subtropical gyres (e.g. Figs. 5.6, 5.8 and 5.9), where oceanic and atmospheric CO_2 are close to equilibrium, or constrain the strength of CO_2 sources and sinks at high latitudes or in upwelling regions (Fig. 5.10).

The ocean carbonate system is most simply described by the 6 components alkalinity, DIC, CO_2, HCO_3^-, CO_3^{2-}, and pH, which are linked by 4 independent equations; the carbonate system thus has two degrees of freedom, so knowledge of two components allows the system to be completely determined (Box 1 and Fig. 5.7). The next most important acid-base system in seawater after CO_2 is boron, which adds 3 extra components $\left(B_{SW},\ B(OH)_3,\ \text{and}\ B(OH)_4^-\right)$ and 2 independent equations. Thus with knowledge of B_{SW}, the system continues to have 2 degrees of freedom. Complete determination of the carbonate system from $\delta^{11}B$-derived pH therefore requires a second carbonate system parameter to be known.

In any determination of the carbonate system it is crucially important to consider the relationships—and associated error propagation—between different carbonate system components (Fig. 5.7). For instance due to the sub-parallel relationship between pH and [CO_3^{2-}] in ALK-DIC space, the use of these parameters to determine a less closely-related component of the system (such as DIC) results in large propagated uncertainties. Rae et al. (2011) show that even with errors of only ~0.03 in pH from $\delta^{11}B$ and ~10 μmol/kg in [CO_3^{2-}] from benthic B/Ca, propagated uncertainty on DIC is ~300 μmol/kg, so although the DIC

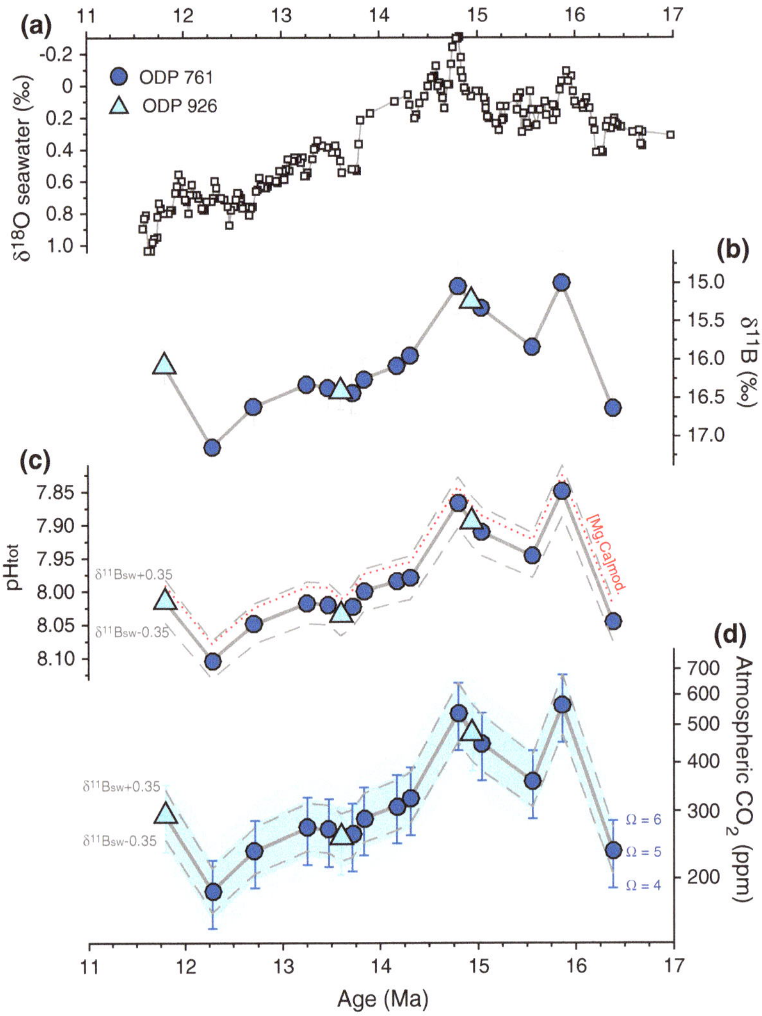

Fig. 5.8 Estimates of Miocene pH and atmospheric CO_2 from foraminiferal $\delta^{11}B$ for the Miocene, illustrating the influence of different $\delta^{11}B_{SW}$ and $\Omega_{calcite}$ values. **a** $\delta^{18}O_{SW}$ from benthic foraminiferal $\delta^{18}O$ and Mg/Ca (Lear et al. 2010), reflecting changes in global ice volume. **b** $\delta^{11}B$ of the planktic foraminifera *T. sacculifer* (formerly known as *G. sacculifer* and also known as *G. trilobus* in the Miocene; Spezzaferri et al. 2015) from ODP761 (*circles*) and ODP926 (*triangles*); (Foster et al. 2012). **c** pH calculated from $\delta^{11}B$ using the *T. sacculifer* calibration of Sanyal et al. (2001), as adjusted by Foster et al. (2012), $\delta^{11}B_{SW}$ of 37.82‰ (Foster et al. 2012), temperatures derived from Mg/Ca on these samples, and Miocene [Ca] = 14 mmol/kg and [Mg] of 48 mmol/kg (Horita et al. 2002; Hain et al. 2015). The influence of changing $\delta^{11}B_{SW}$ by ±0.35‰ (Foster et al. 2012; *grey dashed lines*) and of using modern [Ca] and [Mg] (*red dotted line*) is also shown. **d** Atmospheric CO_2 derived from the pH values in (**c**) and assuming $\Omega_{calcite} = 5 \pm 1$ (*error bars* on each point). The influence of changing $\delta^{11}B_{SW}$ by ±0.35‰ (Foster et al. 2012; *grey dashed lines*) is also shown. Confidence intervals on CO_2 of 68% and 95% (*blue shaded bands*) are derived from 1000 Monte Carlo simulations taking into account errors of ±0.2‰ on $\delta^{11}B$ measurements, ±1 °C on SST, ±1 psu on salinity, 2 mmol/kg on [Ca] and [Mg], and ±1 on $\Omega_{calcite}$. These errors are quoted at ~95% confidence and are assumed to have a normal distribution, except for $\Omega_{calcite}$ where a uniform distribution is used. Note that even without any calculation of pH or pCO_2, the close correspondence between $\delta^{11}B$ measurements and $\delta^{18}O_{SW}$ implies that CO_2 and climate are closely coupled during this interval. CO_2 calculations were carried out with a modified version of csys.m from Zeebe and Wolf-Gladow (2001). Code that reproduces this figure is available in the online materials accompanying this article

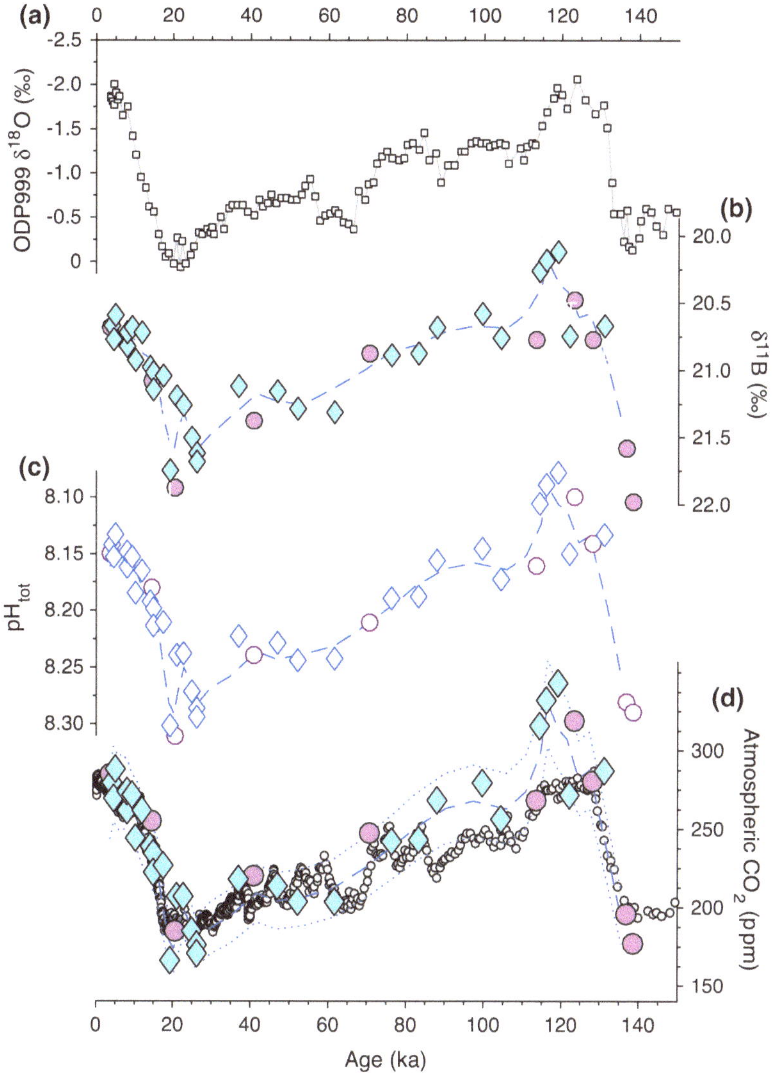

Fig. 5.9 Estimates of pH and atmospheric CO_2 for the last glacial cycle derived from foraminiferal boron isotopes. **a** $\delta^{18}O$ of *G. ruber* from ODP999 (Schmidt et al. 2004; Schmidt et al. 2006). **b** $\delta^{11}B$ of *G. ruber* from ODP999 (Foster 2008; *diamonds*) and $\delta^{11}B$ of *T. sacculifer* from ODP668 (Hönisch and Hemming 2005; *circles*). *T. sacculifer* values have been corrected to account for analytical offsets in early NTIMS measurements (see Hönisch et al. 2009), using the offset between *T. sacculifer* and *G. ruber* in these records in samples <5 kyr. The *blue dashed line* shows a 3-point moving average through the combined data. **c** $\delta^{11}B$-derived pH (*open symbols*), using the *G. ruber* calibration of Henehan et al. (2013), corrected for size fraction as in Martínez-Botí et al. (2015a), and the *T. sacculifer* calibration of Sanyal et al. (2001). **d** $\delta^{11}B$-dervied atmospheric CO_2 compared to the composite ice core record (Bereiter et al. 2015). *T. sacculifer*-derived estimates are as in the original publication (Hönisch and Hemming 2005). *G. ruber*-derived estimates are recalculated using modern alkalinity of 2330 μmol/kg and SST from Mg/Ca on the same samples (Foster 2008), and account for a modern offset of 21 ppm CO_2 between the surface ocean and atmosphere at this core site (Henehan et al. 2013). The dashed line shows a 3-point moving average through the combined data, with the *dotted lines* showing the influence of ±175 μmol/kg alkalinity (close to the whole range of the modern surface ocean)

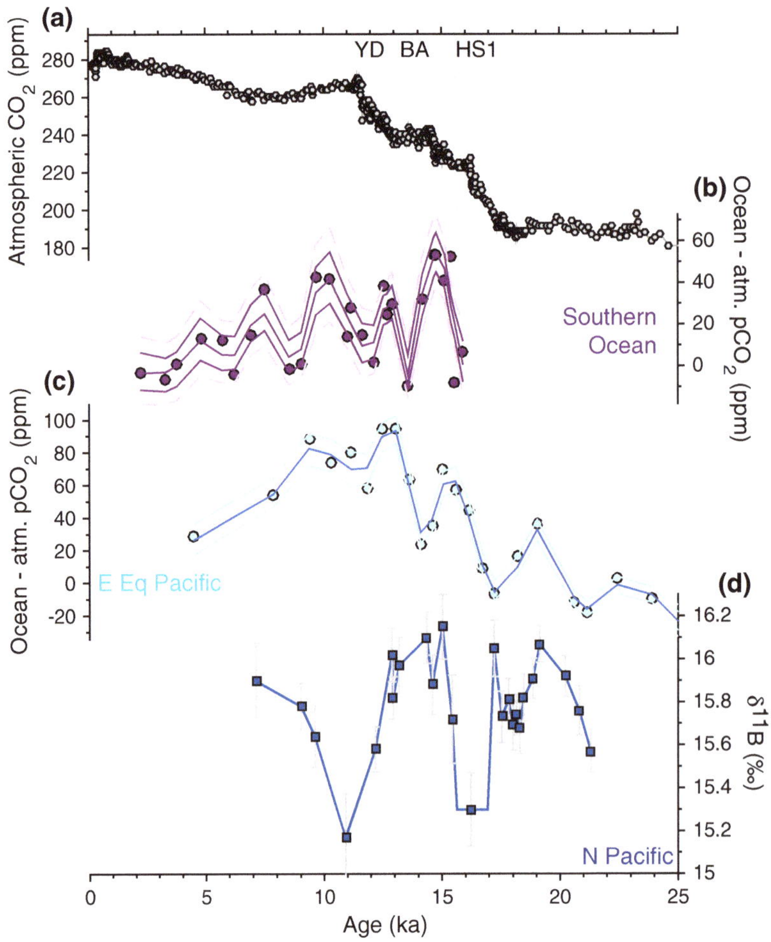

Fig. 5.10 Boron isotope records, that constrain CO_2 release from upwelling regions and the deep ocean over the last deglaciation. **a** Atmospheric CO_2 from Antarctic ice cores (Bereiter et al. 2015). **b** Ocean-atmosphere pCO_2 difference, derived from $\delta^{11}B$ on *G. bulloides* from Southern Ocean sediment core PS2498 and assuming modern alkalinity ±125 μmol/kg (Martínez-Botí et al. 2015b). *Lines* show the 68 and 95% confidence intervals calculated from 10,000 Monte Carlo simulations. **c** As in (**b**) but with $\delta^{11}B$ data from *T. sacculifer* from sediment core ODP1238 from the East Equatorial Pacific (Martínez-Botí et al. 2015b). **d** Boron isotope data from *C. wuellerstorfi* from the deep North Pacific, with error bars equivalent to 2SD on replicate measurements (Rae et al. 2014). Note the pulses of high surface water CO_2 and low deep water pH during intervals of atmospheric CO_2 rise

determination is accurate, the uncertainty spans a larger range than the whole modern ocean! However these carbonate system relationships also present opportunities, particularly in the determination of CO_2 from pH. As pH and CO_2 are very closely coupled, a well-constrained pH value will largely control the calculated CO_2 (Fig. 5.7). Therefore a reasonable estimate of a second carbonate system parameter, with generous error bars, should still result in a relatively accurate and precise determination of CO_2. Indeed it is far preferable to estimate a second carbonate system parameter with low precision that is likely to encompass the right value, than to use an estimate that is precise but inaccurate: provided the pH determination is accurate, the former approach will yield the correct CO_2 within error, while the latter will produce spurious results with the impression of high precision.

The two most commonly used second carbonate system parameters in $\delta^{11}B$-based CO_2 reconstructions are alkalinity and Ω-derived $[CO_3^{2-}]$. Alkalinity estimates in early studies (Hönisch and Hemming 2005; Foster 2008) were based on the close relationship between alkalinity and salinity in modern surface seawater: reconstructions of salinity, either from paired $\delta^{18}O$ and Mg/Ca measurements or from scaling with global sea level change, were thus used to estimate alkalinity. However alkalinity may vary independently of salinity due to changes in $CaCO_3$ cycling (Broecker and Peng 1987). Recent studies have thus used modern alkalinity but with a large range (e.g. ±175 μmol/kg; Martínez-Botí et al. 2015a) and a uniform rather than normal distribution. CO_2 is then determined via Monte-Carlo simulations, and the resulting CO_2 estimates include every possible alkalinity within the given range with equal likelihood.

On timescales beyond the last ~3 Myr the assumption that alkalinity has remained close to modern values is unlikely to hold. On these timescales several studies have used assumptions or models of calcite saturation state (Ω_{ca})

$$\Omega_{ca} = \frac{[Ca][CO_3^{2-}]}{K_{spc}} \tag{5.13}$$

to estimate $[CO_3^{2-}]$ (Pearson et al. 2009; Anagnostou et al. 2016). It is thought that carbonate compensation maintains the global mean surface Ω_{ca} in the range 4–6 on long timescales (Ridgwell 2005; Hönisch et al. 2012; Hain et al. 2015), so given estimates of [Ca] (Horita et al. 2002) and K_{sp} (from temperature and major ion composition; Hain et al. 2015), $[CO_3^{2-}]$ can be estimated.

The ultimate test of any proxy-based reconstruction of atmospheric CO_2 is comparison to the ice core CO_2 record. This is illustrated in Fig. 5.9, demonstrating that despite the various potential sources of uncertainty, foraminiferal boron isotopes are able to provide accurate CO_2 reconstruction, with fully propagated uncertainty of ~20 ppm (2SD). Although application further back in the geological may present other challenges, the close match to ice core CO_2 demonstrates the potential of foraminiferal $\delta^{11}B$ for pH and CO_2 reconstruction.

5.5 Proxy Application: Examples

Spurred on by the development work described above, application of boron isotopes in foraminifera to the geological record has grown rapidly in recent years. Here I give a brief overview of some highlights from this work.

5.5.1 Glacial-Interglacial CO_2

Over the glacial cycles of the last 800 kyr, atmospheric CO_2 levels are well known, but the cause of CO_2 change between glacial and interglacial periods remains a mystery (EPICA 2004; Kohfeld and Ridgwell 2009). All leading hypotheses invoke changes in CO_2 partitioning between the deep ocean and the atmosphere, likely mediated by changes in circulation and productivity at high latitudes where deep waters are ventilated (Sigman et al. 2010). Boron isotope records can be used to test these hypotheses, by constraining CO_2 (dis)equilibrium between the surface ocean and the atmosphere, and CO_2 storage and release in deep waters.

A notable first application of $\delta^{11}B$ over glacial cycles was provided by Sanyal et al. (1995), who used planktic foraminifera and mixed benthic foraminifera to suggest that the pH of the ocean was ~0.3 units higher during the last glacial maximum compared to the Holocene. This was taken as support for glacial CO_2 drawdown by elevated ocean alkalinity (e.g. Archer and Maier-Reimer 1994). However reconciling this with changes in $CaCO_3$ preservation in the deep ocean is problematic (Sigman et al. 1998), and more recent work from monospecific benthics suggests much smaller changes in pH in the deep glacial ocean (Hönisch et al. 2008; Yu et al. 2010; Rae et al. 2014).

Several studies have used $\delta^{11}B$ in planktic foraminifera to reconstruct the extent of disequilibrium (ΔpCO_2) between CO_2 in surface waters and the atmosphere (Sanyal et al. 1997; Sanyal and Bijma 1999; Palmer and Pearson 2003; Foster and Sexton 2014; Naik et al. 2015; Martínez-Botí et al. 2015b; see Fig. 5.10). Recent work shows a pronounced increase in CO_2 outgassing in the East Equatorial Atlantic during the LGM (Foster and Sexton 2014), indicative of enhanced upwelling, while surface water CO_2 in the West Equatorial Atlantic stays close to equilibrium with the atmosphere, suggesting that the upwelled CO_2 and nutrients from the East are progressively drawn-down by biological productivity. In the East (Martínez-Botí et al. 2015b) and West (Palmer and Pearson 2003) Equatorial Pacific, ΔpCO_2 is similar to or slightly lower than modern during the LGM, then shows a dramatic increase during the deglaciation, interpreted as an increase in the CO_2 content of upwelled waters. In the Indian Ocean, $\delta^{11}B$ records have been used to show variability in CO_2 outgassing related to different monsoon conditions (Naik et al. 2015; Palmer et al. 2010), with a peak in outgassing in the Bølling-Allerød interstadial, associated with an invigorated SW/Summer monsoon.

At high latitudes Yu et al. (2013) show that North Atlantic surface waters remained a sink of CO_2 from the atmosphere for the last 18 kyr. In contrast Southern Ocean surface waters saw pulses of high CO_2 during the last deglaciation (Martínez-Botí et al. 2015b), coincident with increases in atmospheric CO_2. This provides the first direct evidence for the role of the Southern Ocean in driving deglacial CO_2 rise, likely as a result of enhanced upwelling of CO_2-rich deep waters; the advection of these waters to the subsurface of the East Equatorial Pacific may also explain the pulses of CO_2 outgassing seen in this region (Martínez-Botí et al. 2015b; Fig. 5.10).

Boron isotope records from benthic foraminifera have also been used to examine the deep ocean carbonate system over the last deglaciation. These might be expected to show a pattern of increased deep ocean pH during intervals of CO_2 release from the deep ocean to the atmosphere. However Rae et al. (2014)'s data from the deep North Pacific show the opposite (Fig. 5.10d), with pulses of low pH during CO_2 rise in HS1 and the Younger Dryas. These are interpreted to result from mixing of CO_2-rich water from mid depths through the water column during pulses of local deep water formation (supported by comparison to radiocarbon and $\delta^{13}C$ data). This emphasizes the importance of considering changes in circulation, as well biogeochemistry, in interpreting benthic $\delta^{11}B$ records.

5.5.2 pH and CO_2 Beyond the Ice Cores

The reconstruction of atmospheric CO_2 change beyond the reach of the ice cores has been a major goal for studies using foraminiferal $\delta^{11}B$. Applications of such data include understanding the major climate transitions as Earth evolved from greenhouse to icehouse over the Cenozoic (Pearson et al. 2009; Bartoli et al. 2011; Fig. 5.6), to determining climate sensitivity in a warmer world (Anagnostou et al. 2016; Martínez-Botí et al. 2015a).

Early applications of foraminiferal $\delta^{11}B$ to reconstruct long term changes in surface ocean pH and atmospheric CO_2 indicated substantial decrease in CO_2 over the Cenozoic (Spivack et al. 1993; Pearson and Palmer 1999, 2000). However Pearson and Palmer (2000)'s iconic $\delta^{11}B$-CO_2 record also shows high amplitude spikes in the Palaeogene, and low and constant values in the Neogene (Fig. 5.6), leading to some suggestions that CO_2 and climate were decoupled (Shevenell et al. 2004; Pagani et al. 2005b; Mosbrugger et al. 2005).

More recent work shows high CO_2 values (~ 1500 ppm) during the early Eocene climatic optimum (EECO), which then decrease through the rest of the Eocene (Anagnostou et al. 2016). At higher resolution, Penman et al. (2014) use $\delta^{11}B$ to provide the first direct evidence of ocean acidification during the PETM, while Pearson et al. (2009) show decreasing CO_2 levels over the Eocene-Oligocene boundary, though with an intriguing re-bound following Antarctic ice growth in the early Oligocene. In the Miocene, Foster et al. (2012) show a close coupling between atmospheric CO_2 and global climate and sea level (Fig. 5.8), which is also observed at higher temporal resolution (Fig. 5.6; Greenop et al. 2014; Badger et al. 2013).

A strong link between CO_2 and climate has also been shown using foraminiferal $\delta^{11}B$ in the Pliocene and early Pleistocene (Fig. 5.6). Seki et al. (2010), Bartoli et al. (2011) and Martínez-Botí et al. (2015a) show that the Pliocene warm period had CO_2 levels of ~400 ppm, similar to modern day anthropocene values. Martínez-Botí et al. (2015a) further quantify the relationship between CO_2 and climate at high resolution over this interval, and show that Earth System climate sensitivity in this warmer climate was half that of the Pleistocene. This can be attributed to a reduced ice-albedo feedback in the Pliocene, with equilibrium climate sensitivity found to be the same in both intervals when ice volume changes are considered. Between the Pliocene and the Pleistocene CO_2 falls (Seki et al. 2010; Bartoli et al. 2011), and Honisch et al. (2009) show that further intensification of glaciation during the mid-Pleistocene transition was linked to lower CO_2 in glacial periods, while CO_2 in interglacials remained relatively similar (Fig. 5.6).

5.6 Summary and Outlook

Recent developments in boron isotope analyses have spurred on an exciting era in the development and application of the boron isotope pH proxy in foraminifera. Boron isotope analysis by column chromatography and MC-ICPMS now gives reproducibility of ~0.2‰, equivalent to ~0.02 pH units, on around 5–20 benthic or 50–200 planktic foraminifera. Calibration studies show that foraminiferal $\delta^{11}B$ lies close to that of borate ion at seawater pH, consistent with simple models of $\delta^{11}B$-pH systematics based on sole incorporation of the borate ion into foraminiferal carbonate. Offsets between foraminifera and borate ion are largely attributable to modification of the local microenvironment by photosynthesis, respiration, and calcification. pH elevation during biomineralisation might be expected to drive substantial elevation of foraminiferal $\delta^{11}B$ above seawater borate, but this is not observed in the majority of species. The majority of the boron incorporated into foraminiferal calcite appears to be derived from borate ion at or close to seawater pH, providing an important constraint on models of biomineralisation.

Given the close coupling between seawater pH and CO_2, foraminiferal $\delta^{11}B$ measurements may provide relatively precise constraints on past CO_2 change, even with conservative estimates of a second carbonate system parameter. Boron isotope-based CO_2 reconstructions show a close match to the ice core CO_2 record where these overlap. On glacial-interglacial timescales, $\delta^{11}B$ records from regions of CO_2 disequilibrium in the surface ocean and in the deep ocean provide evidence of deglacial CO_2 release from the Southern Ocean and North Pacific, and changes in upwelled CO_2 in the tropics. Beyond the reach of the ice cores, foraminiferal $\delta^{11}B$ records increasingly demonstrate coupling between CO_2 and climate over key intervals of the last 60 Myr.

Future work should continue to refine boron isotope calibration in foraminifera, capitalize on boron's constraints on biomineralisation, and further constrain the evolution of seawater $\delta^{11}B$ to improve long-term pH and CO_2 reconstructions.

Acknowledgements This chapter was greatly improved thanks to detailed and thoughtful reviews by Marcus Gutjahr and an anonymous reviewer, editorial comments by Gavin Foster, and informal reviews by David Evans and Rosanna Greenop. Other members of the STAiG lab boron group, including Will Gray, Ben Taylor, Jessica Crumpton-Banks, Eloise Littley, and Matt Kaminski, along with the "B-team" at Southampton, are thanked for their hard work developing boron isotopes in foraminifera. Discussions over the last ten years with Michael Henehan, Gavin Foster, Richard Zeebe, Tim Elliott, Wally Broecker, Andy Ridgwell, Daniela Schmidt, Bärbel Hönisch, Kat Allen, Jess Adkins, and Jonathan Erez have all greatly contributed to my thinking about boron isotopes in foraminifera and CO_2 reconstruction; though any omissions or mistakes are mine alone. This work was partially supported by a research fellowship from the University of St Andrews, $100 (pending) from Richard Zeebe, and NERC grants NE/N003861/1 and NE/N011716/1.

References

Al-Ammar AS, Gupta RK, Barnes RM (2000) Elimination of boron memory effect in inductively coupled plasma-mass spectrometry by ammonia gas injection into the spray chamber during analysis. Spectrochim Acta Part B 55(6):629–635

Allen KA et al (2011) Controls on boron incorporation in cultured tests of the planktic foraminifer Orbulina universa. Earth Planet Sci Lett 309(3–4):291–301

Allen KA et al (2016) Trace element proxies for surface ocean conditions: a synthesis of culture calibrations with planktic foraminifera. Geochem Cosmochim Acta 193:197–221

Allison N, Finch AA (2010) $\delta^{11}B$, Sr, Mg and B in a modern porites coral: the relationship between calcification site pH and skeletal chemistry. Geochim Cosmochim Acta 74(6):1790–1800

Allison N et al (2014) Corals concentrate dissolved inorganic carbon to facilitate calcification. Nat Commun 5:1–6.

Anagnostou E et al (2012) Evaluation of boron isotope ratio as a pH proxy in the deep sea coral Desmophyllum dianthus: evidence of physiological pH adjustment. Earth Planet Sci Lett 349–350:251–260

Anagnostou E et al (2016) Changing atmospheric CO_2 concentration was the primary driver of early Cenozoic climate. Nature 533(7603):380–384

Archer D, Maier-Reimer E (1994) Effect of deep-sea sedimentary calcite preservation on atmospheric CO_2 concentration. Nature 367:260–263

Badger MPS et al (2013) CO_2 drawdown following the middle Miocene expansion of the Antarctic ice sheet. Paleoceanography 28(1):42–53

Balan E et al (2016) First-principles study of boron speciation in calcite and aragonite. Geochem Cosmochim Acta 193:119–131

Barker S, Greaves M, Elderfield H (2003) A study of cleaning procedures used for foraminiferal Mg/Ca paleothermometry. Geochem Geophys Geosyst 4 (9):8407

Bartoli G, Hönisch B, Zeebe RE (2011) Atmospheric CO_2 decline during the Pliocene intensification of Northern Hemisphere glaciations. Paleoceanography 26(4):n/a

Bentov S, Erez J (2006) Impact of biomineralization processes on the Mg content of foraminiferal shells: a biological perspective. Geochem Geophys Geosyst 7 (1):1–11

Bentov S, Brownlee C, Erez J (2009) The role of seawater endocytosis in the biomineralization process in calcareous foraminifera. PNAS 106(51):21500–21504

Bereiter B et al (2015) Revision of the EPICA Dome C CO_2 record from 800 to 600 kyr before present. Geophys Res Lett 42(2):542–549

Bijma J, Faber WW, Hemleben C (1990) Temperature and salinity limits for growth and survival of some planktonic foraminifers in laboratory cultures. J. Foraminiferal Res 20(2):95–116

Bijma J, Hönisch B, Zeebe RE (2002) Impact of the ocean carbonate chemistry on living foraminiferal shell weight: comment on "Carbonate ion concentration in glacial-age deep waters of the Caribbean Sea" by W.S. Broecker and E. Clark. Geochem Geophys Geosyst 3 (11):1–7

Boyle EA (1981) Cadmium, zinc, copper, and barium in foraminifera tests. Earth Planet Sci Lett 53(1):11–35

Branson O et al (2015) The coordination and distribution of B in foraminiferal calcite. Earth Planet Sci Lett 416 (C):67–72

Broecker WS (2005) The role of the ocean in climate yesterday, today and tomorrow. Eldigio Press

Broecker WS, Peng TH (1987) The role of $CaCO_3$ compensation in the glacial to interglacial atmospheric CO_2 change. Global Biogeochem Cycles 1:15–29

Brumsack HJ, Zuleger E (1992) Boron and boron isotopes in pore waters from ODP Leg 127, Sea of Japan. Earth Planet Sci Lett 113(3):427–433

Corliss BH (1985) Microhabitats of benthic foraminifera within deep-sea sediments. Nature 314(6010):435–438

de Nooijer LJ, Langer G et al (2009a) Physiological controls on seawater uptake and calcification in the benthic foraminifer Ammonia tepida. Biogeosciences 6:2669–2675

de Nooijer LJ, Toyofuku T, Kitazato H (2009b) Foraminifera promote calcification by elevating their intracellular pH. Proc Nat Acad Sci 106(36):15374–15379

de Nooijer LJ et al (2014) Biomineralization in perforate foraminifera. Earth Sci Rev 135:48–58

Deyhle A, Kopf A (2004) Possible influence of clay contamination on B isotope geochemistry of carbonaceous samples. Appl Geochem 19(5):737–745

Dickson AG (1990) Thermodynamics of the dissociation of boric acid in synthetic seawater from 273.15 to 318.15 K. Deep Sea Res Part A Oceanogr Res Pap 37 (5):755–766

Dickson AG, Sabine CL, Christian JR (eds) (2007) Guide to best practices for ocean CO_2 measurements. PICES special publication 3. North Pacific Marine Science Organization, Sidney, p 176

Edgar KM et al (2015) Assessing the impact of diagenesis on d11B, $\delta^{13}C$, $\delta^{18}O$, Sr/Ca and B/Ca values in fossil planktic foraminiferal calcite. Geochim Cosmochim Acta 166(C):189–209

EPICA (2004) Eight glacial cycles from an Antarctic ice core. Nature 429(6992):623–628

Erez J (2003) The source of ions for biomineralization in foraminifera and their implications for paleoceanographic proxies. Rev Min Geochem 54(1):115–149

Farmer J et al (2015) Outside the pH box: boron isotopes in synthetic calcite precipitated under varying solution chemistry. In: AGU fall meeting abstracts. AGU

Farmer JR, Hönisch B, Uchikawa J (2016) Single laboratory comparison of MC-ICP-MS and N-TIMS boron isotope analyses in marine carbonates. Chem Geol 447:173–182

Feldmeijer W et al (2013) The effect of chemical pretreatment of sediment upon foraminiferal-based proxies. Geochem Geophys Geosyst 14(10): 3996–4014

Fietzke J et al (2010) Boron isotope ratio determination in carbonates via LA-MC-ICP-MS using soda-lime glass standards as reference material. J Anal At Spectrom 25 (12):1953–1957

Fietzke J et al (2015) Century-scale trends and seasonality in pH and temperature for shallow zones of the Bering Sea. Proc Natl Acad Sci 112(10):2960–2965

Foster GL (2008) Seawater pH, pCO_2 and $[CO_3{=}]$ variations in the Caribbean Sea over the last 130 kyr: a boron isotope and B/Ca study of planktic foraminifera. Earth Planet Sci Lett 271(1–4):254–266

Foster GL, Rae JWB (2016) Reconstructing ocean pH with boron isotopes in foraminifera. Ann Rev Earth Planet Sci 44(1):207–237. doi:10.1146/annurev-earth-060115-012226

Foster GL, Sexton PF (2014) Enhanced carbon dioxide outgassing from the eastern equatorial Atlantic during the last glacial. Geology 42(11):1003–1006

Foster GL, Pogge von Strandmann PAE, Rae JWB (2010) Boron and magnesium isotopic composition of seawater. Geochem Geophys Geosyst 11(8):Q08015

Foster GL, Lear CH, Rae JWB (2012) The evolution of pCO_2, ice volume and climate during the middle Miocene. Earth Planet Sci Lett 341–344:243–254

Foster GL et al (2013) Interlaboratory comparison of boron isotope analyses of boric acid, seawater and marine $CaCO_3$ by MC-ICPMS and NTIMS. Chem Geol 358:1–14

Gaillardet J, Lemarchand D, Göpel C (2001) Evaporation and sublimation of boric acid: application for boron purification from organic rich solutions. Geostand Geoanal Res 25(1):67–75

Gagnon AC, Adkins JF, Erez J, Eiler JM (2013) Sr/Ca sensitivity to aragonite saturation state in cultured subsamples from a single colony of coral: mechanism of biomineralization during ocean acidification. Geochem Cosmochim Acta 105:240–254

Glas MS, Fabricius KE et al (2012a) The O_2, pH and Ca^{2+} microenvironment of benthic foraminifera in a high CO_2 world. PLoS ONE 7(11):e50010

Glas MS, Langer G, Keul N (2012b) Calcification acidifies the microenvironment of a benthic foraminifer (*Ammonia* sp.). J Exp Mar Biol Ecol 424–425:53–58

Greenop R et al (2014) Middle Miocene climate instability associated with high-amplitude CO_2 variability. Paleoceanography 29(9):845–853

Greenop R et al (2017) A record of Neogene seawater $\delta^{11}B$ reconstructed from paired $\delta^{11}B$ analyses on benthic and planktic foraminifera. Clim Past 13:149–170

Gutjahr M et al (2014) Boron isotope intercomparison project (BIIP): development of a new carbonate standard for stable isotopic analyses. In: EGU general assembly conference abstracts, vol 16

Hain MP et al (2015) The effects of secular calcium and magnesium concentration changes on the thermodynamics of seawater acid/base chemistry: implications for eocene and cretaceous ocean carbon chemistry and buffering. Global Biogeochem Cycles 29(5):517–533

Hemleben C, Bijma J (1994) Foraminiferal population dynamics and stable carbon isotopes. In: Carbon cycling in the glacial ocean: constraints on the ocean's role in global change. Springer, Berlin, Heidelberg, pp 145–166

Hemleben C, Spindler M, Anderson OR (2012) Modern planktonic foraminifera. Springer Science & Business Media, New York, NY

Hemming NG, Hanson GN (1992) Boron isotopic composition and concentration in modern marine carbonates. Geochim Cosmochim Acta 56:537–543

Hemming NG, Hanson GN (1994) A procedure for the isotopic analysis of boron by negative thermal ionization mass spectrometry. Chem Geol 114(1–2):147–156

Hemming NG, Reeder RJ, Hanson GN (1995) Mineral-fluid partitioning and isotopic fractionation of boron in synthetic calcium carbonate. Geochim Cosmochim Acta 59(2):371–379

Henehan MJ et al (2013) Calibration of the boron isotope proxy in the planktonic foraminifera Globigerinoides ruber for use in palaeo-CO_2 reconstruction. Earth Planet Sci Lett 364:111–122

Henehan MJ et al (2016) A new boron isotope-pH calibration for Orbulina universa, with implications for understanding and accounting for "vital effects". Earth Planet. Sci. Lett. 454(C):282–292

Hershey JP et al (1986) The ionization of boric acid in NaCl, NaCaCl and NaMgCl solutions at 25 °C. Geochim Cosmochim Acta 50(1):143–148

Hönisch B, Hemming NG (2004) Ground-truthing the boron isotope-paleo-pH proxy in planktonic foraminifera shells: partial dissolution and shell size effects. Paleoceanography 19:PA4010

Hönisch B, Hemming NG (2005) Surface ocean pH response to variations in pCO_2 through two full glacial cycles. Earth Planet Sci Lett 236(1–2):305–314

Hönisch B et al (2003) The influence of symbiont photosynthesis on the boron isotopic composition of foraminifera shells. Mar Micropaleontol 49(1–2):87–96

Hönisch B et al (2004) Assessing scleractinian corals as recorders for paleo-pH: empirical calibration and vital effects. Geochim Cosmochim Acta 68(18):3675–3685

Hönisch B, Bickert T, Hemming NG (2008) Modern and Pleistocene boron isotope composition of the benthic foraminifer Cibicidoides wuellerstorfi. Earth Planet Sci Lett 272(1–2):309–318

Hönisch B et al (2009) Atmospheric carbon dioxide concentration across the mid-pleistocene transition. Science 324(5934):1551–1554

Hönisch B et al (2012) The geological record of ocean acidification. Science 335(6072):1058–1063
Horita J, Zimmermann H, Holland HD (2002) Chemical evolution of seawater during the Phanerozoic: implications from the record of marine evaporites. Geochim Cosmochim Acta 66(21):3733–3756
Jorgensen BB et al (1985) Symbiotic photosynthesis in a planktonic foraminifera, globigerinoides sacculifer (Brady), studied with microelectrodes. Limnol Oceanogr 30:1253–1267
Kaczmarek K et al (2015) Boron incorporation in the foraminifer Amphistegina lessonii under a decoupled carbonate chemistry. Biogeosciences 12:1753–1763
Kaczmarek K et al (2016) Investigating the effects of growth rate and temperature on the B/Ca ratio and $\delta^{11}B$ during inorganic calcite formation. Chem Geol 421(C):81–92
Kakihana H, Kotaka M (1977) Equilibrium constants for boron isotope-exchange reactions. Bull Res Lab Nucl React 2:1–12
Kasemann SA et al (2009) In situ boron isotope analysis in marine carbonates and its application for foraminifera and palaeo-pH. Chem Geol 260(1–2):138–147
Katz ME et al (2010) Traditional and emerging geochemical proxies in foraminifera. J Foramin Res 40(2):165–192
Kiss E (1988) Ion-exchange separation and spectrophotometric determination of boron in geological materials. Anal Chim Acta 211:243–256
Klochko K et al (2006) Experimental measurement of boron isotope fractionation in seawater. Earth Planet Sci Lett 248(1–2):276–285
Klochko K et al (2009) Re-evaluating boron speciation in biogenic calcite and aragonite using ^{11}B MAS NMR. Geochim Cosmochim Acta 73(7):1890–1900
Kohfeld KE, Ridgwell A (2009) Glacial-interglacial variability in atmospheric CO_2. In: Saltzman E, Quere CL (eds) Surface ocean-lower atmospheres processes. Geophysical monograph series. AGU, Washington DC
Köhler-Rink S, Kühl M (2000) Microsensor studies of photosynthesis and respiration in larger symbiotic foraminifera. I The physico-chemical microenvironment of Marginopora vertebralis, Amphistegina lobifera and Amphisorus hemprichii. Mar Biol 137 (3):473–486
Kotaka M, Kakihana H (1977) Thermodynamic isotope effect of trigonal planar and tetrahedral species. Bull Res Lab Nucl React 2:13–29
Krief S et al (2010) Physiological and isotopic responses of scleractinian corals to ocean acidification. Geochem Cosmochim Acta 74(17):4988–5001
Lear CH, Mawbey EM, Rosenthal Y (2010) Cenozoic benthic foraminiferal Mg/Ca and Li/Ca records: Toward unlocking temperatures and saturation states. Paleoceanography 25(4):PA4215
Lemarchand D et al (2000) The influence of rivers on marine boron isotopes and implications for reconstructing past ocean pH. Nature 408(6815):951–954
Lemarchand D et al (2002) An optimized procedure for boron separation and mass spectrometry analysis for river samples. Chem Geol 182(2–4):323–334
Liu Y, Tossell JA (2005) Ab initio molecular orbital calculations for boron isotope fractionations on boric acids and borates. Geochim Cosmochim Acta 69 (16):3995–4006
Lombard F et al (2010) Effect of carbonate ion concentration and irradiance on calcification in planktonic foraminifera. Biogeosciences 7:247–255
Lutze GF, Thiel H (1987) Cibicidoides wuellerstorfi and Planulina ariminensis, elevated epibenthic Foraminifera. In: Altenbach AV, Lutze GF, Weinholz, P (eds) Beobachtungen an Benthos-Foraminiferen (Teilprojekt A3). Berichte aus dem Sonderforschungsbereich 313 "Sedimentation im Europäischen Nordmeer". Sonderforschungsbereich 313, Universität Kiel, Keil, pp 17–30
Martínez-Botí MA, Foster GL et al (2015a) Plio-Pleistocene climate sensitivity evaluated using high-resolution CO_2 records. Nature 518(7537):49–54
Martínez-Botí MA, Marino G et al (2015b) Boron isotope evidence for oceanic carbon dioxide leakage during the last deglaciation. Nature 518(7538):219–222
Mavromatis V et al (2015) Characterization of boron incorporation and speciation in calcite and aragonite from co-precipitation experiments under controlled pH, temperature and precipitation rate. Geochim Cosmochim Acta 150:299–313
McCulloch M, Falter J et al (2012a) Coral resilience to ocean acidification and global warming through pH up-regulation. Nat Clim Change 2(8):623–627
McCulloch M, Trotter J et al (2012b) Resilience of cold-water scleractinian corals to ocean acidification: boron isotopic systematics of pH and saturation state up-regulation. Geochim Cosmochim Acta 87:21–34
Misra S et al (2014) Determination of $\delta^{11}B$ by HR-ICP-MS from mass limited samples: application to natural carbonates and water samples. Geochem Cosmochim Acta 140:531–552
Morard R et al (2009) Morphological recognition of cryptic species in the planktonic foraminifer Orbulina universa. Mar Micropaleontol 71(3–4):148–165
Mosbrugger V, Utescher T, Dilcher DL (2005) Cenozoic continental climatic evolution of Central Europe. Proc Natl Acad Sci U S A 102(42):14964–14969.
Naik SS et al (2015) Tracing the strength of the southwest monsoon using boron isotopes in the eastern Arabian Sea. Geophys Res Lett 42(5):1450–1458
Nehrke G et al (2013) A new model for biomineralization and trace-element signatures of Foraminifera tests. Biogeosciences 10(10):6759–6767
Ni Y et al (2007) A core top assessment of proxies for the ocean carbonate system in surface dwelling foraminifers. Paleoceanography 22(3):PA3212
Ni Y, Foster GL, Elliott T (2010) The accuracy of $\delta^{11}B$ measurements of foraminifers. Chem Geol 274(3–4):187–195

(sinistral) to constrain CO_2 uptake in the subpolar North Atlantic during the last deglaciation. Paleoceanography 28:237–252
Zachos JC et al (2001) Trends, rhythms, and aberrations in global climate 65 Ma to present. Science 292:686–693
Zachos JC, Dickens GR, Zeebe RE (2008) An early Cenozoic perspective on greenhouse warming and carbon-cycle dynamics. Nature 451(7176):279–283
Zeebe RE (2005) Stable boron isotope fractionation between dissolved $B(OH)_3$ and $B(OH)_4$. Geochim Cosmochim Acta 69(11):2753–2766
Zeebe RE, Sanyal A (2002) Comparison of two potential strategies of planktonic foraminifera for house building: Mg^{2+} or H^+ removal? Geochim Cosmochim Acta 66(7):1159–1169
Zeebe RE, Wolf-Gladow DA (2001) CO_2 in seawater: equilibrium, kinetic, isotopes. Elsevier, Amsterdam
Zeebe RE et al (2001) A theoretical study of the kinetics of the boric acid-borate equilibrium in seawater. Mar Chem 73(2):113–124
Zeebe RE et al (2003) Vital effects in foraminifera do not compromise the use of $\delta^{11}B$ as a paleo-pH indicator: evidence from modeling. Paleoceanography 18(2):1043

Boron Isotopic Systematics in Scleractinian Corals and the Role of pH Up-regulation

6

Malcolm T. McCulloch, Juan P. D'Olivo, James Falter, Lucy Georgiou, Michael Holcomb, Paolo Montagna and Julie A. Trotter

Abstract

The boron isotopic composition ($\delta^{11}B$) of scleractinian corals has been used to track changes in seawater pH and more recently as a probe into the processes controlling bio-calcification. For corals that precipitate aragonite skeletons, up-regulation of pH appears to be a general characteristic, typically being ~0.3 to ~0.6 pH units higher than ambient seawater. The relationship between the pH of the corals calcifying-fluid (pH_{cf}) and seawater pH_T (total scale) is shown to be dependent on both physiological as well environmental factors. In laboratory experiments conducted on symbiont-bearing (zooxanthellate) corals under conditions of constant temperature and seawater pH, changes in the $\delta^{11}B$ derived calcifying fluid pH_{cf} is typically $^1/_3$ to $^1/_2$ of that of ambient seawater. Similar linear relationships are found for cold water corals that live in relatively stable, cold, deep-water environments but at significantly elevated levels of pH_{cf} (~0.5–1 pH units above seawater), a likely response to the lower pH of their deep-sea environments. In contrast, zooxanthellae-bearing corals living in shallow-water reef environments that experience significant natural variations in temperature, light, nutrients and seawater pH, show different types of responses. For example, over seasonal time scales

M.T. McCulloch (✉) · J.P. D'Olivo · J. Falter · L. Georgiou · M. Holcomb · J.A. Trotter
Oceans Institute Graduate School and School of Earth Sciences, The University of Western Australia, Crawley, WA 6009, Australia
e-mail: Malcolm.McCulloch@uwa.edu.au

M.T. McCulloch · J.P. D'Olivo · J. Falter · L. Georgiou
ARC Centre of Excellence for Coral Reef Studies, The University of Western Australia, Crawley, WA 6009, Australia

P. Montagna
Institute of Marine Sciences, CNR, Via Gobetti, 101, Bologna, Italy

H. Marschall and G. Foster (eds.), *Boron Isotopes*, Advances in Isotope Geochemistry,
https://doi.org/10.1007/978-3-319-64666-4_6

Porites corals from the Great Barrier Reef (GBR) have a large range in pH_{cf} of ~8.3 to ~8.5, significantly greater (~×2 to ~×3) than that of reef-water (pH_T ~8.01 to ~8.08), and an order of magnitude greater than that expected from 'static' laboratory experiments. Strong physiological controls, but of a different character, are found in corals grown in a Free Ocean Carbon Enrichment Experiment (FOCE) conducted in situ within the Heron Island lagoon (GBR). These corals exhibit near constant pH_{cf} values regardless of external changes in temperature and seawater pH. This pattern of strong physiologically controlled '*pH-homeostasis*', with elevated but constant pH_{cf} has been found despite large natural seasonal variations in the pH (±0.15 pH units) of the lagoon waters, as well as the even larger super-imposed decreases in seawater pH (~0.25 pH units) designed to simulate year 2100 conditions. In natural reef environments we thus find that the processes influencing the up-regulation of pH_{cf} in symbiont-bearing corals are subject to strong physiological controls, behaviour that is not well simulated in the current generation of aquaria-based experiments with fixed seawater pH and temperature. Conversely, cold-water corals that lack symbionts and inhabit the relatively stable deep-sea environments hold the best prospects for providing reliable reconstructions of seawater pH. Clearly, further studies utilising the $\delta^{11}B$-pH_{cf} proxy combined with other DIC/carbonate-ion proxies (e.g. B/Ca), but conducted under realistic 'natural' conditions, are required to elucidate the processes controlling coral bio-calcification and to better understand the vulnerability of scleractinian corals to anthropogenic driven warming and ocean acidification.

Keywords
pH up-regulation · Boron isotopes · Corals

6.1 Introduction

Scleractinian corals, particularly those that precipitate aragonite skeletons, have been subject to a growing number of boron isotopic ($\delta^{11}B$) studies (Allison et al. 2014; Blamart et al. 2007; D'Olivo et al. 2015; Honisch and Hemming 2004; Krief et al. 2010; McCulloch et al. 2012a; Pelejero et al. 2005; Reynaud et al. 2004; Rollion-Bard et al. 2011; Trotter et al. 2011; Wei et al. 2009). Initial $\delta^{11}B$ studies (Hemming and Hanson 1992; Vengosh et al. 1991) showed that to first order, boron is incorporated in the carbonate skeleton of corals as the borate species, hence raising the possibility of utilising changes in the boron isotopic composition of corals as a proxy for seawater pH. These and subsequent studies (Douville et al. 2010; Honisch and Hemming 2005; Liu et al. 2009; Pelejero et al. 2005; Wei et al. 2009) were mainly focused on quantifying changes in seawater pH. These studies have been conducted over both geological timescales [e.g. (Douville et al. 2010) during the last deglaciation] as well more recent timescales, particularly over the modern industrial era (Pelejero et al. 2005; Wei et al. 2009) where rapidly increasing levels of anthropogenic CO_2 are causing ocean acidification (Caldeira et al. 2007; Ciais et al. 2014). These studies, however, assume that the calcifying fluid from which corals precipitate their calcium carbonate skeleton has the same pH as ambient seawater. At that time this was a reasonable premise since there was seemingly good agreement between direct measurements of seawater pH and those calculated from $\delta^{11}B$ systematics, the latter derived from the then widely utilised boron isotopic speciation curve of Kakihana et al. (1977).

More recent studies (Allison et al. 2014; Krief et al. 2010; McCulloch et al. 2012a; Trotter et al. 2011) utilising the updated curve of Klochko et al. (2006) to define the isotopic fractionation between the two main boron species, boric acid ($B(OH_3)$) and borate ion ($B(OH)_4^-$), revealed that corals up-regulate the pH of the coral calcifying fluid (pH_{cf}). As we will show, while this represents a significant limitation in the use of $\delta^{11}B$ as a seawater pH proxy in symbiont (zooxanthellae) bearing corals especially, it now provides a powerful means to interrogate the processes controlling bio-calcification in general. This latter aspect is of increasing importance given that corals are now facing the dual challenges of declining seawater pH due to CO_2 driven ocean acidification, combined with rapidly increasing sea surface temperatures and associated coral bleaching (Hughes et al. 2003). Therefore, understanding the processes controlling the pH of the corals calcifying fluid (pH_{cf}) is critical to understanding the vulnerability of corals to the combined effects of ocean acidification and global warming.

6.2 Calcification in Scleractinian Corals

Before examining the application of boron isotopic studies, it is useful to review some of the basic factors controlling the calcification of scleractinian corals. Biological controls of the calcifying environment occur at a several levels. Firstly, with the synthesis and incorporation of biological templates upon which calcification is directed, and secondly, with the partial isolation of a fluid whose composition can be biologically manipulated to induce bio-calcification (Allemand et al. 2004, 2011). In scleractinian corals, precipitation of their calcium carbonate skeleton occurs from a partially isolated extracellular calcifying medium (Allemand et al. 2004), located at the interface between the coral polyp's basal cell layer and the underlying skeleton (Fig. 6.1). Although formation of the coral's skeleton is a strongly biologically-mediated process (Allemand et al. 2004, 2011), precipitation of aragonite is still ultimately determined by the composition and conditions of the calcifying medium (Al-Horani et al. 2003; Allemand et al. 2004). The initial source of the calcifying-fluid is ambient seawater (Erez 2003), with up-regulation of its pH thought to mainly occur via Ca-ATPase pumping of Ca ions into the calcifying region in exchange for protons (Allemand et al. 2004; Cohen and McConnaughey 2003).

Manipulation of the corals internal pH_{cf} at the site of calcification shifts the equilibrium composition of dissolved inorganic carbon (DIC) in favour of CO_3^{2-} relative to HCO_3^-. This increases the carbonate saturation state of the calcifying

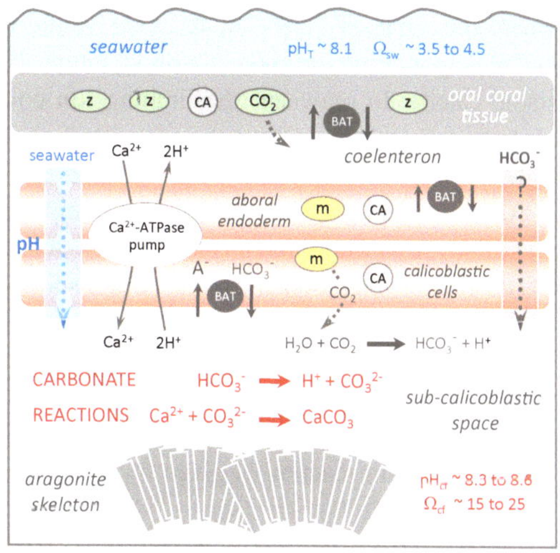

Fig. 6.1 Schematic diagram showing some of the basic elements of the calcification process in zooxanthellae-bearing corals Allemand et al. (2004), Zoccola et al. (2015). *Z* zooxanthellae, *BAT* bicarbonate anion transporters, *CA* carbonic anhydrases, *m* mitochondrion. Removal of protons from the calcification site occurs primarily via Ca^{2+}-ATPase exchangers that pump $2H^+$ ions from the calcifying medium into the coelenteron in exchange for each Ca^{2+} ions. The carbonic anhydrases catalyse the forward reactions converting CO_2 into HCO_3^- ions (Moya et al. 2008), with pH up-regulation shifting the carbonate equilibria to favour CO_3^{2-} ions, the latter essential for calcification. Diffusion of CO_2 into the sub-calicoblastic space may also occur thus increasing the DIC of the calcifying fluid relative to seawater (Erez 2003)

fluid (Ω_{cf}), not only enhancing the kinetics of calcification, but also has the potential to counter the effects (Cohen and Holcomb 2009; Holcomb et al. 2009) of reduced carbonate saturation in seawater. The latter effect is most clearly evident in deep-water corals that can exist at and in some cases below the carbonate saturation horizon for aragonite (Fautin et al. 2009; Thresher et al. 2011). The impact of ocean acidification (Caldeira et al. 2007), hence decreased CO_3^{2-} concentrations, on the rate of calcification is seemingly straightforward since calcification is ultimately controlled by the reaction of:

$$Ca^{2+} + CO_3^{2-} \Leftrightarrow CaCO_3 \quad (6.1)$$

with carbonate saturation state given by:

$$\Omega = [Ca^{2+}][CO_3^{2-}]/K_{spx} \quad (6.2)$$

where the solubility constant K_{spx} is specific to either aragonite or the more thermodynamically stable calcite. In seawater, the limiting ion concentration for this reaction is the concentration of CO_3^{2-} ion (~200–250 μmol kg^{-1} compared to ~10,000 μmol kg^{-1} for Ca^{2+}). The complication arises from the different pathways by which CO_2/carbonate/bicarbonate might be transported to and utilised within the calcifying environment. The most obvious avenue is via the direct seawater bicarbonate (HCO_3^-) route since this is the dominant carbonate species in seawater. In this case the relevant reaction is given by:

$$HCO_3^- \Leftrightarrow CO_3^{2-} + H^+ \quad (6.3)$$

hence the importance of Ca^{2+} pumping to maintain pH levels (Cohen and McConnaughey 2003) by removing H^+ ions from the site of calcification. As indicated, removal of protons from the calcification site occurs primarily via Ca^{2+}-ATPase exchangers that pump $2H^+$ ions from the calcifying medium into the coelenteron in exchange for each Ca^{2+} ion (Fig. 6.1). Importantly, the borate and boric species that makeup seawater are also likely to be transported via the same route, with their speciation showing a similar type of dependence on pH as in seawater. The alternate route for CO_3^{2-} transport is via diffusion of CO_2 into the extra-cellular medium, presumably sourced from CO_2 enriched, low pH domains, possibly related to the activity of zooxanthellae. In this case the carbonic anhydrase (CA)-catalysed reaction (Moya et al. 2008) is given by:

$$H_2O + CO_2 \Leftrightarrow HCO_3^- + H^+ \quad (6.4)$$

with HCO_3^- being an intermediate product and again emphasising the importance of Ca^{2+} pumping to facilitate this reaction. Finally we note there has been some suggestion of a specific HCO_3^- pump (Fig. 6.1) but there is no direct evidence to support this (Zoccola et al. 2015).

The dependence of inorganic calcification rates on Ω is given by the empirical rate dependence law for abiotic calcification (Burton and Walter 1987):

$$R_{calcif} = k(\Omega - 1)^n \quad (6.5)$$

where k is the rate law constant and n is the order of the reaction, both of which are temperature dependent. The role of carbonate saturation state was examined by McCulloch et al. (2012a) who applied this equation to the carbonate saturation state of the calcifying-fluid (Ω_{cf}). Thus, determining the processes controlling both the internal, biologically-mediated pH_{cf}, the DIC (dissolved inorganic carbon) and hence the carbonate saturation-state Ω_{cf} at the site of calcification, are likely to be key to understanding how biogenic calcifiers will respond to ocean acidification and ocean warming in both cold, deep waters as well as tropical reef environments. This, however, requires knowledge of the DIC of the calcifying fluid, an important and still poorly constrained parameter, an issue that we will address later.

6.3 Boron Isotopic Systematics

Boron exists in seawater as two different molecular species, boric acid ($B(OH)_3$) and borate ion ($B(OH)_4^-$), with their relative abundance being pH dependent. Changes in the isotopic ratio of ^{11}B (~80%) and ^{10}B (~20%), are commonly expressed in delta notation (in per mil, ‰) as:

$$\delta^{11}B_{coral} = [(^{11}B/^{10}B_{coral}/^{11}B/^{10}B_{NIST951}) - 1] \times 1000 \tag{6.6}$$

where $^{11}B/^{10}B_{coral}$ is the isotopic ratio measured in the coral and $^{11}B/^{10}B_{NIST951}$ is the isotopic ratio of the NIST SRM 951 boric acid standard. As shown in Fig. 6.2, there is a pronounced isotopic fractionation between the two dissolved boron species. As the relative abundance of each species is pH dependent, as defined by the acid–base relationship of $B(OH)_3 + 2H_2O \Leftrightarrow B(OH)_4^- + H_3O^+$, the isotopic composition of each species also changes with pH, and can be expressed by the reaction:

$$^{10}B(OH)_3 + ^{11}B(OH)_4^- \Leftrightarrow ^{11}B(OH)_3 + ^{10}B(OH)_4^- \tag{6.7}$$

Using these relationships, the pH_{cf} of the calcifying fluid can be calculated from the $\delta^{11}B$ composition of the coral carbonate ($\delta^{11}B_{carb}$). The equation used to convert the measured boron isotopic composition of the coral carbonate skeleton to a pH value (Zeebe and Wolf-Gladow 2001) of the calcifying fluid (pH_{cf}) is given by:

$$pH_{cf} = pK_B - \log[-(\delta^{11}B_{sw} - \delta^{11}B_{carb})/(\delta^{11}B_{sw} - \alpha_{(B3-B4)}\delta^{11}B_{carb} - 1000(\alpha_{(B3-B4)} - 1))] \tag{6.8}$$

where $\delta^{11}B_{sw}$ and $\delta^{11}B_{carb}$ represent the $\delta^{11}B$ composition of seawater ($\delta^{11}B_{sw}$ = 39.61‰) (Foster et al. 2010) and carbonate respectively, and $\alpha_{(B3-B4)}$ defines the boron isotope fractionation in seawater ($\alpha_{(B3-B4)}$ = 1.0272) (Klochko et al. 2006). The dissociation constant of boric acid pK_B has a well-established value of 8.597 at 25 °C and salinity of 35 (Dickson 1990). For deep-water corals, temperature and pressure corrections (Rae et al. 2011; Zeebe and Wolf-Gladow 2001) are also applied using coefficients from CO2SYS Matlab version 1.1 (Lewis and Wallace 1998). This equation also assumes that the calcifying fluid has the same $\delta^{11}B$ composition as seawater, given it is the primary source of boron and due to the low K_D of B/Ca the bulk of the boron remains in the calcifying fluid during calcification.

The assumption that only the borate ion is partitioned into the calcium carbonate skeleton of biogenic calcifiers has been questioned, based on the significant quantities of boric acid species observed from nuclear magnetic resonance and electron-loss spectroscopy (Klochko et al. 2009; Rollion-Bard et al. 2011). These studies reported variable proportions (12–48%) of the trigonally co-ordinated $B(OH)_3$ in different skeletal components of the carbonate skeletons of corals. Conversely, more recent experiments (Mavromatis et al. 2015; Noireaux et al. 2015) conducted on inorganic aragonite indicate that borate is the only species incorporated into the aragonite structure, with the small percentage (<15%) of trigonal detected boric acid probably being the result of a secondary (post crystallisation) change of coordination. Importantly, in aragonite no additional boron isotopic fractionation occurs during the incorporation of the borate species (Noireaux et al. 2015), hence in this respect the $\delta^{11}B$ systematics of aragonitic corals is relatively straightforward.

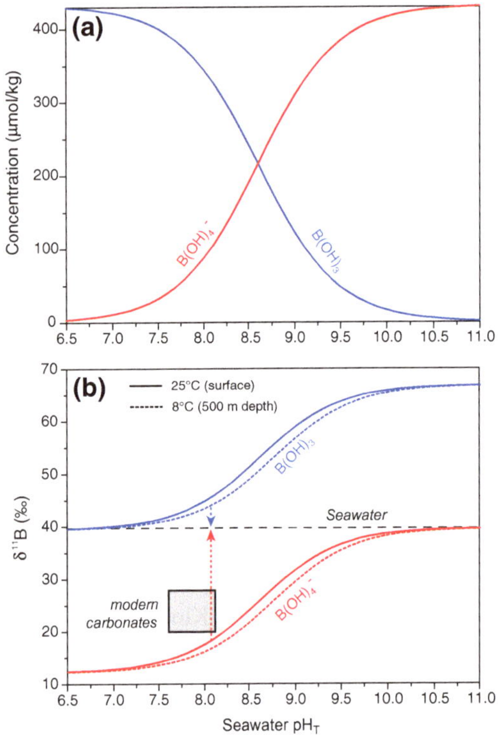

Fig. 6.2 **a** Boron speciation in seawater as a function of seawater pH_T (total scale) with the borate ion ($B(OH)_4^-$) favoured at high seawater pH_T. **b** Boron isotope composition ($\delta^{11}B$) of the boric ($B(OH)_3$) versus borate species (Klochko et al. 2006) as a function of seawater pH_T. The *blue* and *red arrows* show the relative contribution of each species to the overall seawater $\delta^{11}B$ composition. *Grey box* shows the typical $\delta^{11}B$ compositions of aragonitic marine carbonates, which generally lie above the borate curve. *Dashed lines* show the dependence on temperature at typical surface (25 °C) and deeper water depths (8 °C)

6.4 Boron Isotopic Compositions of Scleractinian Corals

The $\delta^{11}B$ isotopic compositions for a number of coral species subject to a range of experimentally fixed values of ambient seawater pH are shown in Fig. 6.3. Most shallow-water symbiont-bearing corals that inhabit tropical to sub-tropical reef-waters have a typical range in seawater pH from ~8.0 to 8.1, and $\delta^{11}B$ compositions from ~22 to ~25‰. Higher $\delta^{11}B$ values are present in temperate corals, such as *Cladocora caespitosa*, found in the Mediterranean (Trotter et al. 2011). The highest $\delta^{11}B$ compositions of ~24–28‰ are found in corals inhabiting the much colder (~0–14 °C) generally deep-water environments (Anagnostou et al. 2012; McCulloch et al. 2012b) where zooxanthellae symbionts are absent in the coral host. In these deep-water environments, there is a large depth dependent decrease in seawater pH with particular species (e.g. *Desmophyllum dianthus*) having concomitantly lower $\delta^{11}B$ values as seawater pH decreases. Likewise, under experimental conditions designed to simulate future higher pCO_2 conditions, shallow water corals also generally exhibit the same trend of declining $\delta^{11}B$ with lower seawater pH (Fig. 6.3).

The importance of an accurate, reliable calibration for the pH dependence of the inorganic borate ion [$B(OH)_4^-$] and hence $\delta^{11}B$ composition is also illustrated in Fig. 6.3. Prior to the experimental re-determination of the borate curve by Klochko et al. (2006), the commonly accepted and widely used curve was that of Kakihana et al. (1977). The *Kakihana* curve (Fig. 6.3), however, has systematically higher $\delta^{11}B$ values (~5‰ at pH_{sw} ~8.0) and tropical shallow-water corals have, coincidentally, similar $\delta^{11}B$ compositions. This led to the reasonable assumption that this group of corals calcified with approximately seawater-like pH values with little or no pH up-regulation. This initial explanation, however, was not viable for calcitic foraminifer (Fig. 6.4) since they lie substantially below the *Kakihana* curve, on the now accepted *Klochko* curve. This revision to the inorganic borate $\delta^{11}B$ versus pH curve (Klochko et al. 2006) thus has important implications for the pH dependence of coral bio-calcification, which we now examine in more detail below.

6.5 Experimental Constraints on the Relationship Between Calcifying Fluid pH and Seawater pH

Determining the relationship between the pH of ambient seawater and that of the coral's calcifying-fluid (pH_{cf}) is an essential prerequisite in the application of $\delta^{11}B$ as a seawater pH proxy,

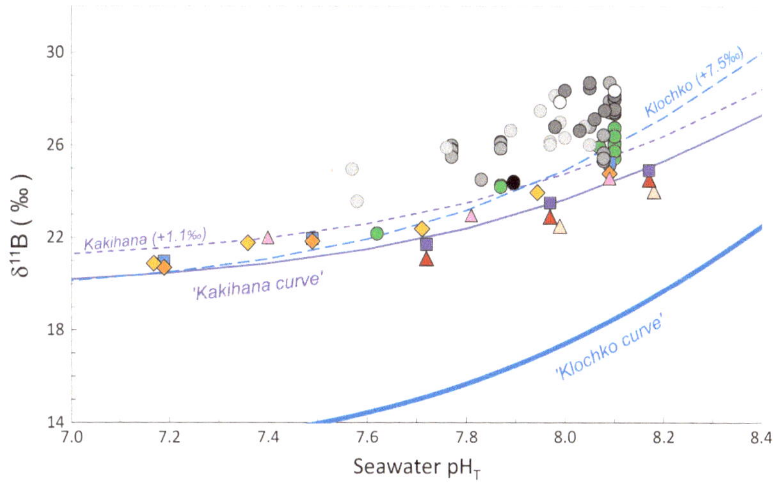

Fig. 6.3 Plot of $\delta^{11}B$ isotopic compositions for shallow water zooxanthellae-bearing corals (*coloured symbols* see Fig. 6.4) and cold-water azooxanthellate scleractinian (i.e. aragonitic) corals (*grey circles*). Aragonitic corals lie substantially above the *Klochko* borate $\delta^{11}B$-pH calibration curve (Klochko et al. 2006) indicative of significant up-regulation of their calcifying-fluid pH_{cf}. Also shown is the obsolete *Kakihana curve* (Kakihana et al. 1977) on (or near) which many zooxanthellae-bearing corals plot. This unfortunate coincidence led to the erroneous assumption that coral calcification occurred under essentially abiotic conditions with seawater-like pH. *Dashed lines* show the differing slopes and hence sensitivities of the *Klochko* versus *Kakihana curves*

but also in understanding the pH related controls on bio-calcification. This latter process is of particular importance since it is integral to the coral's ability to withstand the effects of decreasing seawater pH due to ocean acidification. Numbers of experiments have therefore been conducted to determine the relationship between $\delta^{11}B$ preserved in the corals carbonate skeleton relative to its ambient seawater pH (Honisch and Hemming 2004; Honisch et al. 2004; Reynaud et al. 2004). As already discussed, the interpretation of these earlier experiments (Honisch and Hemming 2004; Honisch et al. 2004; Reynaud et al. 2004; Rollion-Bard et al. 2003) was based on the *Kakihana* curve, with minor differences or offsets in plots of $\delta^{11}B$ versus seawater pH accounted by what were thought to be mainly species-specific effects (Fig. 6.3). A significant advance was made by Trotter et al. (2011) who instead calculated offsets in the pH_{cf} relative to ambient seawater pH. This pH_{cf} versus pH_T approach utilising the *Klochko* relationships gave significantly improved correlations compared to the *Kakihana* curve due to the non-linear nature of the $\delta^{11}B$–pH borate curve (Fig. 6.3). This pH_{cf} versus pH_T based approach, thus provided convincing evidence supporting the validity of the *Klochko* borate curve as well as the important role of pH up-regulation in scleractinian corals.

The relationship between pH_{cf} versus seawater pH is shown in Fig. 6.4, as sub-parallel species-specific linear arrays. At seawater pH_T of 8.1, the temperate Mediterranean coral *Cladocora caespitosa* exhibits the greatest elevation in pH_{cf} relative to seawater with $\Delta pH \sim 0.5$ ($\Delta pH = pH_{cf} - pH_T$). The lowest ΔpH values (~ 0.3) are found in the more rapidly growing *A. nobilis, A. digitifera* and other tropical *Acropora* species, whereas the tropical corals *Porites* spp., *P. cylindrica,* and *S. pistillata* have intermediate ΔpH values of ~ 0.4. The remarkable feature of these relationships is the exceptional strong species-dependent correlations with r^2 of ~ 0.99 being commonly found for experiments conducted over a range of pH values across different laboratories.

Furthermore, independent measurements of calcifying pH_{cf} compositions generally show very

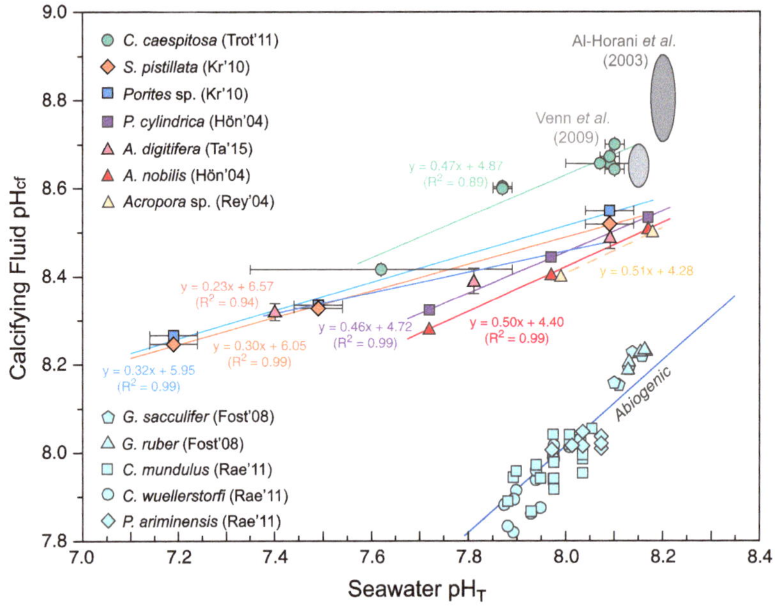

Fig. 6.4 Calcifying-fluid pHcf compositions determined from boron isotopic systematics of scleractinian (aragonitic) corals grown under controlled laboratory conditions for a range of seawater pH_T (Honisch et al. 2004; Krief et al. 2010; Reynaud et al. 2004; Tanaka et al. 2015; Trotter et al. 2011). Corals systematically elevate their pH_{cf} by ~0.3–0.5 pH units and under controlled 'static' conditions form species dependent *linear arrays.* Also shown (*grey ellipses*) are the average pH_{cf} values measured over daily cycles using micro-electrodes (Al-Horani et al. 2003) and pH sensitive dyes (Venn et al. 2009), both of which are broadly consistent with the longer duration (weeks to month) pH_{cf} in the coral samples. For comparison, calcitic foraminifera (*light blue symbols*) are shown which plot on or close to the abiogenic borate curve (Foster et al. 2008; Rae et al. 2011)

good agreement when allowance is made for intrinsic differences between the methods. This is illustrated (Fig. 6.5) for the coral species *Stylophora pistillata* where colonies were grown under different seawater pH treatments, but with each treatment having constant conditions (Venn et al. 2013). The pH_{cf} values were determined using both the pH dye indicator SNARF (Venn et al. 2013) as well as the direct determinations of $\delta^{11}B$ compositions (Holcomb et al. 2014) from different skeletal regions (see Fig. 6.5). Importantly, both approaches (Holcomb et al. 2014; Venn et al. 2013) gave similar slopes (0.32–0.37) between seawater pH and pH_{cf} and form highly correlated linear arrays. The only major difference between methods is the lower magnitude of pH_{cf} up-regulation detected by the SNARF experiment, being ~0.2 pH units compared to ~0.4 pH units (at seawater pH_T ~8) for the $\delta^{11}B$ determinations of pH_{cf}. SNARF measurements require imaging the dyes through a glass mount by confocal microscopy, hence only the very thin transparent coral tissue at the lateral growth edge (Fig. 6.5) can be analysed. Thus, the lower pH values determined within this very thin tissue region is consistent with it being more directly exposed to seawater. Interestingly, coral apexes were found to have the highest pH_{cf} and experienced the smallest changes in pH_{cf} in response to acidification (Holcomb et al. 2014). Lateral growth was found to be associated with both lower pH_{cf} and relatively greater sensitivity to acidification. Thus up-regulation of pH_{cf} is a spatially variable process within corals (Holcomb et al. 2014), with its response being critical to determining the sensitivity of calcification to ocean acidification.

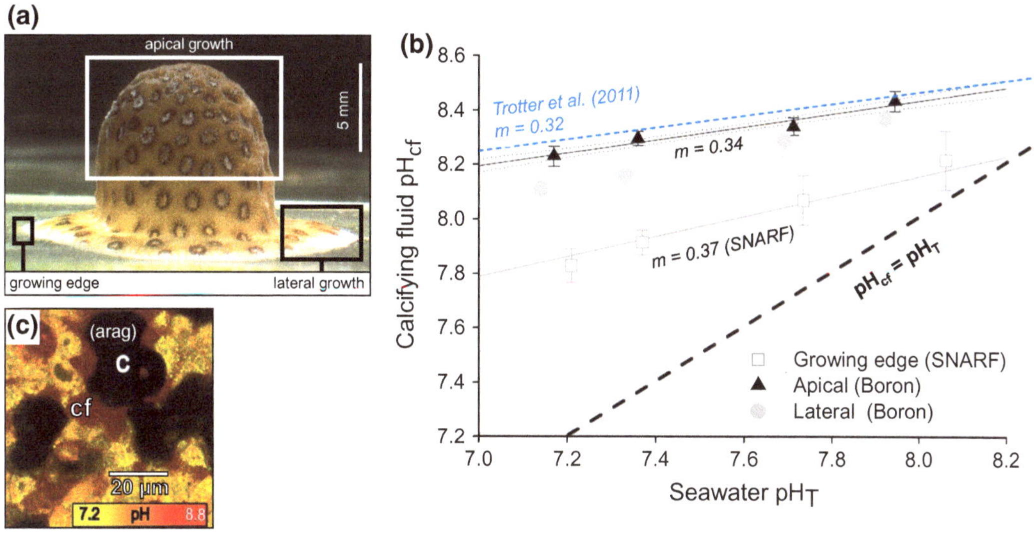

Fig. 6.5 **a** Image of cultured specimen of *Stylophora pistillata* and the regions sampled for comparative pH_{cf} determinations using pH sensitive SNARF dye (Venn et al. 2013) and $\delta^{11}B$ measurements (Holcomb et al. 2014). SNARF pH measurements undertaken by confocal microscopy image the outermost growing edge, whereas $\delta^{11}B$ compositions were determined from the lateral growth and main apical growth regions. **b** Plot of SNARF pH_{cf} of the lateral growing edge (*open squares*) and $\delta^{11}B$ values measured in apical (*black triangles*) and lateral (*grey circles*) skeletal regions versus seawater pH_T (Holcomb et al. 2014). The *blue dashed line* represents pH_{cf} calculated (Trotter et al. 2011) from $\delta^{11}B$ compositions of *Stylophora pistillata* (Krief et al. 2010). **c** Confocal microscopy image (Venn et al. 2013) showing the small micron-scale variability in pH of the irregularly distributed calcifying fluid

6.6 Cold-Water Corals

Cold-water corals have a wide global distribution and inhabit a very diverse range of environments, from the shallow, cold, nutrient-rich waters of fjords, such as those for example in Chile and Norway (Försterra et al. 2005; Roberts et al. 2006), down to thousands of metres reaching (and below) the carbonate saturation horizon. The major characteristic of these corals is an absence of light-harvesting symbiotic dinoflagellates (Allemand et al. 2004; Gattuso et al. 1999), which in regard to $\delta^{11}B$ systematics avoids some potential complexities common to zooxanthellae-bearing corals. In Fig. 6.6, the pH_{cf} data derived from $\delta^{11}B$ measurements is plotted against the ambient seawater pH_T. To first order, it is clear that the pH_{cf} values all lie significantly above the tropical zooxanthellae-bearing tropical corals, thus with a greater elevation in pH_{cf} relative to seawater (i.e. pH = $pH_{cf} - pH_T$) that ranges from ~0.6 to ~0.8 pH units, the largest offset being in corals from low seawater pH_T environments. This general pattern of higher pH is attributed to a requirement to elevate pH_{cf} in order to maximise Ω_{cf} to facilitate calcification. This suggests a minimum or threshold value for Ω_{cf} required to sustain calcification in this challenging low temperature environment, where temperatures may range from below 0 °C (e.g. Antarctic waters) to ~14 °C (e.g. Mediterranean Sea). However, it is noted that there is no obvious dependence of pH_{cf} on seawater temperature since *D. dianthus* (Fig. 6.6) live within a temperature range from ~4 to 14 °C (Anagnostou et al. 2012; McCulloch et al. 2012b).

To apply the $\delta^{11}B$ seawater pH proxy to azooxanthellate cold-water corals, it is also necessary to establish a correlation between the measured pH_{cf} and inferred seawater pH.

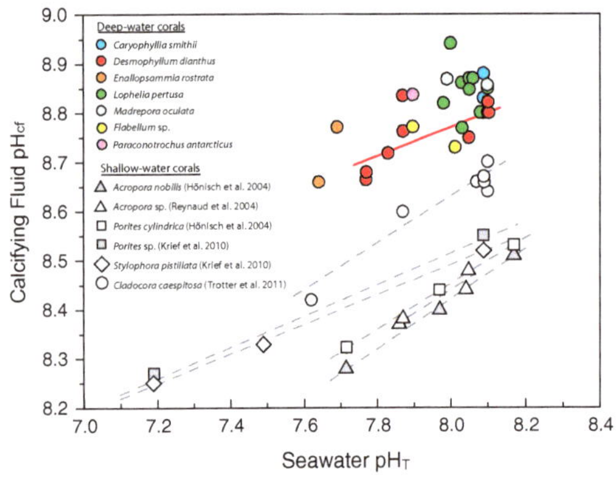

Fig. 6.6 Cold-water coral calcifying fluid pH_{cf} calculated from $\delta^{11}B$ measurements (Anagnostou et al. 2012; McCulloch et al. 2012b) versus seawater pH_T. Cold-water corals (*coloured circles*) have pH_{cf} values up to ~0.5–0.8 pH unit higher than tropical corals (*grey lines*). The cold-water coral, *Desmophyllum dianthus*, defines an approximate *linear array* of pH_{cf} versus seawater pH_T ($r^2 = 0.79$). The less coherent data for some of the other species may be due to non-representative seawater pH_T estimates and/or the presence of physiological species dependent effects. Data from Anagnostou et al. (2012), McCulloch et al. (2012b)

To date, for cold-water corals this calibration is derived from samples collected in situ, and whose ambient seawater pH has been determined either directly or more commonly from databases (e.g. Anagnostou et al. 2012). This latter procedure still represents a significant limitation given that many sample collection sites either lack this information or are attributed interpolated pH values that might not reflect conditions at the time of coral growth. Furthermore, there is an inherent limitation in calibrating the relationship between pH_{cf} versus seawater pH_T because the natural range in pH_T during the life of the corals is often limited. Hence, the necessity to compare samples from disparate locations, such as the relatively warm high salinity waters of the Mediterranean Sea (high pH) versus the much colder and low pH deep-waters of, for example, the Atlantic and Southern Ocean. Despite these limitations there is a reasonable correlation between ambient seawater and pH_{cf} for the solitary deep-sea species *D. dianthus*, where $pH_{cf} = 0.29\ pH_T + 6.4$ ($r^2 = 0.79$, $n = 8$). Thus it is argued that, with further more rigorous calibrations, deep-sea azooxanthellate corals hold much promise as palaeo-seawater pH archives.

6.7 pH Up-Regulation in the Natural Environment

While the above findings suggest an extremely coherent relationship between the calcifying fluid pH_{cf} and seawater pH, the critical test is the actual in situ or 'real-world' response. In natural shallow marine environments especially, there can be considerable variability in not only seawater pH but also temperature, light, and nutrients, across a wide range of time-scales from diurnal to seasonal cycles through decades to millennia. On seasonal timescales, the most significant parameter is temperature, with the closely associated parameter of light as well as nutrients also varying. Changes in these ambient conditions will affect not only the kinetics of coral calcification but also importantly the metabolism of symbionts that likely play a key role in the supply of DIC to the coral host. The major exception is for cold-water corals inhabiting deep-sea environments where light does not penetrate, and where there are generally relatively stable conditions of temperature and pH on annual to decadal timescales. Given that these deep-sea corals are azooxanthellate, hence lack symbionts, they arguably have less complex pathways for the supply of DIC to the calcifying fluid.

6.7.1 Corals Under Free Ocean Carbon Enrichment (FOCE) Conditions

Free Ocean Carbon Enrichment (FOCE) experiments have been undertaken to provide realistic environmental conditions (Gattuso et al. 2014; Kline et al. 2012) that incorporate natural variability in the key parameters of temperature, pH, and as far as possible light and water flow. In addition, by modulating the seawater pH, the FOCE system enables realistic simulations of the effects of ocean acidification within natural reef-environments, at levels of atmospheric pCO_2 and seawater pH that are predicted to occur by the end of this century according to RCP4.5 (van Vuuren et al. 2011). Such experiments were recently conducted within the Heron Island reef flat of the Great Barrier Reef (GBR) covering a range of pCO_2 scenarios (Kline et al. 2012). Here the FOCE experiment was thus designed to raise the dissolved pCO_2 to ~600 ppm, hence lowering the seawater pH (and hence Ω_{arag}) of the treatment flumes by ~0.25 pH units below ambient levels. These changes within the FOCE flumes were undertaken progressively over an approximately eight month period (March to mid-December 2010), and superimposed upon the natural daily variations in seawater pH as well as light and temperature. The response of calcifying fluid pH_{cf} compositions in the branching coral *P. cylindrica* cultivated within the FOCE flumes over this period was thus determined using $\delta^{11}B$ compositions (Georgiou et al. 2015).

The findings from these experiments are shown in Fig. 6.7 with several surprising outcomes. Firstly, there were substantial differences in $\delta^{11}B$ between colonies (22–26‰) that translate to substantial differences in pH_{cf} (8.4–8.6). Secondly and most surprisingly, the compositions of individual growth segments (i.e. separate branches) of *P. cylindrica* along their respective central growth axis show essentially no or, at the very most, minor (<0.02) declines in pH_{cf} despite the very large decrease in seawater pH from ~8.25 to 7.7. These data show that these corals exhibited a pattern of pH-homeostasis, with elevated but constant pH_{cf}, despite the large natural seasonal variations (~0.3 pH units) and upon which an additional reduction of seawater pH of ~0.25 pH units was added during the FOCE treatment.

The cause of this unexpected behaviour requires further investigation (see also Wall et al. 2016), but while up-regulation of pH_{cf} is a requirement for calcification of aragonitic corals, there are clearly other factors that influence its response to changes in the ambient reef-water pH. It is also noted that the Heron Island lagoon is somewhat unusual in being a semi-enclosed

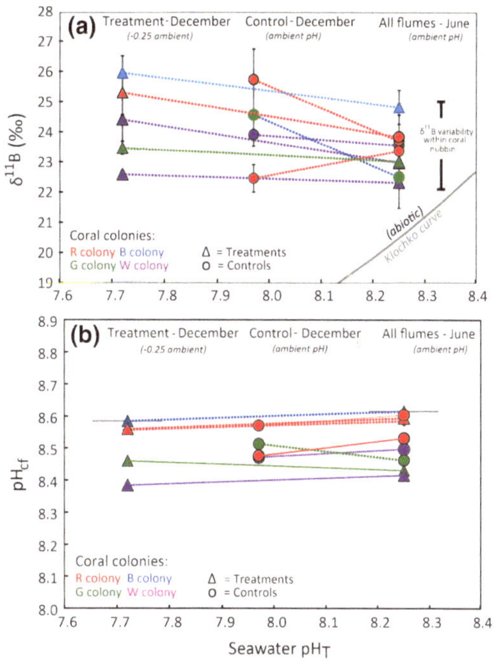

Fig. 6.7 **a** Boron isotopic compositions (Georgiou et al. 2015) from colonies (*coloured symbols*) of *Porites cylindrica*, subject to a Free Ocean Carbon Enrichment (FOCE) experiment conducted in the Heron Island lagoon of the GBR (Kline et al. 2012). At the commencement of the experiment (June 2010) the control and treatments colonies were at same the ambient lagoon water pH_T (~8.26). In December FOCE treatment colonies are at ~0.25 pH units below ambient (control) lagoon conditions. The $\delta^{11}B$ determinations were undertaken using samples representative (~5 mg) of coral growth during these periods (Georgiou et al. 2015). **b** Calcification fluid (pH_{cf}) derived from the $\delta^{11}B$ measurements (*above*) plotted against FOCE and lagoon water pH_T for the months of June and December. Despite the large natural and FOCE induced variations in seawater pH (Kline et al. 2015), for each coral branch there is a near constant pH_{cf}. Differences between colonies is significantly greater (>×5) than from polyps from the same branch, indicative of strong physiological controls (i.e. homeostasis) on the calcifying fluid pH_{cf}

and relatively shallow lagoon that experiences very large daily and seasonal fluctuations in seawater pH (Kline et al. 2015). The highest pH values of lagoon waters occur in the winter, due in part to cooling but also as a result of reduced calcification (a source of CO_2), whereas in summer, increased calcification together with temperature acts to reduce the seawater pH. Given these extreme natural fluctuations (Kline et al. 2015), we surmise that local adaptations (e.g. homeostasis of pH_{cf}) by corals growing within the highly variable Heron Island lagoon played a significant role in their neutral response to the enhanced conditions of ocean acidification simulated in the FOCE experiment.

6.7.2 Corals Under Natural Conditions of Seasonal Forcing

Corals growing under natural conditions provide another means to examine the relationship between pH_{cf} and seasonally driven fluctuations in seawater pH. A prerequisite is that the sclerochronology of coral growth must be well defined, such as occurs in *Porites*, which usually have well-defined annual growth bands (e.g. Knutson et al. 1972; Buddemeier et al. 1974) that can be sampled at better than monthly resolution. Also important is the availability of long-term in situ records of both reef-water temperature and pH, a requirement to relate changes in the δ^{11}B-derived pH_{cf} to conditions in the reef environment. This often represents a significant limitation, and although monitoring of reef-water pH is now becoming more routine, reliable 'long-term' data sets are still limited, especially given that multi-year records are ideally required. Thus, here we use *Porites* cores collected in 2013 from Davies Reef in the central GBR at the site of the Australian Institute of Marine Science (AIMS) reef-water monitoring station where long-term in situ sea surface temperature (SST) records are available, together with more limited seasonal records of reef-water pH variability (Albright et al. 2013; Barnes 1983; Dennison and Barnes 1988). This is crucial since many reefs exhibit significant variability in reef-water pH at both diurnal as well as seasonal time-scales, dependent on the combination of productivity/calcification controls and the residence times of water within reef systems. In order to extend the existing pH_T data over multiple years, these factors were quantified using seasonal in situ records of seawater carbonate chemistry (Albright et al. 2013; Barnes 1983; Dennison and Barnes 1988). Although being of limited duration the seasonal records enabled the reef-induced component of seawater pH to be estimated (Falter et al. 2013) and hence extended using the much longer temperature records.

Using the above approach, a time-series (Fig. 6.8) was obtained for reef-water pH, temperature, and δ^{11}B-pH_{cf} using corals from Davies Reef for the period from ~2007 to the end of 2013 (McCulloch et al. 2017). The reef-water pH ranges from ~8.02 in the summer to ~8.09 in the winter, an overall seasonal variation of ~0.07 pH units. This pattern of reef-water pH variability for Davies Reef is similar and only slightly greater than that for the open-ocean since the Davies reef-waters have short residence times (hours), being subject to ongoing open-water exchange. Seasonal δ^{11}B-pH_{cf} records were obtained (Fig. 6.8) for two cores (Davies 2 and 3; McCulloch et al. 2017); both exhibit large variation in pH_{cf} (~0.25 pH units) with a typical pH range from 8.3 to 8.5, the highest values being during winter. Also shown are the pH_{cf} values expected using the aquaria-derived relationships with seawater pH described previously for *Porites*, where $pH^*_{cf} = 0.4\ pH_T + 4.72$ (average for *Porites* calibration from Fig. 6.4). Thus, although seasonal variations in seawater pH and calcifying fluid pH_{cf} remain in phase, the amplitude of pH_{cf} variations are an order of magnitude larger (Fig. 6.8) than that predicted from laboratory (aquaria) derived pH_{cf} versus seawater pH relationship.

The role of temperature in influencing the calcifying fluid pH_{cf} and seawater pH is also illustrated in Figs. 6.9 and 6.10. As already noted, for Davies Reef the seasonal variation in seawater pH is relatively minor (±0.035 pH units), contrasting to the much larger variations in pH_{cf} exhibited by both *Porites* colonies (±0.12 pH units). Importantly, the largest deviation in pH_{cf} (δ^{11}B) compared to those expected from changes in seawater pH alone, occurs during the summer when temperatures are higher and hence physiological forcing is expected to be at its greatest. This is apparent, for example, in Fig. 6.10 where the two co-located *Porites* colonies converge to the theoretical winter value of pH_{cf}. This pattern is thus consistent with

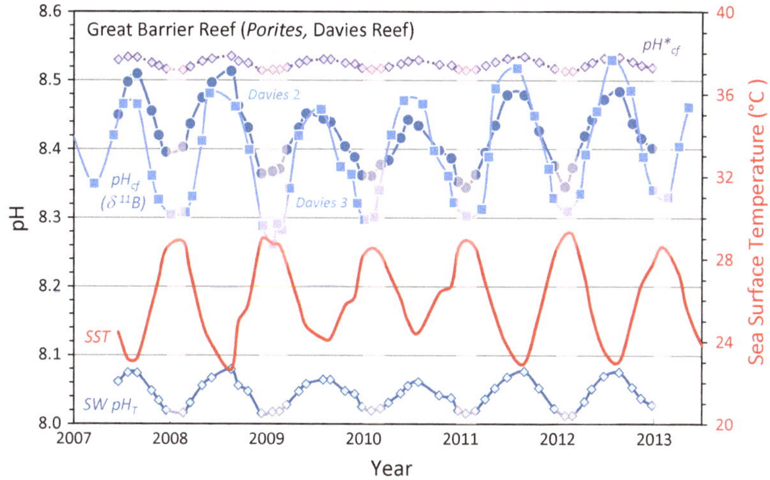

Fig. 6.8 Time-series (2007–2013) of reef water pH (Albright et al. 2013; Falter et al. 2013) and temperature measured at the Davies Reef (GBR) AIMS monitoring station, together with pH_{cf} derived from $\delta^{11}B$ measurements of several *Porites* coral cores (Davies 2 and 3). Reef water pH_T ranges from ~8.02 in summer (vertical shaded zones) to a maximum of ~8.09 in winter, an overall seasonal amplitude of ~0.07 pH units. The $\delta^{11}B$ derived seasonal variations in pH_{cf} are ~0.25 pH units, an order of magnitude larger than that expected from aquaria calibrations (pH^*_{cf}) due to changes in seawater pH_T alone. This indicates major physiological enhancement of the seasonal amplitude in pH_{cf} under natural reef conditions

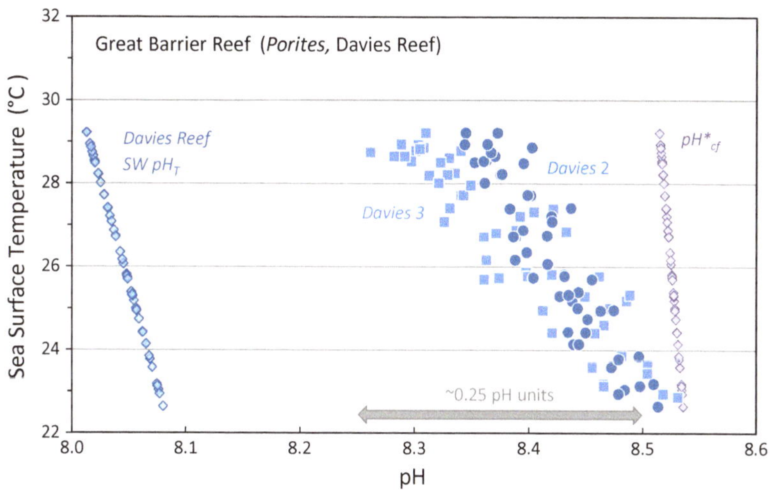

Fig. 6.9 Plot of $\delta^{11}B$ derived pH_{cf} values for *Porites* corals from Davies Reef (GBR) and ambient reef water pH_T, versus reef water temperatures. The range in pH_{cf} for *Porites* corals based on variations in seawater pH alone is shown as pH^*_{cf}. Although the absolute magnitude of pH_{cf} up-regulation relative to reef water pH_T is maintained (~0.45–0.3 pH units), the overall seasonal range in pH_{cf} for the *Porites* colonies is an order of magnitude greater than that expected from changes in reef water pH alone

subtly varying differences in the physiological control on pH_{cf} that are most strongly manifested during the summer months when symbiont activity is at its greatest.

Although there is a strong relationship (Fig. 6.10b) between temperature and $\delta^{11}B$ derived pH_{cf} for both the Davies Reef *Porites* colonies (Davies 2 and 3, $r^2 = 0.8–0.85$), they

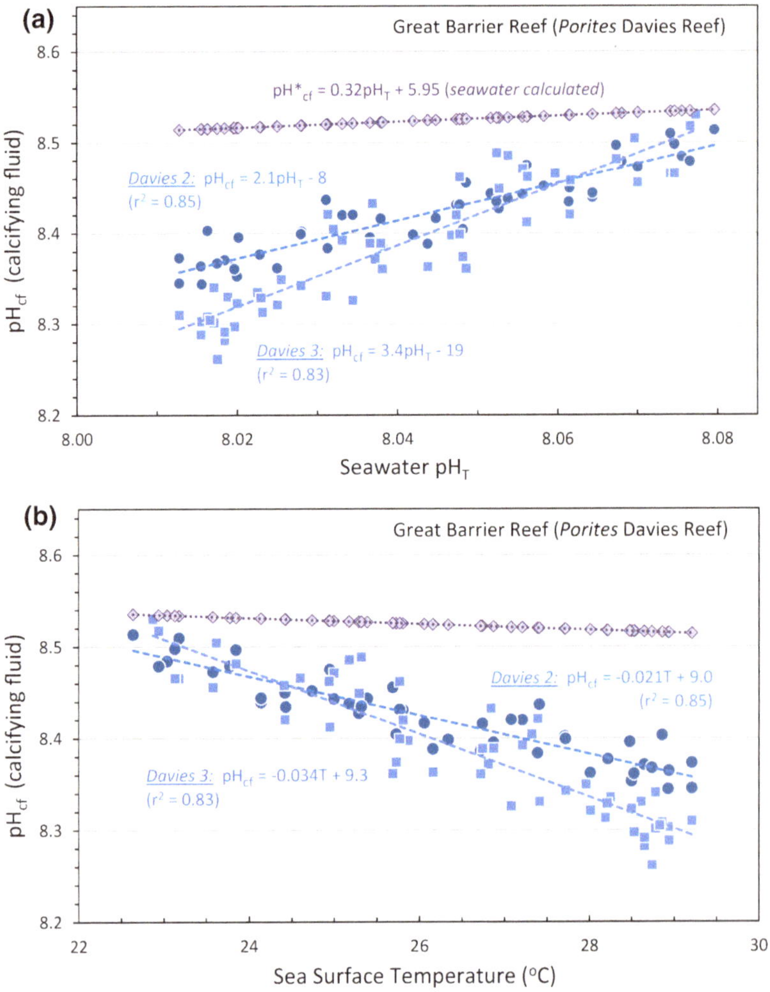

Fig. 6.10 **a** Comparison of $\delta^{11}B$ derived pH_{cf} for *Porites* colonies (Davies 2 and 3) from Davies Reef (GBR) with both the in situ determined reef water pH_T and the pH^*_{cf} values inferred from aquaria calibrations. The *Porites* colonies have an order of magnitude larger range in pH_{cf} than that expected from changes in reef-water pH alone. **b** Calcifying-fluid pH_{cf} versus in situ measured temperatures indicate a sensitivity of pH_{cf} from 0.02 to 0.032 pH units per degree

exhibit different sensitivities in pH_{cf} of ~0.02–0.034 per degree increase in temperature. This strong but variable temperature dependent sensitivity of pH_{cf} up-regulation in zooxanthellae-bearing corals again points to differing physiological controls at the colony scale. This clearly limits the application of $\delta^{11}B$-pH_{cf} by itself as a seawater pH palaeo-proxy given that changes in calcifying fluid pH_{cf} are closely linked to metabolic induced changes in DIC that likely accompany increasing temperatures. It is noted that such findings are not unexpected since the key parameter that ultimately dictates calcification is the saturation state of the calcifying fluid, which is directly dependent on total DIC, temperature, as well as pH of the calcifying fluid.

6.8 Summary and Conclusions

While the initial $\delta^{11}B$ studies of corals were largely motivated by its potential application as a proxy for palaeo-seawater pH, it is becoming increasingly apparent that under 'natural' real-world conditions a Pandora's Box has been opened with respect to physiological-environmental controls that influence coral bio-calcification. This is particularly the case for zooxanthellae-bearing corals, especially in the highly dynamic conditions that characterise many coral reefs. Although further studies are still required, particularly with processes controlling the DIC enrichment of the calcifying-fluid, four broad categories of pH_{cf} versus seawater pH_T responses have been identified (Fig. 6.11). Firstly, under conditions of constant temperature and seawater pH as applied in many experimental treatments, the coral's calcifying-fluid pH_{cf} is closely linked to changes in seawater pH with constant species-specific pH_{cf}/pH_T gradients from 0.3 to 0.5. Under these conditions, zooxanthellate corals typically up-regulate their pH_{cf} by ~0.3–0.6 pH units. Deep-sea azooxanthellate corals constitute a second closely related category, showing a very similar behaviour but with a greater magnitude of pH_{cf} up-regulation (~0.6–1 pH units). We conjecture that these two categories correspond to conditions of quasi-equilibrium between pH_{cf} and DIC, a result of stable seawater pH and possibly relatively constant seawater supply of DIC. Importantly, because of these characteristics, azooxanthellate corals may best satisfy the requirements for utilising the $\delta^{11}B$ palaeo-seawater pH proxy.

The third and fourth categories represent responses of zooxanthellae-bearing corals from tropical and sub-tropical environments. In tropical environments (third category), *Porites* show strong seasonal controls on pH_{cf}, apparently amplifying concomitant changes in seawater pH_T with the pH_{cf}/pH_T seasonal amplitudes ranging from 2 to 3. Here we conjecture that the dominant factor is from seasonal variability in symbiont derived DIC, resulting in higher carbonate-ion concentrations and higher rates of calcification despite the lower pH_{cf} during the summer. The fourth (sub-tropical) category of constant pH_{cf} may thus represent a physiological response in pH_{cf} to optimise calcification under conditions of limited DIC supply, with the

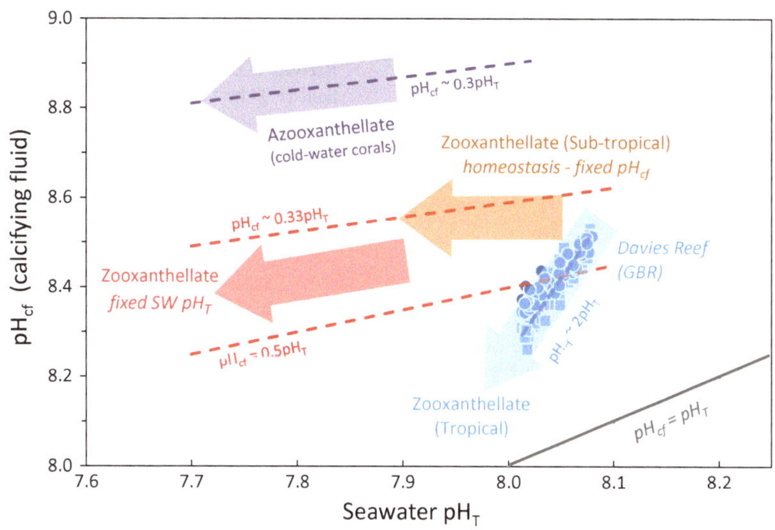

Fig. 6.11 Relationships of calcifying-fluid pH_{cf} ($\delta^{11}B$ derived) for scleractinian (aragonitic) corals versus ambient seawater pH_T. Four main responses are shown: (1) zooxanthellae-bearing corals grown under 'controlled' experimental conditions of fixed seawater pH_T show good correlations between seawater pH_T and the coral pH_{cf}. The gradient of pH_{cf}/pH_T is ~0.3–0.5 (*red*); (2) cold-water azooxanthellate corals with highly elevated pH_{cf}. The gradient of pH_{cf}/pH_T is ~0.3 (*purple*); (3) tropical zooxanthellae-bearing corals grown under 'natural' reef conditions with a gradient of pH_{cf}/pH_T ~2–3 (*blue*); (4) pH-homeostasis in zooxanthellae-bearing corals from highly variable sub-tropical environments (i.e. Heron Island lagoon) that exhibit near constant pH_{cf} independent of seawater and FOCE treatments of $pH_{T)}$ (*orange*). The *black solid line* corresponds to the inorganic $\delta^{11}B$ borate fractionation where $pH_{cf} = pH_T$ (Klochko et al. 2006)

calcifying-fluid pH_{cf} being optimised at a fixed value to maintain year-round calcification despite large variability in the ambient lagoon waters.

The latter findings are in marked contrast to constant-condition experiments, demonstrating more complex sets of interrelated physiological responses to changing environmental conditions. Up-regulation of the calcifying fluid pH_{cf} is clearly a common strategy that optimises coral calcification at an approximately constant rate. Process-based studies of bio-calcification must therefore be conducted under more realistic 'real-world' conditions to identify the critical thresholds of coral calcification to rapidly changing conditions that now increasingly characterise our high CO_2 World.

Acknowledgements This research was supported by funding from an ARC Laureate Fellowship awarded to Professor Malcolm McCulloch and the ARC Centre of Excellence for Coral Reef Studies. Dr Julie Trotter was supported by an ARC Future Fellowship and Dr Michael Holcomb by an ARC Super Science Fellowship. Dr Paolo Montagna is grateful for ongoing support by the Institute of Marine Sciences, CNR, Italy and this is ISMAR-CNR Bologna scientific contribution n. 1915. We thank Anne-Marin Comeau and Dr Kai Rankenburg for their technical assistance with measurements of the $\delta^{11}B$ isotopic and B/Ca elemental ratios being conducted at The University of Western Australia's Advanced Geochemical Facility for Indian Ocean Research (AGFIOR).

References

Al-Horani FA, Al-Moghrabi SM, de Beer D (2003) Microsensor study of photosynthesis and calcification in the scleractinian coral, *Galaxea fascicularis:* active internal carbon cycle. J Exp Mar Biol Ecol 288(1):1–15

Albright R, Langdon C, Anthony K (2013) Dynamics of seawater carbonate chemistry, production, and calcification of a coral reef flat, central Great Barrier Reef. Biogeosciences 10(10):6747–6758

Allemand D, Ferrier-Pagès C, Furla P, Houbrèque F, Puverel S, Reynaud S, Tambutté E, Tambutté S, Zoccola D (2004) Biomineralisation in reef-building corals: from molecular mechanisms to environmental control. CR Palevol 3(6–7):453–467

Allemand D, Tambutté E, Zoccola D, Tambutté S (2011) Coral calcification, cells to reefs. In: Dubinsky Z, Stambler N (eds) Coral reefs: an ecosystem in transition. Springer, Berlin, pp 119–150

Allison N, Finch AA, Eimf (2010) Delta B-11, Sr, Mg and B in a modern Porites coral: the relationship between calcification site pH and skeletal chemistry. Geochim Cosmochim Acta 74(6):1790–1800

Allison N, Cohen I, Finch AA, Erez J, Tudhope AW (2014) Corals concentrate dissolved inorganic carbon to facilitate calcification. Nat Commun 5:5741

Anagnostou E, Huang KF, You CF, Sikes EL, Sherrell RM (2012) Evaluation of boron isotope ratio as a pH proxy in the deep sea coral Desmophyllum dianthus: evidence of physiological pH adjustment. Earth Planet Sci Lett 349:251–260

Barnes D (1983) Profiling coral reef productivity and calcification using pH and oxygen electrodes. J Exp Mar Biol Ecol 66(2):149–161

Blamart D, Rollion-Bard C, Meibom A, Cuif J-P, Juillet-Leclerc A, Dauphin Y (2007) Correlation of boron isotopic composition with ultrastructure in the deep-sea coral *Lophelia pertusa:* implications for biomineralization and paleopH. Geochem Geophys Geosyst 8(Q12001):11

Buddemeier RW, Maragos JE, Knutson DW (1974) Radiographic studies of reef coral exoskeletons: rates and patterns of coral growth. J Exp Mar Biol Ecol 14 (2):179–199

Burton EA, Walter LM (1987) Relative precipitation rates of aragonite and Mg calcite from seawater: temperature or carbonate ion control? Geology 15(2):111–114

Caldeira K, Archer D, Barry JP, Bellerby RGJ, Brewer PG, Cao L, Dickson AG, Doney SC, Elderfield H, Fabry VJ, Feely RA, Gattuso JP, Haugan PM, Hoegh-Guldberg O, Jain AK, Kleypas JA, Langdon C, Orr JC, Ridgwell A, Sabine CL, Seibel BA, Shirayama Y, Turley C, Watson AJ, Zeebe RE (2007) Comment on "modern-age buildup of CO_2 and its effects on seawater acidity and salinity" by Hugo A. Loaiciga. Geophys Res Lett 34(18)

Ciais P, Sabine C, Bala G, Bopp L, Brovkin V, Canadell J, Chhabra A, DeFries R, Galloway J, Heimann M (eds) (2014) Carbon and other biogeochemical cycles. Climate change 2013: the physical science basis. Contribution of Working Group I to the fifth assessment report of the Intergovernmental Panel on Climate Change. Cambridge University Press, Cambridge, pp 465–570

Cohen AL, Holcomb M (2009) Why corals care about ocean acidification: uncovering the mechanism. Oceanography 22(4):118–127

Cohen AL, McConnaughey T (2003) Geochemical perspectives on coral mineralization. In: Dove P, Weiner S, Yoreo J (eds) Biomineral Rev Mineral Geochem pp. 151–187

Dennison WC, Barnes DJ (1988) Effect of water motion on coral photosynthesis and calcification. J Exp Mar Biol Ecol 115(1):67–77

Dickson AG (1990) Thermodynamics of the dissociation of boric acid in synthetic seawater from 273.15 to 318.15 K. Deep Sea Res Part A 37:755–766
Douville E, Paterne M, Cabioch G, Louvat P, Gaillardet J, Juillet-Leclerc A, Ayliffe L (2010) Abrupt sea surface pH change at the end of the Younger Dryas in the central sub-equatorial Pacific inferred from boron isotope abundance in corals (Porites). Biogeosciences 7(8):2445–2459
D'Olivo JP, McCulloch MT, Eggins SM, Trotter J (2015) Coral records of reef-water pH across the central Great Barrier Reef, Australia: assessing the influence of river runoff on inshore reefs. Biogeosciences 12(4):1223
Erez J (2003) The source of ions for biomineralization in foraminifera and their implications for paleoceanographic proxies. In: Dove PM, Yoreao JJD, Weiner S (eds) Rev Mineral Geochem 54:115–149
Falter JL, Lowe RJ, Zhang ZL, McCulloch M (2013) Physical and biological controls on the carbonate chemistry of coral reef waters: effects of metabolism, wave forcing, sea level, and geomorphology. PLoS ONE 8(1)
Fautin DG, Guinotte JM, Orr JC (2009) Comparative depth distribution of corallimorpharians and scleractinians (Cnidaria: Anthozoa). Mar Ecol Prog Ser 397:63–70
Försterra G, Beuck L, Häussermann V, Freiwald A (eds) (2005) Shallow-water *Desmophyllum dianthus* (Scleractinia) from Chile: characteristics of the biocoenoses, the bioeroding community, heterotrophic interactions and (paleo)-bathymetric implications. In: Freiwald A, Roberts JM (eds) Cold-water corals and ecosystems. Springer, Berlin, pp 937–977
Foster GL, Rae J, Elliott T (2008) Boron isotope measurements of marine carbonate using MC-ICPMS. Geochim Cosmochim Acta 72(12):A279
Foster GL, Pogge von Strandmann, PAE, Rae JWB (2010) Boron and magnesium isotopic composition of seawater. Geochem Geophys Geosyst 11(8)
Gattuso J-P, Allemand D, Frankignoulle M (1999) Photosynthesis and calcification at cellular, organismal and community levels in coral reefs: a review on interactions and control by carbonate chemistry. Am Zool 39:160–183
Gattuso JP, Kirkwood W, Barry JP, Cox E, Gazeau F, Hansson L, Hendriks I, Kline DI, Mahacek P, Martin S, McElhany P, Peltzer ET, Reeve J, Roberts D, Saderne V, Tait K, Widdicombe S, Brewer PG (2014) Free-ocean CO_2 enrichment (FOCE) systems: present status and future developments. Biogeosciences 11(15):4057–4075
Georgiou L, Falter J, Trotter J, Kline DI, Holcomb M, Dove SG, Hoegh-Guldberg O, McCulloch M (2015) pH homeostasis during coral calcification in a free ocean CO_2 enrichment (FOCE) experiment, Heron Island reef flat, Great Barrier Reef. Proc Natl Acad Sci 112(43):13219–13224
Hemming NG, Hanson GN (1992) Boron isotopic composition and concentration in modern marine carbonates. Geochim Cosmochim Acta 56(1):537–543
Holcomb M, Cohen AL, Gabitov RI, Hutter JL (2009) Compositional and morphological features of aragonite precipitated experimentally from seawater and biogenically by corals. Geochim Cosmochim Acta 73 (14):4166–4179
Holcomb M, Venn AA, Tambutte E, Tambutte S, Allemand D, Trotter J, McCulloch M (2014) Coral calcifying fluid pH dictates response to ocean acidification. Sci Rep 4:5207–5211
Honisch B, Hemming NG (2004) Ground-truthing the boron isotope-paleo-pH proxy in planktonic foraminifera shells: partial dissolution and shell size effects. Paleoceanography 19(4)
Honisch B, Hemming NG (2005) Surface ocean pH response to variations in pCO_2 through two full glacial cycles. Earth Planet Sci Lett 236:305–314
Honisch B, Hemming NG, Grottoli AG, Amat A, Hanson GN, Bijma J (2004) Assessing scleractinian corals as recorders for paleo-pH: empirical calibration and vital effects. Geochim Cosmochim Acta 68:3675–3685
Hughes TP, Baird AH, Bellwood DR, Card M, Connolly SR, Folke C, Grosberg R, Hoegh-Guldberg O, Jackson JBC, Kleypas J, Lough JM, Marshall P, Nystrom M, Palumbi SR, Pandolfi JM, Rosen B, Roughgarden J (2003) Climate change, human impacts, and the resilience of coral reefs. Science 301(5635):929–933
Kakihana H, Kotaka M, Satoh S, Nomura M, Okamoto M (1977) Fundamental studies on the ion-exchange separation of boron isotopes. Chem Soc Jpn B50:158–163
Kline DI, Teneva L, Schneider K, Miard T, Chai A, Marker M, Headley K, Opdyke B, Nash M, Valetich M (2012) A short-term in situ CO_2 enrichment experiment on Heron Island (GBR). Sci Rep 2
Kline DI, Teneva L, Hauri C, Schneider K, Miard T, Chai A, Marker M, Dunbar R, Caldeira K, Lazar B, Rivlin T, Mitchell BG, Dove S, Hoegh-Guldberg O (2015) Six month in situ high-resolution carbonate chemistry and temperature study on a coral reef flat reveals asynchronous pH and temperature anomalies. PLoS ONE 10(6):e0127648
Klochko K, Kaufman AJ, Yoa W, Byrne RH, Tossell JA (2006) Experimental measurement of boron isotope fractionation in seawater. Earth Planet Sci Lett 248:261–270
Klochko K, Cody GD, Tossell JA, Dera P, Kaufman AJ (2009) Re-evaluating boron speciation in biogenic calcite and aragonite using ^{11}B MAS NMR. Geochim Cosmochim Acta 73(7):1890–1900
Knutson DW, Buddemeier RW, Smith SV (1972) Coral chronometers: seasonal growth bands in reef corals. Science 177(4045):270–272
Krief S, Hendy EJ, Fine M, Yam R, Meibom A, Foster GL, Shemesh A (2010) Physiological and isotopic responses of scleractinian corals to ocean acidification. Geochim Cosmochim Acta 74 (17):4988–5001
Lewis E, Wallace DWR (1998) Program developed for CO_2 system calculations. In: C.D.I.A. Center (ed).

Oak Ridge National Laboratory, U.S. Department of Energy, Oak Ridge, p 21
Liu Y, Liu W, Peng Z, Xiao Y, Wei G, Sun W, He J, Liu G, Chou C-L (2009) Instability of seawater pH in the South China Sea during the mid-late Holocene: evidence from boron isotopic composition of corals. Geochim Cosmochim Acta 73(5):1264–1272
Mavromatis V, Montouillout V, Noireaux J, Gaillardet J, Schott J (2015) Characterization of boron incorporation and speciation in calcite and aragonite from co-precipitation experiments under controlled pH, temperature and precipitation rate. Geochim Cosmochim Acta 150:299–313
McCulloch M, Falter J, Trotter J, Montagna P (2012a) Coral resilience to ocean acidification and global warming through pH up-regulation. Nat Clim Change 2(8):623–627
McCulloch MT, Trotter JA, Montagna P, Falter J, Dunbar R, Freiwald A, Försterra G, López Correa M, Maier C, Rüggeberg A, Taviani M, Thresher R (2012b) Boron isotope systematics of cold-water scleractinian corals: internal pH up-regulation and response to ocean acidification. Geochim Cosmochim Acta 87:21–34
McCulloch MT, D'Olivo JP, Falter J, Holcomb M, Trotter JA (2017) Coral calcification in a changing World and the interactive dynamics of pH and DIC upregulation. Nat Commun 8
Moya A, Tambutté S, Bertucci A, Tambutté E, Lotto S, Vullo D, Supuran CT, Allemand D, Zoccola D (2008) Carbonic anhydrase in the scleractinian coral *Stylophora pistillata* characterization, localization, and role in biomineralization. J Biol Chem 283 (37):25475–25484
Noireaux J, Mavromatis V, Gaillardet J, Schott J, Montouillout V, Louvat P, Rollion-Bard C, Neuville D (2015) Crystallographic control on the boron isotope paleo-pH proxy. Earth Planet Sci Lett 430:398–407
Pelejero C, Calvo E, McCulloch MT, Marshall JF, Gagan MK, Lough JM, Opdyke BN (2005) Preindustrial to modern interdecadal variability in coral reef pH. Science 309(5744):2204–2207
Rae JWB, Foster GL, Schmidt DN, Elliott T (2011) Boron isotopes and B/Ca in benthic foraminifera: proxies for the deep ocean carbonate system. Earth Planet Sci Lett 302(3–4):403–413
Reynaud S, Hemming NG, Juillet-Leclerc A, Gattuso J-P (2004) Effect of pCO_2 and temperature on the boron isotopic composition of the zooxanthellate coral *Acropora* sp. Coral Reefs 23:539–546
Roberts JM, Wheeler AJ, Freiwald A (2006) Reefs of the deep: the biology and geology of cold-water coral ecosystems. Science 312(5773):543–547
Rollion-Bard C, Chaussidon M, France-Lanord C (2003) pH control on oxygen isotopic composition of symbiotic corals. Earth Planet Sci Lett 215(1–2):275–288
Rollion-Bard C, Blamart D, Trebosc J, Tricot G, Mussi A, Cuif J-P (2011) Boron isotopes as pH proxy: a new look at boron speciation in deep-sea corals using ^{11}B MAS NMR and EELS. Geochimica et Cosmochimicha Acta 75:1003–1012
Tanaka K, Holcomb M, Takahashi A, Kurihara H, Asami R, Shinjo R, Sowa K, Rankenburg K, Watanabe T, McCulloch M (2015) Response of Acropora digitifera to ocean acidification: constraints from $\delta^{11}B$, Sr, Mg, and Ba compositions of aragonitic skeletons cultured under variable seawater pH. Coral Reefs 34 (4):1139–1149
Thresher RE, Tilbrook B, Fallon S, Wilson NC, Adkins J (2011) Effects of chronic low carbonate saturation levels on the distribution, growth and skeletal chemistry of deep-sea corals and other seamount megabenthos. Mar Ecol Prog Ser 442:87–99
Trotter JA, Montagna P, McCulloch MT, Silenzi S, Reynaud S, Mortimer G, Martin S, Ferrier-Pageè C, Gattuso J-P, Rodolfo-Metalpa R (2011) Quantifying the pH 'vital effect' in the temperate zooxanthellate coral *Cladocora caespitosa*: validation of the boron seawater pH proxy. Earth Planet Sci Lett 303:163–173
van Vuuren DP, Edmonds J, Kainuma M, Riahi K, Thomson A, Hibbard K, Hurtt GC, Kram T, Krey V, Lamarque J-F, Masui T, Meinshausen M, Nakicenovic N, Smith SJ, Rose SK (2011) The representative concentration pathways: an overview. Clim Change 109(1–2):5–31
Vengosh A, Kolodny Y, Starinsky A, Chivas AR, McCulloch MT (1991) Coprecipitation and isotopic fractionation of boron in modern biogenic carbonates. Geochim Cosmochim Acta 55(10):2901–2910
Venn AA, Tambutte E, Lotto S, Zoccola D, Allemand D, Tamubtte S (2009) Imaging intracellular pH in a reef coral and symbiotic anemone. In: Falkowski PG (ed) PNAS, pp 16574–16579
Venn AA, Tambutte E, Holcomb M, Laurent J, Allemand D, Tambutte S (2013) Impact of seawater acidification on pH at the tissue–skeleton interface and calcification in reef corals. Proc Natl Acad Sci U S A 110(5):1634–1639
Wall M, Fietzke J, Schmidt GM, Fink A, Hofmann LC, de Beer D, Fabricius KE (2016) Internal pH regulation facilitates in situ long-term acclimation of massive corals to end-of-century carbon dioxide conditions. Sci Rep 6:30688
Wei G, McCulloch MT, Mortimer G, Deng W, Xie L (2009) Evidence for ocean acidification in the Great Barrier Reef of Australia. Geochim Cosmochim Acta 73:2332–2346
Zeebe R, Wolf-Gladow DA (2001) CO_2 in seawater: equilibrium, kinetics, isotopes Elsevier Oceanography Series, vol 65. Elsevier, Amsterdam, p 65
Zoccola D, Ganot P, Bertucci A, Caminiti-Segonds N, Techer N, Voolstra CR, Aranda M, Tambutté E, Allemand D, Casey JR, Tambutté S (2015) Bicarbonate transporters in corals point towards a key step in the evolution of cnidarian calcification. Sci Rep 5:9983

Boron in the Weathering Environment

7

Jérôme Gaillardet and Damien Lemarchand

Abstract

This chapter reviews the state of art of the use of boron isotopes to understand water-rock interaction in the Critical Zone, the thin and reactive layer at the Earth's surface. Because boron isotopes are largely fractionated by adsorption, coprecipitation and evaporation-condensation processes, boron isotopes are well adapted to trace the main processes that convert rocks into soils and sediments on terrestrial surfaces. The difference in affinity of boron isotopes between trigonal and tetrahedral species is the main cause of isotope fractionation of boron at the Earth's surface. Due to the competition between the speciation of boron in solution and the speciation on boron onto or into solids or gas, large isotopic variations are predicted and observed. Measured boron isotopic composition in the weathering environment varies over a considerable range of about 70‰. Precipitation, rivers and biomass are usually enriched in ^{11}B, while a complementary depletion in ^{11}B (enrichment in ^{10}B) is observed in clay minerals and on organic or inorganic surfaces. At the ecosystem scale, boron appears to behave as a micronutrient with a major flux of boron associated with biological recycling. The inputs of boron to ecosystems by chemical weathering or from the atmosphere are minor. When the residence time of water in the critical zone is high, such as in groundwater systems, boron contents increase and are much more dominated by a weathering signal. Boron is mainly added to the ocean by rivers, while the most important sink of boron is adsorption on clay minerals. This makes boron a particularly good tracer of the weathering/erosion balance of terrestrial surfaces in addition to its capacity for tracing the pH and ancient seawater. A lot remains to be

J. Gaillardet (✉)
Institut de Physique du globe de Paris, Sorbonne Paris Cité, Université Paris Diderot, 75005 Paris, France
e-mail: gaillardet@ipgp.fr

J. Gaillardet
Institut Universitaire de France, Paris, France

D. Lemarchand
Université de Strasbourg, EOST, 1 Rue Blessig, 67084 Strasbourg Cedex, France
e-mail: lemarcha@unistra.fr

H. Marschall and G. Foster (eds.), *Boron Isotopes*, Advances in Isotope Geochemistry,
https://doi.org/10.1007/978-3-319-64666-4_7

done to better understand the behavior of boron and boron isotopes at the Earth's surface and on the secular evolution of boron isotopes in the ocean but our review of the available literature shows that this tracer has a great potential at a local (ecosystem) and global (ocean) scale.

Keywords
Critical zone · Weathering · Biological recycling · Clay formation
Nutrient · Denudation

7.1 The Main Fractionating Mechanisms of Boron on Terrestrial Surfaces

7.1.1 The Crystallochemistry of Boron

Boron is the fifth element in the periodic table (atomic number 5) with the ground-state electronic structure [He]$2s^2 2p^1$. Because of its poor stability during nucleosynthesis reactions in stars, its abundance on Earth is very low, comparable to elements of atomic number 30–40 (e.g. Zinc). Boron is highly incompatible during magmatic processes resulting in a large enrichment in the continental crust (about 10 $\mu g\ g^{-1}$, Rudnick and Gao 2003) relative to the mantle (about 0.1 $\mu g\ g^{-1}$, McDonough and Sun 1995).

Boron's neighbors in the periodic table are carbon and aluminium and it is close to silicon, with which it shares common physico-chemical properties. Its very high ionization potential results in the formation of covalent (rather than ionic) bonds. Boron is a strong electron acceptor and is therefore almost always bonded to oxygen in geological materials. In minerals and in aqueous liquids, boron is found in coordination III (BO_3, H_3BO_3) or IV (BO_4, $B(OH)_4^-$). This difference in coordination is the principal driving force controlling the chemical reactivity of boron, overshadowing other possible effects like temperature, and therefore often reduces the interpretation of the boron distribution in minerals and solutions to a simple question of boron sources and speciation.

In granitic and volcanic rocks, boron substitutes for Si or Al and therefore belongs to the group of structure-forming elements, by opposition of charge-compensator elements that occupy the other crystallographic sites to balance the intracrystalline electric charges. Boron concentrations in silicic intrusive rock range between 10 and 30 $\mu g\ g^{-1}$, with a large variability between minerals: only traces of boron are found in quartz, feldspar and plagioclases usually have boron concentrations in the lower range ($\sim 10\ \mu g\ g^{-1}$) and much higher concentrations are observed in phyllosilicates like micas (up to 100 $\mu g\ g^{-1}$). In secondary weathering products like clay minerals, boron concentrations can be very high (up to 1000 $\mu g\ g^{-1}$) mostly depending on the nature of the surrounding fluid from which they form. Boron is variably distributed between the crystal structure and the interlayer sites depending on the mineral transformation and the succession of fluids in contact (Williams et al. 2001a; Voinot et al. 2013).

In carbonates, incorporation of boron concentrations and isotopes is pH-dependent and has been used in many studies to reconstruct seawater paleo-pH from marine records (see Chaps. 5 and 6). However, the actual mechanism of boron incorporation into the carbonate lattice is still debated and the initial model of boron incorporation as borate ion (Vengosh et al. 1991a; Hemming and Hanson 1992) is challenged by new results provided by nuclear magnetic resonance analyses, ^{11}B NMR (Klochko et al. 2009; Rollion-Bard et al. 2011; Mavromatis et al. 2015; Noireaux et al. 2015; see Chap. 4 for full discussion). These studies indicate that boron is found in both trigonal and tetrahedral forms and may occupy different crystallographic sites depending on the calcium carbonate polymorph (calcite or aragonite). So far, the relationship between water chemistry and boron speciation in carbonates remains unclear and calls for a more complex mechanism of boron incorporation.

7.1.2 Interaction of Dissolved Boron with Mineral Surfaces

Early studies of Raman spectroscopy (Edwards et al. 2002) and ^{11}B NMR spectroscopy (Onak et al. 2005) have demonstrated that boron in solution exists in trigonal (boric acid, $B(OH)_3$) and tetrahedral (borate ion, $B(OH)_4^-$) forms. Monomeric species are only found in solutions with boron concentrations below 0.025 mol/kg whereas polymeric forms prevail at higher concentrations and high pH (Ishihara et al. 1991). Dissolved boron reacts as a Lewis acid with solid surfaces to form mostly inner-sphere complexes (ligand exchange) with (poly)hydroxyl groups (e.g. Pizer and Selzer 1984; Shao et al. 2000; Lemarchand et al. 2007). Formation of outer-sphere complexes (physical adsorption) has however been reported at the surface of some minerals like birnessite (Lemarchand et al. 2007) or hydrous ferric oxides (Peak et al. 2003). Some reactions of dissolved boron with surfaces are presented in Fig. 7.1. The surface protonation/deprotonation reactions and the boron weak acidity in aqueous solutions explain the pH-dependence of boron interactions with solids. The absence of ions other than protons or hydroxides involved in boron complexation also explains the almost independence of boron surface reactions to the solution ionic strength (e.g. Goldberg 2005; Lemarchand et al. 2007).

The main boron adsorbing surfaces in soils are clay minerals, metal oxides, carbonate and organic matter (Goldberg 1997). The boron partition coefficient ($Kd = [B]_{ads}/[B]_{sol}) = (m_{Bads}/m_{solid})/(m_{Bsol}/m_{sol})$ with $[B]_{ads}$ and $[B]_{sol}$ the concentrations of adsorbed and dissolved boron and m_{Bads}, m_{solid}, m_{Bsol}, m_{sol}, the mass of adsorbed boron, the mass of solid, the mass of boron in solution, and the mass of solution, respectively) between solution and surface depends on the specific surface area of the solid exposed to water and on the density of the surface hydroxyl groups. It may therefore vary over several orders of magnitude. Tectosilicates, characterized by a high level of polymerization like quartz and feldpars, show very small Kd values (typically below 1), whereas phyllosilicates and organic compounds show much higher values (typically in the order of 10 or 100 with a maximum value of 40,000 reached by the synthetic boron specific resin Amberlite IRA 743, Lemarchand et al. 2002). Boron shows moderate affinity for carbonate surface with Kd values typically comprised between 1 and 6 (Goldberg and Forster 1991; Ranjbar and Jalali 2014).

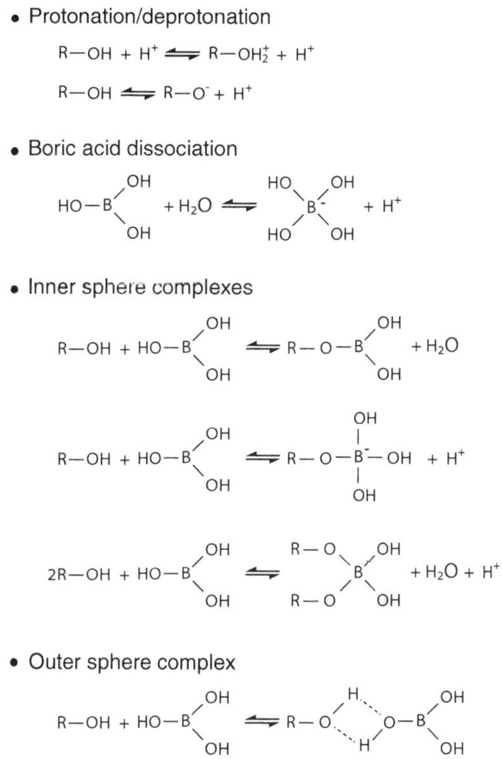

Fig. 7.1 Examples of chemical reactions of boron complexation onto solid surfaces and speciation in aqueous solutions. *R* represents a cation on a mineral surface and is commonly Si, Al or any other major or minor cation (Ca, Fe, Mg,…) in silicates or C in organic materials

The combination of a generally large Kd value with common mineral and organic surfaces (Fig. 7.2a) with the relatively low water/rock ratio of porous media like soils or groundwaters lead boron to be largely in adsorbed form in the weathering environment. Solid surfaces therefore

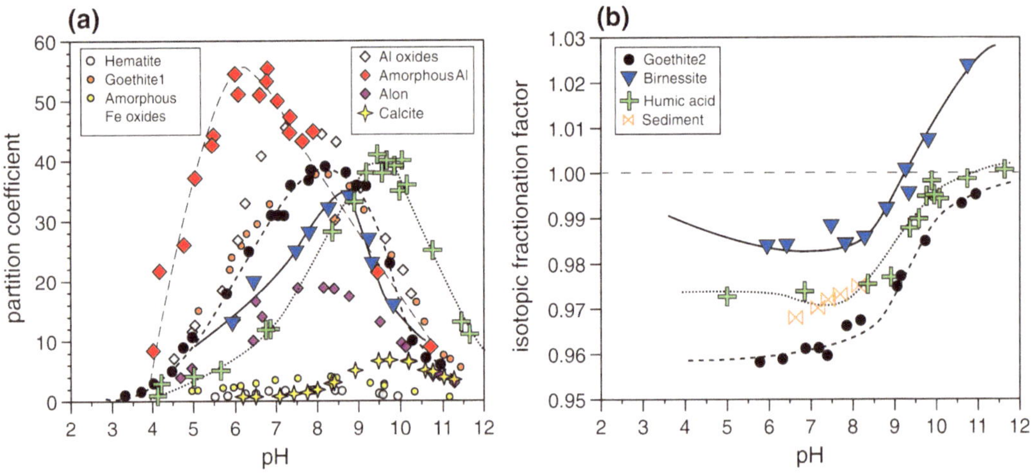

Fig. 7.2 Relationships between pH and **a** the boron partition coefficient (concentration on the solid/concentration in solution) and **b** the solid-solution boron isotopic fractionation factor. For hematite, goethite1, Fe-oxides, Al-oxides, amorphous-Al, data are from Goldberg and Glaubig (1985). Goethite2 and birnessite are from Lemarchand et al. (2007). Humic acid is from Lemarchand et al. (2005). Calcite data are from Goldberg (1991). Sediment data are from Palmer et al. (1987). Solid and dashed lines for humic acids goethite and birnessite correspond to the curves modeled using two to four surface complexes as proposed by Lemarchand et al. (2005, 2007)

constitute an important pool of exchangeable boron that can be mobilized by changes to the ambient hydro-chemical conditions. For instance, a modest partition coefficient of 5 in a porous media with a typical water/rock ratio of 0.1 leads to 98% of boron being adsorbed. In rivers or lakes, where the water/rock ratio is much larger (on the order of 10^5), boron is found mostly in the dissolved phase. It is therefore expected that the transport of particles from soils to rivers leads to a massive desorption of boron.

The chemical stability of the boron-ligand complex depends on the acid dissociation constant of both the boric acid and the ligand (Pizer and Selzer 1984 and references therein). Adsorption on surfaces therefore reflects the competition for the surface reactive sites (that may be pH-dependent) between protons, hydroxides, boric acid and borate ions. A compilation of the pH-dependence of boron partition coefficients with various solid surfaces is shown in Fig. 7.2a. Bell-shaped curves with maximum partition coefficients reached between 8 and 10 (close to the acid dissociation constant of boric acid, pK_A = 9.01 in diluted waters, Dickson, 1990) are reported for the considered compounds.

Various approaches have been proposed to model the interactions with solid surfaces. Keren and Mezuman (1981) proposed a phenomenological equation, but this empirical approach fails to provide a physical significance to the parameter determined. Freundlich and Langmuir adsorption isotherms were used to satisfactorily fit experimental data (Lemarchand et al. 2005, 2007; Ranjbar and Jalali 2014). However, the determined parameter values are difficult to extrapolate to other conditions than the experimentally tested ones. In parallel, more sophisticated chemical models were adapted to boron, like the total capacitance (Goldberg et al. 2000) and the triple layer model (Goldberg 2005), with mixed success. So far, the total capacitance model requires smaller number of adjustable parameters than other methods and can be applied to satisfactorily predict the boron distribution in soils, but still assumes that boron only forms inner-sphere complexes and is not adapted to investigate with precision the nature of the boron complexation reactions (see review by Goldberg and Su 2007).

Only a few studies have focused on the boron isotopic fractionation during boron adsorption/desorption reactions. Schwarcz et al. (1969) were the first to investigate the fractionation of boron isotopes during adsorption onto clays. Palmer et al. (1987) reported boron isotopic fractionation during adsorption experiments on marine clay and first proposed that changes of boron speciation may be the cause of the observed isotopic fractionation. Later, Lemarchand et al. (2005 and 2007) investigated the boron isotopic fractionation during interactions with humic acids and Fe- or Mn-oxides. In these studies, coupling boron isotope data with Diffuse-Reflectance Infrared Fourier-Transformed (DRIFT) spectroscopy clarified the nature of the boron reactions with solid surfaces and proposed the formation of monodentate or bidentate B-complexes or outer-sphere B-complexes. Large isotopic fractionations are observed in relation to the water pH, which can mainly be explained by the boron coordination in water and adsorbed on the surface but also, to a smaller extent, by the nature of the surface as well as the complex formed. Figure 7.2b shows the modeled curves proposed by Lemarchand et al. (2005, 2007) for the surface-solution fractionation factor for humic acids and oxides. The boron isotopic fractionation factor between solution and surface is highest at acidic pH (down to $\alpha = 0.945$, pH < 7) and close to 1 at basic pH (pH > 9). In order to fit the sigmoid dependence of the isotope data, several complexes have to be postulated depending on the pH range, each having different isotopic fractionation and different complexation constant. The origin of the large isotopic fractionation is easily explained by the difference of steric structure between B-surfaces complexes and B species in solution (both pH-dependent). For example, steric strains induced by the formation of trigonal bidendate complexes are responsible for the strong enrichment in ^{10}B at the surface of goethite, whereas the formation of a B trigonal outer-sphere complex causes the enrichment of birnessite surfaces in ^{11}B at pH 9 (Lemarchand et al. 2007). The structure of the surface complexes formed during the adsorption of boron onto goethite and birnessite are shown in Fig. 7.3.

7.1.3 Coprecipitation of Boron into Solids

The behavior of boron isotopes during reactions of mineral precipitation has been less studied than during adsorption reactions. So far, published studies have mainly focused on carbonates and rarely on silicates. They are based on experimental precipitations conducted at moderate to high temperature, to ensure high reaction rates.

Studies of boron coprecipitation in carbonates are motivated by the use of boron isotopes in geological archives to reconstruct ancient seawater pH, on the basis that the borate ion is the only boron species to be incorporated into the carbonate lattice. This hypothesis is however challenged by ^{11}B NMR observations (Rollion-Bard et al. 2011; Noireaux et al. 2015). Recent inorganic precipitation experiments conducted by Mavromatis et al. (2015) and Noireaux et al. (2015) showed that, in addition to pH, there is a crystallographic control on the incorporation of boron isotopes in carbonate crystals. In aragonite, boron is quantitatively present in its four-coordinated form, whereas both the tetrahedral and trigonal forms are observed in calcite. These authors proposed that tetrahedral boron is the only form incorporated in aragonite by substitution of CO_3^{2-} anions but that boron actually enters the calcite lattice in both tetrahedral and trigonal forms. The possibility of forming trigonal species in calcite was further proposed from first principle calculations by Balan et al. (2016) in the form of $BO_2(OH)^{2-}$. In parallel, Uchikawa et al. (2015) and Mavromatis et al. (2015) recently proposed that the boron incorporation in carbonates can be controlled by a diversity of mechanisms (see Chap. 4 for more discussion of this topic).

Published studies on B coprecipitation in silicates at moderate to high temperatures, include mineral-silicate melt isotope partitioning (e.g. Tonarini et al. 2003), the behavior of boron in coexisting hydrous solutions and silicate melts in subduction zones (e.g. Peacok and Hervig 1999) and diagenetic environments, in particular those involving maturation of organic matter

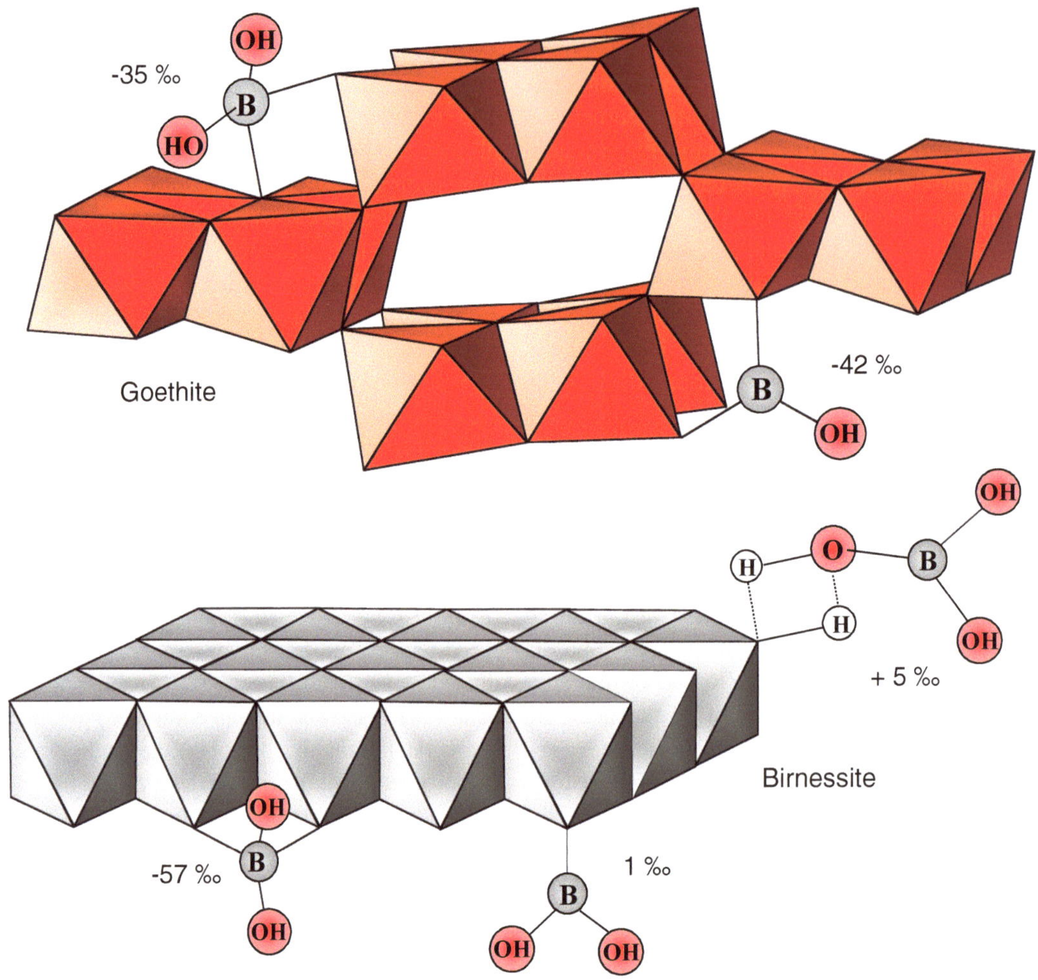

Fig. 7.3 Structure of the boron complexes formed at the surface of goethite (α-FeOOH) and birnessite ($K_{0.1}Mn_{0.22}$, $0.9H_20$) and corresponding isotopic fractionation factors between the surface and H_3BO_3 form in solution. Results from experiments at 25 °C and I = 0.1 M. Complexes surfaces were constrained by infrared spectroscopic analysis and modeling of speciation at the oxide-solution interface

(e.g. Williams et al. 2001a, b). This later study showed that boron initially adsorbed on clay mineral surfaces at low temperature is subsequently incorporated into the silicate lattice by substitution of Si in the form of tetrahedral B species if temperature exceeds 60 °C, corresponding to the transformation of smectite to illite. As the tetrahedral boron is preferentially enriched in the ^{10}B isotope, the illitization process of the smectite was shown to enrich the (trans)formed clay minerals in ^{10}B. These studies allowed the determination of an empirical relationship between temperature and the magnitude of the fluid-solid boron isotope fractionation, shown in Fig. 7.4. The interesting point is that the temperature dependence of the boron isotopic fractionation appears similar for all reactions involving silicates and water regardless of the conditions under which the reaction has occurred (i.e. aqueous fluid/silicate melts, smectite/illite transformation and adsorption onto detrital clay minerals).

In the absence of a thorough investigation at ambient temperature of the boron isotopic fractionation during reactions of mineral precipitation, it is still speculated that the incorporation of

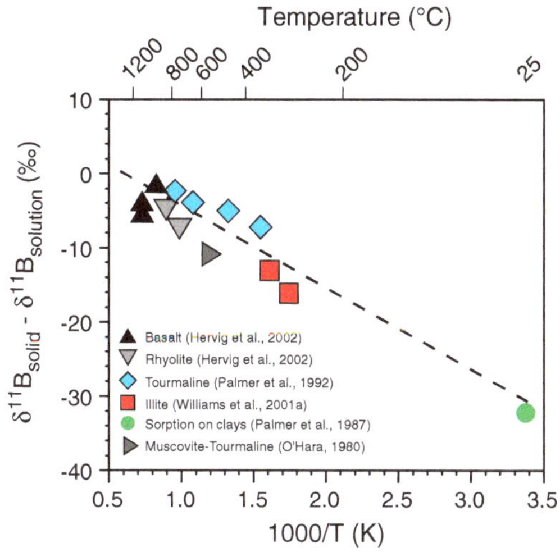

Fig. 7.4 Compilation of boron isotopic fractionation, expressed as the solid-fluid isotopic shift, associated with B coprecipitation in different solids and B sorption onto clays as a function of temperature. The dashed line corresponds to the least-square fitting of the data for reactions between water and silicates and corresponds to $\delta^{11}B_{solid}-\delta^{11}B_{solution} = -11280.T^{-1} + 6.49$ with T in K

boron isotopes into solids should primarily proceed through the formation of the same surface complexes that control adsorption followed by a quantitative incorporation into the mineral lattice (Sen et al. 1994; Lemarchand et al. 2007; Lu et al. 2011). If so, precipitation and adsorption reactions should generate similar boron isotopic fractionations and the difference would only depend on the rate of boron exchange with the fluid.

7.1.4 Behavior of Boron During Evaporation and Condensation Processes

In addition to its ability to sorb onto surfaces, boron is a volatile element. Boron is partitioned into a vapor phase and volcanic rocks as indicated by its abundance in fumaroles (i.e. Leeman et al. 2005). Experimental studies conducted at high temperature (Spivack et al. 1990; Palmer et al. 1992) showed that boron is partitioned between a vapor phase and brine/solid phase in hydrothermal conditions, with the vapor slightly enriched in ^{11}B. The presence of gaseous boric acid in the atmosphere has been recognized from the first evidence of an increase by a factor of hundreds of the boron-chlorinity ratio in rainwaters relative to seawater and preferential removal of chlorine as the rain continued (Sugawara 1948). In 1959, Gast and Thomson conducted seawater evaporation experiments and concluded that seawater is a major source of atmospheric boron. Fogg and Duce (1985) used a filter sampling system and found a large proportion of atmospheric boron present in the gas phase as boric acid. In addition, a number of studies have used experimental approaches to determine boron volatility and associated isotopic fractionations, motivated by the issue (or advantage) of boron volatility during its separation from geological matrices (Ishikawa and Nakamura 1990; Xiao et al. 1997; Gaillardet et al. 2001). Available studies show that the behavior of boron and boron isotopes during the evaporation of seawater is strongly dependent upon experimental conditions. If only boric acid evaporates from a solution at neutral pH, it is expected, in the absence of kinetic effects, that the isotopic composition of the gaseous form will be that of the boric acid species. In that case, the $\delta^{11}B$ of the gas ($\delta^{11}B_{gas}$) is simply a function of seawater pH at equilibrium:

$$\delta^{11}B_{gas} = \delta^{11}B_{sw} + 1000(\alpha_{3-4} - 1)\frac{1}{1 + 10^{pH-pKA}}$$

where $\delta^{11}B_{sw}$ is the isotopic composition of seawater, α_{3-4} is the boron isotopic fractionation factor between boric acid and borate ion in solution ($\alpha_{3-4} = 1.027$, Klochko et al. 2006) and pK_A is the boric acid dissociation constant ($pK_A = 8.6$ in seawater at 25 °C and 35 psu, Dickson 1990). As a result, the modeled boron isotopic composition of the atmospheric gaseous form in equilibrium with seawater (pH = 8.2 should be close to 50‰), which is higher than the value for seawater ($\delta^{11}B = 39.6$‰). Experimental studies using seawater solutions at different pH confirm that the vapor phase from a seawater solution is enriched in ^{11}B (Chetelat et al. 2005, Xiao et al. 2007). However, re-interpreted using the fractionation factor of Klochko et al. (2006), these data show that the $\delta^{11}B$ of the vapor phase is much lower than the $\delta^{11}B$ of aqueous boric acid. This important conclusion implies either fractionation of boric acid during its transfer from the solution to the gas, which is unlikely, or kinetic fractionation during evaporation with preferential escape of the ^{10}B-enriched boric acid. Using B-doped seawater, Xiao et al. (2007) made the interesting observation that the apparent isotopic shift between seawater and gaseous boron increases with B concentration, which they interpreted as indicating a less important kinetic effect at high B concentration. By contrast, Xiao et al. (2001) performed airflow experiments and showed that the vapor phase was enriched in ^{10}B compared to seawater. This result was used by Miyata et al. (2000) and Rose-Koga et al. (2006) to interpret natural data of $\delta^{11}B$ in rainwater and air by proposing that the evaporation of boron at the seawater-atmosphere interface and condensation of boron in clouds were characterized by a seawater-vapor kinetic fractionation of +25.5‰ (the vapor is enriched in ^{10}B compared to seawater) and a rain-vapor fractionation of +32‰.

This rather confusing situation regarding the behavior of B isotopes during evaporation and condensation clearly shows that the experimental conditions are important and can easily change the direction and magnitude of the B isotope fractionation. Experiments of Chetelat et al. (2005) and Xiao et al. (2007) were conducted in stagnant air systems while Xiao et al. (2001) reported evaporation experiments under airflow conditions. It is plausible that kinetic effects are only visible in airflow systems. More experimental work is needed to address both the kinetic and the equilibrium fractionation factors associated with B evaporation from solution or B condensation from a gas phase.

7.1.5 Behavior of Boron in Biological Processes

Boron is an important micronutrient involved in multiple biological processes. In line with its great chemical affinity for polyol groups, boron is involved in all reactions of sugar transport (particularly studied in sugar beets), but it also pairs with calcium to consolidate the cell structure (for example, a change in size and shape of leaves is the first symptom of boron deficiency). Boron is also involved in reproductive growth in regulating metabolic processes similarly to plant hormones (see review by Blevins and Lukaszewski 1998). Despite worldwide problems of boron deficiency or toxicity in agricultural soil, boron isotopes in plants has not received a great deal of interest, especially from an isotopic point of view. The main reason for this is the difficulty in separating boron from the organic matrix as well as problems with the analysis of boron isotopes by mass spectrometry due to isobaric interferences caused by residual organic impurities (Lemarchand et al. 2002; Rosner et al. 2011; Roux et al. 2015). This has led some authors to use ^{10}B labeling approaches to determine the source of boron and its pathways between plant organs instead (e.g. Boaretto et al. 2011; He et al. 2015). Based on the idea that the boron uptake by plant roots occurs without significant isotopic fractionation, the boron isotopic signature of plants has been used as tracer of provenance (e.g. Wieser et al. 2001; Chang et al. 2016). These studies assume no isotopic fractionation between the crop and the soil, a hypothesis that is challenged by recent works showing that boron isotopes in tree leaves are indeed close to the soil solution value but that a large isotopic fractionation exists between leaves, branches, trunk and roots (Cividini et al. 2010; Noireaux et al. 2017). The fractionation of boron isotopes during plant uptake is still unknown but data published in these latter studies indicate that excreted boron from leaves is enriched in ^{11}B compared to the plant, thus generating a boron isotopic fractionation by the biological cycle at the ecosystem scale. The release of ^{11}B-rich boron from leaves is consistent with the preferential complexation of boron by polysaccharide constituents of plant tissues. It is clear that the behavior of boron isotopes in biological processes still requires thorough exploration.

7.2 Biogeochemistry of Boron in the Critical Zone[1]

7.2.1 Boron Isotopes in Precipitation

Boron geochemistry in precipitation and more generally in the atmosphere is a long-standing topic that has attracted a significant amount of work and is however still much debated. Boron exists in the atmosphere in two forms, particulate and gaseous. The gaseous form represents 90–95% of the total, according to the pioneering paper of Fogg and Duce (1985). Gaseous boron is thought to be mostly derived from the evaporation of boron from seawater, although the possibility of boron released by vegetation, biomass burning and anthropogenic activities exists. A series of papers have reported boron isotopes in precipitation. Both boron concentrations and isotope compositions are very variable. B concentrations range from a 0.1 to several 100 μg L^{-1} and $\delta^{11}B$ spans the whole range of natural values (−20–50‰). A first order conclusion is that boron is enriched in rainwater compared to seawater, which is reflected by higher B/Na ratios (or B/Cl) ratios in precipitation as compared to seawater ratios (Chetelat et al. 2005; Rose-Koga et al. 2006; Millot et al. 2010). The contribution of boron derived from seasalt aerosols is significant but rarely dominant and the excess of boron in precipitation is attributed by several authors to both the contribution of gaseous H_3BO_3 and the dissolution of non-marine aerosols (e.g. Chetelat et al. 2009). The existence of a boron meteoric line linking $\delta^{11}B$ and δD ($\delta^{11}B = 2.6.\delta D - 133$) in rainwaters has been proposed by Rose-Koga et al. (2006), suggesting that the dominant process explaining the large range of $\delta^{11}B$ of rainwater is Rayleigh distillation in which a vapor phase, evaporated from the surface ocean, is progressively condensed in precipitation and loses its gaseous boron. According to Rose-Koga et al. (2006), the vapor formed by the evaporation of seawater is enriched in ^{10}B ($\delta^{11}B$ = 14‰). When B condenses in cloud droplets, a reverse isotopic fractionation of about 31‰ occurs, producing rainwaters with $\delta^{11}B$ close to that of seawater (39–40‰) or slightly higher (45‰). This model of evaporation-condensation was also invoked by Roux et al. (2017). With progressive condensation, $\delta^{11}B$ in precipitation decreases from 45‰ values down to negative values. As shown above, this interpretation, however, relies on an imperfect knowledge of the isotopic fractionation between H_3BO_3 in solution (seawater or rain) and the vapor phase (Miyata et al. 2000; Chetelat et al. 2005; Rose-Koga et al. 2006). Other interpretations for explaining low $\delta^{11}B$ measured in precipitation have been proposed including the dissolution of atmospheric dusts (Millot et al. 2010; Rose-Koga et al. 2006), the contribution of biomass burning, anthropogenic emissions (Chetelat et al. 2009; Zhao et Liu, 2010) or fertilizer application (Roux et al. 2017). Anthropogenic emissions are particularly evident when the boron isotopic data of rainwaters are fitted on a Rayleigh distillation pattern (Rose-Koga et al. 2006) or based on the correlation of $\delta^{11}B$ values with NO_3/Na ratios (Chetelat et al. 2009; Millot et al. 2010; Zhao and Liu 2010). While in a Rayleigh distillation process, low $\delta^{11}B$ must be associated with low B concentrations, when additional sources are involved, low $\delta^{11}B$ values can be associated with high B concentrations (Fig. 7.5).

[1]The Critical Zone is the zone between the top of the lithosphere and the lower atmosphere. In this zone, all compartments are connected and respond to external climatic, tectonic or anthropogenic forcings (Anderson et al. 2004). Critical Zone Observatories (CZO) are field sites deployed along environmental gradients where stream, vegetation, soil pore water, precipitation and groundwater are monitored.

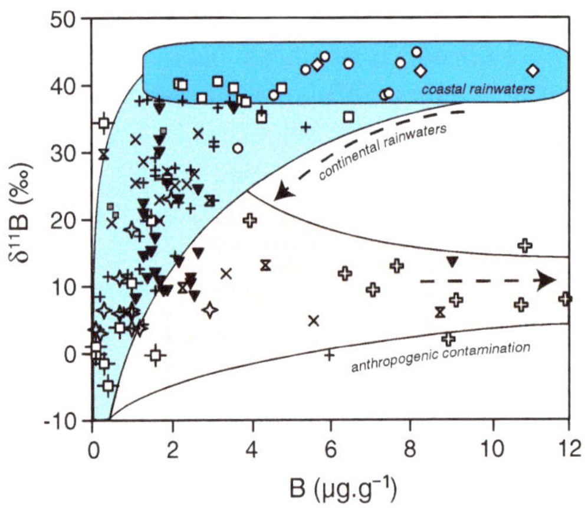

Fig. 7.5 $\delta^{11}B$ values plotted as a function of B concentration ($\mu g\ L^{-1}$) in rainwaters of Montiers (Eastern France, Roux et al. 2017), Brest (West France, Millot et al. 2010), Clermont Ferrand (Center France, Millot et al. 2010), Dax (South-West France, Millot et al. 2010), Orleans (Center France, Millot et al. 2010), Paris (Center France, Chetelat et al. 2009; Guiana, Chetelat et al. 2005; Reunion, Négrel et al. 2007 Nepal, Rose et al. 2000; and Quebec, Rose-Koga et al. 2006). Also shown is the continental rainwater array as defined by a Rayleigh distillation model for α = 1.015 and using coastal rains as the origin of air masses. The effect of biogenic and/or anthropogenic contributions are also shown (Figure modified from Roux et al. 2017)

In addition to wet deposition, the contribution of dry deposition to the terrestrial surface may be important. Roux et al. (2017) estimated that, in Eastern France, around half of the atmospheric flux of boron to the ground of a forested plot was derived from dry deposition and measured relatively high B concentrations (100–500 ppm) associated with crustal-like $\delta^{11}B$ values derived from dust.

Understanding the origin of boron in precipitation is an important issue for the purpose of extracting the mineral weathering contribution in continental waters as precipitation may represent a large input flux to watersheds. It is also critical for constraining the share between the "external" and "internal" boron inputs to hydrosystems and to better understand the biogeochemical cycle of boron and ecosystem sustainability, boron being an essential micronutrient for plant growth. So far, the contribution of atmospheric boron to terrestrial surface is poorly understood and requires urgent experimental and field studies.

7.2.2 Boron Isotopes in Soil Profiles

To date, there are only few studies of boron isotopes in soils. The pioneering measurements of Spivack et al. (1987) on a soil profile (oxisol type) in the Upper Orinoco watershed developed on granitic rock under wet tropical conditions showed that the weathering residue (kaolinite and oxides) was enriched in light boron (around 2‰ in the bulk soils) and that boron was lost in the solid phase from the bottom to the top of the profile. Subsequent studies have revealed that this behavior of B and B isotopes is however not general and appears to be more complex (Fig. 7.6). In South India, the dry tropical soil from the Mule Hole experimental watershed is enriched in ^{10}B (by 10‰ compared to the bedrock) but the boron concentration in the bulk soil is enriched compared to the bedrock, rather than depleted, as found in the Upper Orinoco watershed (Noireaux et al. 2017). In the temperate soil profile of the Strengbach catchment in the Vosges mountains, Western Europe, boron is depleted compared to insoluble elements but shows an enrichment in the heavy isotope (by more than 5‰), attributed by Lemarchand et al. (2012) to the incorporation of ^{11}B derived from the vegetation and dissolved in the soil solution into secondary minerals. In the Strengbach soil profile, all granite minerals have similar $\delta^{11}B$ and clay minerals have a constant but fractionated $\delta^{11}B$ value (5–10‰ higher that the bedrock) along the profile making $\delta^{11}B$ a good proxy for solving mass budget equations. In this particular soil profile, it is estimated by Lemarchand et al. (2012) that 70% of boron into the upper soil solid phases is of biological origin. This biological control of boron isotopes by vegetation in soils is not observed in Mule Hole despite the importance of biological cycling in the tropics. The boron enrichment in Mule Hole is attributed by Noireaux et al. (2017) to the downward migration of boron and to its fixation onto fine clay minerals while erosion processes removed the complementary depleted upper soil profile. Bulk soil data from Mule Hole show that $\delta^{11}B$ (Fig. 7.6) is strongly related to the Al/Si ratio, which is a proxy of the relative enrichment of clays versus quartz and

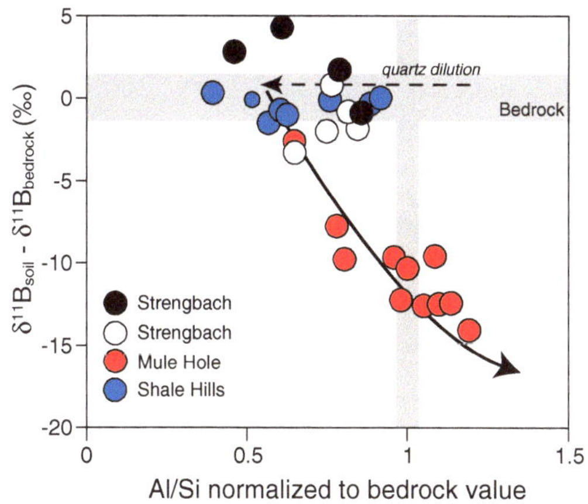

Fig. 7.6 Published data of boron isotopic composition in bulk soils as a function of Al/Si ratio (g g^{-1}). The Al/Si ratio is a tracer for gain size and discriminates Al-rich (clays and phylosilicates) from Si-rich (mostly quartz) minerals. Data are from Mule Hole, Strengbach and Shale Hills Critical Zone Observatories. ^{11}B enrichments are likely due to the incorporation of biogenic boron

unweathered minerals. In a soil profile developed in Shale Critical Zone Observatory in Pennsylvania, it is shown by Noireaux et al. (2014) that the loss of boron along the soil profile mimics that of Al and is attributed to the migration of fine particles with no isotope fractionation. All these studies suggest that grain size is important for understanding boron isotope geochemistry in soils. This effect was addressed in detail by Lemarchand et al. (2012) in the Strengbach soil profile. The analysis of different soil granulometric fractions showed that the finest component (clay) is 2–3 times enriched in boron compared to the coarsest fraction highlighting the important role of phyllosilicates on the budget and mobility of boron in soil profiles. While the granitic bedrock in the Strengbach basin has boron concentrations of 30–40 ppm, the finest clay component can have 100 ppm of boron. In this catchment, the finest clays appear to be in equilibrium with the modern day soil solution ($\delta^{11}B_{clay}-\delta^{11}B_{solution} = -30‰$). In Mule Hole, by contrast, the clay fraction appears to be fractionated by about −50‰ and cannot be in equilibrium with the soil solution without employing an unrealistic solid-fluid fractionation factor. These few preliminary studies on the behavior of boron isotopes in soils need to be generalized and the data incorporated into evolution models taking into account the long-term geomorphological history. Nonetheless, the existing data clearly demonstrate that light boron is preferentially incorporated into secondary soil products and that most of the boron in soils is contained in clay minerals that are generally enriched in boron compared to the bedrock. The idea that the solid phase of the upper soil profiles can incorporate biologically processed boron is very interesting and may have a number of possible paleoenvironmental applications, in particular for tracing the feedbacks between life and soil formation in the past.

7.2.3 Boron Isotopes at the Catchment Scale

A few recent studies have investigated highly instrumented catchments (Critical Zone Observatories, CZO) in which, in addition to stream water time series, vegetation, soil pore water, precipitation and groundwater were also investigated. Although preliminary, these studies have revealed the great potential of boron isotopes to understand Critical Zone (CZ) processes and feedbacks, particularly with regard to the involvement of boron in the biological cycle.

Cividini et al. (2010) investigated the Strengbach Basin in Eastern France and Noireaux et al. (2017) investigated the Mule Hole catchment in South India. Both basins are forested crystalline catchments. The Strengbach catchment experiences a cold temperate climate condition while the Mule Hole catchment has a dry tropical climate. These studies showed that the behavior of boron isotopes in the two systems obeys common rules (Figs. 7.7 and 7.8) despite important differences. In both CZOs, boron inputs to the soil + vegetation system are characterized by a significant contribution of atmospheric wet deposition. As indicated earlier, it is still currently difficult to determine whether boron from the atmosphere is external (i.e. of marine or

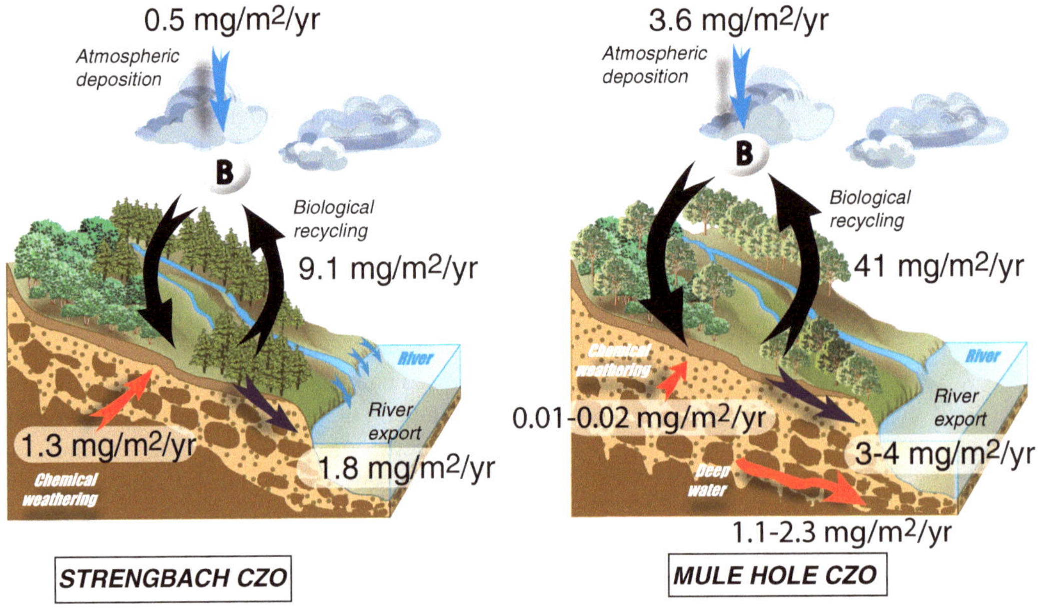

Fig. 7.7 Fluxes of boron in the two Critical Zone Observatories of Mule Hole (tropical climate) and Strengbach (temperate climate). In both cases, the recycling flux of boron is more important that the input and output fluxes

anthropogenic origin) or internally recycled from the catchment (biological aerosols, biomass burning, etc.).

A predominant feature of boron in CZ observatories is the large flux associated with vegetation cycling (Fig. 7.7). Although different in magnitude, in the Strengbach and Mule Hole watersheds, the recycling flux of boron by vegetation is 5–10 times that of boron export by streams. This turnover of boron in the ecosystem is similar to that of the major nutrients (e.g. N, P). The boron flux from the vegetation to the ground in both cases is almost equally shared between boron contained in the litterfall and soluble boron exudated from leaves, leached by rainfall and reaching the ground (throughfall) in a soluble form (Noireaux et al. 2017). Boron issued from soil weathering reactions is negligible in the Mule Hole watershed while in the Strengbach, its contribution is almost double that of the atmosphere.

Interestingly, boron isotopes are highly fractionated between the different CZ compartments. Figure 7.8 shows that the range of $\delta^{11}B$ in the different reservoirs of Mule Hole and Strengbach CZOs is extremely large and spans 80‰. Overall, the picture emerging from CZO studies is that biomass is significantly enriched in ^{11}B compared to the soil or bedrock minerals and local rainwater. In the Mule Hole watershed, while the contribution of mineral weathering is not significant, the throughfall is 10–15‰ enriched in ^{11}B compared to rainfall and litter samples (Fig. 7.8), showing that a major boron fractioning mechanism occurs between boron exudates from leaves and boron in plant tissue.

In order to assess the importance of biological recycling on the $\delta^{11}B$ of runoff water in the Strengbach basin, soil solutions were sampled at different depths. They show a characteristic decrease of boron concentration with depth indicating absorption in the root zone and confirming that boron is a limiting nutrient for plant growth. Exploiting the very large isotopic shift between vegetation (30–35‰) and soil minerals (−25‰), a simple model was developed by Cividini et al. (2010) to calculate the amount of boron in the soil solution that is derived from weathering reactions or involved in vegetation cycling. In the model, B in the soil solution is

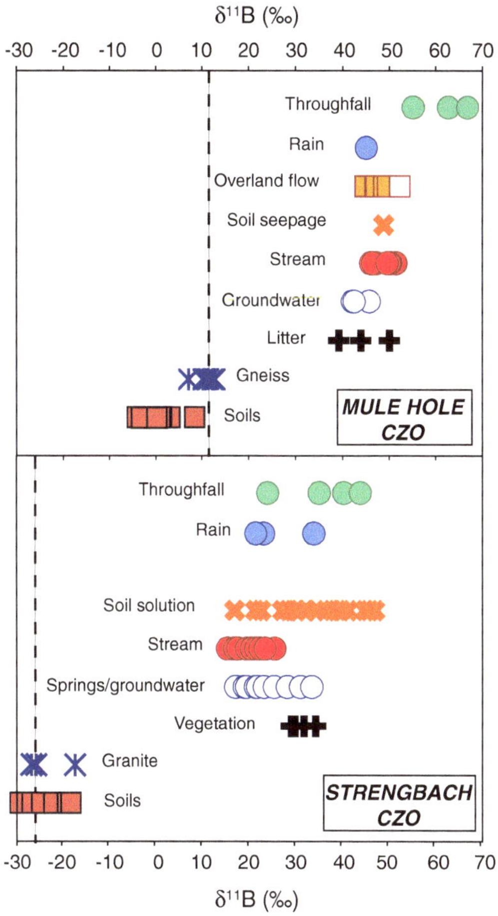

Fig. 7.8 Comparison of $\delta^{11}B$ values in the different compartments of two contrasted critical zone observatories, Mule Hole basin, India (tropical climate) and Strengbach basin (France). Both show the same trend of ^{11}B enrichment in the vegetation

controlled by weathering reactions, atmospheric input and biological recycling. Mineral dissolution is assumed to release boron with no isotopic fractionation while neoformation of secondary minerals strongly fractionates B isotopes (30‰) by enriching the solution in ^{11}B. Vegetation uptake results in limited discrimination between ^{10}B and ^{11}B (−5‰). The results from this simple approach applied to different depths of the upper part of the soil profile indicate that the flux of boron uptake by roots is higher than that released by weathering at all depths but particularly in the upper soil layers, where it is recycled from leaf litter. Both fluxes are coupled along the depth profiles.

In both the Mule Hole and Strengbach watersheds, more important contributions of chemical weathering to deep soils and groundwater solutions are found (Fig. 7.7). The contribution of the pool of boron released by mineral dissolution and fractionated by its incorporation in secondary minerals is variable and depends upon the hydrological setting. In the Strengbach, 50% of the boron at the outlet of the catchment is found to originate from groundwater/rock interactions. By contrast, in Mule Hole, most of the boron exported by the stream is derived from overland flow, essentially derived from vegetation that entirely controls the $\delta^{11}B$ at the outlet. The boron isotopic composition of boron derived from groundwater/rock interaction is variable and regulated by source mineral $\delta^{11}B$. Overall, in the Strengbach, the stream is 45‰ higher in $\delta^{11}B$ that the bedrock while in Mule Hole CZO, this difference is about 60‰, reflecting the much greater importance of biomass recycling (Fig. 7.8).

Finally, available studies suggest that $\delta^{11}B$ in the soil pore water depends on depth with surface water more affected by biological cycling than deeper waters which are more strongly controlled by weathering reactions. Therefore river $\delta^{11}B$ could serve as an interesting tracer of the hydrological cycle at the catchment scale distinguishing between surficial and deeper water paths both characterized by different residence time. There may be a great interest at coupling at the catchment scale boron isotopes with oxygen and hydrogen isotopes from the water molecule, or with gaseous tracers of residence times.

7.2.4 Boron Isotopes in Groundwaters

Boron isotopes have been determined in a wide range of groundwater types ranging from shallow to deep water bodies, in combination with source tracers (Sr isotopes) or other stable isotopic systems (N, O isotopes). The use of boron in coastal aquifers is motivated by the contrast in concentration and isotopic ratio between continental waters and seawater (Pennisi et al. 2006) to trace salinization processes (Vengosh et al. 2005). The mobility of boron in hydrothermal systems also makes it a good proxy for assessing the hydrothermal contribution to groundwaters (Millot and Negrel 2007; Millot et al., 2011; Louvat et al. 2011). In general, groundwaters are B-rich and concentrations increase as a function of total dissolved solids. Boron isotopic compositions are extremely variable, from low crustal values (0‰) to δ^{11}B up to 40–50‰ (Negrel et al. 2002; Lemarchand et al. 2015). There is a broad agreement that the boron composition of groundwaters is controlled both by the host bedrock mineralogy and processes fractionating boron isotopes, mainly the sorption onto mineral and organic surfaces and coprecipitation of boron in secondary minerals. Evaporites of marine origin are a major source of boron, as well as organic matter rich-layers (Williams and Hervig 2004). Once released from primary minerals by dissolution, boron is removed from the solution by sorption and coprecipitation. Two basic equations governing the exchange of boron between solution and mineral or organic surfaces can be written. Firstly, if a volume of water containing dissolved boron with an initial concentration of dissolved boron $[B]_T$ equilibrates with mineral or organic surfaces:

$$[B]_{sol} = \frac{\omega_R}{K_d + \omega_R}[B]_T$$

where $[B]_{sol}$ is the soluble B concentration, K_d is the partition coefficient of boron ($Kd = [B]_{ads}/[B]_{sol}$) and ω_R is the water rock ratio. Typical values for a porosity of 20–30% and a K_d of 20 (Lemarchand and Gaillardet 2006) give the first order result that 98% of the boron stock is adsorbed. As a consequence, in groundwaters, the dissolved B and its isotopic composition are buffered by the solid phase and subtle changes in Kd, which vary as function of pH and temperature for example, can have important effects on the dissolved phase. This also explains the chemostatic behavior of B in groundwater-fed rivers when runoff variations occur (Cividini et al. 2010). A second equation can be written for isotopes:

$$\delta^{11}B_{sol} = \frac{K_d + \omega_R}{\alpha K_d + \omega_R}\delta^{11}B_T + 1000\frac{K_d.(1-\alpha)}{\alpha K_d + \omega_R}$$

where $\delta^{11}B_{sol}$ and $\delta^{11}B_T$ (in ‰) are respectively the boron isotopic composition of the solution at equilibrium and initially, and α is the averaged boron isotopic fractionation factor between solids and solution.

The fractionation factors of boron isotopes due to adsorption are still poorly known. As shown in Sect. 7.1, boron sorption onto mineral or organic surfaces can be viewed as a competitive process between boron species in solution and complexes of boron forming at the surface. The simplest model would be to assume that the borate ion in solution is the only species to sorb onto mineral or organic surfaces. However, as shown above, laboratory experiments (Lemarchand et al., 2005, 2007) have shown that this hypothesis is not tenable and that different surface complexes form depending on the solution pH and have different partition coefficients (Kd) and solid-solution fractionation factors.

A model of water-rock interaction, including water advection and two types of reactions (fast exchange reactions with surfaces and slow dissolution reactions of silicate minerals) was proposed by Lemarchand and Gaillardet (2006) in the Mackenzie river basin in order to explain the high δ^{11}B measured in the river and its tributaries. The model shows that boron isotopic ratios are sensitive to weathering rates and groundwater advection velocity (hence residence time) and evolve after a given perturbation in two steps as a result of exchange reactions involving boron and fractionating its isotopes (Fig. 7.9). Firstly, ion exchange reactions introduce a retardation factor by scavenging (or releasing) boron on (or from) mineral surfaces with associated isotopic fractionation. Secondly, after the perturbation front has passed though the system, a steady state can be established where the

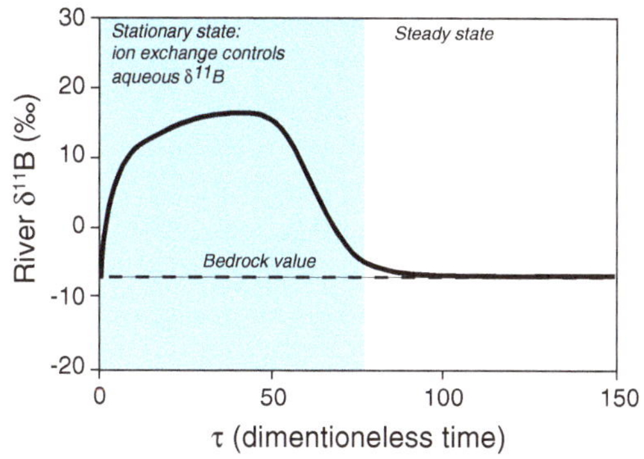

Fig. 7.9 Modeled boron isotopic composition in shallow groundwaters of the Mackenzie river basin illustrating the competition between ion exchange (*fast*) and dissolution (*slow*) reactions, and time, here expressed by the dimensionless τ number (time normalized to the residence time of water in the groundwater). High ^{11}B enrichment are generated when fast ion exchange reaction are dominant

isotopic ratio of the solution is no longer controlled by exchange reactions and is only driven by dissolution reactions. This theoretical model shows that over a limited period of time, ^{11}B can become strongly enriched in solution (for example, in response to an increase of boron by rock dissolution caused by an abrupt change in climate conditions), if all other parameters such as pH, source minerals and temperature remain constant. The duration of this transient state depends on the residence time of the solution and on the adsorption properties of the host rocks. For a typical groundwater-rock system with residence times of water of 500 yr in the Mackenzie glacial tills, the dissolved $\delta^{11}B$ will be dominated by ion exchange over 10–25 kyr. It was proposed by Lemarchand and Gaillardet (2006) that the high $\delta^{11}B$ values measured in the lowland of the Mackenzie river basin reflect an ancient perturbation in the weathering rates (associated with the last deglaciation) that has still not reached its steady state (Fig. 7.9). Such a scenario strongly depends upon the local hydrology, water flowpaths and host rock exchange properties, but it clearly highlights the importance of exchange reactions in controlling the boron isotopic composition of continental waters and the potentially powerful application of boron isotopes in water-rock systems.

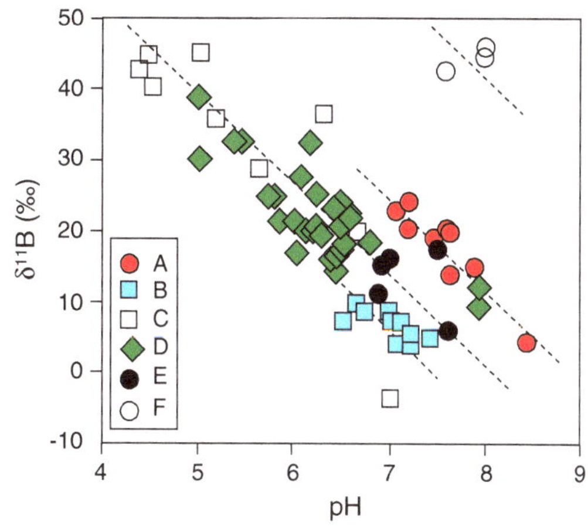

Fig. 7.10 Relationship between dissolved $\delta^{11}B$ in groundwaters and pH (from Lemarchand et al., 2015). Vertical shifts are mainly due to variations in the hosting bedrocks. Data from *A* Wyoming, USA; *B* Cornia Plain, Italy; *C* French Guyana; *D* Strengbach CZO, France; *E* Shale Hills CZO, USA; *F* Mule Hole CZO

In the Wyodak-Anderson coal bed aquifer (USA), Lemarchand et al. (2015) have modeled the large range of $\delta^{11}B$ in the aquifer (15–25‰) by a simpler model of water-rock interaction including a pH-dependency of the fractionation factor between adsorbed and dissolved boron. Plausible variations in the pH values of the aquifer between 7 and 8.5 explain the range of $\delta^{11}B$ measured in the aquifer with a solid-liquid fractionation factor of 15‰. The inverse dependence of $\delta^{11}B$ in aquifers with pH is observed in a number of aquifers, dominated by exchange reactions (Fig. 7.10).

This use of B isotopes in groundwater systems is however still limited by our poor understanding of B sorption mechanisms and associated isotope fractionation.

7.2.5 Boron Isotopes in River Systems

The few investigated Critical Zone Observatories for boron isotopes allows us to disentangle some of the relevant elementary processes in the critical zone but are not providing any global information. The behavior of boron in large river systems has been investigated by a few studies.

Due to low concentrations, the analysis of boron isotopes in river waters has long been a challenge. Early measurements used pyrohydrolysis (Spivack et al. 1987) or ion probe (Rose et al. 2000) following evaporation of large volumes of river water. More recently, a pre-concentration step has been developed using the B-specific resin Amberlite IRA 743 (Lemarchand et al. 2002a) in order to purify the necessary amount of boron for isotopic analysis.

The boron isotopic composition of dissolved boron in large rivers (Lemarchand et al. 2002; Spivack et al. 1987; Rose et al. 2000; Chetelat and Gaillardet 2005; Lemarchand and Gaillardet 2006, Chetelat et al. 2009) shows a wide range of values (from −6 to 44‰). Figure 7.11 gives a global picture of the boron input to the ocean and associated $\delta^{11}B$. The flux of B to the ocean comprises an atmospheric (so called "cyclic") component and a probable anthropogenic component (Schlessinger and Vengosh 2016). Using chloride concentration as a proxy for atmospheric inputs, it has been shown by Rose et al. (2000) for the Himalayan rivers, Lemarchand and Gaillardet (2006) for the Mackenzie, Chetelat et al. (2009) for the Changjiang river and more generally for the 22 largest rivers of the world that the atmospheric input to the global boron budget is probably <10–20% of the total dissolved boron, except in coastal rivers where this proportion is much higher (Chetelat et al. 2005). The anthropogenic boron content of rivers is more difficult to assess. It is dominant in the polluted rivers such as the Seine river (Chetelat and Gaillardet 2005), France, where most of the boron at the river mouth is anthropogenic. In the Changjiang river, China, Chetelat et al. (2009) estimated that about 20% of the dissolved boron was derived from industrial and farming activities.

In the Ganges-Brahamapura, Mackenzie and Changjiang river systems, lithogenic boron in rivers is mainly derived from silicate weathering and evaporite dissolution when such lithologies

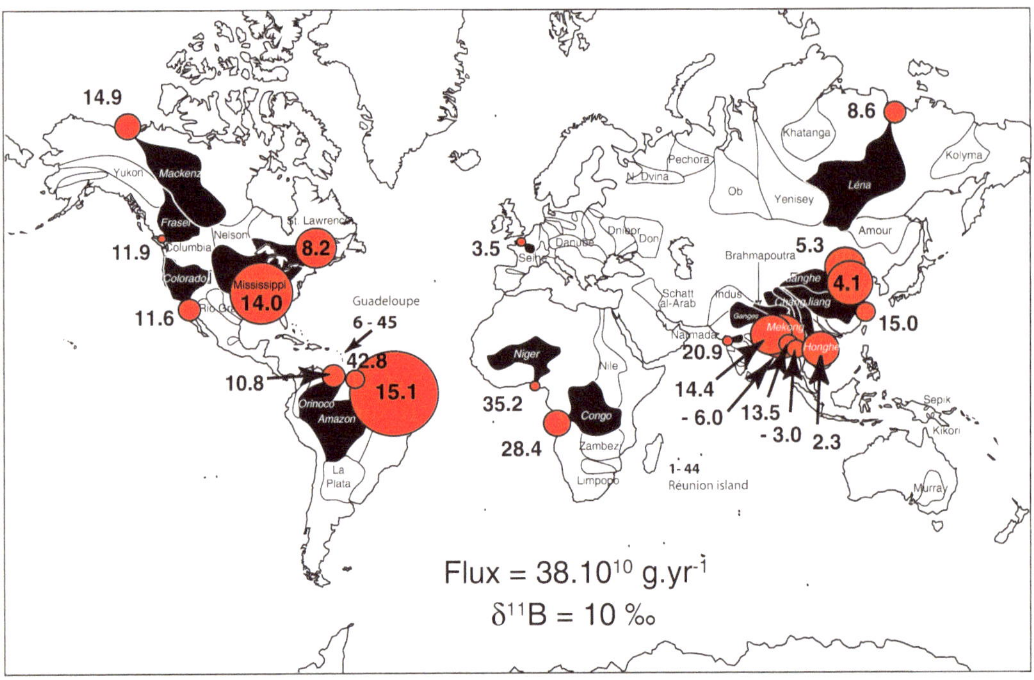

Fig. 7.11 Global riverine input of boron to the ocean. Database from Lemarchand et al. (2000) and Spivack et al. (1987) updated by more recent studies of Lemarchand and Gaillardet (2006), Chetelat et al. (2009), Louvat et al. (2011, 2014) for Lesser Antilles and Réunion Islands. Note the ^{11}B enrichment in tropical surface waters

are present in the catchment. The contribution of evaporites and hydrothermal springs to the Ganges ranges between 10 and 60% of total dissolved boron (Rose et al. 2000), while this proportion is lower in the Changjiang river (20%, Chetelat et al. 2009) and in the Mackenzie river (<15%, Lemarchand and Gaillardet 2006). In all cases, carbonate dissolution provides a negligible contribution to the dissolved B budget, making boron an excellent tracer for understanding silicate weathering mechanisms and tracing silicate weathering rates or associated atmospheric CO_2 consumption rates.

Due to the abilities of boron to interact with mineral surfaces and to be incorporated into secondary minerals, both of which highly fractionate B isotopes, it is expected that hydrological conditions (runoff/soil water/groundwater inputs) and weathering regimes (weathering-limited vs. transport-limited) will have a significant, if not dominant, effect on setting the dissolved $\delta^{11}B$ of rivers. In the Ganges-Brahmaputra river system, once corrected for evaporite dissolution, Rose et al. (2000) have shown that the range of $\delta^{11}B$ in Himalayan rivers can be interpreted as indicating the incongruent weathering of Himalayan bedrock in which dissolved boron is partly retained in secondary clay minerals. This interpretation assumes that the fractionation of B isotopes between the soil solution and the neoformed clay minerals is pH-dependent as only the borate species (whose $\delta^{11}B$ is pH-dependent) is incorporated. The high $\delta^{11}B$ values measured in some Himalayan rivers are therefore interpreted as being influenced by a high proportion of B uptake by secondary minerals and high soil pH values. In order to delineate the factors influencing the fractionation of boron in rivers, Louvat et al. (2011, 2014) investigated rivers draining andesite in the Lesser Antilles and basaltic rocks in La Réunion Island. They showed that riverine boron was influenced by hydrothermal inputs in active volcanic regions with low $\delta^{11}B$ and confirmed that a wide range of $\delta^{11}B$ could be produced in rivers not impacted by hydrothermal activity depending on the weathering regimes and probably the biological productivity. The highest $\delta^{11}B$ values found in the Lesser Antilles and Réunion are between +30 and 45‰, remarkably similar to the highest values found in other large tropical rivers (Congo, Amazon lowland rivers, Niger, Guiana: in the global river database of Lemarchand et al. 2000). Interestingly, tropical rivers from South Taiwan (Liu et al. 2012) characterized by weathering-limited weathering regimes (short residence time in the soils) do not show such high values, again suggesting the importance of weathering regimes. In transport- limited regimes, both active precipitation of soil clay minerals or oxides and active biological recycling are possible mechanisms for producing the high $\delta^{11}B$ values observed in tropical rivers but more work remains to be done, in particular using the CZO facilities to fully understand what controls the $\delta^{11}B$ of dissolved boron in rivers.

7.2.6 Partitioning of Boron Isotopes Between Water and Modern Day River Sediments

There are few reported data on the boron isotopic composition of river sediments (Spivack et al. 1987; Chetelat et al. 2009). In the Changjiang river (Chetelat et al. 2009), 70% of boron is transported in solid form as suspended sediments (sand fraction not included) and 30% in solution. These authors showed that these phases have distinctly different $\delta^{11}B$ values (Fig. 7.12), and that the $\delta^{11}B$ value of the suspended sediments is close to the crustal value (around −7‰). In the Changjiang, Al/B, used as a proxy for the mobility of boron during chemical weathering (dissolution vs. reincorporation in secondary phases) shows a positive correlation with the difference between the isotopic composition of the dissolved and suspended loads, suggesting that the reincorporation of boron in secondary minerals is controlling the boron isotopic composition of the dissolved load. These preliminary conclusions need to be confirmed and the effects of grain size and sedimentary recycling carefully assessed, but available data clearly show that boron isotopes are fractionated by large scale continental weathering processes. Following

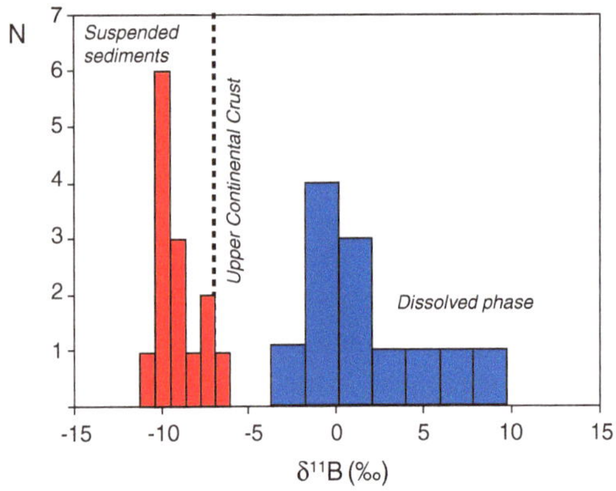

Fig. 7.12 Boron isotopes in the dissolved phase and suspended sediments of the Changjiang Basin, China (Chetelat et al. 2009), showing the overall fractionation of boron isotopes due to chemical weathering processes. UCC is the Upper Continental Crust (Chaussidon and Albarède 1992)

Chaussidon and Albarède (1992), boron isotopes need to be explored as a proxy of paleo-weathering conditions or to trace the long-term evolution of the continental crust.

7.2.7 Boron Isotopes in Lakes

Lakes have been investigated for boron isotopes in a few studies aiming at constraining their geochemical evolution (Vengosh et al. 1991b, 1991c, 1995; Xiao et al. 1992; Leslie et al. 2014). The marine versus non marine origin of evaporites in the sedimentary archives have also motivated the investigation of boron isotopes in modern salt lakes. Boron in lakes has very variable concentrations, ranging from typical values of 1 μg L^{-1} to a 0.5 g L^{-1} in hypersaline lakes associated with borate deposits and essentially correlated with alkalinity. The boron isotopic composition of lakes is also very variable ranging between 0 and 60‰ and overall enriched in ^{11}B relative to continental crust (−7‰). Most lakes analyzed to date show a positive correlation between Na/Cl and B/Cl (Leslie et al. 2014), with the exception of the Australian salt lakes, which are depleted in B compared to Cl and some lakes of the Great Basin, Northern America, which show a strong B enrichment. The boron geochemistry of lakes reflects both the inputs of boron to the lakes (riverine and groundwater inputs, evaporite dissolution, hotsprings in active tectonic settings) and in situ lake processes, mainly boron adsorption on mineral surfaces. Depending on the local geology, proximity to the sea, anthropogenic inputs and lake dynamics (evaporation, pH, water-sediment ratio, sediment lithology), large variations of $\delta^{11}B$ can occur within an individual lake. In the Great Australian salt Lakes, Vengosh et al. (1991b) have shown that the dominant process affecting lake $\delta^{11}B$ is the adsorption of boron into clay minerals with the preferential uptake of ^{10}B, thus causing the increase of lake $\delta^{11}B$. Boron coprecipitation of saline minerals (^{10}B enriched) also contributes to increase lake water $\delta^{11}B$ when evaporation becomes important (Vengosh et al. 1995). According to Leslie et al. (2014), two global trends exist. In most cases, lake $\delta^{11}B$ values increase as dissolved boron decreases as ^{10}B is preferentially lost by adsorption at high solution concentrations (Antarctica, Dead Sea, Australia). Other lakes do not show or show a little change in $\delta^{11}B$ with B concentration, indicating that B removal onto surfaces or precipitates is not an active process (China and Great Basin).

Beyond this diversity of situations, boron geochemistry in lakes confirms the general idea that boron is a reactive element in solution, likely to be sorbed on mineral surfaces and coprecipitated in neoformed minerals, both processes preferentially enriching the solution in ^{11}B. In detail, boron residence times in lakes should be better constrained. There is still an important lack of experimental data on the determination of the fractionation factors for the different phases that scavenge boron in lakes. A theoretical treatment of how $\delta^{11}B$ evolves as a function of lake chemistry and dynamics is currently lacking although lakes could be used as an interesting analogue for understanding how seawater acquired its composition.

7.3 The Riverine Input of Boron to the Ocean and Secular Evolution of the Ocean

7.3.1 The Riverine Input of Boron to the Ocean

The discharge-weighted average riverine boron flux based on the 22 largest rivers of the world (representing 39% of the water discharge to the ocean) by Lemarchand et al. (2000) is 0.38 Tg yr^{-1} with a $\delta^{11}B$ value of 10‰. Thus, river waters are enriched in ^{11}B compared to the continental crust (around −7‰, Chaussidon and Albarède 1992). A revised value of global riverine $\delta^{11}B$ value calculated using the most recent data from the Changjiang river (Chetelat et al. 2009) is within error of the Lemarchand et al. (2000) estimate. Rivers draining active volcanic areas have high B concentrations and low $\delta^{11}B$ due to hydrothermal input. Data from Louvat et al. (2011) and Louvat et al. (2014) extrapolated to the global scale indicate that this continental hydrothermal component may be significant on the global boron budget. In Basse-Terre Island (Lesser Antilles), boron concentration is hydrothermal springs can be 10–100 times enriched compared to rivers. We lack precise estimates of the boron content of rivers from volcanic islands, but if we assume a boron concentration of only 5 times that of continental rivers, then, with 10% of the global runoff to the ocean derived from volcanic areas (Allègre et al. 2010), the seawater input of boron due to volcanic islands should be half of the global boron input to the ocean.

The global sediment flux to the ocean is 12,600 Tg yr^{-1} (Syvitski et al. 2005). Based on data from large rivers, the mean boron content of suspended sediments is 70 ppm (Chetelat et al. 2009), the dissolved flux of boron to the ocean is augmented by a particulate flux of 0.9 TgB yr^{-1}, twice the estimate by Schlessinger and Vengosh (2016). However, this pool of boron is unlikely to be solubilized in seawater, so only the dissolved riverine boron probably affects the secular evolution of boron in seawater.

The important question for determining the net input of boron to the ocean and its residence time is how much of the riverine flux is cyclic (transiting rapidly through the atmosphere), compared to how much is anthropogenic and lithogenic. As shown above, the proportion of cyclic boron in rivers depends on knowledge of boron in precipitation, which is hard to assess because boron exists in the atmosphere both in gaseous and particulate forms, and to distinguish between sources and fractionation effects due to evaporation- condensation processes is difficult. However, as far as large rivers are concerned, an upper limit on the contribution of rainwater to surface waters can be estimated based on Cl concentrations in rivers, corrected for evaporite dissolution, where appropriate (Lemarchand et al. 2002; Lemarchand and Gaillardet 2006; Chetelat et al. 2009; Rose et al. 2000). Except for coastal rivers in which most of the dissolved boron may be of atmospheric origin (Chetelat et al. 2005), the atmospheric input of B to the global river flux does not exceed 15%. The input of anthropogenic B is variable and only a few rivers have been investigated in detail. In the Seine river, both B concentration and isotopic signatures of B show that most of the dissolved boron is anthropogenic (Chetelat and Gaillardet 2005) while in the Mackenzie river, this component is negligible (Lemarchand and Gaillardet 2006). In the Changjiang river, Chetelat et al. (2009) have estimated that 20% of the dissolved boron in the river is of anthropogenic origin. These conclusions are not consistent with the idea proposed by Schlessinger and Vengosh (2016) that 80% of the boron transferred from land to sea each year is anthropogenic and the riverine signal from chemical weathering is minor. More work is needed to estimate the global anthropogenic contribution of boron in rivers and thus to the ocean but the global flux of 0.38 TgB yr^{-1} proposed by Lemarchand et al.

(2000) is in good agreement with the Na flux to the ocean derived from rock weathering (Gaillardet et al. 1999) and estimates of the crustal B/Na crustal ratio (Li et al. 2000).

7.3.2 Boron as an Integrated Tracer of Global Weathering and Erosion

Two main processes of boron removal in the ocean are adsorption onto sedimentary clays carried to the ocean by rivers and derived from the physical erosion of continents and coprecipitation with clays during burial diagenesis (Schwarz et al. 1969; Palmer et al. 1987). As adsorption and coprecipitation of boron preferentially incorporate the light isotope (Palmer et al. 1987; Williams et al. 2001a), these mechanisms have first been invoked to explain why the ocean has a high $\delta^{11}B$ (39.61‰, Foster et al. 2010). Lemarchand et al. (2000) have proposed an ocean model for boron isotopes in which the inputs of boron to the ocean are rivers (0.38 TgB yr^{-1}), hydrothermal fluids at the sea floor (0.04 TgB yr^{-1}) and the expulsion of diagenetic fluids from accretionary prisms (0.02 TgB yr^{-1}). The oceanic output fluxes of boron considered by Lemarchand et al. (2000), in addition to boron sorption and coprecipitation with clays (0.13 TgB yr^{-1}), are the consumption of boron during the low-temperature interaction of seawater with basalts and gabbros from the oceanic crust (0.27 TgB yr^{-1}) and the coprecipitation in carbonates (0.06 TgB yr^{-1}). Since this study, there have not been major changes in the oceanic B flux estimates or any new suggestions for other sources and sinks. More recent data from Staudigel et al. (2013) have confirmed that weathering of the oceanic crust is a net sink of boron. The new estimate of the sediment flux to the ocean from Syvitski et al. (2005) of 12,600 Tg yr^{-1} also has no implication for the value of the oceanic sink of boron due to sorption onto clay surfaces.

It is important at this stage to emphasize that the boron oceanic cycle is closely linked to terrestrial weathering processes. The dominant flux of boron to the ocean is the release of boron by chemical weathering of terrestrial surfaces and the dominant removal fluxes are linked to mechanical erosion of terrestrial rocks. Both processes have been shown to be coupled (Gaillardet et al. 1999; West et al. 2005) and therefore this makes boron isotopes in the ocean a very good integrated proxy of water-rock interaction on Earth.

Based on the above fluxes, Lemarchand et al. (2000) showed that boron in the ocean is at steady state (within error), which means that the sum of the input fluxes to the ocean equals that the output fluxes. They then proposed a very long residence time value for boron in the ocean, 14 Myr, consistent with the homogeneity of boron concentration and isotopic composition in the modern ocean (Foster et al. 2010) and implying that the boron isotopic composition of seawater could have only varied if significant variations of the main input (chemical erosion) or output fluxes (physical erosion and oceanic crust production) occurred. Simon et al. (2006) developed a model evaluating the sensitivity of seawater $\delta^{11}B$ to oceanic crust production and the water/rock ratio during seawater-crust interaction and concluded, like Lemarchand et al. (2000), that the $\delta^{11}B$ of the ocean has probably reached a steady state with the modern crustal production rate of 3 cm yr^{-1}, a seawater-crust ratio between 1 and 3 and a B residence time in the ocean of 10 Myr.

It is important to stress here, however, that the oceanic cycle of boron is still under-constrained and that future investigations should particularly focus on a better understanding on boron sorption mechanisms onto clays and oxides and on the behavior of boron in sedimentary basin during burial diagenesis. The importance of clay recrystallization during diagenesis and the importance of sedimentary organic matter as a source for boron in sedimentary fluids have been suggested by Williams et al. (2001b) and Williams and Hervig (2004). The importance of organic matter for the oceanic boron cycle mirrors its role in the boron cycling at the watershed scale as stressed above.

7.3.3 The Secular Evolution of Boron Isotopes in the Ocean

As boron in the ocean is mostly influenced by continental weathering processes, the secular evolution of boron isotopes in the ocean is of central interest for exploring the denudation history of the Earth. In addition, boron isotopes in marine carbonates are widely used as a tracer of paleo-acidity (see Chap. 5), which requires the knowledge of the $\delta^{11}B$ of seawater at the time of carbonate precipitation. Based on modeling the $\delta^{11}B$ in ancient carbonates, several attempts have been made to reconstruct the secular evolution of seawater $\delta^{11}B$. Pearson and Palmer (2000) used the difference in $\delta^{11}B$ in planktonic forams sampled at different water depths to circumvent the problem of having two unknowns (ocean $\delta^{11}B$ and pH) for one equation. Assuming no significant changes in ocean productivity over time, they deduced the secular evolution of $\delta^{11}B$ in seawater (Fig. 7.13). Foster et al. (2012) used the changes in the difference between carbonate $\delta^{13}C$ and $\delta^{11}B$ to constrain the pH gradient with depth and infer $\delta^{11}B$ seawater values at the Eocene Oligocene transition. Raitzsch and Hönisch (2013) analyzed benthic forams covering the last 50 Myr and clearly showed that the deep-water pH calculated from these data is unrealistic if the $\delta^{11}B$ of the seawater remained constant. By using modeled pH values of ocean deep water, they proposed a $\delta^{11}B$ secular curve of the ocean (Fig. 7.13). Recently, Greenop et al. (2017) applied the method of Foster et al. (2012) for the last 23 Myr and produced a new curve of secular seawater $\delta^{11}B$ evolution. All these methods, based on carbonate $\delta^{11}B$, suffer from uncertainties (see Greenop et al. 2017 for a review of these uncertainties) and are model-dependent but are nethertheless similar to the theoretical curve calculated by Lemarchand et al. (2000). Using a very simple box model, these authors deduced the $\delta^{11}B$ of seawater over the last 50 Myr based on plausible secular evolutions of physical erosion, continental runoff, seafloor spreading rates and carbonate precipitation over the last 100 Myr. They showed that seawater $\delta^{11}B$ variations of the ocean should be expected as a result of the formation of the Himalaya and Tibetan plateau and the high sediment flux generated by the India-Eurasia collision. The re-estimation of oceanic crust production by Cogné and Humler (2004) reinforced the idea that the $\delta^{11}B$ of the ocean is mainly controlled by terrestrial weathering and erosion processes. Since 50 Myr, calculated seawater $\delta^{11}B$ is estimated to have increased from 36.5 to 39.6‰ due increased scavenging of ^{10}B in the ocean due to enhanced sediment production (Fig. 7.13). Increased denudation (D) implies an increase in both the dissolved (W) and particulate (P) river fluxes, which means, from a B isotope perspective, an increase of ^{11}B delivery to the ocean and greater ^{10}B removal within the ocean, both mechanisms contributing to an increase in the $\delta^{11}B$ of seawater. It is interesting to note that such an increase is compatible with the decrease of global W/W + P ratios over the Cenozoic inferred by the evolution of Li isotopes in the ocean (Hathorne and James 2006; Misra and Froelich

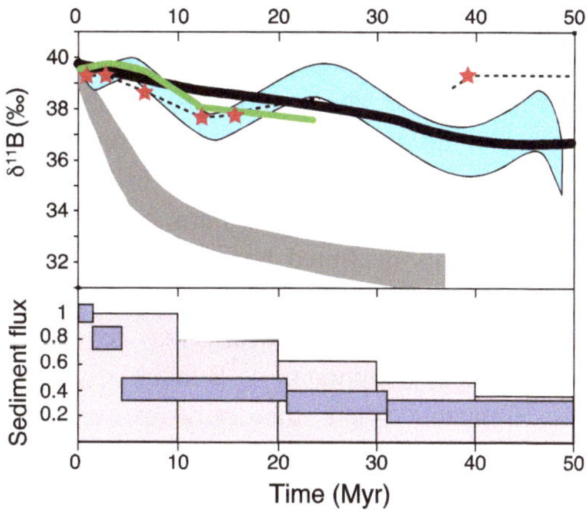

Fig. 7.13 Estimated boron isotopic composition of the Cenozoic seawater. Different approaches were used including measurements and models. *Blue zone* is from Raitzsch and Hönisch 2013; *green line* from Greenop et al. (2017); *red stars* from Pearson and Palmer (2000), *gray zone* from Paris et al. (2010) and *black line* from Lemarchand et al. (2000). The increase of the global sediment flux to the ocean over the last 60 Myr is shown for comparison, taken from Li and Elderfield (2013) in deep purple and Lemarchand et al. (2002) in *light purple*. These curves represent the global continental denudation and influence the boron isotopic budget in the ocean via adsorption onto solid surfaces (preferentially sequestrating ^{10}B)

2012; Dellinger et al. 2015; Pogge von Strandmann and Henderson 2015, Greenop et al. 2017).

None of the above indirect approaches is devoid of uncertainties. So far, the only study that have attempted to measure the $\delta^{11}B$ directly in ancient seawater is from Paris et al. (2010). These authors measured boron isotopes in seawater fluid inclusions in carefully selected halite from the geological record. They reported a much steeper evolution of $\delta^{11}B$ in the ocean over the Cenozoic than the previous estimates (Fig. 7.13), increasing from 32‰ to the modern value (40‰) over the last 40 Myr. The hypotheses sustaining this work are that fluid inclusions sampled seawater unaffected by evaporation processes (a hypothesis checked on modern halite samples) and that boron did not react within the fluid inclusion after deposition. According to Raitzsch and Hönisch (2013), the evolution of seawater $\delta^{11}B$ proposed by Paris et al. (2010) is not compatible with reconstructions of Cenozoic deep water pH as this would have implied massive precipitation of carbonates at the Eocene-Oligocene boundary which is at odds with observations of the paleo-CCD (Carbonate Compensation Depth).

7.4 Conclusion

This review of the behavior of boron and boron isotopes in the terrestrial weathering environment demonstrates their great potential for understanding the processes that transform rock into soil and ecosystem dynamics. Boron is a nutrient and has an important affinity for mineral and organic surfaces. Although data sets are still limited, biological cycling and soil formation are the principal processes that fractionate B isotopes, making boron a useful tool for understanding the processes that shape the Critical Zone. More work is urgently needed to better constrain experimentally the fractionation factors of boron isotopes in the weathering environment and to better assess the atmospheric and biological sub-cycles.

Given the long residence time of B in the ocean, the secular evolution of boron isotopes in the seawater records terrestrial evolution of life and soil processes and so, indirectly, the consumption of atmospheric CO_2 by chemical weathering. As seawater $\delta^{11}B$ is highly sensitive to global weathering and denudation (though rivers) and because of the importance of boron isotopes in carbonates for reconstructing ocean paleo-acidity and atmospheric paleo-CO_2 levels, the reconstruction of past seawater $\delta^{11}B$ is really a research priority.

Acknowledgements We thank Gavin Foster for the invitation to write this chapter. Jacques Schott, Pascale Louvat, Claire Rollion-Bard, Johanna Noireaux and Julien Bouchez are thanked for useful comments.

References

Allègre CJ, Louvat P, Gaillardet J, Meynadier L, Rad S, Capmas F (2010) The fundamental role of island arc weathering in the oceanic Sr isotope budget. Earth Planet Sci Lett 292(1):51–56

Anderson SP, Blum J, Brantley SL, Chadwick O, Chorover J, Derry LA, Richter D (2004) Proposed initiative would study earth's weathering engine. Eos 85:265–272

Balan E, Pietrucci F, Gervais C, Blanchard M, Schott J, Gaillardet J (2016) First-principles study of boron speciation in calcite and aragonite. Geochim Cosmochim Acta 193:119–131

Blevins DG, Lukaszewki KM (1998) Boron in plant strucutre and function. Annu Rev Plant Physiol Plant Mol Biol 49:481–500

Boaretto RM, Quaggio JA, Mattos DJ, Muraoka T, Boaretto AE (2011) Boron uptake and distribution in field grown citrus trees. J Plant Nutr 34:839–849

Chaussidon M, Albarède F (1992) Secular boron isotope variations in the continental crust: an ion microprobe study. Earth Planet Sci Lett 108(4):229–241

Chang C-T, You C-F, Aggarwal SK, Chung C-H, Chao H-C, Liu H-C (2016) Boron and strontium isotope ratios and major/trace elements concentrations in tea leaves at four major tea growing gardens in Taiwan. Environ Geochem Health 38:737–748

Chetelat B, Liu CQ, Gaillardet J, Wang QL, Zhao ZQ, Liang CS, Xiao YK (2009a) Boron isotopes geochemistry of the Changjiang basin rivers. Geochim Cosmochim Acta 73(20):6084–6097

Chetelat B, Gaillardet J, Freydier R, Négrel P (2005) Boron isotopes in precipitation: experimental constraints and field evidence from French Guiana. Earth Planet Sci Lett 235(1):16–30

Chetelat B, Gaillardet J (2005) Boron isotopes in the Seine River, France: a probe of anthropogenic contamination. Environ Sci Technol 39(8):2486–2493

Chetelat B, Gaillardet J, Freydier R (2009b) Use of B isotopes as a tracer of anthropogenic emissions in the atmosphere of Paris. France Appl Geochem 24 (5):810–820

Cividini D, Lemarchand D, Chabaux F, Boutin R, Pierret MC (2010) From biological to lithological control of the B geochemical cycle in a forest watershed (Strengbach, Vosges). Geochim Cosmochim Acta 74(11):3143–3163

Cogné JP, Humler E (2004) Temporal variation of oceanic spreading and crustal production rates during the last 180 My. Earth Planet Sci Lett 227(3):427–439

Dellinger M, Gaillardet J, Bouchez J, Calmels D, Louvat P, Dosseto A, Maurice L (2015) Riverine Li isotope fractionation in the Amazon River basin controlled by the weathering regimes. Geochim Cosmochim Acta 164:71–93

Dickson AG (1990) Thermodynamics of the dissociation of boric acid in synthetic seawater from 273.15 to 318.15 K. Deep-Sea Res 37:755–766

Edwards JO, Morrison GC, Ross VF, Schultz JW (2002) The structure of the aqueous borate ion. J Am Chem Soc 77:266–268

Fogg TR, Duce RA (1985) Boron in the troposphere: distribution and fluxes. J Geophy Res Atmos 90 (D2):3781–3796

Foster GL, Pogge von Strandmann PAE, Rae JWB (2010) Boron and magnesium isotopic composition of seawater. Geochem Geophys Geosyst 11(8)

Foster GL, Lear CH, Rae JW (2012) The evolution of pCO2, ice volume and climate during the middle Miocene. Earth Planet Sci Lett 341:243–254

Gaillardet J, Dupré B, Louvat P, Allegre CJ (1999) Global silicate weathering and CO_2 consumption rates deduced from the chemistry of large rivers. Chem Geol 159(1):3–30

Gaillardet J, Lemarchand D, Göpel C, Manhès G (2001) Evaporation and sublimation of boric acid: application for boron purification from organic rich solutions. Geostand Newsl 25:67–75

Gast JA, Thompson TG (1959) Evaporation of boric acid from sea water. Tellus 11:344–347

Goldberg S, Glaubig RA (1985) Boron adsorption on aluminium and iron oxide minerals. Soil Sci Soc Amer 49:1374–1379

Goldberg S, Forster HS (1991) Boron sorption on calcareous soils and reference calcites. Soil Sci 152:304–310

Goldberg S (1997) Reactions of boron with soils. Plant Soil 193:35–48

Goldberg S, Lesch SM, Suarez DL (2000) Predicting boron adsorption by soils using soil chemical parameters in the constant capacitance model. Soil Sci Soc Am J 64:1356–1363

Goldberg S (2005) Inconsistency in the triple layer model description of ionic strength dependent boron adsorption. J Colloid Interface Sci 285:509–517

Goldberg S, Su C (2007) New advances in boron soil chemistry. Advances in plant and Animal Boron Nutrition 313–330

Greenop R, Foster GL, Sosdian SM, Hain MP, Oliver KIC, Goodwin P, Chalk TB, Lear CH, Wilson PA (2017) A record of Neogene seawater δ11B reconstructed from paired δ11B analyses on benthic and planktic foraminifera. Clim Past 13:149–170. doi:10.5194/cp-13-149-2017

Hathorne EC, James RH (2006) Temporal record of lithium in seawater: a tracer for silicate weathering? Earth Planet Sci Lett 246(3):393–406

He J, Cui J, Zhu D, Zhou W, Liao S, Geng M (2015) The use of boron isotopes to evaluate boron uptake by rape grown in acid soil treated with boron containing goethite. J Plant Nutr Soil Sci 178:935–943

Hemming NG, Hanson GN (1992) Boron isotopic composition and concentration in modern marine carbonates. Geochim Cosmochim Acta 56:537–543

Hervig RL, Moore GM, Williams LB, Peacock SM, Holloway JR, Roggensack K (2002) Isotopic and elemental partitioning of boron between hydrous fluid and silicate melt. Am Miner 87:769–774

Ishihara K, Mouri Y, Funahashi S, Tanaka M (1991) Mechanistic study of the complex formation of boric acid. Inorg Chem 30:2356–2360

Ishikawa T, Nakamura E (1990) Suppression of boron volatilization from a hydrofluoric acid solution using a boron-mannitol complex. Anal Chem 62(23):2612–2616

Keren R, Mezuman U (1981) Boron adsorption by clay minerals using a phenomenological equations. Clays Clay Miner 29:198–204

Klochko K, Cody GD, Tossell JA, Dera P, Kaufman AJ (2009) Re-evaluating boron speciation in biogenic calcite and aragonite using 11B MAS NMR. Geochim Cosmochim Acta 73:1890–1900

Klochko K, Kaufman AJ, Yao W, Byrne RH, Tossell JA (2006) Experimental measurement of boron isotope fractionation in seawater. Earth Planet Sci Lett 248 (1):276–285

Leeman WP, Tonarini S, Pennisi M, Ferrara G (2005) Boron isotopic variations in fumarolic condensates and thermal waters from Vulcano Island, Italy: implications for evolution of volcanic fluids. Geochim Cosmochim Acta 69(1):143–163

Lemarchand D, Gaillardet J, Göpel C, Manhes G (2002a) An optimized procedure for boron separation and mass spectrometry analysis for river samples. Chem Geol 182(2):323–334

Lemarchand D, Gaillardet J, Lewin E, Allegre CJ (2000) The influence of rivers on marine boron isotopes and implications for reconstructing past ocean pH. Nature 408(6815):951–954

Lemarchand D, Gaillardet J, Lewin E, Allegre CJ (2002b) Boron isotope systematics in large rivers: implications for the marine boron budget and paleo-pH reconstruction over the Cenozoic. Chem Geol 190(1):123–140

Lemarchand D, Gaillardet J (2006) Transient features of the erosion of shales in the Mackenzie basin (Canada), evidences from boron isotopes. Earth Planet Sci Lett 245(1):174–189

Lemarchand D, Cividini D, Turpault MP, Chabaux F (2012) Boron isotopes in different grain size fractions: exploring past and present water–rock interactions from two soil profiles (Strengbach, Vosges Mountains). Geochim Cosmochim Acta 98:78–93

Lemarchand D, Jacobson AD, Cividini D, Chabaux F (2015) The major ion, 87 Sr/86 Sr, and δ 11 B geochemistry of groundwater in the Wyodak-Anderson coal bed aquifer (Powder River Basin, Wyoming, USA). CR Geosci 347(7):348–357

Lemarchand E, Schott J, Gaillardet J (2005) Boron isotopic fractionation related to boron sorption on humic acid and the structure of surface complexes formed. Geochim Cosmochim Acta 69(14):3519–3533

Lemarchand E, Schott J, Gaillardet J (2007) How surface complexes impact boron isotope fractionation: evidence from Fe and Mn oxides sorption experiments. Earth Planet Sci Lett 260(1):277–296

Leslie D, Lyons WB, Warner N, Vengosh A, Olesik J, Welch K, Deuerling K (2014) Boron isotopic geochemistry of the McMurdo dry valley lakes, Antarctica. Chem Geol 386:152–164

Li YH (2000) A compendium of geochemistry: from solar nebula to the human brain. Princeton University Press: USA

Li G, Elderfield H (2013) Evolution of carbon cycle over the past 100 million years. Geochim Cosmochim Acta 103:11–25

Liu YC, You CF, Huang KF, Wang RM, Chung CH, Liu HC (2012) Boron sources and transport mechanisms in river waters collected from southwestern Taiwan: isotopic evidence. J Asian Earth Sci 58:16–23

Louvat P, Gaillardet J, Paris G, Dessert C (2011) Boron isotope ratios of surface waters in Guadeloupe, Lesser Antilles. Appl Geochem 26:S76–S79

Louvat P, Gayer E, Gaillardet J (2014) Boron behavior in the rivers of Réunion island, inferred from boron isotope ratios and concentrations of major and trace elements. Procedia Earth Planet Sci 10:231–237

Lu P, Nuhfer NT, Kelly S, Li Q, Konishi H, Elswick E, Zhu C (2011) Lead coprecipitation with iron oxyhydroxide nano-particles. Geochim Cosmochim Acta 75:4547–4561

Mavromatis V, Montouillout V, Noireaux J, Gaillardet J, Schott J (2015) Characterization of boron incorporation and speciation in calcite and aragonite from co-precipitation experiments under controlled pH, temperature and precipitation rate. Geochim Cosmochim Acta 150:299–313

McDonough WF, Sun SS (1995) The composition of the earth. Chem Geol 120:223–253

Millot R, Négrel P (2007) Multi-isotopic tracing (δ 7 Li, δ 11 B, 87 Sr/86 Sr) and chemical geothermometry: evidence from hydro-geothermal systems in France. Chem Geol 244(3):664–678

Millot R, Guerrot C, Innocent C, Négrel P, Sanjuan B (2011) Chemical, multi-isotopic (Li–B–Sr–U–H–O) and thermal characterization of Triassic formation waters from the Paris Basin. Chem Geol 283(3):226–241

Millot R, Petelet-Giraud E, Guerrot C, Négrel P (2010) Multi-isotopic composition (δ 7 Li–δ 11 B–δD–δ 18 O) of rainwaters in France: origin and spatio-temporal characterization. Appl Geochem 25(10):1510–1524

Misra S, Froelich PN (2012) Lithium isotope history of Cenozoic seawater: changes in silicate weathering and reverse weathering. Science 335(6070):818–823

Miyata Y, Tokieda T, Amakawa H, Uematsu M, Nozaki Y (2000) Boron isotope variations in the atmosphere. Tellus B 52(4):1057–1065

Negrel P, Petelet-Giraud E, Kloppmann W, Casanova J (2002) Boron isotope signatures in the coastal groundwaters of French Guiana. Water Resources Research 38(11)

Négrel P, Guerrot C, Millot R (2007) Chemical and strontium isotope characterization of rainwater in France: influence of sources and hydrogeochemical implications. Isot Environ Health Stud 43(3):179–196

Noireaux J, Gaillardet J, Sullivan PL, Brantley SL (2014) Boron isotope fractionation in soils at shale hills CZO. Procedia Earth Planet Sci 10:218–222

Noireaux J, Mavromatis V, Gaillardet J, Schott J, Montouillout V, Louvat P, Rollion-Bard C, Neuville DR (2015) Crystallographic control on the boron isotope paleo-pH proxy. Earth Planet Sci Lett 430:398–407

Noireaux J, Riotte J, Gaillardet J, Louvat P, Bouchez J, Sekkar M, Kumar MS, Candaudap F, Braun J-J MS, et al (2017) Vegetation cycling and chemical weathering in a tropical catchment: constraints from B isotope geochemistry. Submitted to Geochim Cosmochim Acta

O'Hara PF (1980) Metamorphic and structural geology of the Northern Bradshaw Mountains, Yavapai County, Arizona. Geology Arizona State University, Tempe, p 146

Onak T, Landesman H, Williams R, Shapiro I (2005) The B11 nuclear magnetic resonance chemical shifts and spin coupling values for various compounds. J Phys Chem 63:1533–1535

Palmer MR, Spivack AJ, Edmond JM (1987) Temperature and pH controls over isotopic fractionation during adsorption of boron on marine clay. Geochim Cosmochim Acta 51:2319–2323

Palmer MR, London D, Morgan GB, Babb HA (1992) Experimental determination of fractionation of 11B/10B between tourmaline and aqueous vapor: A temperature- and pressure-dependent isotopic system. Chem Geol Isot Geosci Sect 101:123–129

Peacok S, Hervig R (1999) Boron isotopic composition of subduction-zone metamorphic rocks. Chem Geol 160:281–291

Peak D, Luther Iii GW, Sparks DL (2003) ATR-FTIR spectroscopic studies of boric acid adsorption on hydrous ferric oxide. Geochim Cosmochim Acta 67:2551–2560

Pearson PN, Palmer MR (2000) Atmospheric carbon dioxide concentrations over the past 60 million years. Nature 406(6797):695–699

Paris G, Gaillardet J, Louvat P (2010) Geological evolution of seawater boron isotopic composition recorded in evaporites. Geology 38(11):1035–1038

Pennisi M, Bianchini G, Muti A, Kloppmann W, Gonfiantini R (2006) Behaviour of boron and strontium isotopes in groundwater–aquifer interactions in the Cornia Plain (Tuscany, Italy). Appl Geochem 21 (7):1169–1183
Pizer R, Selzer R (1984) The boric acid/lactic acid system. Equilibria and reaction mechanism. Inorg Chem 23:3023–3026
Pogge von Strandmann PAP, Henderson GM (2015) The Li isotope response to mountain uplift. Geology 43 (1):67–70
Raitzsch M, Hönisch B (2013) Cenozoic boron isotope variations in benthic foraminifers. Geology 41 (5):591–594
Ranjbar F, Jalali M (2014) Surface complexation model of boron adsorption by calcareous soils. Int J Environ Sci Technol 11:1317–1326
Rollion-Bard C, Blamart D, Trebosc J, Tricot G, Mussi A, Cuif J-P (2011) Boron isotopes as pH proxy: A new look at boron speciation in deep-sea corals using 11B MAS NMR and EELS. Geochim Cosmochim Acta 75:1003–1012
Rose EF, Chaussidon M, France-Lanord C (2000) Fractionation of boron isotopes during erosion processes: the example of Himalayan rivers. Geochim Cosmochim Acta 64(3):397–408
Rose-Koga EF, Sheppard SMF, Chaussidon M, Carignan J (2006) Boron isotopic composition of atmospheric precipitations and liquid-vapour fractionations. Geochim Cosmochim Acta 70:1603–1615
Rosner M, Pritzkow W, Vogl J, Voerkelius S (2011) Development and validation of a method to determine the boron isotopic composition of crop plants. Anal Chem 83:2562–2568
Roux P, Lemarchand D, Hughes HJ, Turpault M-P (2015) A rapid method for determining boron concentration (ID-ICP-MS) and δ11B (MC-ICP-MS) in vegetation samples after microwave digestion and cation exchange chemical purification. Geostand Geoanal Res. doi:10.1111/j.1751-908X.2014.00328.x
Roux P, Turpault M-P, Kirchen G, Redon P-O, Lemarchand D (2017) Boron and $d^{11}B$ atmospheric inputs to a forest ecosystem (in rev. ES&T)
Rudnick RL, Gao S (2003) Composition of the continental crust. Treatise Geochem 3: 1–64. doi:10.1016/B0-08-043751-6/03016-4
Schlesinger WH, Vengosh A (2016) Global boron cycle in the Anthropocene. Glob Biogeochem Cycles 30. doi:10.1002/2015GB005266
Schwarcz HP, Agyei EK, McMullen CC (1969a) Boron isotopic fractionation during clay adsorption from Sea-Water. Earth Planet Sci Lett 6:1–5
Sen S, Stebbins JF, Hemming NG, Ghosh B (1994) Coordination environments of B impurities in calcite and aragonite polymorphs: a11B MASS NMR study. Am Miner 79:819–825
Shao C, Miyazaki Y, Matsuoka S, Yoshimura K, Sakashita H (2000) Complexation of borate with cross-linked polysaccharide anion exchanger: 11B NMR and adsorption properties studies. Macromolecules 33:19–25
Simon L, Lécuyer C, Maréchal C, Coltice N (2006) Modelling the geochemical cycle of boron: implications for the long-term δ 11 B evolution of seawater and oceanic crust. Chem Geol 225(1):61–76
Staudigel H (2013) Chemical fluxes from hydrothermal alteration of the oceanic crust. Treatise Geochem Holland 4:583–606
Spivack AJ, Palmer MR, Edmond JM (1987) The sedimentary cycle of the boron isotopes. Geochim Cosmochim Acta 51(7):1939–1949
Spivack AJ, Berndt ME, Seyfried WE Jr (1990) Boron isotope fractionation during supercritical phase separation. Geochim Cosmochim Acta 54:2337–2339
Sugawara K (1948) Chemistry of precipitation (Rain and snow). Kugaku (Science, Japan) 18: 485–492
Syvitski JP, Vörösmarty CJ, Kettner AJ, Green P (2005) Impact of humans on the flux of terrestrial sediment to the global coastal ocean. Science 308(5720):376–380
Schwarcz HP, Agyei EK, McMullen CC (1969) Boron isotopic fractionation during clay adsorption from sea-water. Earth Planet Sci Lett 6(1):1–5
Tonarini S, Forte C, Petrini R, Ferrara G (2003) Melt/biotite $^{11}B/^{10}B$ isotopic fractionation and the boron local environment in the structure of volcanic glasses. Geochim Cosmochim Acta 67:1863–1873
Uchikawa J, Penman DE, Zachos JC, Zeebe RE (2015) Experimental evidence for kinetic effects on B/Ca in synthetic calcite: implications for potential B(OH) 4 − and B(OH)3incorporation. Geochim Cosmochim Acta 150:171–191
Vengosh A, Kolodny Y, Starinsky A, Chivas AR, McCulloch MT (1991a) Coprecipitation and isotopic fractionation of boron in modern biogenic carbonates. Geochim Cosmochim Acta 55:2901–2910
Vengosh A, Chivas AR, McCulloch MT, Starinsky A, Kolodny Y (1991b) Boron isotope geochemistry of Australian salt lakes. Geochim Cosmochim Acta 55 (9):2591–2606
Vengosh A, Starinsky A, Kolodny Y, Chivas AR (1991c) Boron isotope geochemistry as a tracer for the evolution of brines and associated hot springs from the Dead Sea. Israel Geochimica et Cosmochimica Acta 55(6):1689–1695
Vengosh ACHIVAS, Chivas AR, Starinsky A, Kolodny Y, Baozhen Z, Pengxi Z (1995) Chemical and boron isotope compositions of non-marine brines from the Qaidam Basin, Qinghai. China Chem Geol 120(1):135–154
Vengosh A, Kloppmann W, Marei A, Livshitz Y, Gutierrez A, Banna M, Guerrot C, Pankratov I, Raanan H (2005) Sources of salinity and boron in the Gaza strip: natural contaminant flow in the southern Mediterranean coastal aquifer. Water Resources Research 41(1)
Voinot A, Lemarchand D, Collignon C, Granet M, Chabaux F, Turpault M-P (2013) Experimental dissolution vs. transformation of micas under acidic soil

conditions: clues from boron isotopes. Geochim Cosmochim Acta 117:144–160
West AJ, Galy A, Bickle M (2005) Tectonic and climatic controls on silicate weathering. Earth Planet Sci Lett 235(1):211–228
Wieser ME, Iyer SS, Krouse HR, Cantagallo MI (2001) Variations in the boron isotope composition of Coffea arabica beans. Appl Geochem 16:317–322
Williams LB, Hervig RL, Holloway JR, Hutcheon I (2001a) Boron isotope geochemistry during diagenesis. Part I. Experimental determination of fractionation during illitization of smectite. Geochim Cosmochim Acta 65:1769–1782
Williams LB, Hervig RL, Hutcheon I (2001b) Boron isotope geochemistry during diagenesis. Part II. Applications to organic-rich sediments. Geochim Cosmochim Acta 65:1783–1794
Williams LB, Hervig RL (2004) Boron isotope composition of coals: a potential tracer of organic contaminated fluids. Appl Geochem 19(10):1625–1636
Xiao Y, Sun D, Wang Y, Qi H, Jin L (1992) Boron isotopic compositions of brine, sediments, and source water in Da Qaidam Lake, Qinghai, China. Geochimica et Cosmochimica Acta 56(4)
Xiao YK, Vocke RD, Swihart GH, Xiao Y (1997) Boron volatilization and its isotope fractionation during evaporation of boron solution. Anal Chem 69(12):5203–5207
Xiao Y, Swihart GH, Xiao Y, Vocke RD Jr (2001) A preliminary experimental study of the boron concentration in vapor and the isotopic fractionation of boron between seawater and vapor during evaporation of seawater. Sci China, Ser B 44(5):540–551
Xiao YK, Li SZ, Wei HZ, Sun AD, Liu WG, Zhou WJ, Swihart GH (2007) Boron isotopic fractionation during seawater evaporation. Mar Chem 103(3):382–392
Zhao ZQ, Liu CQ (2010) Anthropogenic inputs of boron into urban atmosphere: evidence from boron isotopes of precipitations in Guiyang City. China Atmos Environ 44(34):4165–4171

8 Boron Isotopes in the Ocean Floor Realm and the Mantle

Horst R. Marschall

Abstract

This chapter reviews the boron isotopic composition of the ocean floor, including pristine igneous oceanic crust such as mid-ocean ridge basalts and ocean island basalts and their implications for the B isotopic composition of the mantle. The chapter further discusses the B isotopic effects of assimilation of altered crustal materials in mantle-derived magmas. The systematics of seawater alteration on oceanic rocks are discussed, including sediments, igneous crust and serpentinization of ultramafic rocks and the respective marine hydrothermal vent fluids. The chapter concludes with a discussion of the secular evolution of the B isotopic composition of seawater.

Keywords

Boron isotopes · Oceanic crust · Alteration · Vent fluids · Mantle
Seawater

8.1 Introduction

The oceanic crust plays a key role in global geochemical cycles in several ways. Oceanic crust formation is the prime process of magma extraction from the mantle. Intense interaction of seawater with the lithosphere, particularly along the world's 65,000 km long mid-ocean ridge system, affects both seawater chemistry and the composition of the crustal rocks, as well as sections of the oceanic mantle (e.g., German and Lin 2004; Karson et al. 2015). The oceanic crust and part of the lithospheric mantle modified by

H.R. Marschall (✉)
Department of Geology and Geophysics, Woods Hole Oceanographic Institution, Woods Hole, MA 02543, USA
e-mail: hmarschall@whoi.edu; marschall@em.uni-frankfurt.de

H.R. Marschall
Institut Für Geowissenschaften, Goethe Universität Frankfurt, Altenhöferallee 1, 60438 Frankfurt Am Main, Germany

H. Marschall and G. Foster (eds.), *Boron Isotopes*, Advances in Isotope Geochemistry,
https://doi.org/10.1007/978-3-319-64666-4_8

interaction with the hydrosphere plus its sediment load are eventually subducted along the convergent plate margins, introducing chemical and isotopic heterogeneities into the mantle (e.g., Elliott et al. 1997; Jackson et al. 2007; Marschall and Schumacher 2012). Subduction of oceanic crust triggers the production of convergent-margin magmas that determines the composition of the newly produced continental crust today and probably throughout parts of Earth's history (e.g., Reymer and Schubert 1984; Dhuime et al. 2012). Ideal geochemical tracers that can be employed to detect and quantify these exchange processes should show large disparities in abundances among seawater, crustal rocks, sediments, altered rocks of the seafloor and the mantle, and should possess isotopic signatures that are strongly fractionated among the mantle, crust and hydrosphere.

Boron fulfils these criteria better than most other elements. It is a quintessentially crustal element with high concentrations in rocks of continental affinity and in rocks that interacted with the hydrosphere. Oceanic sediments, low-temperature altered oceanic basalts, and serpentinites show very high B abundances (10–200 μg/g), whereas the depleted mantle is characterised by very low B contents (<0.1 μg/g; Leeman and Sisson 1996). Fractionation of the two stable isotopes of boron (^{10}B and ^{11}B) at low temperatures is responsible for surface reservoirs that are enriched in boron with a ^{11}B-rich isotopic composition, and seawater is at the high end of the isotopic scale (Palmer and Swihart 1996).

Boron isotope analyses of pristine, unmetasomatised mantle samples has not been achieved yet, due to analytical limitations relative to the level of precision required for a geologically meaningful interpretation. However, fresh basaltic glasses erupted at mid-ocean ridges have been used to indirectly determine the boron elemental and isotopic composition of the convecting mantle, and ocean-island basalts have been investigated to identify possible recycled crustal materials in the mantle (Chaussidon and Jambon 1994; Chaussidon and Marty 1995; Marschall et al. 2017).

This chapter summarizes the boron isotope characteristics of the ocean-floor realm, including fresh and altered igneous crust and its connection to the mantle, as well as oceanic sediments and serpentinized ultramafic rocks exposed at the seafloor. Boron in hydrothermal vent fluids are reviewed for their bearing on seafloor alteration, and a brief summary of the secular change in the boron isotopic composition of seawater is given.

8.2 The Oceanic Crust

Approximately two thirds of the Earth are covered by large ocean basins submerged under, on average, a 4000 m deep layer of water. The oceanic crust is dominated by basalts, gabbros and exposed mantle rocks and has a limited age of approximately 0–200 Ma due to constant recycling (e.g., Müller et al. 2008). This contrasts with the elevated continents that comprise a diverse rock record spanning a very long history over billions of years (e.g., Dhuime et al. 2012). The oceanic crust is constantly formed at the mid-ocean ridges from where it drifts away, ages and is eventually subducted into the mantle at convergent plate margins. The processes operating at mid-ocean ridges are, therefore, central to the formation of the crust and key to understanding of its general structure and diverse types of crustal accretion process that have been distinguished (e.g., Reynolds et al. 1992; Dick et al. 2003; Escartín et al. 2008).

The textbook model for the structure of fast-spreading oceanic crust is the classic "Penrose model", named after the GSA 1972 Penrose Field Conference, where it was defined for the idealized structure of ophiolites (e.g., Dilek 2003). It consists of mantle and ultramafic cumulates at the bottom, overlain by successive layers of magmatic rocks, including gabbro, sheeted dikes and lava flows (sheets and pillows), and a top layer of siliceous and carbonaceous sediments. At slow-spreading ridges the spreading has a much larger tectonic component with disruption of the magmatic layers and exposure of serpentinized mantle at the seafloor (e.g., Dick et al. 2003).

Boron abundances and isotopic compositions have been analysed for all the above-mentioned rock types and provide a picture of the boron isotopic inventory of the ocean floor. Hydrothermal activity, low-temperature alteration and weathering effectively mobilize boron and strongly influence the B isotopic composition of the affected rocks. These alteration processes have been studied through the investigation of altered rocks, hydrothermal vent fluids, and through laboratory experiments.

8.3 Mid-Ocean Ridge Basalts

Basaltic melts erupted along the global mid-ocean ridges are readily quenched to glass at contact with seawater, thus preserving the geochemical patterns of the melt that may otherwise be disturbed by crystallization and alteration processes in slower-cooled, crystalline rocks. Mid-ocean ridge basalt (MORB) glasses have, thus, been extensively used to decipher the composition of mantle-derived magmas and the composition of their sources, i.e. the upper, depleted, convecting mantle (e.g., Sun et al. 1979; Hofmann 1988; O'Neill and Jenner 2012).

The boron isotopic composition of unaltered MORB glasses has been investigated in several studies (Fig. 8.1) starting with Spivack and Edmond (1987). These authors analysed two samples from the East-Pacific Rise (EPR) resulting in a $\delta^{11}B$ of $-3.0 \pm 2.0‰$. Ishikawa and Nakamura (1992) investigated a number of basaltic rocks from ODP Hole 504B (Galapagos Spreading Center) that showed variable degrees of hydrothermal alteration. They extrapolated the alteration trend back to the least altered sample and argued that $\delta^{11}B = +0.2‰$ was representative of fresh MORB.

Chaussidon and Jambon (1994) analysed 17 MORB glasses from the EPR, the Mid-Atlantic Ridge (MAR) and the Red Sea and found $\delta^{11}B$ to range from -6.5 to $-1.2‰$ with a mean of $-3.9 \pm 3.3‰$. These authors observed a similar range of values in back-arc basin basalts and ocean-island basalts (OIB; see below). Based on

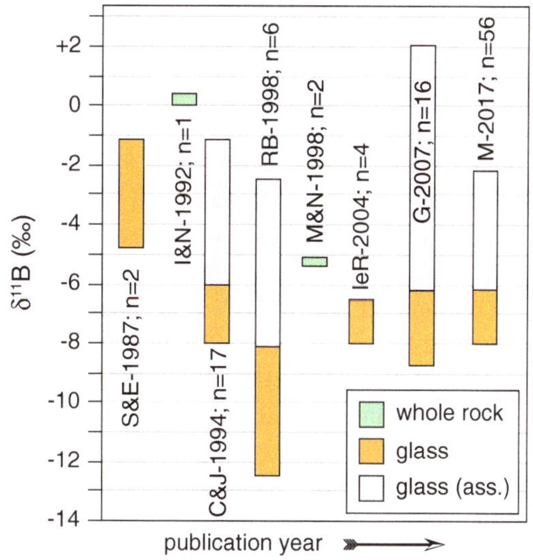

Fig. 8.1 Boron isotopic composition of unaltered, uncontaminated MORB as determined through analyses of MORB glass (*yellow*) and whole-rock samples (*green*). Elevated $\delta^{11}B$ values were found in MORB at various localities and were demonstrably related to assimilation of altered crust into the magmas beneath the ridge (*pink* ass. = assimilation). Sources are: S&E-1987 = Spivack and Edmond (1987), I&N-1992 = (Ishikawa and Nakamura 1992), C&J-1994 = Chaussidon and Jambon (1994), RB-1998 = Roy-Barman et al. (1998), M&N-1998 = Moriguti and Nakamura (1998), leR-2004 = le Roux et al. (2004), G-2007 = Gannoun et al. (2007), M-2017 = Marschall et al. (2017). 'n' refers to the number of unaltered MORB samples analyzed in each study

K_2O, H_2O, MgO and B content, as well as $^{87}Sr/^{86}Sr$ and $^{2}H/^{1}H$ ratios, they argued that the B isotope range observed in the oceanic basalts were not reflecting mantle source heterogeneities, but were due to the assimilation of seawater-altered materials by the magmas beneath the ridge prior to eruption. Chaussidon and Jambon (1994) argued that a value of $\delta^{11}B = -7.0 \pm 1.0‰$ is most representative of the upper mantle.

Moriguti and Nakamura (1998) analyzed two rock samples from ODP Hole 648B (MAR) with $\delta^{11}B = -5.3 \pm 0.2‰$, and le Roux et al. (2004) analyzed four glass samples from the EPR resulting in $\delta^{11}B = -7.3 \pm 0.8‰$ (Fig. 8.1).

Roy-Barman et al. (1998) investigated Os and B isotopes of six MORB glasses from the MAR,

Central Indian Ridge and the EPR. They showed that radiogenic Os is found in samples that also show isotopically heavy B, most likely introduced by assimilation of altered crust. Their two MORB samples with $^{187}Os/^{188}Os < 0.135$ have a $\delta^{11}B$ value of −10.3 ± 2.2‰, which they take as representative of the uncontaminated mantle, following Chaussidon and Marty (1995).

Gannoun et al. (2007) also completed a combined Os and B isotope study and analysed 16 MORB glasses from the EPR, MAR and the South-West Indian Ridge (SWIR). They too found a covariation of radiogenic Os with isotopically heavy B, interpreted as a sign of assimilation of seawater-altered crust by the magma. This led to $\delta^{11}B$ values of as high as +2.1‰. The glasses with the lowest $^{187}Os/^{188}Os (< 0.135)$ have $\delta^{11}B$ values between −8.7 and −6.2‰.

Marschall et al. (2017), the most recent and comprehensive study so far, investigated 56 MORB samples from the northern and southern MAR, the SWIR and from three sections of the EPR. Magma assimilation of altered crust (or seawater, or brines) was identified through elevated chlorine contents and Cl/K ratio. A subset of 40 samples with $Cl/K < 0.08$ was identified to represent uncontaminated MORB and was used to determine the mantle-derived boron isotopic compositions of the various ridge segments. The mean of the six different investigated ridge sections for samples with $Cl/K < 0.08$ was −7.1 ± 0.9‰ (2 standard deviation; Fig. 8.1). The majority of the low-Cl/K samples show $\delta^{11}B$ values between −8.5 and −6.0‰ with no resolvable differences among the various investigated ridge sections, and without any correlation with Cl, Li or B contents (Marschall et al. 2017). Samples from diverse localities, such as the southern MAR, the SWIR and the various sections of the EPR all showed indistinguishable normal distributions around the mean value of −7.1‰. Furthermore, no correlation was found between $\delta^{11}B$ and geochemical parameters that indicate the degree of depletion or enrichment of the mantle source of the MORB magmas, such as La/Sm, Ba/TiO_2, or Zr/Y. Boron isotopes also did not show any systematic variation with radiogenic isotope ratios, such as those of Sr, Nd or Pb (Marschall et al. 2017). Assimilation of altered crust, brines or seawater into the magmas at the ridge produced elevated Cl and B contents and $\delta^{11}B$ values of as high as −2.2 ± 1.7‰.

Shaw et al. (2012) presented H and B isotopes of 7 MORB glasses from the Manus basin with $\delta^{11}B$ value from −10.8 ± 1.0‰ to −3.1 ± 1.0‰ that correlate positively with δD. However, the Manus basin is very complex with possible influences of a plume, subducted slab, and a back-arc system, so that the isotope values of these samples are not representative of MORB produced at an open-ocean spreading center (Shaw et al. 2012).

Any comparison of all these $\delta^{11}B$ values published by a number of authors from different laboratories and determined by various analytical techniques need to take the analytical limitations into account. Well-established silicate reference materials for B isotope analysis only became available relatively recently (e.g., Jochum et al. 2006), and analytical protocols have been improved over the past two decades (e.g., Aggarwal et al. 2009; Foster et al. 2013, 2017; Marschall and Monteleone 2015). Discrepancies among the different studies from the 1980s and 1990s cited above and displayed in Fig. 8.1 at the level of 5‰ or less are not likely to be significant, given the level of accuracy, inter-laboratory comparability and lack of internationally distributed B isotope reference materials. Nonetheless, these studies established that the $\delta^{11}B$ value of fresh, uncontaminated MORB most likely had to be between approximately −12 and 0‰. Today, the best estimate for the $\delta^{11}B$ of uncontaminated MORB is −7.1 ± 0.9‰, and it is homogenous in boron isotopes on the level of current analytical precision and accuracy. It is further concluded that assimilation of seawater or seawater-altered materials produces elevated $\delta^{11}B$ values as observed in a number of MORB glass samples (Chaussidon and Jambon 1994; Marschall et al. 2017; Fig. 8.1).

8.4 Ocean Island Basalts

Ocean island basalts (OIB) exhibit chemical and isotopic characteristics that distinguish them from the basalts typically erupted at mid-ocean ridges (Hofmann 1997, 2003). MORB represent the composition of the convecting upper mantle, and the excursions of the OIB from the MORB field require the existence of chemically, isotopically and possibly mineralogically distinct domains in the Earth's mantle (i.e., in the source of these OIB) that persist over hundreds of millions to billions of years (e.g., McKenzie and O'Nions 1983; Zindler and Hart 1986; Stracke et al. 2005).

The genesis of the enriched components in the mantle is a long-standing matter of debate, and various hypotheses have been put forward to explain their origin. One possibility is that radiogenic parent-daughter trace elements were fractionated in the mantle itself or by interaction of mantle and core (e.g., Vidal and Dosso 1978; Allègre et al. 1980; Menzies and Wass 1983; Halliday et al. 1990; Collerson et al. 2010). However, most studies favor subducted slabs as the origin of the enriched components in OIB (e.g., Chase 1981; Hofmann and White 1982; Weaver 1991; Jackson et al. 2007; Nebel et al. 2013).

The establishment of the connection between recycled surface materials and OIB and its quantification in terms of timescales and mass fractions is of foremost importance to the geosciences, as it would enable us to use the geochemical tracers in OIB to inform models on mantle convection and the long-term evolution of the crust-mantle system (c.f., van Keken et al. 2002).

The interpretation of radiogenic isotope data are complicated by the fact that isotope signatures naturally evolve differently with time depending on the parent/daughter abundance ratios in any particular mantle domain. This generally leaves a range of possible interpretations for the measured data with respect to residence time and mixing and convection mechanisms. The possibility of parent/daughter fractionation in the mantle adds ambiguity to radiogenic isotope signatures as tracers of recycled material.

An alternative tool to radiogenic isotopes are stable isotope ratios. These do not change over time and, hence, remove the uncertainty of time-dependent evolution models. Also, stable isotope fractionation is strong in crustal materials at the surface, but small or negligible in the hot mantle and in differentiating magmas. Stable isotope signatures of erupted lavas can, therefore, be interpreted to reflect mantle heterogeneities produced by recycled surface materials.

Boron stable isotopes in OIB have been employed by a number of studies in an attempt to identify recycled materials in their mantle sources. Analyses of B isotopes in OIB are less difficult and slightly more precise than B isotope analyses of MORB glasses, because many have higher B abundances by a factor of two to three (e.g., Ryan and Langmuir 1993). However, this does not apply to all OIB samples. In addition, two major obstacles have caused difficulties for the interpretation of the data, or demand cautiousness towards the interpretations proposed in the published studies discussed in more detail below:

(1) The boron isotopic composition of the upper mantle was not well defined in the past, and some studies have discussed the OIB data relative to a δ^{11}B value for the depleted upper mantle higher than −7‰; such elevated δ^{11}B values in MORB have been identified as produced by shallow magma assimilation of seawater-altered materials and are not accepted as representative of the MORB mantle (see discussion above). An incorrect geochemical "baseline" for the mantle, i.e., for the mantle that does not contain recycled materials, inevitably leads to misinterpretations of data on possibly exotic mantle domains.
(2) Assimilation of even small amounts of hydrothermally altered materials by the OIB magmas during their ascent through the crust can influence their B isotopic composition, in which case the measured signal is no longer

representative of the composition of the mantle source. This, on the other hand, makes B isotopes a sensitive indicator for crustal assimilation in basalts and they have been employed in OIB studies for this purpose.

The largest number of studies were completed on samples from Hawai'i and Iceland, with only a few published data from other ocean islands. Chaussidon and co-workers presented boron isotope data from Hawai'i and Iceland, as well as Galapagos, St. Helena, MacDonald seamount, and Afar (Fig. 8.2; Chaussidon and Jambon 1994; Chaussidon and Marty 1995; Gurenko and Chaussidon 1997; Roy-Barman et al. 1998). These studies consistently show positive correlations of $\delta^{11}B$ values and indicators of assimilation of low-temperature altered crust, such as increasing δD values or radiogenic Os. Decreasing MgO contents that correlate with increasing $\delta^{11}B$ indicate combined assimilation and fractional crystallization (Chaussidon and Jambon 1994). A second trend is observed in samples from subaerial volcanoes with $\delta^{18}O$ values below that of uncontaminated mantle indicating assimilation of crust altered by high-temperature meteoric fluids: these samples show lower $\delta^{11}B$ values than MORB-source mantle, as low as −14.6 ± 3.0‰ (2SE) in samples from Afar (Fig. 8.2; Chaussidon and Marty 1995). The least contaminated OIB with the highest $^{3}He/^{4}He$ ratios are taken as representative of the OIB-source mantle with a relatively homogeneous $\delta^{11}B$ value for all investigated localities of −10 ± 2‰ (Fig. 8.2; Chaussidon and Marty 1995). No boron isotope mantle heterogeneities are indicated in these studies and all excursions to high or low $\delta^{11}B$ values are demonstrated to be caused by assimilation of altered crust.

Hawai'i whole-rock isotope data from three different shield stages (Kilauea, Mauna Loa and Koolau) showed a very narrow range in $\delta^{11}B$ values from −5.4 to −3.0‰ for 25 samples (Fig. 8.2; Tanaka and Nakamura 2005). The three stages show slightly different $\delta^{11}B$ values of −5.1 ± 0.6‰ for Koolau, −4.0 ± 0.6‰ for Kilauea, and −3.4 ± 0.6‰ for Mauna Loa. Seawater-alteration and assimilation of altered crust were excluded as causes of these isotopic variations, because they do not correlate with B content nor with fractionation indicators (Tanaka and Nakamura 2005). Radiogenic isotopes show variations among these three Hawai'ian volcanoes, and the authors interpret the variation as a range between MORB-source mantle and deeply subducted sediments with a low $\delta^{11}B$ value. However, they take a value of −3.3‰ for the unmodified MORB-source mantle and argue that values between approximately −4 and −5.4‰, as observed in the Kilauea and Koolau lavas, would require a recycled sediment

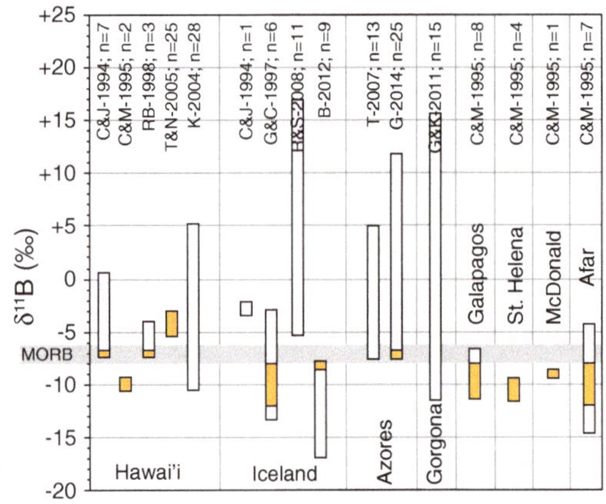

Fig. 8.2 Boron isotopic composition of unaltered OIB samples from various localities. *Orange fields* depict estimates of B isotopic composition of the OIB mantle source, whereas *pink fields* display samples with evidence for assimilation of hydrothermally altered crust. The *green fields* show samples that were interpreted to represent the heterogenous mantle sources of the OIB, but a reconsideration in the light of crustal assimilation has been published or is suggested here. Sources are: C&J-1994 = Chaussidon and Jambon (1994), C&M-1995 = Chaussidon and Marty (1995), RB-1998 = Roy-Barman et al. (1998), T&N-2005 = Tanaka and Nakamura (2005), K-2004 = Kobayashi et al. (2004), R&S-2008 = Rose-Koga and Sigmarsson (2008), B-2012 = Brounce et al. (2012), T-2007 = Turner et al. (2007), G-2014 = Genske et al. (2014), G&K-2011 = Gurenko and Kamenetsky (2011). 'n' refers to the number of samples analyzed in each study, including analyses of matrix glass, melt inclusions, and whole-rock samples. MORB range from Marschall et al. (2017)

component. This model has to be reversed, however, if a $\delta^{11}B$ value for the MORB-source mantle of approximately −7.1‰ is accepted, and a component with isotopically heavy B is required, instead, to explain the data for all three Hawai'ian volcanic centers.

In a very similar way, Kobayashi et al. (2004) interpreted data from melt inclusions in orthopyroxene and plagioclase from Hawai'i (28 analyses from 5 different samples, most of which were also analyzed by Tanaka and Nakamura 2005). Their data range from −10.5 ± 2.5‰ to +5.2 ± 2.6‰ (Fig. 8.2). The authors use a $\delta^{11}B$ reference value of −3 ± 3‰ for MORB-source mantle and interpret all observed values higher than that as due to assimilation of altered Hawai'ian crust into the magma (Kobayashi et al. 2004). Values lower than −6‰ are taken as evidence for deeply recycled ancient subducted crust; however, there are no data that are significantly lower than the more recently suggested $\delta^{11}B$ value for the MORB-source mantle (−7.1 ± 0.9‰; Marschall et al. 2017), suggesting that all data may be explained by melts derived from mantle without an unusual B isotopic composition, and that the observed range may be due to various degrees of assimilation of altered crust. Kobayashi et al. (2004) also report a wide range of Li isotopes in the melt inclusions, but these would have to be revisited in the light of kinetic isotope fractionation during diffusion of Li in and out of the melt inclusions after entrapment.

A parallel story can be told about the Azores hot-spot lavas; initial boron isotope data from the Azores were interpreted to indicate deeply recycled oceanic crust based on an assumed MORB-source mantle value of −4.6‰ and the idea that the lowest observed $\delta^{11}B$ values of −7.6 ± 1.0‰ would require a mantle component enriched in isotopically light B relative to the MORB source (Turner et al. 2007). A more recent study on the Azores (that involved some of the same authors) that included a larger number of samples from a larger number of islands in the archipelago also did not report any $\delta^{11}B$ lower than MORB source mantle (as defined by Marschall et al. 2017), but a range of very high values of up to +11.8‰ without any correlation to the observed range in radiogenic isotopes (Genske et al. 2014). The authors conclude that boron isotopes do not indicate mantle heterogeneities in the Azores, but that they are in general a very good tool to distinguish between different crustal assimilation processes in ocean-island basalts (Genske et al. 2014).

Gurenko and Kamenetsky (2011) found a range of $\delta^{11}B$ values from −11.5 ± 5.4‰ (2SD) to +15.6 ± 4.8‰ in olivine melt inclusions from Gorgona island komatiites, Columbia. The lowest $\delta^{11}B$ values are not significantly lower than MORB. The strongly elevated values that were observed were suggested to be due to hydrous metasomatism in the mantle source, rather than assimilation of altered crustal materials or brines by the ascending magma. This conclusion supports the model of komatiite formation by hydrous melting of the mantle rather than dry melting under extremely high temperatures (Gurenko and Kamenetsky 2011). The authors suggest that the mantle source was metasomatized by fluids derived from a subducting slab, which delivered isotopically heavy boron to the mantle source of the komatiites.

Boron isotope studies on various volcanic eruption products from Iceland revealed that assimilation of altered crust, as well as partial melting of altered crust may produce a large range of B isotopic compositions. Rose-Koga and Sigmarsson (2008) found the lowest $\delta^{11}B$ values (approximately −5‰; Fig. 8.2) in basalts with mantle-like O isotopic composition. In contrast, strongly elevated $\delta^{11}B$ values (up to +17‰) were found in rhyolites with low $\delta^{18}O$ values and low $^{230}Th/^{232}Th$ ratios. The series of rocks are interpreted as mixing between mantle-derived basalts and crustal rhyolites derived from melting of hydrothermally-altered basalts (Rose-Koga and Sigmarsson 2008). Brounce et al. (2012) investigated plagioclase-hosted basaltic melt inclusions and basaltic matrix glass from the Laki Fissure eruption. They found $\delta^{11}B$ values in both groups of samples that range from mantle-like

(−7.8 ± 4.2‰) to enriched in the light isotope (δ^{11}B = −16.9 ± 4.1‰; Fig. 8.2). The latter were interpreted to be affected by assimilation of crustal materials into the magma that were altered by deeply-circulating meteoric water (Brounce et al. 2012). This interpretation is supported by volatile abundance analyses and by reports of low-δ^{11}B altered crust from the Icelandic Deep Drilling Project reported by Raffone et al. (2010).

In summary, the range of OIB samples investigated to date show a relatively wide range of δ^{11}B values. These excursions from the B isotopic composition of uncontaminated MORB is, to a large degree, generated by assimilation of crustal materials into the ascending magma. These crustal materials were altered by seawater or meteoric fluids and are enriched in isotopically heavy or light boron, respectively. Consequently, the identification of B isotopic heterogeneities in the mantle source of OIB is not easy, but may be achievable through combinations of various isotope and trace-element systems. Such uncontaminated mantle sources were suggested for Hawai'i with δ^{11}B values of approximately −4 ± 1‰ and for a range of other localities with values of −10 ± 2‰ (Fig. 8.2). It can, therefore, be concluded from all studies completed to date that the global δ^{11}B range of uncontaminated OIB is restricted to within four or five per mil of the δ^{11}B value of uncontaminated MORB, i.e., between −12 and −3‰ (Fig. 8.2).

8.5 The Mantle

The boron isotopic composition of the mantle has not been analyzed directly. The first reason for this lack of data is that the abundances of B in unmetasomatized mantle rocks are too low to be analyzed with currently available methods to yield a precision and accuracy that would be geologically meaningful; the second reason is that it is very difficult to identify mantle samples that lack any metasomatic overprint, hydrothermal alteration, weathering or interaction with basaltic magmas. Any of these processes are prone to alter the abundance and isotopic composition of B in mantle samples.

As an alternative strategy, basalts and basaltic glasses have been used to unravel the B isotopic composition of the mantle, based on the idea that the boron isotope fractionation between rocks and silicate melt is negligible at magmatic temperatures (e.g., Chaussidon and Jambon 1994; Chaussidon and Marty 1995). Uncontaminated MORB and OIB should, therefore, faithfully record the unfractionated isotopic composition of their mantle sources. Marschall et al. (2017) discussed this connection between basalts and their mantle sources in more detail, based on experimentally determined boron partition coefficients between peridotitic and basaltic minerals and silicate melt, as well as estimated boron isotope fractionation factors. The latter are not well constraint by experiments and had to be extrapolated from experiments on other fluid, mineral and melt phases conducted at lower temperatures, and from a limited number of natural rocks (Marschall et al. 2017). Nevertheless, the authors concluded that uncontaminated mantle-derived basalts accurately reflect the δ^{11}B value of their mantle source within ±0.30‰ and that crystal fractionation and magmatic differentiation do not change the B isotopic composition of magmas to a measurable degree, unless assimilation of altered crustal material occurs.

Heterogeneity of the global MORB-source mantle has been identified through radiogenic isotopes and trace elements, but no such variation is detectable for B isotopes at the current level of analytical precision (±0.9‰, 2 standard error; see discussion above; Marschall et al. 2017). The mantle domains sampled by OIB magmas may show a small variation with δ^{11}B values by up to 4‰ higher and up to 5‰ lower than the MORB source mantle (−10 ± 2 to −4 ± 1‰; see discussion above).

Marschall et al. (2017) argued that the δ^{11}B value of the depleted upper mantle is best represented by the global range of MORB with low Cl/K that show a normal distribution around the mean of −7.1 ± 0.9‰ in their study (Fig. 8.3). Chaussidon and Jambon (1994) estimated a δ^{11}B

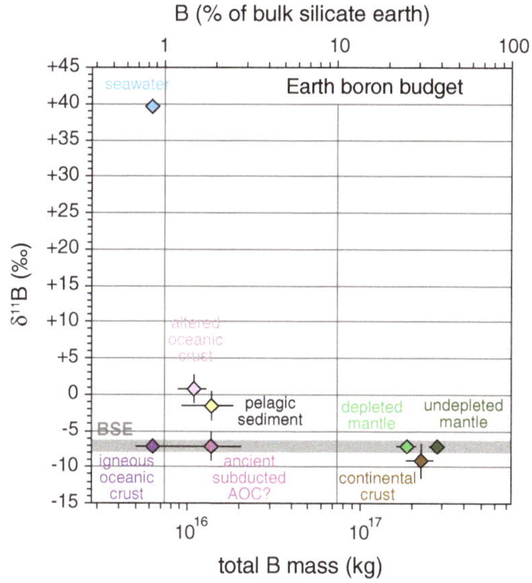

Fig. 8.3 Geochemical budget of B in Earth's major silicate reservoirs. Isotopic compositions versus the total mass of boron on the *bottom* x-axis, and fraction of boron of the bulk silicate earth budget on the *top* x-axis. The bulk continental crust is estimated to be enriched in isotopically light B compared to the mantle. Ancient subducted AOC (*altered oceanic crust*) is estimated to be indistinguishable from the mantle in B isotopes (Figure from Marschall et al. 2017)

value of −7 ± 1‰ for the MORB-source mantle, but discussed the possibility that recycled materials derived from subducted altered oceanic crust could have enriched the upper mantle in isotopically heavy boron, and that the primitive mantle had a lower δ^{11}B than the one observed through uncontaminated MORB. However, Marschall et al. (2017) rejected this hypothesis based on the lack of variation in δ^{11}B among MORB samples from various global ridge sections, some of which show clear enriched or depleted geochemical signatures in radiogenic isotope space and trace elements. They argue that different portions of recycled materials into various parts of the global ridge system would express itself in a measurable boron isotopic variation, if the recycled material had sufficient amounts of exotic boron; but this has not been observed to date.

Chaussidon and Marty (1995) took the δ^{11}B value of −10 ± 2‰ that they identified as the uncontaminated endmember of a range of OIB from various localities as representative of the primitive mantle. A subset of their OIB samples show high ^{3}He/^{4}He ratios indicating an undegassed (primitive) mantle source, and these samples are taken as representative of the primitive mantle δ^{11}B value. However, it should be noted that the boron isotopic composition of OIB samples are prone to contamination (as discussed above), and that the primitive mantle value estimated by Chaussidon and Marty (1995) overlaps within error with the MORB-source mantle value estimated by Marschall et al. (2017). Furthermore, interlaboratory comparison and accuracy limitations among the various studies that have been conducted over a time span of several decades additionally blur the picture. Future research should focus on the least contaminated OIB samples and unravel the variation in the OIB mantle sources and a possible difference between the OIB and MORB-source mantle.

In contrast to the complications seen in MORB and OIB, clear variation and mostly strongly elevated δ^{11}B values exist in basalts erupted along subduction zones, and these are interpreted to represent addition of isotopically heavy boron to the mantle source of arc magmas (de Hoog and Savov 2017). Supra-subduction mantle, therefore, experiences the addition of isotopically distinct boron ultimately derived from seawater (de Hoog and Savov 2017).

8.6 Hydrothermal Alteration and Weathering

Unaltered mafic igneous and ultramafic rocks show low concentrations of boron combined with a relatively well-constrained boron isotopic composition, as discussed above for MORB, OIB and the mantle. Modern seawater, in contrast, shows higher boron concentrations than most unaltered basalt, gabbro, and peridotite samples (4.5 µg/g; Uppstroem 1974; Spivack and Edmond 1987), and its isotopic composition is strongly enriched in the heavy isotope (+39.61 ± 0.04‰; Table 8.1; Fig. 8.3; Spivack and Edmond 1987; Foster et al. 2010).

Fluid-rock interaction in rocks of the oceanic crust is frequently identified and quantified using O and Sr isotopes (e.g., Alt 1995; Hart et al. 1999; Bach et al. 2001), and these tracers work very reliably in hydrothermal systems with relatively high fluid-rock ratios. The trace element B with its distinct isotopic seawater–mantle dichotomy is an ideal tracer for alteration processes at the seafloor, and it can be used to detect even the smallest influence of seawater or seawater-derived fluids on rocks and magmas (e.g., Chaussidon and Jambon 1994; Marschall 2017). Low- and high-temperature altered oceanic crust and serpentinized mantle rocks show significant alterations in their B contents and B isotopic compositions.

8.6.1 Alteration of the Igneous Crust

The B isotopic composition of hydrothermally overprinted and weathered rocks is variable and depends on temperature, fluid/rock ratio, pH of the fluid and the type of secondary minerals (in particular their crystallographic coordination of B). Weathering of basalt at the seafloor produces smectite with tetrahedrally coordinated B in contact to seawater $\left(\delta^{11}\mathrm{B} = +39.6\ ‰,\ {\sim}80\%\right.$ B in trigonal coordination). Boron isotope fractionation under these conditions is large and weathered basalts are expected to display low δ^{11}B values relative to seawater. Experimentally and theoretically determined B isotope fractionation is approximately 30–35‰ for temperatures of the seafloor (~0 °C) (Schwarcz et al. 1969; Palmer et al. 1987; Kowalski and Wunder 2017), and a δ^{11}B value of +5 to +10‰ could be expected for alteration in contact to seawater. Spivack and Edmond (1987) determined a B isotope fractionation of 32‰ between seawater and alteration products of basalt, with a δ^{11}B value of +8‰ for the alteration products. Theoretical and experimental work further predicts that alteration at a temperature of 100 °C would result in a δ^{11}B value of +18‰ in the secondary minerals (see Kowalski and Wunder 2017).

Low-temperature alteration (<150 °C) and seafloor weathering generally lead to a strong enrichment of boron in the alteration products (mainly in clays) with abundances of up to 140 µg/g (Thompson and Melson 1970; Donnelly et al. 1979; Spivack and Edmond 1987). Weathering and low-T alteration by seawater leads to a B isotopic signature that is enriched in the heavy isotope (^{11}B) compared to fresh MORB, and δ^{11}B values that range from −4 to as high as +25‰ have been found in low-temperature altered MORB (Table 8.1; Fig. 8.4; Spivack and Edmond 1987; Ishikawa and Nakamura 1992; Smith et al. 1995). Nonetheless, the majority of data for low-temperature altered MORB shows δ^{11}B values varying between 0 and +5‰, whereas higher δ^{11}B values are found in basalts and gabbros that were altered at higher temperatures (Fig. 8.4). The B isotopic composition of these rocks show a relatively large scatter as a function of boron concentrations; nevertheless, their is a tendency towards higher δ^{11}B in low-B samples in altered gabbros and basalts (Fig. 8.4a), whereas the ultramafic rocks are invariably enriched in the heavy isotope at high B concentrations (Fig. 8.4 b). There is a negative correlation (albeit weak) between δ^{11}B and δ^{18}O for the combined data from altered ophiolite rocks and MORB (Fig. 8.4c). This demonstrates that high alteration temperatures in basalts and gabbros lead to a moderate enrichment of isotopically very heavy boron or, in some cases, a loss of isotopically light boron, whereas low-temperature alteration leads to a strong enrichment of boron with a moderately elevated δ^{11}B value. This is consistent with a decrease of fluid–mineral boron isotope fractionation as temperature increases, and with a preferential up-take of B by clay minerals at very low temperatures, contrasted with leaching of boron from the rocks at high temperatures.

The evaluation of the imprint of high-T alteration is complicated by the fact that rocks that experienced high-T alteration subsequently cooled prior to sampling and are likely to also record lower-T alteration. Whole-rock analyses of high-T altered rocks are, therefore, prone to

Table 8.1 Compilation of published boron data for different rocks, fluids, and reservoirs

Material	$\delta^{11}B$(‰)	[B] (μg/g)	References
MORB-source mantle	−7 ± 1	0.065	Chaussidon and Jambon (1994)
	−10 ± 2	0.010–0.015	Chaussidon and Marty (1995)
	−7.1 ± 0.9	0.060	Marschall et al. (2017)
Primitive mantle	−7 ± 1	0.25 ± 0.10	Chaussidon and Jambon (1994)
	−10 ± 2	0.090	Chaussidon and Marty (1995)
	−7.1 ± 0.9	0.173	Marschall et al. (2017)
Continental crust	−10 ± 3	10	Chaussidon and Albarède (1992)
	−9.4 ± 0.4	11	Marschall et al. (2017)
Upper continental crust	−8.8	43	Kasemann et al. (2000)
Modern seawater	+ 39.5 ± 0.3	4.5	Spivack and Edmond (1987)
	+ 39.61 ± 0.04	n.d.	Foster et al. (2010)
Vent fluids			
Mid-ocean ridge, basalt-hosted (sediment-starved)	+30.0 to +36.8	4.9–5.9	Spivack and Edmond (1987)
	+26.7 to +36.8	3.8–6.1	Palmer (1991)
	+24.3 to +26.0	4.5–5.1	James et al. (1995)
	+28.5 to +35.2	5.4–6.9	You et al. (1994)
Back arc basin (sediment starved)	+22.5 to +29.8	8.1–8.8	Palmer (1991)
	+17.8 to +38.0	4.4–11	Yamaoka et al. (2015a)
Mid-ocean ridge (sediment hosted)	+10.1 to +23.2	17–23	Palmer (1991)
	−2.2 to +22.6	2.9–105	James et al. (1999)
Back arc basin (sediment hosted)	−1.0 to +9.2	22–37	You et al. (1994)
	+2.3 to +19.1	7.4–51	Yamaoka et al. (2015a)
Ultramafic hosted	+25 to +30	0.3–3.6	e.g. Foustoukos et al. (2008)
Marine evaporite setting	+29.7 to +39.0	6.8–8.8	Palmer (1991)
Altered MORB			
<150 °C[a]	+0.1 to +9.2	8.9–69	Spivack and Edmond (1987)
	−4.3 to +24.9	1.1–104	Smith et al. (1995)
>150 °C[a]	−0.1 to +1.0	0.17–0.52	Ishikawa and Nakamura (1992)
Serpentinite	+8.3 to +12.6	50–81	Spivack and Edmond (1987)
	+7.0 to +9.9	n.d.	Lécuyer et al. (2002)
	+11.4 to +16.3	34–91	Boschi et al. (2008)
	+29.6 to +40.7	10–65	Vils et al. (2009)
	+9.3 to +19.6	19–36	Harvey et al. (2014)
Ophiolites			
Oman pillows (<60 °C[a])	−1.1 to +11.9	1.4–29.1	Yamaoka et al. (2012)
Troodos pillows	+0.2 to +15.6	3.8–207	Yamaoka et al. (2015b)
Oman sheeted dikes (<200–350 °C[a])	+1.1 to +17.5	1.5–11.6	Yamaoka et al. (2012)
Oman sheeted dikes	−1.6 to +16.9	5.0–11.1	Smith et al. (1995)

(continued)

Table 8.1 (continued)

Material	$\delta^{11}B$(‰)	[B] (μg/g)	References
Troodos sheeted dikes	+3.3 to +10.6	0.6–18	Yamaoka et al. (2015b)
Troodos sheeted dikes ($\delta^{18}O > +6$‰)	−0.9 to +7.2	3.1–8.1	Smith et al. (1995)
Troodos sheeted dikes ($\delta^{18}O \leq +5$‰)	−0.1 to +7.8	2.4–3.4	Smith et al. (1995)
Oman upper gabbros (350–450 °C[a])	+8.3 to +18.6	1.6–5.0	Yamaoka et al. (2012)
Oman lower gabbros (>450 °C[a])	+7.3 to +17.7	0.25–3.8	Yamaoka et al. (2012)
Troodos gabbros	−1.7 to +18.5	0.3–8.4	Yamaoka et al. (2015b)
Oman serpentinites	−5.7 to +10.0	n.d.	Yamaoka et al. (2012)
Marine sediments			
Carbonate sediment	−5.0 to +23.0	0.3–7	Spivack and You (1997)
	+8.9 to +26.2	10–17	Vengosh et al. (1991)
	+4.8 to +10.5	13–26	Ishikawa and Nakamura (1993)
Calcareous sediment	+18.4 to +18.9	n.d.	Lécuyer et al. (2002)
Pelagic clay	−4.3 to +2.8	64–157	Spivack et al. (1987)
	−6.6 to −1.8	96–132	Ishikawa and Nakamura (1993)
	−13.1 to +4.0	101–163	Tonarini et al. (2011)
Chert	−9.3 to +7.7	49–97	Kolodny and Chaussidon (2004)
Siliceous ooze and biogenic silica	−3.8 to +4.5	55–77	Ishikawa and Nakamura (1993)
	−11.8 to +5.4	86–148	Tonarini et al. (2011)
Turbidite	−6 to −2	75–105	You et al. (1995)
Borates	+18.2 to +31.7	several wt %	Swihart and Moore (1986)
Marine sediment pore fluids	+27.8 to +37.8	3.2–21	James and Palmer (2000), ODP 1037
	+34.6 to +43.9	3.6–6.0	Kopf et al. (2000), ODP 1039B
	+34.4 to +40.5	0.1–5	Spivack and You (1997), ODP 851
	+35 to +49	4–12	You et al. (1995), ODP 808
Biogenic carbonate			
Corals (aragonite)	+23.0 to +24.7	50–68	Hemming and Hanson (1992)
	+26.7 to +31.9	51–80	Vengosh et al. (1991)
	+23.3 to +27.0	49–58	Gaillardet and Allegre (1995)
	+23.9 to +26.2	46–54	Hemming et al. (1998)
	+23.1 to +26.1	n.d.	Lécuyer et al. (2002)
	+27.9 to +38.5	53–114	Blamart et al. (2007)
	+20.9 to +29.3	28–56	Allison et al. (2010)
Ancient corals	+23.6 to +27.3	39–52	Gaillardet and Allegre (1995)
Bivalves, algae and ooids (aragonite)	+21.2 to +22.2 +20.1 to +25.3	14–28 11–15	Hemming and Hanson (1992), Vengosh et al. (1991)
Gastropoda (aragonite)	+19.8 to +31.5	2–3	Vengosh et al. (1991)
Foraminifera (calcite)	+4.9 to +32.2	9–54	Vengosh et al. (1991)
	+21.6 to +25.9	n.d.	Pearson and Palmer (2000)

(continued)

Table 8.1 (continued)

Material	$\delta^{11}B$(‰)	[B] (μg/g)	References
	+12.0 to +16.8	3–26	Rae et al. (2011)
	+19.6 to +20.7	8–13	Foster (2008)
Brachiopoda, gastropoda (calcite)	+15.0 to +19.7	n.d.	Lécuyer et al. (2002)
	+18.5 to 20.8	19–26	Hemming and Hanson (1992)
Algae, echinoids (high-Mg calcite)	+22.3 to 23.0	44–55	Hemming and Hanson (1992)
	+22.1 to 27.9	65 ± 5	Fietzke et al. (2015)
Forearc			
Serpentinite seamount	+5.4 to +25.3	6.6–126	Benton et al. (2001)
Mud volcanoes	−7.7 to +39.5	2–870	Kopf and Deyhle (2002)
Trench fluids	+20 to +50	3.6–42	E.g., You et al. (1993), Kopf et al. (2003)

[a]The temperatures refer to estimates for the conditions of hydrothermal alteration of these samples at the seafloor. *n.d.* not determined

record a contaminated signal. In the case of boron this contamination can be severe, due to the large difference in concentrations between high- and low-*T* alteration products, respectively. No boron concentration or boron isotope analyses of mineral separates or in situ analyses of minerals from high-temperature altered oceanic crust have been published. Whole-rock analyses are available for samples from two ODP drill cores and from ophiolites (Oman and Troodos; Fig. 8.4).

High-temperature altered samples from ODP Hole 504B (Costa Rica Rift) show low B concentrations (0.17 0.52 μg/g) and $\delta^{11}B$ values close to 0‰ (Ishikawa and Nakamura 1992). High-temperature altered samples from Hole 735B (Atlantis Bank, Indian Ocean) show higher abundances of B (1.1–7.1 μg/g) and a wide range of mostly high $\delta^{11}B$ values between −4.3 and +24.9‰ (Smith et al. 1995; Hart et al. 1999; see Fig. 8.4) High-temperature altered Layer-3 samples from the Troodos and Oman ophiolites range from slightly depleted to strongly enriched in boron (0.25–11.6 μg/g; Table 8.1) with respect to fresh MORB, and they show significantly elevated $\delta^{11}B$ values of −1.7 to +18.6‰, but mostly between +8 and +18‰ (Fig. 8.4; Smith et al. 1995; Yamaoka et al. 2012, 2015b).

The interpretation of the ophiolite data is, however, complicated by the tectonic setting in which the exposed crust was formed and where the hydrothermal alteration took place, i.e., a supra-subduction zone rather than a mid-ocean ridge (e.g., Searle and Cox 1999; Fonseca et al. 2017). For example, glass orbicules from the upper pillow lava section in the Troodos ophiolite show geochemical evidence for the contribution of sediment melts and subduction-zone fluids to the depleted-mantle derived magmas that formed these lavas, as demonstrated by Fonseca et al. (2017). These authors found that $\delta^{11}B$ in these glass orbicules range from low values (−8.2 ± 0.5‰) indistinguishable from the depleted mantle to strongly elevated values (+5.9 ± 1.1‰), with a negative correlation of $\delta^{11}B$ and B concentrations (≈3–10 μg/g). These boron isotopic signatures were observed in fresh glass and are not the result of seafloor alteration. Instead, they are interpreted as contributions to the Troodos ophiolite magmas in a subduction setting, in which isotopically heavy B was likely contributed from the slab (Fonseca et al. 2017; see also Chap. 9 of this volume). Further

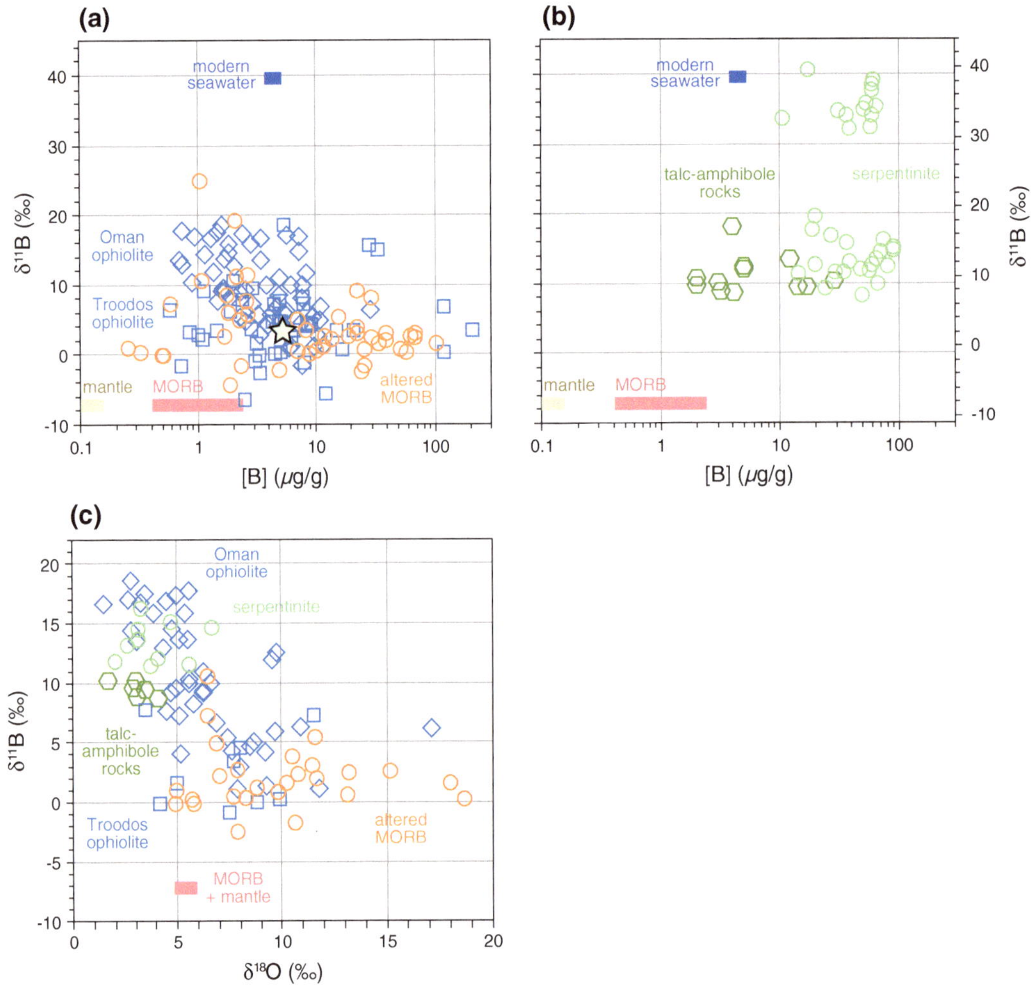

Fig. 8.4 **a** Boron isotopic composition versus B concentration of altered mafic rocks sampled from the ocean floor (*orange circles*) and from ophiolites (*blue squares and diamonds*). The altered-oceanic crust composite for ODP Hole 735B of Smith et al. (1995) is shown by the *green star*. **b** Boron isotopic composition versus B concentration of serpentinites (*green circles*) and talc-amphibole-rich metasomatic rocks recovered from the ocean floor (*green hexagons*). **c** Boron isotopic composition versus oxygen isotopic composition for a subset of the samples displayed in (**a**) and (**b**) for which data were available. See Table 8.1 for data sources

complication of the interpretation of ophiolite data arises from the possibility of post-obduction low-temperature hydrous alteration involving meteoric fluids (e.g., Chavagnac et al. 2013).

8.6.2 Serpentinization

Serpentinites formed at the ocean floor by hydration of peridotites are strongly enriched in boron with concentrations between 10 and 91 µg/g (Fig. 8.4; Table 8.1). Most samples display δ^{11}B values between +7.0 and +19.9‰ (Fig. 8.4; Table 8.1; Spivack and Edmond 1987; Lécuyer et al. 2002; Boschi et al. 2008; Harvey et al. 2014). Exceptionally high δ^{11}B values were found in serpentinites from ODP Leg 209 (Mid-Atlantic Ridge) of approximately +30 to +41‰ (Fig. 8.4; Table 8.1; Vils et al. 2009). The latter were explained by serpentinization by seawater-derived fluids that were processed through interaction with mantle rocks prior to serpentinization of the investigated samples, leading to a strong enrichment of isotopically heavy boron. Vils et al. (2009) predict evolved serpentinization fluids with δ^{11}B values between +50 and +60‰, which have yet to be discovered. Lower δ^{11}B values between −6 and +10‰ were reported for serpentinites from the Oman ophiolite (Table 8.1; Lécuyer et al. 2002).

Serpentinized ultramafic rocks, therefore, have the ability to incorporate very high concentrations of B that is enriched or strongly enriched in the heavy isotope. This enrichment also occurs at high alteration temperatures, conditions at which mafic rocks tend to show loss of boron or only very minor enrichment (Fig. 8.4).

Metasomatic rocks produced by the interaction of serpentinites with silica-rich fluids derived from gabbroic intrusions have been investigated along with coexisting serpentinites (Boschi et al. 2008; Harvey et al. 2014). These rocks are rich in talc, as well as tremolite or chlorite, the formation of which postdates serpentinization (Boschi et al. 2008; Harvey et al. 2014). This steatitization process led to boron loss from the samples compared to the serpentinite precursor with concentrations ranging from 2.0 to 28.7 µg/g (Fig. 8.4). This is accompanied by a small isotopic shift to lower δ^{11}B values between +8.8 and +13.5‰ for all but one sample (Boschi et al. 2008; Harvey et al. 2014).

8.6.3 Hydrothermal Vent Fluids

Boron concentrations and B isotope ratios have been reported for hydrothermal vent fluids from sediment-starved systems at the Mid-Atlantic Ridge (MARK, 23°N; TAG, 26°N; Broken Spur, 29°N), the Juan de Fuca Ridge, and various vents between 11°N and 21°N on the EPR (Table 8.1; Spivack and Edmond 1987; Palmer 1991; You et al. 1994; James et al. 1995). Additional data come from sediment-starved back-arc basins, including the Manus Basin, the Mariana Trough, and the North Fiji Basin (Palmer 1991; Yamaoka et al. 2015a).

Boron concentrations for the endmember hydrothermal fluids for all of these sites range from seawater-like to slightly enriched ([B] = 0.8 to 2 times seawater concentration). This enrichment in boron correlates with an enrichment in the light isotope, i.e., with a decrease in the δ^{11}B values from near-seawater (as high as +38.0‰) to values as low as +17.8‰ for the most B-rich fluids (Fig. 8.5; Table 8.1). This trend is consistent with interaction of the incoming seawater with the mafic crust and extraction of isotopically light B from the rocks at high temperatures.

Such a leaching process is also predicted from estimates of high-T (300 °C) partition coefficients of B between greenschist- and amphibolite-facies minerals and hydrous fluid (D_{B}[mineral/fluid]), which are between 0.001 and 0.03 for quartz, albite, epidote, chlorite, talc, clinopyroxene and Ca-amphibole, predicting high fluid mobility during fluid–rock interaction under greenschist-facies conditions (Marschall et al. 2006).

Sediment-hosted vent fluids from mid-ocean ridge systems (Guyamas Basin and Escanaba Trough) and from back-arc basins (Okinawa

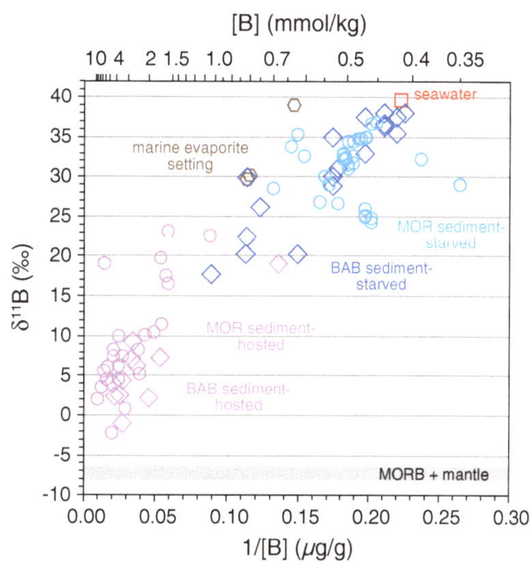

Fig. 8.5 Boron isotopic composition and B concentration of vent fluids sampled from hydrothermal vents located at mid-ocean ridges (*MOR*) and back-arc basins (*BAB*) with or without sediment cover on the oceanic crust. See Table 8.1 for data sources

Trough) continue the B – δ^{11}B trend observed for the sediment-starved vent fluids (Fig. 8.5). They show higher B contents than the fluids from sediment-starved systems, typically higher than 15 μg/g (1.4 mmol/kg) and ranging up to 105 μg/g (9.7 mmol/kg = 27 times seawater concentration; Table 8.1). The increase in B contents is accompanied by a decrease in δ^{11}B, and while some samples show values as high as +23.2‰, most are below +12‰, and some reach negative values as low as −2.2‰ (Fig. 8.5).

Ultramafic-hosted hydrothermal systems have been sampled along the Mid-Atlantic Ridge, including boron contents from Lost City (~100 °C), Rainbow (~350 °C) and Logatchev (~350 °C). These are consistently lower than the B content of seawater and endmember hydrothermal fluids have been determined to 0.34–0.55 μg/g (31–51 μmol/kg), 2.57 μg/g (238 μmol/kg), and 3.62 μg/g (335 μmol/kg) for Lost City, Rainbow, and Logatchev, respectively (Schmidt et al. 2007; Foustoukos et al. 2008; Boschi et al. 2008; Seyfried et al. 2011, 2015). Boron isotopic data for ultramafic-hosted vent fluids are not yet available, but have been estimated to +25 to +30‰ (Foustoukos et al. 2008; Boschi et al. 2008).

It is, hence, concluded from the high boron abundances in serpentinite and the lower-than-seawater abundances in ultramafic-hosted vent fluids that boron is sequestered from seawater during serpentinization at the seafloor. This sequestration may be strongest at lower temperatures (Lost City), but is still efficient at high temperatures (Rainbow and Logatchev). The latter contrasts with mafic systems, in which boron is leached from the rocks at high temperatures of alteration.

8.6.4 Subaerial Hydrothermal Alteration

Seawater contains relatively high concentrations of boron (4.5 μg/g or 418 μmol/kg) with an isotopically very heavy composition (δ^{11}B = +39.6‰; Table 8.1), which leads to the strong increase in boron abundances and δ^{11}B values of rocks altered by interaction with seawater. Meteoric waters, in contrast, have B concentrations that are typically two orders of magnitude lower than those of seawater. For example, freshwater lakes from Iceland have B contents of 23–49 ng/g (2.1–4.5 μmol/kg) and δ^{11}B values between −1.8 and +18.2‰ (Aggarwal et al. 2000). High-temperature hydrothermal fluids from Iceland show negative δ^{11}B values (−6.7 to −1.5‰) and B concentrations generally lower than seawater (0.7–5.7 μg/g; Aggarwal et al. 2000). Low-temperature well fluids show a much larger range in δ^{11}B values (−4.7 to +25.0‰) and lower B concentrations (0.05–1.2 μg/g; Aggarwal et al. 2000).

Basaltic rocks from Iceland altered by meteoric fluids show elevated B contents compared to MORB (3.3–12.4 μg/g) and a shift to low δ^{11}B values, as low as −18.3‰ (Raffone et al. 2010). This isotopically light B is also apparent in Icelandic rhyolites that were formed from or have assimilated fluid-altered crust (see above; Brounce et al. 2012). This shift towards isotopically light B is unique to subaerial hydrothermal alteration by meteoric water and is not observed in seawater-altered crust. Consistent with this fractionation behavior, hydrothermal fluids from subaerial systems supplied by meteoric water tend to show δ^{11}B values that are lower than the values observed in submarine hydrothermal vents (Fig. 8.5), i.e., they range from −22 to +14‰ for fluids from Yellowstone, Etna, Iceland, and Alberta oil sands (Palmer and Sturchio 1990; Aggarwal et al. 2000; Pennisi et al. 2000; Williams et al. 2001b).

8.7 Oceanic Sediments

Oceanic sediments generally include terrigenous sediments derived from continents, dominantly delivered as detritus by wind, rivers and currents to the shelves, slopes and basins near the continents, and marine sediments including clay, biogenic carbonate and silica, which are more evenly and slowly deposited in the ocean basins.

Pelagic clay is typically composed of smectite produced from weathering of volcanic material and illite delivered from the continents.

Pelagic clays show high B concentrations of approximately 60–160 µg/g (Table 8.1; Spivack et al. 1987; Ishikawa and Nakamura 1993). Boron isotopic compositions of smectite formed by weathering of basalt at the seafloor are different from those of the terrigenous illite. Submarine smectite displays δ^{11}B values between $+2$ and +9‰, whereas terrigenous clay shows significantly lower δ^{11}B values of -13 to -8‰ (Ishikawa and Nakamura 1993). The resulting B isotopic composition of pelagic clays is a combination of these clay components in addition to biogenic carbonate and biogenic silica and ranges from -13.1 to +5.4‰ (Fig. 8.6; Table 8.1; Spivack et al. 1987; Ishikawa and Nakamura 1993; Tonarini et al. 2011). Turbidites, which are dominated by continental detritus in addition to clay derived from weathered volcanic material, also show high B contents and moderately negative δ^{11}B values (Fig. 8.6; Table 8.1). Marine cherts show B concentrations almost as high as turbidites and pelagic clays and a very similar range in B isotopic composition (Fig. 8.6; Table 8.1; Kolodny and Chaussidon 2004). Chemical sediments that are rich in salts and marine borates generally show high δ^{11}B that reflect their seawater origin, and concentrations may be very high, in particular in the case of borates (Swihart and Moore 1986; Vengosh et al. 1992; Paris et al. 2010).

Carbonate sediments and calcareous sediments show much lower B contents than pelagic clays, cherts and turbidites, typically <20 µg/g with a wide range in isotopic composition from $\delta^{11}B = -5$ to +26‰ (Fig. 8.6; Table 8.1; Spivack and You 1997; Vengosh et al. 1991; Ishikawa and Nakamura 1993; Lécuyer et al. 2002). This variation may be related to variable amounts of clay contained in these sediments and to diagenetic processes. Pristine biogenic carbonate shows a much smaller range in isotopic composition, with B concentration and B isotopic composition depending on species, skeletal mineralogy (aragonite, calcite, or high-Mg

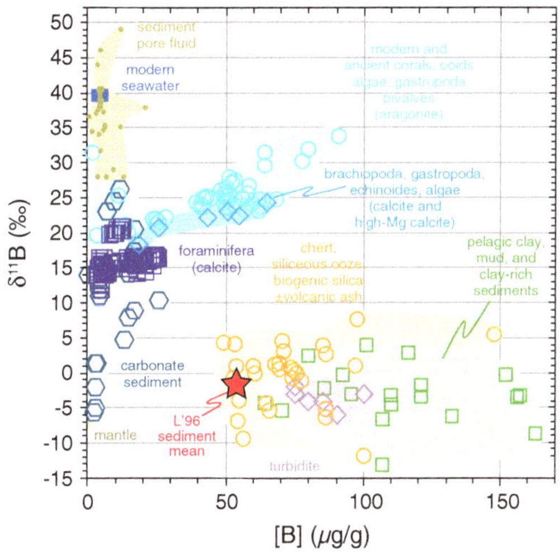

Fig. 8.6 Boron isotopic composition and B concentration of marine sediments, pore fluids, and biogenic carbonate. See Table 8.1 for data sources. The *red star* marks the weighted average for marine sediments proposed by Leeman and Sisson (1996)

calcite), as well as seawater pH and seawater δ^{11}B during growth. The details of these parameters and their impact on the B isotopic composition of biogenic carbonate are discussed in separate chapters of this book (Rae 2017; McCulloch et al. 2017). Concentrations of B in aragonite corals and high-Mg calcite coralline algae are generally high (typically 50 µg/g), whereas foraminifera and other calcite-shell organisms show lower B contents in the range of 3–26 µg/g (Fig. 8.6; Table 8.1). The δ^{11}B values are typically $+13$ to +20‰ in foraminifera and approximately +25‰ in corals and coralline algae (Fig. 8.6; Table 8.1).

Boron in clay minerals is crystallographically hosted in different positions, namely adsorbed onto the mineral surface, hosted in interlayer sites of the clay, and incorporated into tetrahedral sites where it replaces Si or Al (Williams et al. 2001a; Williams and Hervig 2005). These different portions of B are released into pore fluids at different stages of diagenesis and metamorphism with the adsorbed B released at the earliest stage (Spivack et al. 1987). In general, recrystallization of clay minerals, and the transformation from smectite to illite during deeper burial and diagenesis is predicted to lead to fluid release accompanied by the preferential loss of isotopically heavy B and the consequential decrease in the sediment $\delta^{11}B$ with increasing grade of diagenesis (Williams et al. 2001a; Williams and Hervig 2005). Shales and slates show lower $\delta^{11}B$ values than comparable modern sediments, which is compatible with loss of isotopically heavy B during diagensis (Vengosh et al. 1991; Ishikawa and Nakamura 1993). However, in detail the exchange of boron and the fractionation of its isotopes between clay minerals and pore fluid in compacting sediments may be quite complex with various stages of adsorption and desorption, depending for example on temperature, burial depth, fluid pH, and the concentration of ammonium released from decomposing organic matter (Hüpers et al. 2016).

During hydrothermal alteration, sediments behave similar to basaltic rocks with respect to B. Boron is leached from the rocks and $\delta^{11}B$ values decrease during high-temperature fluid/rock interaction. Spivack et al. (1987), for example, investigated sediments containing 32–62 μg/g B with $\delta^{11}B$ values between −4.5 and −1.2‰. These sediments were intruded by a basaltic sill generating hydrothermal fluid circulation within the sediments. The sediments in the vicinity of the sill showed lower B concentrations of 14–33 μg/g, and $\delta^{11}B$ values of approximately −8‰. In zones of greenschist facies hydrothermal metamorphism, B concentrations were decreased to ~ 1.3 μg/g (B isotopic compositions of these B-poor samples are not reported).

A weighted average for seafloor sediments on a global scale was estimated by Leeman and Sisson (1996) to be $[B] = 53\,\mu g/g$ and $\delta^{11}B = -1.6$‰ (Fig. 8.6).

8.8 Paleo-Ocean Chemistry of Boron

Boron has a long residence time in the ocean of approximately 10–20 million years and is isotopically very well mixed (Lemarchand et al. 2002; Foster et al. 2010). The B isotopic composition of modern ocean water is well defined at +39.61 ± 0.04‰ and the long residence time suggests that the rate of change was low in the Cenozoic at approximately 0.1‰/Ma (Fig. 8.7; Lemarchand et al. 2000; Foster et al. 2010). The reconstruction of the B isotopic composition of ocean water is, however, a matter of debate, and especially the more distant history of its secular evolution is poorly constraint.

Accurate data on the isotopic composition of the paleo-ocean is required for reconstruction of paleo-ocean pH, which is linked to the CO_2 content of the atmosphere. The boron isotope record of biogenic carbonate is used to reconstruct the pH of the ocean water at the time of growth of the organism that built the carbonate shell (Rae 2017; Branson 2017; McCulloch et al. 2017). This procedure requires knowledge on the pH-dependent B isotope fractionation between $CaCO_3$ and seawater for the particular species that is investigated, but it also requires a well-defined B isotopic composition of the seawater at the time of growth.

Estimates on geochemical cycles in subduction zones and the deep mantle also require knowledge of the long-term evolution of the isotopic composition of ocean water, because strong variations in its composition would have affected the isotope budget of altered oceanic crust, serpentinites and sediments. This in turn would cause variations in the subducted and deeply recycled B isotope signal throughout Earth's history and would have to be taken into account in the geochemical interpretation of the rock record.

Initial work by Pearson and Palmer (2000) estimated that the $\delta^{11}B$ of seawater varied by less than 2‰ since the mid Miocene, and it was further assumed that it was close to the modern value throughout the Paleogene. More recent reconstructions of the seawater $\delta^{11}B$ value are based on box models that take into account the sources and sinks of boron in the oceans (Lemarchand et al. 2000, 2002; Joachimski et al. 2005; Simon et al. 2006). The sources are riverine input, hydrothermal vents, and fluids expelled from continental margins, while the sinks are low-T alteration of the oceanic crust,

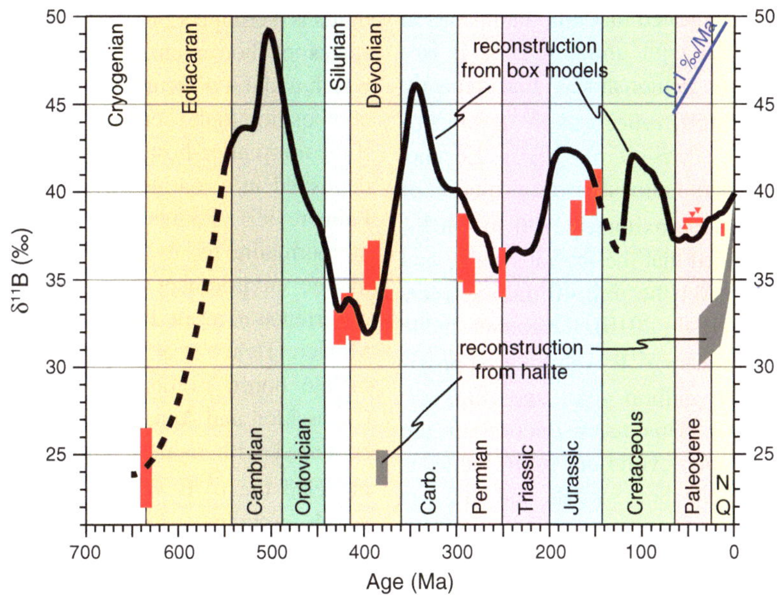

Fig. 8.7 Boron isotopic evolution of seawater during the late Neoproterozoic and Phanerozoic. Estimates marked by the *thick black line* are based on box models based on ocean spreading rates, major orogenies and erosion (Lemarchand et al. 2002; Joachimski et al. 2005). The *stippled line* is interpolated for periods where no model exists. Reconstruction of seawater δ^{11}B based on data from marine carbonate are marked in *red boxes* (Joachimski et al. 2005; Kasemann et al. 2010; Foster et al. 2012; Clarkson et al. 2015; Anagnostou et al. 2016). Minimum and maximum seawater values for Eocene estimates are marked by upward- and downward-pointing *triangles*, respectively. Reconstruction of seawater δ^{11}B based on data from halite is marked by the *gray* field (Paris et al. 2010). The latter is inconsistent with the carbonate reconstruction and with the box model. A more detailed discussion of the Paleogene–Neogene evolution of seawater δ^{11}B is presented by Gaillardet and Lemarchand (2017)

formation of clay, and carbonate precipitation. All of these estimates are afflicted with increasing uncertainties the further back in Earth history they are projected. However, estimates of the element flux to the oceans are to some degree constrained through other element and isotope systems, such as seawater ^{87}Sr/^{86}Sr, that are well defined through investigation of the rock record.

In the case of boron isotopes, the investigation of the rock record for seawater reconstruction is progressing only slowly. It is difficult to identify pristine samples that have preserved the B isotopic record of seawater at the time of their formation, due to the high susceptibility of low-T alteration phases for this element. In addition, the dependence of carbonate δ^{11}B on the pH and δ^{11}B of seawater complicates the interpretation of such samples, especially in the case of biogenic carbonate formed by biological species for which B isotope fractionation factors are unknown. However, several studies have used marine carbonate to reconstruct seawater δ^{11}B for various stages during the late Neoproterozoic and the Phanerozoic (Fig. 8.7; Joachimski et al. 2005; Kasemann et al. 2010; Foster et al. 2012; Clarkson et al. 2015; Anagnostou et al. 2016).

The estimates include seawater δ^{11}B variations of ± 2‰ since the Jurassic, and larger variations of ± 9‰ in the Paleozoic with excursions to very high and very low values of +32 to +49‰ (Fig. 8.7). Even more extreme values have been suggested for the end of the Cryogenian (635 Ma) with seawater δ^{11}B values as low as +25‰ (Fig. 8.7; Kasemann et al. 2010).

Alternative to biogenic carbonates, halite was used to reconstruct the B isotopic composition of seawater. It was suggested that fluid inclusions in halite trap seawater with unfractionated B isotopes and it was demonstrated that recently formed halite faithfully reflects the δ^{11}B value of modern seawater (Paris et al. 2010). Eocene, Miocene and modern halite show an evolution of δ^{11}B values from approximately +30 to +40‰, and it was concluded that the seawater value had risen by 8–10‰ over the past 40 million years (Fig. 8.7; Paris et al. 2010). Devonian halite suggested a seawater δ^{11}B value of approximately +25‰ 380 million years ago (Fig. 8.7; Paris et al. 2010). This reconstruction of the seawater B isotopic evolution from halite is inconsistent with the box models and the carbonate record discussed above (Fig. 8.7), and it is unclear how to reconcile these contrasting records. Paris et al. (2010). suggest that the box models may have underestimated the deposition of siliciclastic material (that acts as a sink of isotopically light B) provided by the erosion of the Himalaya since the Eocene. The box models have large uncertainties, and simply varying the amount of oceanic crust that is altered along the mid-ocean ridges alone introduces a 10‰ uncertainty for the seawater δ^{11}B value without changing any of the other sources and sinks (Simon et al. 2006). Yet, uncertainties in the box models do not explain the discrepancies between the halite and carbonate records, and the possibility remains that the fluid inclusions trapped in ancient halite are not pristine fossil seawater, but were diagenetically processed instead.

Reconstructions of the B isotope composition of any pre-Neoproterozoic seawater that would represent the global oceans has not been completed convincingly. Attempts to do so include investigation of tourmaline from supracrustal rocks from Greenland, South Africa or Australia (Chaussidon and Albarède 1992; Chaussidon and Appel 1997; Grew et al. 2015). Tourmaline is a reliable and robust recorder of boron and other chemical and isotopic systems that will reflect the environment in which it formed (see for example reviews by van Hinsberg et al. 2011; Marschall and Jiang 2011). However, formation of this mineral from marine sediments or hydrothermal fluids is a complex process that involves multiple stages of boron enrichment and fractionation, which make it difficult to connect any tourmaline composition to the composition of seawater, even if the tourmaline-hosting sediments were initially deposited in a marine environment (Byerly and Palmer 1991; Palmer and Swihart 1996). Also, no tourmaline has ever been found in the modern marine environment, with the exception of an occurrence in a salt dome cap rock in the Gulf of Mexico (Henry et al. 1999). The seawater–tourmaline boron isotope connection proposed by Chaussidon and Appel (1997) can, therefore, not be tested with modern samples. These authors modeled the δ^{11}B value of the marine clay in which authigenic Eoarchean tourmaline from Isua (Greenland) would have formed to be -8 ± 7‰, and they argue that this requires a δ^{11}B of early Archean seawater of $+27 \pm 11$‰. However, the estimated value for the marine clay is within the range of modern marine clay (Fig. 8.6; Table 8.1), demonstrating that the Archean clay could as well have formed in contact with seawater with a present-day B isotopic composition. The authors assumed that smectite/illite ratios were much lower in the Archean sediments, because of smaller continents and consequently a much lower input of continental detritus. Since it is really the smectite that reflects the B isotopic composition of the seawater and illite carries isotopically light B derived from the continents, a bulk sediment with higher smectite/illite ratio would reflect lower δ^{11}B values in the smectite and, hence, in the seawater which led to its formation (Chaussidon and Appel 1997).

In a more recent study, Grew et al. (2015) investigated a large number of tourmalines from various rock types in the Isua supracrustal belt and found a range of δ^{11}B values ranging from -29.2 to -1.8‰. The authors present a box model for the evolution of B isotopes from seawater to clay-rich sediments followed by diagenesis and metamorphism to gneisses in which finally the tourmaline formed. This box model follows the approach of Chaussidon and Appel (1997), but results in a lower δ^{11}B estimate for seawater of +14 ± 15‰ based on their new sample set and modern parameters for B isotope fractionation during the various mineral-fluid fractionation steps (Grew et al. 2015). However, Grew et al. (2015) argue that this value probably does not reflect the B isotopic composition of global seawater in the early Archean, but rather the composition of water in a restricted ocean basin in which the Isua sediments were likely deposited, diagenetically altered and enriched in boron. In conclusion, the Precambrian boron isotopic evolution of seawater remains unconstrained up to the late Neoproterozoic.

8.9 Summary and Outlook

Boron isotopes have an important role in high-temperature geochemistry in the investigation of mantle sources of oceanic basalts, in the quantification of assimilation processes along mid-ocean ridges and on ocean islands, and in the investigation of hydrothermal alteration processes as expressed in the composition of vent fluids and altered crustal and mantle rocks, as well as sediments. The B isotopic composition of pristine, uncontaminated MORB has been determined as $\delta^{11}B = -7.1 \pm 0.9‰$, and no variation was detected between N-MORB and E-MORB, or as a function of degree of melting or spreading rate. This value also represents the B isotopic composition of the depleted upper mantle, and probably that of the primitive mantle. Mantle-source variations for ocean island basalts detected so far show deviations of less than 5‰ from the MORB value, and this spread may still be affected by crustal assimilation processes. Assimilation of seawater-altered materials or brines into magmas lead to elevated $\delta^{11}B$ values in MORB and OIB. Assimilation of crust altered by meteoric water leads to low $\delta^{11}B$ values in OIB.

The B concentrations and B isotopic variations in pristine, uncontaminated MORB and mantle samples are too low to be resolved with currently available analytical tools. Future analytical development may make these samples and their variability accessible. OIB samples already show a certain variability that is currently underexploited.

Alteration of the oceanic crust and of the exposed mantle is a major boron sink in the oceans. Low-temperature alteration and weathering of basalts lead to a strong enrichment in boron with a moderate increase in the $\delta^{11}B$ value of the rocks (Fig. 8.8). High-temperature alteration of the mafic igneous crust leads to only moderate enrichment or even leaching of boron from the rocks, but with a strongly elevated $\delta^{11}B$ value in the altered rocks (Fig. 8.8). High-temperature vent fluids are consequently enriched in boron compared to seawater, and their isotopic composition is enriched in isotopically light boron leached from the rocks (Fig. 8.8). Sediments show a very similar behavior, whereas serpentinization of mantle rocks leads to a strong enrichment of B in the serpentinite even at high temperatures.

Boron isotopes also play a vital role in low-temperature geochemistry, and foremost in the reconstruction of paleo-seawater pH related to the evolution of the CO_2 content of the atmosphere. The application of this tool hinges

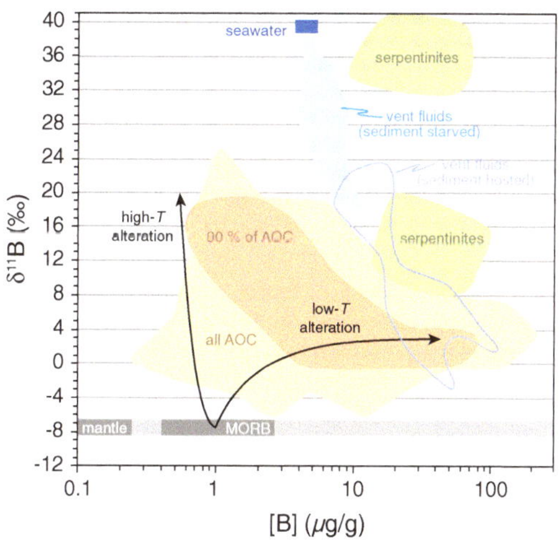

Fig. 8.8 Boron isotopic composition and B concentration of altered oceanic crust, serpentinites and vent fluids. The *black arrows* highlight the approximate evolution of basalts during high-*T* and low-*T* alteration, resepctively. Vent fluids show an evolution away from seawater towards enrichment in isotopically light B (*blue field*). The individual data are displayed in Figs. 8.4 and 8.5

on the accurate knowledge of the secular evolution of the B isotopic composition of seawater, which is currently poorly constrained. Models for the evolution of seawater are afflicted with large uncertainties, and carbonate and halite records are in conflict with each other for the Phanerozoic. No reliable estimates are available for the first 85% of Earth history.

Acknowledgements I would like to thank Cees-Jan de Hoog and Angela Helbling for discussion, Simone Kasemann and Martin Palmer for constructive reviews, and Gavin Foster for editorial handling. Preparation of this review chapter was financially supported by a WHOI Independent Study Award from the Andrew W. Mellon Foundation (#GG15920) to the author.

References

Aggarwal JK, Palmer MR, Bullen TD, Arnórsson S, Ragnarsdóttir KV (2000) The boron isotope systematics of Icelandic geothermal waters: 1. Meteoric water charged systems. Geochimica Cosmochimica Acta 64:579–585

Aggarwal J, Böhm F, Foster G, Halas S, Hönisch B, Jiang SY, Košler J, Liba A, Rodushkin I, Sheehan T, Shen JJS, Tonarini S, Xie Q, You CF, Zhao ZQ, Zuleger E (2009) How well do non-traditional stable isotope results compare between different laboratories: results from interlaboratory comparison of boron isotope measurements. J Analyt Atom Spectrom 24:825–831

Allègre CJ, Brevart O, Dupré B, Minster JF (1980) Isotopic effects produced in a continously differentiating earth mantle. Phil Trans Royal Soc 297:447–477

Allison N, Finch AA, EIMF (2010) δ^{11}B, Sr, Mg and B in a modern *Porites* coral: the relationship between calcification site pH and skeletal chemistry. Geochim Cosmochim Acta 74:1790–1800

Alt JC (1995) Subseafloor Processes in Mid-Ocean Ridge Hydrothermal Systems. In: Humphris SE (ed) Seafloor hydrothermal systems: physical, chemical, biological, and geological interactions, Geophysical Monograph Series, vol 91, American Geophysical Union, Washington, DC, pp 85–114

Anagnostou E, John EH, Edgar KM, Foster GL, Ridgewell A, Ingis GN, Pancost RD, Lunt DJ, Pearson PN (2016) Changing atmospheric CO_2 concentration was the primary driver of early Cenozoic climate. Nature 533:380–384

Bach W, Alt JC, Niu Y, Humphris SE, Erzinger J, Dick HJB (2001) The geochemical consequences of late-stage low-grade alteration of lower ocean crust at the SW Indian Ridge: Results from ODP Hole 735B (Leg 176). Geochim Cosmochim Acta 65:3267–3287

Benton LD, Ryan JG, Tera F (2001) Boron isotope systematics of slab fluids as inferred from a serpentine seamount. Mariana forearc Earth Planet Sci Lett 187:273–282

Blamart D, Rollion-Bard C, Meibom A, Ciuf JP, Juillet-Leclerc A, Dauphin Y (2007) Correlation of boron isotopic composition with ultrastructure in the deep-sea coral *Lophelia pertusa*: implications for biomineralization and paleo-pH. Geochem Geophys Geosys (G^3) 8, Q12001

Boschi C, Dini A, Früh-Green GL, Kelley DS (2008) Isotopic and element exchange during serpentinization and metasomatism at the Atlantis Massif (MAR 30° N): insights from B and Sr isotope data. Geochim Cosmochim Acta 72:1801–1823

Branson O (2017) Boron incorporation into marine $CaCO_3$. In: Marschall HR, Foster GL (eds) Boron Isotopes—The Fifth Element, Advances in Isotope Geochemistry, vol 7. Springer, Heidelberg, 71–105

Brounce M, Feineman MD, LaFemina P, Gurenko AA (2012) Insights into crustal assimilation by Icelandic basalts from boron isotopes in melt inclusions from the 1783–1784 Lakaggar eruption. Geochim Cosmochim Acta 94:164–180

Byerly GR, Palmer MR (1991) Tourmaline mineralization in the Barberton greenstone belt, South Africa: early Archean metasomatism by evaporite-derived boron. Contrib Mineral Petrol 107:387–402

Chase CG (1981) Oceanic island Pb—2-stage histories and mantle evoltuion. Earth Planet Sci Lett 52:277–284

Chaussidon M, Albarède F (1992) Secular boron isotope variations in the continental crust: an ion microprobe study. Earth Planet Sci Lett 108:229–241

Chaussidon M, Appel PWU (1997) Boron isotopic composition of tourmalines from the 3.8-Ga-old Isua supracrustals, West Greenland: implications on the δ^{11}B value of early Archean seawater. Chem Geol 136:171–180

Chaussidon M, Jambon A (1994) Boron content and isotopic composition of oceanic basalts: geochemical and cosmochemical implications. Earth Planet Sci Lett 121:277–291

Chaussidon M, Marty B (1995) Primitive boron isotope composition of the mantle. Science 269:383–386

Chavagnac V, Monnin C, Ceuleneer G, Boulart C, Hoareau G (2013) Characterization of hyperalkaline fluids produced by low-temperature serpentinization of mantle peridotites in the Oman and Ligurian ophiolites. Geochem Geophys Geosys G3(14):2496–2522

Clarkson MO, Kasemann SA, Wood RA, Lenton TM, Daines SJ, Richoz S, Ohnemueller F, Meixner A, Poulton SW, Tipper ET (2015) Ocean acidification and the Permo-Triassic mass extinction. Science 348:229–232

Collerson KD, Williams Q, Ewart AE, Murphy DT (2010) Origin of HIMU and EM-1 domains sampled by ocean island basalts, kimberlites and carbonatites: the role of CO_2-fluxed lower mantle melting in thermochemical upwellings. Phys Earth Planet Inter 181:112–131

De Hoog CJ, Savov IP (2017) Subduction zones, dehydration, metasomatism, mud and serpentinite volcanoes, and arc magmatism. In: Marschall HR, Foster GL (eds) Boron isotopes—The fifth element, Advances in Isotope Geochemistry, vol 7, Springer, Heidelberg, 219–249

Dhuime B, Hawkesworth CJ, Cawood PA, Storey CD (2012) A change in the geodynamics of continental growth 3 billion years ago. Science 335:1334–1336

Dick HJB, Jian L, Schouten H (2003) An ultraslow-spreading class of ocean ridge. Nature 426:405–412

Dilek Y (2003) Ophiolite concept and its evolution. In: Dilek Y, Newcomb S (eds) Ophiolite concept and the evolution of geological thought, Special Paper Geological Society of America, vol 373. Boulder, Colorado, pp 1–16

Donnelly T, Francheteau J, Bryan W, Robinson P, Flower M, Salisbury M (eds) (1979) The chemistry of altered basalts at site 417, Deep Sea Drilling Project Leg 51, Initial Reports of the Deep Sea Drilling Project, vol. 51–53, U.S. Government Printing Office, Washington, D.C

Elliott T, Plank T, Zindler A, White W, Bourdon B (1997) Element transport from slab to vocanic front at the Mariana arc. J Geophys Res 102:14991–15019

Escartn J, Smith DK, Cann J, Schouten H, Langmuir CH, Escrig S (2008) Central role of detachment faults in accretion of slow spread oceanic lithosphere. Nature 455:790–794

Fietzke J, Ragazzola F, Halfar J, Dietze H, Foster LC, Hansteen TH, Eisenhauer A, Steneck RS (2015) Century-scale trends and seasonality in pH and temperature for shallow zones of the Bering Sea. Proc Nat Acad Sci 112:2960–2965

Fonseca ROC, Kirchenbaur M, Ballhaus C, Münker C, Zirner A, Gerdes A, Heuser A, Botcharnikov R, Lenting C (2017) Fingerprinting fuid sources in Troodos ophiolite complex orbicular glasses using high spatial resolution isotope and trace element geochemistry. Geochim Cosmochim Acta 200:145–166

Foster GL (2008) Seawater pH, pCO_2 and $[CO_3^{2-}]$ variations in the Caribbean Sea over the last 130kyr: a boron isotope and B/Ca study of planktic foraminifera. Earth Planet Sci Lett 271:254–266

Foster GL, Pogge von Strandmann PAE, Rae JWB (2010) Boron and magnesium isotopic composition of seawater. Geochem Geophys Geosys (G^3)11, Q08015

Foster GL, Lear CH, Rae JWB (2012) The evolution of pCO_2, ice volume and climate during the middle Miocene. Earth Planet Sci Lett 341–344:243–254

Foster GL, Hönisch B, Paris G, Dwyer GS, Rae JWB, Elliott T, Gaillardet J, Hemming NG, Louvat P, Vengosh A (2013) Interlaboratory comparison of boron isotope analyses of boric acid, seawater and marine $CaCO_3$ by MC-ICPMS and NTIMS. Chem Geol 358:1–14

Foster GL, Marschall HR, Palmer MR (2017) Boron isotope analysis of geologic materials. In: Marschall HR, Foster GL (eds) Boron Isotopes—The Fifth Element, Advances in Isotope Geochemistry, vol. 7, Springer, Heidelberg, 13–31

Foustoukos DI, Savov IP, Janecky DR (2008) Chemical and isotopic constraints on water/rock interactions at the Lost City hydrothermal field, 30 N Mid-Atlantic Ridge. Geochim Cosmochim Acta 72:5457–5474

Gaillardet J, Allegre CJ (1995) Boron isotopic compositions of corals: seawater or diagenesis record? Earth Planet. Sci Lett 136:665–676

Gaillardet J, Lemarchand D (2017) Boron isotopes in riverine systems and the weathering environment. In: Marschall HR, Foster GL (eds) Boron Isotopes—The Fifth Element, Advances in Isotope Geochemistry, vol. 7, Springer, Heidelberg, 165–190

Gannoun A, Burton KW, Parkinson IJ, Alard O, Schiano P, Thomas LE (2007) The scale and origin of the osmium isotope variations in mid-ocean ridge basalts. Earth Planet Sci Lett 259:541–556

Genske FS, Turner SP, Beier C, Chu MF, Tonarini S, Pearson NJ, Haase KM (2014) Lithium and boron isotope systematics in lavas from the Azores islands reveal crustal assimilation. Chem Geol 373:27–36

German CR, Lin J (2004) The thermal structure of the oceanic crust, ridge spreading and hydrothermal circulation: how well do we understand their interconnections? In: German CR, Lin J, Parson LM (eds) Mid-Ocean ridges: hydrothermal interactions between the lithosphere and Oceans, Geophysical Monograph Series, vol 148, American Geophysical Union, pp 1–18

Grew ES, Dymek RF, de Hoog CJ, Harley SL, Boak J, Hazen RM, Yates MG (2015) Boron isotopes in tourmaline from the ca. 3.7–3.8 a Isua supracrustal belt, Greenland: sources for boron in Eoarchean continental crust and seawater. Geochim Cosmochim Acta 163:156–177

Gurenko AA, Chaussidon M (1997) Boron concentrations and isotopic composition of the Icelandic mantle: evidence from glass inclusions in olivine. Chem Geol 135:21–34

Gurenko AA, Kamenetsky VS (2011) Boron isotopic composition of olivine-hosted melt inclusions from Gorgona komatiites, Colombia: new evidence supporting wet komatiite origin. Earth Planet Sci Lett 312:201–212

Halliday AN, Davidson JP, Holden P, Dewolf C, Lee DC, Fitton JG (1990) Trace- Element fractionation in plumes and the origin of HIMU mantle beneath the Cameroon Line. Nature 347:523–528

Hart SR, Blusztajn J, Dick HJB, Meyer PS, Muehlenbachs K (1999) The fingerprint of seawater circulation in a 500-meter section of ocean crust gabbros. Geochim Cosmochim Acta 63:4059–4080

Harvey J, Savov IP, Agostini S, Cliff RA, Walshaw R (2014) Si-metasomatism in serpentinized peridotite: the effects of talc-alteration on strontium and boron

isotopes in abyssal serpentinites from Hole 1268a, ODP Leg 209. Geochim Cosmochim Acta 126:30–48
Hemming NG, Hanson GN (1992) Boron isotopic composition and concentration in modern marine carbonates. Geochim Cosmochim Acta 56:537–543
Hemming NG, Guilderson TP, Fairbanks RG (1998) Seasonal variations in the boron isotopic composition of coral: a productivity signal? Global Biogeochem. Cycles 12:581–586
Henry DJ, Kirkland BL, Kirkland DW (1999) Sector-zoned tourmaline from the cap rock salt dome. Eur J Mineral 11:263–280
Hofmann AW (1988) Chemical differentiation of the Earth: the relationship between mantle, continental crust, and oceanic crust. Earth Planet Sci Lett 90:297–314
Hofmann AW (1997) Mantle geochemistry: the message from oceanic volcanism. Nature 385:219–229
Hofmann AW (2003) Sampling mantle heterogeneity through oceanic basalts: isotopes and trace elements. In: Holland HD, Turekian KK (eds) The mantle and core, Treatise on Geochemistry, vol 2, Elsevier-Pergamon, pp 61–101
Hofmann AW, White WM (1982) Mantle plumes from ancient oceanic crust. Earth Planet Sci Lett 57:421–436
Hüpers A, Kasemann SA, Kopf AJ, Meixner A, Toki T, Shinjo R, Wheat CG, You CF (2016) Fluid flow and water–rock interaction across the active Nankai Trough subduction zone forearc revealed by boron isotope geochemistry. Geochim Cosmochim Acta 193:100–118
Ishikawa T, Nakamura E (1992) Boron isotope geochemistry of the oceanic crust from DSDP/ODP Hole 504B. Geochim Cosmochim Acta 56:1633–1639
Ishikawa T, Nakamura E (1993) Boron isotope systematics of marine sediments. Earth Planet Sci Lett 117:567–580
Jackson MG, Hart SR, Koppers AAP, Staudigel H, Konter J, Blusztajn J, Kurz M, Russel JA (2007) The return of subducted continental crust in Samoan lavas. Nature 448:684–687
James RH, Palmer MR (2000) Marine geochemical cycles of the alkali elements and boron: the role of sediments. Geochim Cosmochim Acta 64:3111–3122
James RH, Elderfield H, Palmer MR (1995) The chemistry of hydrothermal fluids from Broken Spur site, 29 N Mid-Atlantic Ridge. Geochim Cosmochim Acta 59:651–659
James RH, Rudnicki MD, Palmer MR (1999) The alkali element and boron geochemistry of the Escanaba Trough sediment-hosted hydrothermal system. Earth Planet Sci Lett 171:157–169
Joachimski MM, Simon L, van Geldern R, Lécuyer C (2005) Boron isotope geochemistry of Paleozoic brachiopod calcite: implications for a seular change in the boron isotope geochemistry of seawater over the Phanerozoic. Geochim Cosmochim Acta 69:4035–4044
Jochum KP, Stoll B, Herwig K, Willbold M, Hofmann A, Amini M, Aarburg S, Abouchami W, Hellebrand E, Mocek B, Raczek I, Stracke A, Alard O, Bouman C, Becker S, Dücking M, Brätz H, Klemd R, de Bruin D, Canil D, Cornell D, de Hoog CJ, Dalpé C, Danyushevsky LV, Eisenhauer A, Gao Y, Snow JE, Groschopf N, Günther D, Latkoczy C, Giullong M, Hauri EH, Höfer HE, Lahaye Y, Horz K, Jacob DE, Kasemann SA, Kent AJR, Ludwig T, Zack T, Mason PRD, Meixner A, Rosner M, Misawa K, Nash BP, Pfänder JA, Premo WR, Sun WD, Tiepolo M, Vannucci R, Vennemann T, Wayne D, Woodhead JD (2006) MPI-DING reference glasses for in situ microanalysis: New reference values for element concentrations and isotope ratios. Geochem Geophys Geosys (G^3) 7, Q02008. doi:10.1029/2005GC001060
Karson JA, Kelley DS, Fornari DJ, Perfit MR, Shank TM (2015) Discovering the deep: a photographic atlas of the seafloor and oceanic crust. Cambridge University Press
Kasemann S, Erzinger J, Franz G (2000) Boron recycling in the continental crust of the central Andes from the Paleozoic to Mesozoic, NW Argentina. Contrib Mineral Petrol 140:328–343
Kasemann SA, Prave AR, Fallick AE, Hawkesworth CJ, Hoffmann KH (2010) Neoproterozoic ice ages, boron isotopes, and ocean acidification: implications for a snowball Earth. Geology 38:775–778
Kobayashi K, Tanaka R, Moriguti T, Shimizu K, Nakamura E (2004) Lithium, boron, and lead isotope systematics of glass inclusions in olivines from Hawaiian lavas: evidence for recycled components in the Hawaiian plume. Chem Geol 212:143–161
Kolodny Y, Chaussidon M (2004) Boron isotopes in DSDP cherts: fractionation and diagenesis. In: Hill RJ, Leventhal J, Aizenshtat Z, Baedecker MJ, Claypool G (eds) Geochemical investigations in earth and space science: a tribute to Isaac R. Kaplan, vol 9, The Geochemical Society Special Publications, pp 1–14
Kopf A, Deyhle A (2002) Back to the roots; boron geochemistry of mud volcanoes and its implications for mobilization depth and global B cycling. Chem Geol 192:195–210
Kopf A, Deyhle A, Zuleger E (2000) Evidence for deep fluid circulation and gas hydrate dissociation using boron and boron isotopes of pore fluids in forearc sediments from Costa Rica (ODP Leg 170). Marine Geol 167:1–28
Kopf A, Germán M, Deyhle A, Frape S, Hesse R (2003) Fluid geochemistry in the Japan trench forearc (ODP Leg 186): a synthesis. Proc Ocean Drill Progr Sci Res 186:1–23
Kowalski P, Wunder B (2017) Boron-isotope fractionation among solids-fluids-melts: experiments and atomic modeling. In: Marschall HR, Foster GL (eds) Boron Isotopes—The Fifth Element, Advances in Isotope Geochemistry, vol 7, Springer, Heidelberg, 33–69

Lécuyer C, Grandjean P, Reynard B, Albarède F, Telouk P (2002) $^{11}B/^{10}B$ analysis of geological materials by ICP-MS Plasma 54: application to the boron fractionation between brachiopod calcite and seawater. Chem Geol 186:45–55

Leeman WP, Sisson VB (1996) Geochemistry of boron and its implications for crustal and mantle processes. In: Grew ES, Anovitz LM (eds) Boron: mineralogy, petrology and geochemistry, Reviews in mineralogy, vol 33, pp 645–695

Lemarchand D, Gaillardet J, Lewin É, Allègre CJ (2000) The influence of rivers on marine boron isotopes and implications for reconstructing past ocean pH. Nature 408:951–954

Lemarchand D, Gaillardet J, Lewin É, Allègre CJ (2002) Boron isotope systematics in large rivers: implications for the marine boron budget and paleo-pH reconstruction over the Cenozoic. Chem Geol 190:123–140

le Roux PJ, Shirey SB, Benton L, Hauri EH, Mock TD (2004) In situ, multiple-multiplier, laser ablation ICP-MS measurement of boron isotopic composition (δ11B) at the nanogram level. Chem Geol 203:123–138

Marschall HR (2017) Boron isotopes in the ocean floor realm and the mantle. In: Marschall HR, Foster GL (eds) Boron Isotopes—The Fifth Element, Advances in Isotope Geochemistry, vol 7, Springer, Heidelberg, 191–217

Marschall HR, Jiang SY (2011) Tourmaline isotopes: no element left behind. Elements 7:313–319

Marschall HR, Monteleone BD (2015) Boron isotope analysis of silicate glass with very-low boron concentrations by secondary-ion mass spectrometry. Geostand Geoanal Res 39:31–46

Marschall HR, Schumacher JC (2012) Arc magmas sourced from mélange diapirs in subduction zones. Nature Geosci 5. 10.1038/NGEO1634

Marschall HR, Altherr R, Ludwig T, Kalt A, Gméling K, Kasztovszky Z (2006) Partitioning and budget of Li, Be and B in high-pressure metamorphic rocks. Geochim Cosmochim Acta 70:4750–4769

Marschall HR, Wanless VD, Shimizu N, Pogge von Strandmann PAE, Elliott T, Monteleone BD (2017) The boron and lithium isotopic composition of mid-ocean ridge basalts and the mantle. Geochim Cosmochim Acta, Geochim 207:102–138. doi:10.1016/j.gca.2017.03.028

McCulloch MT, D'Olivio JP, Falter J, Georgiou L, Holcomb M, Montagna P, Trotter J (2017) Boron isotopes in corals. In: Marschall HR, Foster GL (eds) Boron Isotopes—The Fifth Element, Advances in Isotope Geochemistry, vol 7, Springer, Heidelberg, 145–163

McKenzie D, O'Nions RK (1983) Mantle reservoirs and ocean island basalts. Nature 301:229–231

Menzies MA, Wass SY (1983) CO_2-rich and LREE-rich mantle below eastern Australia—a REE and isotopic study of alkaline magmas and apatite-rich mantle xenoliths from the Southern-Highlands province. Australia Earth Planet Sci Lett 65:287–302

Moriguti T, Nakamura E (1998) Across-arc variation of Li-isotopes in lavas and implications for crust/mantle recycling at subduction zones. Earth Planet Sci Lett 163:167–174

Müller RD, Sdrolias M, Gaina C, Roest WR (2008) Age, spreading rates, and spreading asymmetry of the world's ocean crust. Geochem Geophys Geosys (G^3) 9, Q04006

Nebel O, Arculus RJ, van Westrenen W, Woodhead JD, Jenner FE, Nebel-Jacobsen YJ, Wille M, Eggins SM (2013) Coupled Hf–Nd Pb isotope co-variation of HIMU oceanic island basalts from Mangaia, Cook-Austral islands, suggest an Archean source component in the mantle transition zone. Geochim Cosmochim Acta 112:87–101

O'Neill HSC, Jenner FE (2012) The global pattern or trace-element distributions in ocean floor basalts. Nature 491:698–704

Palmer MR (1991) Boron isotope systematics of hydrothermal fluids and tourmalines: a synthesis. Chem Geol 94:111–121

Palmer MR, Sturchio NC (1990) The boron isotope systematics of the Yellowstone National Park (Wyoming) hydrothermal system: a reconnaissance. Geochim Cosmochim Acta 54:2811–2815

Palmer MR, Swihart GH (1996) Boron isotope geochemistry: an overview. In: Grew ES, Anovitz LM (eds) Boron: mineralogy, petrology and geochemistry, Reviews in Mineralogy, 1st edn. vol 33, Mineralogical Society of America, Washington, DC, pp 709–740

Palmer MR, Spivack AJ, Edmond JM (1987) Temperature and pH controls over isotopic fractionation during adsorption of boron on marine clay. Geochim Cosmochim Acta 51:2319–2323

Paris G, Gaillardet J, Louvat P (2010) Geological evolution of seawater boron isotopic composition recorded in evaporites. Geology 38:1035–1038

Pearson PN, Palmer MR (2000) Atmospheric carbon dioxide concentrations over the past 60 million years. Nature 406:695–699

Pennisi M, Leeman WP, Tonarini S, Pennisi A, Nabelek P (2000) Boron, Sr, O, and H isotope geochemistry of groundwaters from Mt. Etna (Sicily)—hydrologic implications. Geochim Cosmochim Acta 64:961–974

Rae JWB (2017) Boron isotopes in foraminifera. In: Marschall HR, Foster GL (eds) Boron Isotopes—The Fifth Element, Advances in Isotope Geochemistry, vol 7, Springer, Heidelberg, 107–143

Rae JWB, Foster GL, Schmidt DN, Elliott T (2011) Boron isotopes and BCa in benthic foraminifera: proxies for the deep ocean carbonate system. Earth Planet Sci Lett 302:403–413

Raffone N, Ottolini LP, Tonarini S, Gianelli G, D'Orazio M, Friedleifsson GO (2010) An investigation of trace and isotope light elements in mineral phases from well RN-17 (Reykjanes Peninsula, SW Iceland). In: 11th European Workshop on Modern Developments and Applications in Microbeam Analysis, IOP Conference Series: Materials Science and Engineering, vol 7, IOP Publishing, 012026

Reymer A, Schubert G (1984) Phanerozoic addition rates to the continental crust and crustal growth. Tectonics 3:63–77

Reynolds JR, Langmuir CH, Bender JF, Kastens KA, Ryan WBF (1992) Spatial and temporal variability in the geochemistry of basalts from the East Pacific Rise. Nature 359:493–499

Rose-Koga EF, Sigmarsson O (2008) B-O–Th isotope systematics in Icelandic tephra. Chem Geol 255:454–462

Roy-Barman M, Wasserburg GJ, Papanastassiou DA, Chaussidon M (1998) Osmium isotopic compositions and Re–Os concentrations in sulfide globules from basaltic glasses. Earth Planet Sci Lett 154:331–347

Ryan JG, Langmuir CH (1993) The systematics of boron abundances in young volcanic rocks. Geochim Cosmochim Acta 57:1489–1498

Schmidt K, Koschinsky A, Garbe-Schönberg D, de Carvalho LM, Seifert R (2007) Geochemistry of hydrothermal fluids from the ultramafic-hosted Logatchev hydrothermal field, 15 N on the Mid-Atlantic Ridge: temporal and spatial investigation. Chem Geol 242:1–21

Schwarcz HP, Agyei EK, McMullen CC (1969) Boron isotopic fractionation during clay adsorption from sea-water. Earth Planet Sci Lett 6:1–5

Searle M, Cox J (1999) Tectonic setting, origin, and obduction of the Oman ophiolite. Geol Soc Am Bull 111:104–122

Seyfried WE Jr, Pester NJ, Ding K, Rough M (2011) Vent fluid chemistry of the Rainbow hydrothermal system (36 N, MAR): phase equilibria and in situ pH controls on subseafloor alteration processes. Geochim Cosmochim Acta 75:1574–1593

Seyfried WE Jr, Pester NJ, Tutolo B, Ding K (2015) The Lost City hydrothermal system: constraints imposed by vent fluid chemistry and reaction path models on subseafloor heat and mass transfer processes. Geochim Cosmochim Acta 163:59–79

Shaw AM, Hauri EH, Behn MD, Hilton DR, Macpherson CG, Sinton JM (2012) Long-term preservation of slab signatures in the mantle inferred from hydrogen isotopes. Nat Geosci 5:224–228

Simon L, Lécuyer C, Maréchal C, Coltice N (2006) Modelling the geochemical cycle of boron: implications for the long-term δ^{11}B evolution of seawater and oceanic crust. Chem Geol 225:61–76

Smith HJ, Spivack AJ, Staudigel H, Hart SR (1995) The boron isotopic composition of altered oceanic crust. Chem Geol 126:119–135

Spivack AJ, Edmond JM (1987) Boron isotope exchange between seawater and the oceanic crust. Geochim Cosmochim Acta 51:1033–1043

Spivack AJ, You CF (1997) Boron isotopic geochemistry of carbonates and pore waters, Ocean Drilling Program Site 851. Earth Planet Sci Lett 152:113–122

Spivack AJ, Palmer MR, Edmond JM (1987) The sedimentary cycle of the boron isotopes. Geochim Cosmochim Acta 51:1939–1949

Stracke A, Hofmann AW, Hart SR (2005) Fozo, HIMU, and the rest of the mantle zoo. Geochem Geophys Geosys (G^3) 6, Q05007

Sun SS, Nesbitt RW, Sharaskin AY (1979) Geochemical characteristics of mid-ocean ridge basalts. Earth Planet Sci Lett 44:119–138

Swihart GH, Moore PB (1986) Boron isotopic composition of marine and nonmarine evaporite borates. Geochim Cosmochim Acta 50:1297–1301

Tanaka R, Nakamura E (2005) Boron isotopic constraints on the source of Hawai'ian shield lavas. Geochim Cosmochim Acta 69:3385–3399

Thompson G, Melson WG (1970) Boron contents of serpentinites and metabasalts in the oceanic crust: implicatons for the boron cycle in the oceans. Earth Planet Sci Lett 8:61–65

Tonarini S, Leeman WP, Leat PT (2011) Subduction erosion of forearc mantle wedge implicated in the genesis of the South Sandwich Island (SSI) arc: evidence from boron isotope systematics. Earth Planet Sci Lett 301:275–284

Turner S, Tonarini S, Bindemann I, Leeman WP, Schaefer BF (2007) Boron and oxygen isotope evidence for recycling of subducted components over the past 2.5 Gyr. Nature 447:702–705

Uppstroem LR (1974) The boron/chlorinity ratio of deep seawater from the Pacific Ocean. Deep Sea Res 21:161–162

van Hinsberg VJ, Henry DJ, Marschall HR (2011) Tourmaline: an ideal indicator of its host environment. Can Mineral 49:1–16

van Keken PE, Hauri EH, Ballentine CJ (2002) Mantle mixing: the generation, preservation, and destruction of chemical heterogeneity. Ann Rev Earth Planet Sci 30:493–525

Vengosh A, Kolodny Y, Starinsky A, Chivas AR, McCulloch MT (1991) Coprecipitation and isotopic fractionation of boron in modern biogenic carbonates. Geochim Cosmochim Acta 55:2901–2910

Vengosh A, Starinsky A, Kolodny Y, Chivas AR, Raab M (1992) Boron variations during fractional evaporation of seawater: new constraints on the marine vs nonmarine debate. Geology 20:799–802

Vidal P, Dosso L (1978) Core formation—catastrophic or continuous—Sr and Pb isotope geochemistry constraints. Geophys Res Lett 5:169–172

Vils F, Tonarini S, Kalt A, Seitz HM (2009) Boron, lithium and strontium isotopes as tracers of seawater–serpentinite interaction at Mid-Atlantic ridge, ODP Leg 209. Earth Planet Sci Lett 286:414–425

Weaver BL (1991) The origin of ocean island basalt end-member compositions—trace-element and isotopic constraints. Earth Planet Sci Lett 104:381–397

Williams LB, Hervig RL (2005) Lithium and boron isotopes in illite-smectite: the importance of crystal size. Geochim Cosmochim Acta 69:5705–5716

Williams LB, Hervig RL, Holloway JR, Hutcheon I (2001a) a B isotope geochemistry during diagenesis. Part I. Experimental determination of fractionation

during illitization of smectite. Geochim Cosmochim Acta 65:1769–1782
Williams LB, Wieser ME, Fennell J, Hutcheon I, Hervig RL (2001b) b Application of boron isotopes to the understanding of fluid–rock interactions in a hydrothermally stimulated oil reservoir in the Alberta Basin. Can Geofl 1:229–240
Yamaoka K, Ishikawa T, Matsubaya O, Ishiyama D, Nagaishi K, Hiroyasu Y, Chiba H, Kawahata H (2012) Boron and oxygen isotope systematics for a complete section of oceanic crustal rocks in the Oman ophiolite. Geochim Cosmochim Acta 84:543–559
Yamaoka K, Hong E, Ishikawa T, Gamo T, Kawahata H (2015a) a Boron isotope geochemistry of vent fluids from arc/back-arc seafloor hydrothermal systems in the western Pacific. Chem Geol 392:9–18
Yamaoka K, Matsukura S, Ishikawa T, Kawahata H (2015b) b Boron isotope systematics of a fossil hydrothermal system from the Troodos ophiolite, Cyprus: water–rock interactions in the oceanic crust and subseafloor ore deposits. Chem Geol 396:61–73
You CF, Spivack AJ, Smith JH, Gieskes JM (1993) Mobilization of boron in convergent margins: implications for the boron geochemical cycle. Geology 21:207–210
You CF, Butterfield DA, Spivack AJ, Gieskes JM, Gamo T, Campbell AJ (1994) Boron and halide systematics in submarine hydrothermal systems: effects of phase separation and sedimentary contributions. Earth Planet. Sci. Lett. 123:227–238
You CF, Chan LH, Spivack AJ, Gieskes JM (1995) Lithium, boron, and their isotopes in sediments and pore waters of Ocean Drilling Program Site 808, Nankai Trough: implications for fluid expulsion in accretionary prisms. Geology 23:37–40
Zindler A, Hart SR (1986) Chemical geodynamics. Ann Rev Earth Planet Sci 14:493–571

Boron Isotopes as a Tracer of Subduction Zone Processes

9

Jan C.M. De Hoog and Ivan P. Savov

Abstract

This chapter reviews recycling of boron (B) and its isotopes in subduction zones. It discusses the profound changes in B concentrations and B isotope ratios of various materials involved in convergent margin evolution, in particular highlighting the fate and evolution of progressively dehydrating subducting slabs and the behavior of B during burial and subsequent metamorphism. We review various models used to parameterize these complex and often poorly understood processes and critically evaluate the available data from the literature. We proceed by reviewing B isotope data from mafic arc volcanic rocks and explore systematic variations with subduction zone geometry as well as familiar geochemical tracers of subduction processes. Finally, the role of serpentinisation in the mantle wedge is discussed in the light of new geochemical and petrological insights on the importance of serpentinites and subduction erosion. We provide recommendations for further research on B isotopes in subduction zones and directions where we think this exciting stable isotope tracer may make the biggest impact.

Keywords

Boron · B isotopes · Subduction recycling · Metamorphism
Arc volcanism · Volatiles · Fluids

J.C.M. De Hoog (✉)
School of GeoSciences, Grant Institute, The University of Edinburgh, James Hutton Road, Edinburgh EH9 3FE, UK
e-mail: ceesjan.dehoog@ed.ac.uk

I.P. Savov
School of Earth and Environment, The University of Leeds, Leeds LS2 9JT, UK
e-mail: I.Savov@leeds.ac.uk

9.1 Introduction

Boron and its isotopes provide some of the most powerful tools for investigating fluid-mediated processes at subduction zones, as B is extremely depleted in the mantle but strongly enriched (by orders of magnitude) and isotopically fractionated

H. Marschall and G. Foster (eds.), *Boron Isotopes*, Advances in Isotope Geochemistry,
https://doi.org/10.1007/978-3-319-64666-4_9

(by up to 40‰) in subducted sediments, altered oceanic crust (AOC) and serpentinized mantle (subducted lithospheric section or eroded in the forearc wedge; Fig. 9.1). Upon subduction, much of the slab volatile inventory is lost in the forearc region due to diagenesis and compaction, giving rise to vast forearc serpentinite reservoirs, which may still enter magma source regions through subduction erosion. The remaining fraction of B hosted in AOC and sediments is isotopically light as a result of isotope fractionation that occurs during shallow B loss. In contrast, serpentinized oceanic mantle retains a heavy seawater-derived isotopic signature until its dehydration beneath or beyond the volcanic arc front. Thus, B can track the origin of fluid sources to subduction zone magmatism.

The interplay of volatile sources and dehydration processes from forearc to backarc is complex and results in large differences in B isotopic composition of different volcanic arcs. Across-arc trends of diminishing B concentrations and B isotope fractionation occur in some arcs but not in others. Hence, the complexities of B behavior in subduction zones and therefore the source of fluids remain elusive and are still the subject of intense research, and the topic of review of this chapter.

9.2 Metamorphic Processes in the Subducting Slab

The focus of this section is B processing within the progressively changing slab assemblages during subduction. Boron enters subduction zones in the mafic and ultramafic parts of the oceanic lithosphere as well as overlying sediments including pore waters. What happens in the subduction zone is highly complex, as each of these reservoirs has a unique mineralogical makeup as well as a non-unique range of bulk rock B concentrations and isotopic compositions. In addition, each mineral responds differently to increasing pressures and temperatures during subduction. The compositions of the various reservoirs entering and evolving within the subduction zone will be briefly summarized, the reader is referred to Marschall (2017) for more details.

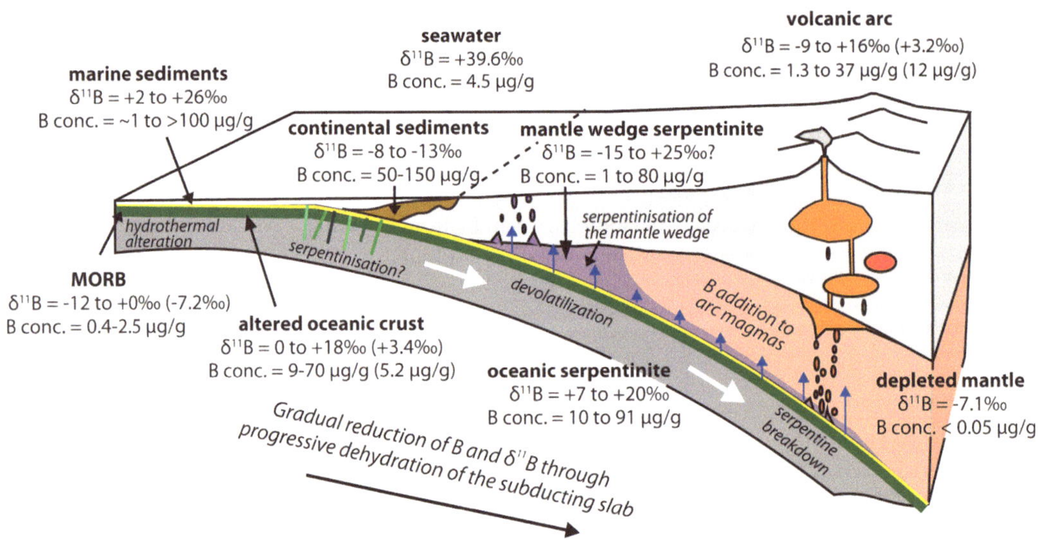

Fig. 9.1 Schematic diagram of the B cycle in subduction zones showing approximate B concentrations and isotopic compositions of various reservoirs (modified from Jones 2014). Values in parentheses are compiled averages. See text and Marschall (2017) for data sources

9.2.1 Physical and Thermal Geometry of Subduction Zones

The physical and thermal layout of subduction zones is of critical importance for the sequence of metamorphic reactions that will occur in the subducting slab as well as for the transport of melts and fluids from the slab into the overlying mantle wedge (e.g., Bebout et al. 1993; Peacock 1993; Van Keken et al. 2002; Cagnioncle et al. 2007; Portnyagin and Manea 2008; Konrad-Schmolke and Halama 2014).

Subduction zones are composed of a subducting slab and an overlying mantle wedge, which itself is overlain by either thick (ca. 30 km) continental or, in the case of intra-oceanic arc systems, thin (ca. 10 km) oceanic crust (Fig. 9.2). The slab is generally composed of, from top to bottom, sediments of various lithological makeup and thickness, ca. 3 km of basalts and diabases/microgabbros, ca. 4 km of gabbroic/cumulate sections of oceanic crust, and a lithospheric root of mantle peridotite (Dilek and Furnes 2011). The altered oceanic crust and the peridotite sections may or may not be partially hydrated (serpentinized, rodingitized, epidotized). The above is the sequence observed at fast-spreading mid-oceanic ridges and well-studied ocean crust sections exposed in ophiolite complexes. In other settings considerable amounts of mantle peridotite can be exposed at the seafloor and much of the igneous material may be missing (Cannat 1993). In addition, fracture zones may be highly altered compared to adjacent crustal sections whereas seamounts may add large amounts of basaltic material, both adding to the diversity of material entering subduction zones (e.g., Wada et al. 2012).

The mantle wedge overlying the slab is harzburgite in mineralogy and highly melt-depleted (Savov et al. 2007; Fryer 2012). It can be subdivided into forearc and sub-arc mantle. The ‘cold nose’ represents the thinnest part of the forearc, which has been chilled by the cold subducting slab and is largely unaffected by mantle corner flow. It may be extensively serpentinized by pore waters and diagenetic fluids derived from the subducting slab, particularly above hot slabs (Mottl et al. 2004; Fryer 2012; Reynard 2013; see Sect. 9.3 for details). The deeper part of the fore-arc is too hot to be serpentinized, as it is prevented from cooling by mantle corner flow in response to the down-dragging forces along the subduction interface. The mantle beneath the volcanic arc, the sub-arc mantle, is hot enough to melt in response to slab-derived fluid fluxing, hence it is here where the onset of melting and arc magma generation take place (Grove et al. 2009; England and Katz 2010).

The crust overlying the mantle wedge can be thick (ca. 20–40 km) continental crust (continental arcs) or thin (ca. 7 km) oceanic crust (intra-oceanic arcs). An accretionary sedimentary wedge may exist at the intersection of subducting slab and overriding plate. However, at many intra-oceanic arcs an accretionary wedge is missing and thus all of the slab assemblages are actively subducted, as in the classic case of the Izu-Bonin-Marianas arc system (Savov et al. 2007; Fryer 2012). The composition and geometry of the subduction interface between the subducting slab and mantle wedge, generally referred to as the subduction channel, is complex and topic of intense research. It is fluxed by large amounts of fluids and subjected to large temperature gradients and high shear stresses, and will be discussed in more detail in Sect. 9.3.

The thermal structure of a subduction zone is dependent on the convergence velocity (V) and age (A) of the subducting slab as well as the angle of subduction δ (Van Keken et al. 2002). These different factors are captured by the thermal parameter ($\Phi = V A \sin \delta$; Kirby et al. 1991; Syracuse et al. 2010), where a high value means a cold subduction zone. Modern thermal models allow the temperature and pressure of the subducting slab to be calculated in sufficient detail to be integrated into thermodynamical models of metamorphic reactions occurring in the slab (Marschall et al. 2007; Konrad-Schmolke and Halama 2014; Walowski et al. 2015). However, thermal models are still being refined and depend strongly on assumptions regarding the depth and extent of coupling between subducting slab and overriding plate (e.g., Syracuse et al. 2010), in

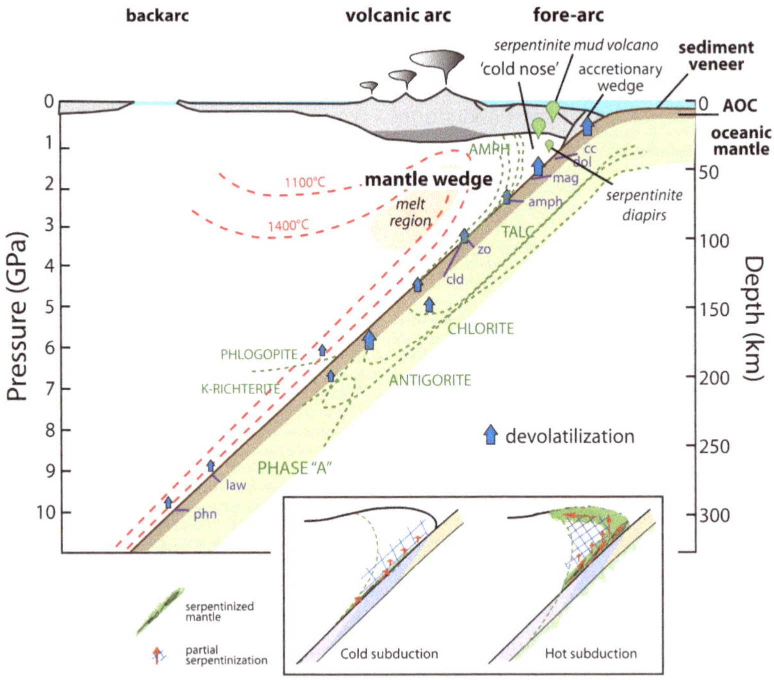

Fig. 9.2 Schematic cross-section of subduction zone indicating major tectonic and geological features as well as mineral stability fields, with minerals in ultramafic lithologies in *green* and those in mafic lithologies in *purple* (after Poli and Schmidt 2002). *Inset* shows detail of serpentinization in the mantle wedge in cold and hot subduction zones, respectively (after Reynard 2013). Mineral abbreviations: *cc* calcite; *dol* dolomite; *mag* magnesite; *amph* amphibole; *zo* zoisite; *cld* chloritoid; *law* lawsonite; *phn* phengite

particular the slab surface temperature being poorly constrained and the topic of much debate (e.g., Penniston-Dorland et al. 2015). This still leaves considerable uncertainty about the temperature profile of the subduction interface and the mantle wedge.

9.2.2 Subduction Metamorphism: Sediments Including Their Pore Waters

Sediments are volumetrically as well as compositionally the most diverse reservoir entering subduction zones. They include biogenic carbonates as well as silica, pelagic clays and continental detritus (Fig. 9.3). Intra-oceanic arc trenches are generally poor in terrigenous sediments and show alternating deep (often volcanic ash-rich) marine pelagic clays (examples: Ryukyu, Japan, Aleutians, Kuriles, Kamchatka, Izu, Java, Tonga, N. Antilles, S. Sandwich Isl.) and/or volcaniclastics (Hikurangi, Vanuatu, Marianas, Costa Rica) from the eroding or erupting nearby arc volcanoes. Trenches near continental arcs receive highly diverse material consisting of larger terrigenous fractions (Nankai, Cascades, Chile, S. Antilles) derived from the erosion of older cratons or eroding younger mountain belts (Himalayas, Andes, Alps). Based on the most recent compilation of the lithology and chemistry of globally subducted sedimentary piles (Plank 2014), the thickness of the sediment sequences entering subduction zones may vary from ~100 m (Tonga Trench) to >3 km (Andaman and Makran Trenches), but on average is in the range of ~500 m, i.e., an order of magnitude less than the thickness of the subducted oceanic crust (Fig. 9.3).

Boron concentration data of subducted sediments is relatively scarce (see Marschall 2017 for a summary), with the majority of the data clustering between 50 and 100 µg/g. The global

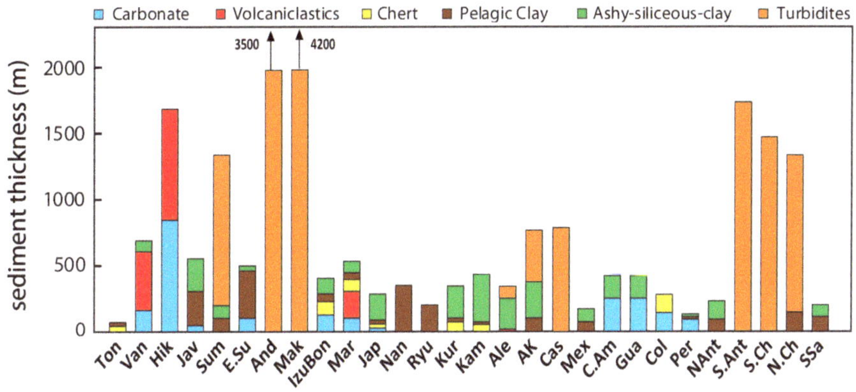

Fig. 9.3 Global sediment compositions and thicknesses at subduction trenches (data from Plank and Langmuir 1998; Plank 2014). Abbreviations: *Ton* Tonga; *Van* Vanuatu; *HIK* N.Zealand; *Jav* Java; *Sum* Sumatra; *E.Su* E.Sunda; *And* Andaman; *Mak* Makran; *IzuBon* Izu-Bonin; *Mar* Marianas; *Jap* Japan; *Nan* Nankai; *Ryu* Ryuku; *Kur* Kurile; *Kam* Kamchatka; *Ale* Aleutians; *AK* Alaska; *Cas* Cascadia; *Mex* Mexico; *C.Am* Central America-Costa Rica; *Gua* Guatemala; *Col* Colombia; *Per* Peru; *NAnt* N. Antilles; *S.Ant* S. Antilles; *S.Ch* S. Chile 45°; *N.Ch* N. Chile 35° S; *SSa* South Sandwich

subducted sediment (GLOSS II) value for B is 68 ± 4 μg/g (Plank 2014).

Only few subducted sediment piles have been methodically evaluated for B and B isotopes; these include examples from the South Sandwich Islands (Tonarini et al. 2011) and Japan (Ishikawa and Nakamura 1993). The South Sandwich sediments have very high B contents (avg. 115 μg/g) and highly variable $\delta^{11}B$ (avg. −4 ± 6.6‰). The Japan trench sediments also have very high B contents (avg. 96 μg/g) but with less variable $\delta^{11}B$ (−3.3 ± 3.7‰), although significant differences exist between sediment types, such as continental detritus ($\delta^{11}B$ = −13 to −8‰), marine clays ($\delta^{11}B$ = +2.3 to +9.2‰), biogenic carbonates ($\delta^{11}B$ = +8.0 to +26.2‰) and biogenic silica ($\delta^{11}B$ = +2.1 to +4.5‰; Ishikawa and Nakamura 1993). Sediment samples from the Aleutian Trench (Leeman et al. 2004) have lower B contents (~80 μg/g) and average $\delta^{11}B$ value of −0.5 ± 1.9‰. Marine carbonates are generally lacking B abundance data but can have very high $\delta^{11}B$ (+14 ± 5‰; Simon et al. 2006). It is important to point out that due to diagenetic processes, B and $\delta^{11}B$ abundances in sediments and their pore fluids vary dramatically with depth (Toki et al. 2015), and thus a large bias of the B and $\delta^{11}B$ values can be expected depending on where in the sedimentary pile the sediments were sampled.

Much of the pore waters and weakly bound water in clays are lost during compaction and diagenesis at shallow depths. As B strongly partitions into such fluids, >80% of the subducted B is recycled in the fore-arc (You et al. 1995, 1996; Savov et al. 2007). Significant B isotope fractionation takes place during diagenesis, as ^{11}B is transported preferentially from sediments to pore fluids during recrystallization and dissolution of clay minerals, carbonates and silica (Ishikawa and Nakamura 1993; Hüpers et al. 2016; Saffer and Kopf 2016). The compositions of these fluids are largely inferred from fluid-mineral fractionation factors (Palmer et al. 1987; Hervig et al. 2002; Sanchez-Valle et al. 2005; Wunder et al. 2005; Kowalski et al. 2013), but are supported by recently recovered forearc materials from the Nankai Trough forearc basin (IODP Exp. 315, 316, 322, 333 and 338) and Mariana forearc (serpentinite) mud volcanoes (ODP Legs 125, 195, and recently completed IODP Exp. 366).

In the Nankai Trough forearc, Hüpers et al. (2016) observed decreasing B concentrations upcore associated with increasingly isotopically heavier B isotopic compositions (up to +49‰). The authors attributed these signatures to volcanic ash alteration at depth and remobilization of B via exchange with NH_4^+. These findings are in accord with the results derived from studies of sediments/fluids and borehole temperature measurements from Barbados, Costa Rica, Taiwan, Mediterranean Ridge, Mariana and the Nankai Trough forearcs (Saffer and Kopf 2016). Compositions of fluids expelled from the shallow decollement zone from these diverse shallow forearc settings become systematically enriched in B with isotopically lighter compositions with progressive heating, burial, and distance from the trench. The Saffer and Kopf (2016) study also revealed that the largest geochemical signatures of deeply sourced fluids should be expected in warm subduction zones, because higher temperatures in the forearcs lead to earlier onset of clay dehydration and desorption of a greater mass of boron. An example is the non-accretionary (warm) Western Pacific Mariana forearc, where multiple expeditions managed to successfully recover a record of depleted peridotite (harzburgite) interactions with B-enriched and slab-derived fluids forming large serpentinite seamounts ("mud volcanoes"; Fryer 2012). These slab-derived fluids have an extremely high alkalinity (30 × seawater), a pH of ~12.5 and very high B concentrations (>3.2 mmol/kg, i.e., 8 × seawater; Mottl 1992; Mottl et al. 2004) with high $\delta^{11}B$ of ca. +9‰ (Savov et al. 2004). These fluids were extracted from serpentinized harzburgite muds and clasts that have also been shown to have very high $\delta^{11}B$ values of ca. +13‰ (Benton et al. 2001).

Large losses of B at shallow depth are also supported by studies of metasediments. For instance, continuous B loss with increasing metamorphic grade was reported for a range of metamorphosed sediments and basalts having protoliths analogous to subducted marine sediments and basalts (Moran et al. 1992). Similar B depletions of up to 70% were noted in metasedimentary rocks from the Catalina Schist in California, USA (Bebout et al. 1993), which already showed large B losses at lawsonite blueschist-facies conditions (ca. 350 °C, 1.0 GPa; Bebout et al. 1993, 1999).

Despite large losses in the shallow fore-arc, the remaining sedimentary B can be transported to great depths by white mica (phengite) ± tourmaline (Bebout et al. 1993, 2007; Domanik et al. 1993; Bebout and Nakamura 2003). Due to its extremely high B content and large stability field, tourmaline is potentially an important reservoir but its abundance is highly variable and its role is not well constrained (Marschall et al. 2009b; Van Hinsberg et al. 2011). Bebout and Nakamura (2003) reported increasingly isotopically lighter prograde mantles of tourmaline from Catalina Schist ($\delta^{11}B$ −7‰ down to −15‰) as well as from Lago di Cignana, W. Alps (UHP cores $\delta^{11}B$ −9 down to −13‰), and concluded that fluids evolve their $\delta^{11}B$ signatures with increasing slab depth. Likewise, Nakano and Nakamura (2001) present evidence for growth of tourmaline during breakdown of muscovite and chlorite in metasediments from the HP Sambagawa Belt (Japan), resulting in little bulk B removal or isotopic fractionation with increasing grade. Prograde subduction-related rims on detrital cores in metasediments from Syros, Greece, have substantially higher $\delta^{11}B$ values (on average +0.9‰; Marschall et al. 2008). Tourmaline may persist to 860 °C at 0.7 GPa (ca. 20 km depth) in B-rich granulite-facies metasediments (MacGregor et al. 2013) and at least 950 °C and 5.5 GPa in silica-undersaturated rocks (Marschall et al. 2009b) and therefore transfer B beyond the arc. On the other hand, tourmaline breakdown accompanied by significant B loss was reported during migmatization of metapelites at ca. 725 °C at 0.5 GPa (ca. 13 km depth; Kawakami 2001a, b), which suggests that B transfer from tourmaline to sediment melt may occur if enough water is present to induce sediment melting.

The role of phengite in subducting sediments remains poorly constrained, even though it is a major host for B. Its great P-T stability suggests it will survive beyond the arc (Domanik and Holloway 1996; Schmidt 1996; Johnson and Plank 1999; Schmidt et al. 2004) and therefore may contribute little to the overlying mantle wedge, but hydrous melts from phengite breakdown are needed for the efficient extraction of K and LILE (Hermann and Spandler 2008).

In summary, due to their low volume and extensive B loss at shallow depths, sediments are unlikely to carry much B to sub-arc depths, and any remaining B is locked up in tourmaline and phengite. Therefore, the contribution of sediments to the B cycle occurs mainly at shallow depths, where B-rich sediment-derived fluids flux the overlying mantle wedge and B is readily fixed in serpentinites. These serpentinites and slab decollement mélanges may be dragged down to greater depths and deliver B to arc magma sources upon breakdown beneath the arc (Hattori and Guillot 2003; Savov et al. 2005, 2007; Marschall and Schumacher 2012). In addition, sediments are likely to contribute significantly to the composition and rheology of the subduction channel, the interface between subducting slab and overlying mantle wedge. Mixtures of aluminiferous sediment and ultramafic mantle material will stabilize chlorite, a water-rich mineral with a much higher thermal stability than serpentine and may therefore persist until sub-arc depth (Bebout 1991; Pawley 2003; Spandler et al. 2008; Padron-Navarta et al. 2010; Marschall and Schumacher 2012; Till et al. 2012; Cannaò et al. 2015). The role of shallow slab-fluid modified mantle wedge serpentinites and the subduction interface will be further discussed in Sect. 9.3.

9.2.3 Subduction Metamorphism: Altered Oceanic Crust

Large sections of the oceanic crust have undergone extensive low to medium temperature hydrothermal alteration near spreading ridges and associated transform and detachment faults, and are referred to as altered oceanic crust (AOC; Smith et al. 1995). Compared to unaltered (dry) oceanic crust, AOC is enriched in hydrous minerals (including chlorite, epidote, zoisite, vesuvianite, prehnite, zeolites, tremolite, talc, brucite, serpentine) and may contain as much as 3–5 wt% H_2O as well as high concentrations of fluid-mobile elements, including B. AOC has a variable B isotopic composition with $\delta^{11}B$ ranging from −4‰ to +25‰ (Marschall 2017) and an average of +3.4‰ (Smith et al. 1995). This is considerably heavier than primitive ocean floor basalts ($\delta^{11}B$ ca. −10 to 0‰; Marschall 2017). The high $\delta^{11}B$ of the AOC is directly correlated with elevated bulk rock and mineral $^{87}Sr/^{86}Sr$ (often up to 0.707) and $\delta^{7}Li$ (up to +20‰; Foustoukos et al. 2008), revealing that seawater (with B ~4.5 µg/g and $\delta^{11}B$ = +39.6‰; Foster et al. 2010) is the ultimate source for the hydrothermal fluids responsible for the hydration reactions.

Ophiolites such as in Oman allow sampling of the crust at depth and have shown downward decreasing B concentrations (from 8 µg/g in pillow lavas down to 1 µg/g in gabbros) with increasing $\delta^{11}B$ (+5.5‰ in pillow lavas to +12‰ in gabbros), which reflects decreasing amounts of hydrothermal alteration at increasing temperature. Importantly, the deep gabbroic sections, which will dehydrate at greatest depth in the subduction zone, have the highest $\delta^{11}B$ and still contain about 30% of the total crust B inventory (Yamaoka et al. 2012). In contrast, lower water-rock ratios in the plutonic complex of the Troodos ophiolite resulted in lower $\delta^{11}B$ values (+4.5‰; Yamaoka et al. 2015), which suggests that the boron isotopic composition of AOC at depth strongly depends on local variations in hydrothermal systems supplying B.

During subduction, pressure and temperature of AOC will increase according to the thermal regime of the subduction zone. In general, the crust will undergo low temperature, high-pressure blueschist to greenschist to high pressure, high temperature eclogite facies metamorphism, during which the main hydrous minerals amphibole, chlorite, micas, epidote and clinozoisite progressively lose water (Schmidt and Poli 1998; Fig. 9.2). Boron, which is mainly hosted by white mica (Domanik et al. 1993; Marschall et al. 2006a), preferentially partitions into the dehydration fluid and is gradually lost from the residue (Mottl 1992; Mottl et al. 2004).

As the subducted slab is generally not available for direct sampling at depth, the B isotopic composition of the crust and its fluids at each particular depth must be estimated by indirect means. Estimates mainly come from three different approaches: (1) the study of ophiolites that represent exhumed formerly subducted terranes; (2) thermodynamic modeling of the subducting slab and its fluids (Sect. 9.4) the composition of arc magmas, particular those from intra-oceanic arcs with little opportunity for crustal contamination or subduction erosion (Sect. 9.5).

The closest equivalent to material from an active subduction zone are blueschist clasts recovered from serpentinite muds during ODP Legs 125 and 195 in the Mariana forearc (Pabst et al. 2012). These blueschists were metamorphosed at about 20–40 km depth and represent material from a subduction mélange, i.e., a mixture of slab-derived mafic material and (hydrated) ultramafic mantle wedge material. In situ B isotopic analysis of silicate minerals from these blueschists gave an average $\delta^{11}B$ value of $-6 \pm 4‰$, which is considerably lower than the average AOC value of $\delta^{11}B = +3.4‰$ and consistent with loss of heavy B to the forearc at shallow depth. This is also in agreement with isotopically heavy $\delta^{11}B$ values of forearc serpentine mud volcanoes and associated cold springs on their summits and flanks (Mottl 1992; Mottl et al. 2004; Savov et al. 2004, 2007; Fryer 2012). In essence, as the heavier isotope ^{11}B prefers the fluid phase (Hervig et al. 2002; Wunder et al. 2005), AOC becomes increasingly isotopically light with low $\delta^{11}B$, with fluids derived from the slab +20‰ (low T) to +5‰ (high T) heavier than the residue.

Examples of exhumed subducted oceanic crust or subduction mélanges from paleo-subduction zones have been reported from Syros, Greece (Marschall et al. 2006a), the Western Alps (Konrad-Schmolke et al. 2011; Lafay et al. 2013; Angiboust et al. 2014; Halama et al. 2014), the Franciscan Complex, California (Bebout 1991; Moran et al. 1992; Bebout et al. 1993, 2007; Domanik et al. 1993; King et al. 2007) and Alaska (Moran et al. 1992). Data from these terranes confirm that the subducting slab becomes increasingly depleted in B but enriched in ^{10}B compared to ^{11}B with depth. For example, Peacock and Hervig (1999) reported $\delta^{11}B$ values of -3 to $-11‰$ in hydrous minerals, mostly white mica, in subduction zone rocks, consistent with earlier loss of heavy B via dehydration reactions. Subduction-related lawsonite-albite to lawsonite−blueschist grade mélange material from the Catalina Schist had $\delta^{11}B$ values of -2.3 to $-12‰$, consistent with slab restite (King et al. 2007). Loss of B during progressive dehydration of meta-igneous rocks was also reported from Syros (Greece), but no B isotope data was presented (Marschall et al. 2009a).

Based on the evidence above, AOC will have little B left when arriving at sub-arc depths, other than what is present in phengite, the amount of which is strongly dependent on the K_2O content of the protolith (Marschall et al. 2007). Any remaining B in dehydrated AOC will have a strongly negative $\delta^{11}B$ signature. In addition, oceanic crust may contain significant areas of fresh or little altered minerals (Cann et al. 2015). These areas may act as a sink of B in percolating fluids if new hydrous minerals form, and delay fluid release till deeper section of the subduction zone,. All of these scenarios leave us with a picture that most B leaves AOC at shallow depths (but deeper than sediments) and little beneath the arc. As most (>90%) of B is transferred to fluids, mass balance requires that the pooled composition of these fluids is close to or slightly higher than the composition of AOC before subduction (+3.4‰; Smith et al. 1995). As was the case for sediments, the heaviest fraction, which is expelled at the shallowest depths, may or may not be recycled into deeper parts of the subduction zone through downward convection in the mantle wedge (see Sect. 9.3).

9.2.4 Serpentinized Oceanic Floor and Mantle

Peridotites are exposed at the seafloor (abyssal peridotites), particularly in slow-spreading ridges and near transform or detachment faults (Cannat 1993), but also make up the lithospheric mantle underlying the oceanic crust. When peridotites react with hydrothermal fluids they become serpentinized, during which process olivine and pyroxenes react to form serpentine, brucite, talc, amphibole and magnesite. Due to their high water contents (up to 13 wt% H_2O), serpentinites are the largest potential reservoir of fluids entering subduction zones (Scambelluri et al. 1995;

Ulmer and Trommsdorff 1995). Serpentine is also an important host for B, which fits well in its crystal structure (Pabst et al. 2011).

Depleted MORB-source mantle has low B (<0.1 μg/g) with $\delta^{11}B$ of ca. −7‰ (Marschall et al. 2017). Interaction with hydrothermal fluids, ultimately derived from seawater ($\delta^{11}B$ = +39.6‰; Foster et al. 2010), results in seafloor serpentinites with a large range of heavy B isotopic compositions of ca. +5 to +40‰ (depending on the water-rock ratios) and very high B contents of up to 90 μg/g (Spivack and Edmond 1987; Boschi et al. 2008, 2013; Vils et al. 2009; Harvey et al. 2014b; see Fig. 8.4 in Marschall (2017) for a summary). Very little is known about the composition and volume of serpentinites in mantle directly underneath the oceanic crust, which may form through deep fluid-infiltration, e.g., during fracturing related to bending of the oceanic plate near the trench (e.g., Ranero et al. 2003; Garth and Rietbrock 2014). The amount of water held by serpentinites in this part of the mantle is estimated to be equivalent to 2 wt% H_2O in the upper 4 km of mantle (Hacker 2008), but is highly uncertain and strongly dependent on the ability of fluids to penetrate to such great depths, which has recently been questioned (Korenaga 2017). The compositions of those serpentinites, if present, are also uncertain. It is assumed by some authors that the B isotopic composition of these oceanic mantle serpentinites is similar to that of abyssal serpentinites (e.g., Konrad-Schmolke and Halama 2014), but the extent to which hydrothermal fluids lose or gain B with depth and if their B isotopic composition changes is essentially unknown (Vils et al. 2009). A possible equivalent can be found in ophiolites, but data is very limited. Serpentinites from the Oman ophiolite have $\delta^{11}B$ values from −4.7 to +10.0‰ (Lécuyer et al. 2002), but little geological background of these samples was provided and it remains unclear if serpentinization happened on the seafloor or during or after ophiolite emplacement.

Abyssal and oceanic mantle serpentinites are dominantly composed of the serpentine polymorphs *lizardite* and *chrysotile* (the main B hosts) plus various amouns of talc, brucite, amphibole (Ca–Mg type: tremolite and/or actinolite), relic peridotite minerals, oxides and complex sulfides and sulfide-metal alloys. Upon subduction, lizardite and chrysotile transform to the serpentine polymorph *antigorite* at about 250–300 °C and will lose 1–2% water. At this stage, 60–80% of B will also be lost (Scambelluri et al. 2004; Deschamps et al. 2011; Kodolanyi and Pettke 2011; Vils et al. 2011; Debret et al. 2013), but the limited available data suggest that no B isotope fractionation occurs during this transition (Scambelluri and Tonarini 2012). Fluid release during the breakdown of brucite to olivine (at ca. 450 °C) is limited by the generally small amount of brucite. Despite the sizeable shallow loss, considerable amounts of B carrying their original heavy isotope signature are retained in antigorite serpentinite (5–30 μg/g; Deschamps et al. 2011; Scambelluri and Tonarini 2012) until extensive dehydration during antigorite breakdown occurs at ca. 650–700 °C. Note that this contrasts with the continuous B loss and isotope fractionation that occurs in subducting sediments and AOC (Sects. 9.2.2, 9.2.3). Depending on the thermal profiles of the subduction zones, the breakdown of serpentinite in the subducted slab occurs at 90–200 km depth, i.e., (partially) underneath the volcanic arc front except in the coldest subduction zones (Van Keken et al. 2011).

As was the case for AOC, much of our understanding of B behavior in serpentinites is based on the study of exhumed formerly subducted terranes, although the setting of these (subducted vs. mantle wedge) is not always clear. The Miocene Cerro del Almirez complex, Spain, is the only known occurrence of a subduction-related serpentinite dehydration front exposed in the field (Trommsdorff et al. 1998; Padrón-Navarta et al. 2011). Here, with increasing metamorphic grade an antigorite serpentinite with heavy B isotope signature ($\delta^{11}B$ = +22‰) breaks down to form chlorite harzburgite with a $\delta^{11}B$ = +6 to −3‰, seemingly indicating

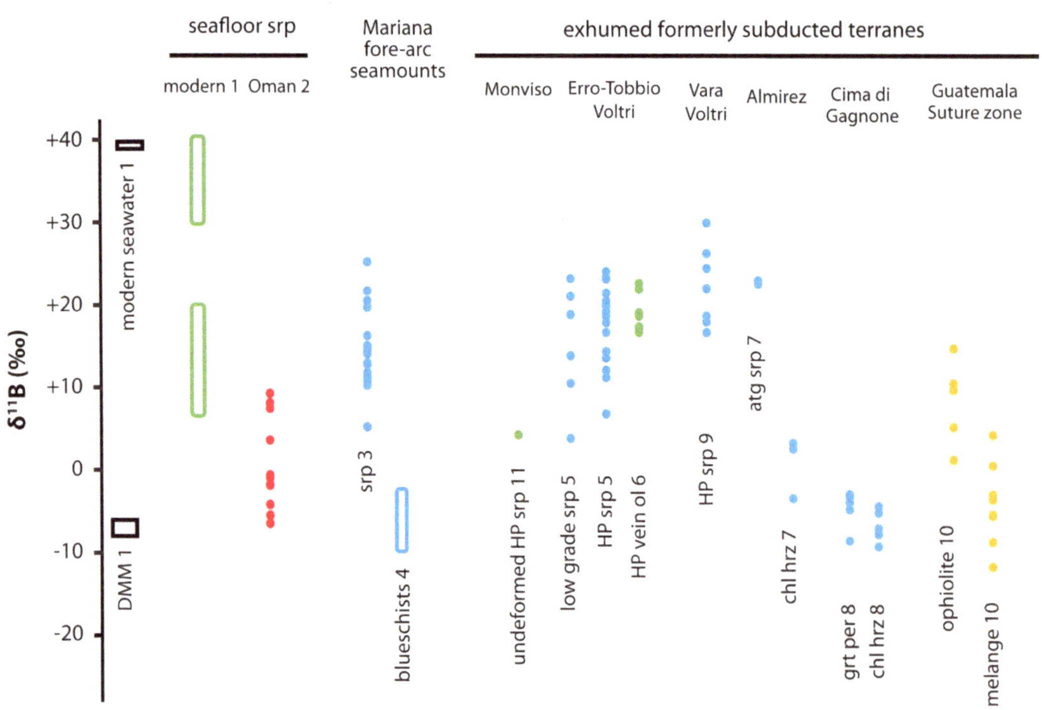

Fig. 9.4 Boron isotopic compositions of subduction-related serpentinites. Color coding according to analytical technique: whole rock TIMS (*blue*), in situ LA-MC-ICP-MS (*yellow*), in situ SIMS (*green*), whole rock ICP-MS (*red*). *Dots* represent individual samples, *open boxes* represent compositional ranges of multiple samples. *DMM* depleted MORB-source mantle; *HP* high pressure; *srp* serpentinite; *atg* antigorite; *chl* chlorite; *ol* olivine; *grt per* garnet peridotite. Data sources: 1 Marschall (2017) and references therein; 2 Lécuyer et al. (2002); 3 Benton et al. (2001); 4 Pabst et al. (2012); 5 Scambelluri and Tonarini (2012); 6 De Hoog et al. (2014); 7 Harvey et al. (2014a); 8 Cannao et al. (2015); 9 Cannao et al. (2016); 10 Martin et al. (2016); 11 Angiboust et al. (2014)

significant B isotopic fractionation during serpentinite dehydration (Harvey et al. 2014a). However, as no accompanying decrease in B concentration was observed, dehydration of the antigorite serpentinite was probably an open-system process assisted by the influx of ^{10}B rich sediment-derived fluids (Harvey et al. 2014a). In the Erro-Tobbio peridotite massif in Voltri (Ligurian Alps, Italy), increasing metamorphic grade (from greenschist facies up to partial serpentinite dehydration in eclogite facies) resulted in little change of $\delta^{11}B$ (ca. +20‰ on average) despite B being lost with increasing metamorphic grade (Scambelluri and Tonarini 2012). However, it remains unclear how much B isotope fractionation occurs during final serpentinite dehydration (antigorite-out isograd). Olivine as a residual phase may contain significant boron (Scambelluri et al. 2004; Tenthorey and Hermann 2004; De Hoog et al. 2014), which is in trigonal coordination (Ingrin et al. 2014) as opposed to B in serpentine (tetrahedral). As ^{11}B favors the trigonal site, equilibrium fluid may have lighter $\delta^{11}B$ than the host serpentinite. In addition, most of the localities described above contain Ti-clinohumite, a hydrous magneisan mineral that may contain significant B (De Hoog et al. 2014). More work is needed to better understand the fate of B during antigorite breakdown and during deep subduction of metamorphic olivine, which also bears relevance to deep B recycling (Sect. 9.6).

9.3 Mantle Wedge Processes and the Subduction Interface

9.3.1 Boron Isotope Composition of Mantle Wedge Serpentinites

Shallow parts of the mantle wedge (10–40 km) are extensively serpentinized by fluids derived from forearc dehydration of the subducting crust (Reynard 2013) and may be transported into deeper parts of the mantle wedge through subduction erosion or mantle wedge corner flow (Tatsumi 1989; Straub and Layne 2002; Hattori and Guillot 2003; Straub and Layne 2003; Deschamps et al. 2010). The extent to which the forearc is serpentinized depends on age of the incoming slabs, as the degree of slab hydration increases with age, whereas temperatures decrease. Hotter slabs like the ones subducted underneath the Cascades, Panama and S. Chile release more fluids at shallow levels (Reynard 2013, and references therein). Significant amounts of B may be fixed in these mantle wedge serpentinites (Deschamps et al. 2010), which will have a heavy B isotope signature as evidenced by the forearc serpentinites drilled in the non-accretionary Mariana arc-basin system (Benton et al. 2001) at Conical Seamount ($\delta^{11}B$ = +5 to +25, average +13‰; Benton et al. 2001) and at South Chamorro ($\delta^{11}B$ = +11 to +18, average +13‰; Savov et al. 2004; Savov, unpublished data). These high $\delta^{11}B$ values are consistent with those predicted by fluid-mineral B isotope fractionation modeling (Spivack et al. 1987; You et al. 1995; Konrad-Schmolke and Halama 2014). Exhumed mantle wedge material supports heavy B isotope signatures. Scambelluri and Tonarini (2012) argued that high $\delta^{11}B$ values from Erro-Tobbio serpentinites (+7 to +23‰, average ca. +15‰) were consistent with an origin as mantle-wedge material. Similarly, antigorite serpentinite from Cerro del Almirez has a heavy B isotope signature ($\delta^{11}B$ = +22‰; Harvey et al. 2014a). Although these have been interpreted to be of oceanic origin (Alt et al. 2012), estimated *P-T* conditions (680–710 °C at 1.6–1.9 GPa; Padrón-Navarta et al. 2011) and metasomatism by sediment-derived fluids accompanying dehydration (Harvey et al. 2014a) are more consistent with a mantle wedge origin. Due to these high $\delta^{11}B$ values reported for forearc serpentinites, the heaviest $\delta^{11}B$ values reported in arc volcanics (up to +20‰; Tonarini et al. 2011) have been attributed to subduction erosion and incorporation of forearc serpentinite mélanges to their sources. Note, however, that some ocean-floor serpentinites as well as sea-floor carbonates may have similarly high $\delta^{11}B$ signatures (Sects. 9.2.2, 9.2.4). In addition, a recent study suggests that not all mantle wedge serpentinites have high $\delta^{11}B$ values. Exhumed mantle wedge serpentinite mélanges from the Guatemala Suture Zone span a wide range from −15.3 to +9.7‰ (Martin et al. 2016), considerably lower than forearc serpentinites presented above. Metasomatism by sediment-derived fluids, which will lower $\delta^{11}B$ values (Cannaò et al. 2015; 2016), was excluded due to the lack of sediments in the area. The authors explain the relatively light $\delta^{11}B$ values by serpentinization of the subduction channel by ^{11}B-depleted AOC-derived fluids at ca. 50–70 km depth, and suggest that mélange serpentinites with negative $\delta^{11}B$ may be quite common. Hence, like the subducting slab, the mantle wedge may become gradually isotopically lighter with depth.

9.3.2 Role of the Subduction Interface in B Recycling

Due to its highly dynamic nature, processes at the subduction interface are necessarily complex (Peacock 2001; Vannucchi et al. 2012). It is thought that subduction channels at the interface play a critical role for upward mass transport and focusing fluid flow (Bebout 2013; Reynard 2013; Angiboust et al. 2014; Hyndman et al. 2015; Harlow et al. 2016). Early thermal models of subduction zones suggested that the mantle wedge directly overlying the slab surface was cold enough for serpentine to be stable (Peacock 1990; Peacock and Hervig 1999; Hattori and Guillot 2003), but more recent models predict a hot slab interface beyond the 'cold nose' (>60–80 km depth) with temperatures in excess of 700 °C, which is too hot for serpentine to persist (Van Keken et al. 2002, 2011; Syracuse et al. 2010). This rules out serpentinite as the reservoir transporting B and other fluid-mobile elements from the front arc to the sub-arc mantle. Instead, an important role for chlorite has been proposed, as chlorite schists are commonly observed in subduction mélanges and recovered from the currently active Mariana Forearc (Savov et al. 2004; Pabst et al. 2011; Fryer 2012) and form when ultramafic material mixed with Al-rich material such as pelitic metasediments or mafic oceanic crust (Bebout and Barton 2002; Spandler et al. 2008; Marschall and Schumacher 2012). However, chlorite typically contains less B than serpentine (Marschall et al. 2006a).

Boron isotopes provide opportunity to unravel subduction interface processes. Several studies provide evidence for complex fluid pathways. Eclogites from oceanic crust making up the Monviso Complex (Western Alps) show evidence of fluid infiltration at high-pressure conditions (Angiboust et al. 2014). Boron isotopic compositions of phengite within metasomatic rinds (0 to +7‰) point to fluids derived from dehydrating serpentine (−4 to 0‰). These fluids probably partially re-equilibrated with mafic and sedimentary lithologies whilst travelling upwards along large subduction-related shear zones in the oceanic crust. In contrast, Marschall et al. (2006b) report very high $\delta^{11}B$ values of +18 to +28‰ for subduction channel fluids in subduction mélange from Syros. Halama et al. (2014) reported B loss from phengite rims in eclogitic micaschists from the Sesia Zone (subducted continental crust) due to retrograde fluid infiltration, but no associated B isotope fractionation, and strongly negative values overall ($\delta^{11}B = -15$ to −9‰). Amphibolite-facies mélange matrix within the Catalina Schist displays $\delta^{11}B$ values of −3.3 to +3.7‰ consistent with infiltration by slab-derived fluids of sedimentary origin (King et al. 2007). Chlorite harzburgites from Cima di Gagnone, Central Alps (Cannaò et al. 2015; Scambelluri et al. 2015), which represent completely dehydrated serpentinites within an ancient subduction mélange, have light $\delta^{11}B$ (−3 to −9‰). These low values were attributed to interaction with sediment-derived fluids during prograde metamorphism. Sediment-derived fluids were also implicated in resetting of B isotope composition of mylonitic serpentinites near Vara in the Voltri Massif, which have $\delta^{11}B$ of +17 to +22‰, whereas undeformed serpentinites have $\delta^{11}B$ of +26 to +30‰ (Cannaò et al. 2016). The variety of processes and B isotope values above demonstrates that B isotopes cannot simply be used to infer the origin of slab interface-related rocks, but instead provide a powerful tool to reconstruct their fluid history.

9.4 Modelling of B Isotope Fractionation During Subduction

The first studies to explain the range of $\delta^{11}B$ in subduction-related young volcanic rocks used binary or ternary mixing models with compositions of AOC ± serpentinite, sediments and mantle as endmember reservoirs with no consideration of (at the time unknown) fluid-solid B isotope fractionation (e.g., Palmer 1991; Ishikawa and Tera 1997; Smith et al. 1997; Ishikawa et al. 2001).

In a ground-breaking study, Peacock and Hervig (1999) reported for the first time that B concentrations and $\delta^{11}B$ of subducted material decreased with increasing metamorphic grade due to B isotope fractionation between dehydration fluids and the slab itself. These authors developed a Rayleigh fractionation model to explain the observed range of $\delta^{11}B$ compositions taking into account temperature-dependent fractionation during B release from 25 to 725 °C ($\Delta^{11}B_{fluid-residue}$ = +20‰ to +5‰). As the slab becomes isotopically lighter upon burial, transfer of B in fluids was required to explain the positive $\delta^{11}B$ values common in nearly all arc volcanic rocks (average $\delta^{11}B = +3.2 \pm 6.2‰$ 1σ; Sect. 9.5.2).

Many studies since have used a similar modeling approach (e.g., Bebout and Nakamura 2003; Rosner et al. 2003; King et al. 2007). These works reinforced the concept that across-arc $\delta^{11}B$ trends can be explained by isotope fractionation during progressive metamorphism of the slab, instead of mixing of various amounts of ^{11}B-enriched and ^{11}B-depleted fluids as in the earlier studies.

A somewhat alternative approach was taken by Rose et al. (2001), who attempted to couple B fractionation to the amount of slab dehydration (water loss). The benefit of this approach is that it also predicts B concentrations in subduction fluids. However, their model used a very low K_D^B $_{rock/fluid}$ of 0.015 and a constant value for $\Delta^{11}B_{fluid-solid}$ of +5‰, which resulted in 99% of B being lost after only 7% dehydration and in unrealistically large isotopic shifts of the residual slab, down to $\delta^{11}B = -15‰$, for only minimal amounts of fluid loss. The same model was adopted by Bouvier et al. (2008, 2010) with the addition of serpentinite as a fluid and B source. The authors assumed the same solid-fluid fractionation for all lithologies (+5‰) and temperatures.

Marschall et al. (2006b) developed a stepwise mass balance-based dehydration model which, like the model by Rose et al. (2001), predicted water loss in combination with B depletions. Prograde metamorphism can be modeled as small steps of batch dehydration, an approach adopted in several B isotope studies aimed at identifying fluid sources in subduction zones (Tonarini et al. 2007, 2011; Scambelluri and Tonarini 2012). The model was refined further by Jones et al. (2014), who incorporated the thermal regime of the subduction zone in their dehydration scenarios in order to calculate more realistic water and B fluxes from different segments of the subducting slab.

All of the above models use a bulk fluid-solid fractionation factor, which ignored the influence of changes in mineralogy of the subducting slab and therefore have limited physical basis for modeling fluid release rates. However, significant progress has been made in recent years in the thermodynamic modeling of mineral reactions in the subducting slab and thermal models of subduction zones. This allows the calculation of the release of fluids or melts from the slab in response to its mineralogical makeup at the dehydration sites (Rüpke et al. 2004; Marschall et al. 2007; Hacker 2008; Van Keken et al. 2011; Wada et al. 2012). By integrating fluid-mineral partitioning and isotope fractionation data with thermodynamic considerations, forward models of fluid-mobile element release from the slab and their isotopic signatures can be derived. For B in AOC this was first done by Marschall et al. (2007), who also pointed out the importance of K_2O content and associated stability of phengite in the slab on the B isotope composition of slab-derived fluids. This forward modeling approach was recently expanded to include additional subduction zone lithologies (sediments, serpentinites, metasomatised mantle wedge) using 2D thermal models instead of single *P-T* paths (Konrad-Schmolke and Halama 2014; Konrad-Schmolke et al. 2016). A B isotope fractionation model for the Kamchatka arc thus developed successfully reproduced the measured across-arc B isotopic variations (Fig. 9.5).

Although extremely capable, these models are still reliant on uncertain *P-T* profiles of subduction zones (e.g., Penniston-Dorland et al. 2015) and largely ignore fluid transport mechanisms

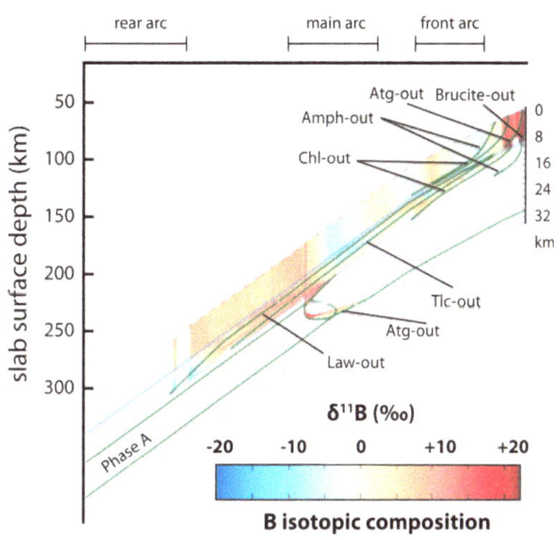

Fig. 9.5 Model-predicted B isotopic composition of various units in the Kamchatka subduction system, including the mantle wedge overlying the subducting slab (after Konrad-Schmolke et al. 2016). The *thick blue line* represents the subduction interface. *Green lines* indicate mineral stability fields. Mineral abbreviations: *Atg* antigorite; *Amph* amphibole; *Chl* chlorite; *Tlc* talc; *Law* lawsonite

from slab to magma source region. Also, the subducted slabs are not homogeneous in volume, composition and physical properties as they may contain fracture zones, oceanic plateaus, etc. which need to be included in the modeling (Wada et al. 2012). As an example, elevated B/Zr in arc volcanoes sampling subducted fracture zones (B isotope data do not yet exist) was interpreted to reflect higher water-rock ratios and involvement of excess hydrous minerals using numerical modeling of the subduction zone thermal structures (Manea et al. 2014). In addition, temporal variations in material entering the subduction zone (Jones et al. 2014), physical erosion of serpentinized forearc serpentinite/blueschist or mélanges (Tonarini et al. 2011) and mantle wedge transport processes such as mélange diapirs (Marschall and Schumacher 2012) also need to be considered. In spite of these complexities, the integrated modeling approach developed by Marschall et al. (2007) and Konrad-Schmolke and Halama (2014) sets a benchmark for future research in fine-tuning our understanding of the B cycle and ultimately the cycle of other volatiles in subduction zones.

9.5 Boron Isotope Signature of Volcanic Arcs

Boron isotopic compositions of volcanic arc rocks have been studied for nearly three decades (Ishikawa and Nakamura 1994; Leeman 1996). This long history is partly due to relatively high B abundances in volcanic arc rocks (usually several to tens of µg/g; Ryan and Langmuir 1993; Ishikawa and Nakamura 1994; see Fig. 9.6) compared to ocean floor basalts (<3 µg/g; Marschall 2017) as well as rapid analytical developments in plasma spectroscopy and mass spectrometry since the 1990s (Nakamura et al. 1992; Tonarini et al. 1997) and was accelerated by the production of volcanic arc rock standards (Tonarini et al. 1997; Gonfiantini et al. 2003; Rosner and Meixner 2004; see also Chaps. 1 and 2 of this volume). The use of boron and its isotopes for the study of volcanic arcs was further driven by the discovery of ^{10}Be as a unique cosmogenic tracer of sediment subduction (Tera et al. 1986) and the subsequent find that B/Be was a similarly powerful tracer of crustal recycling (Morris et al. 1990; Ryan et al. 1995).

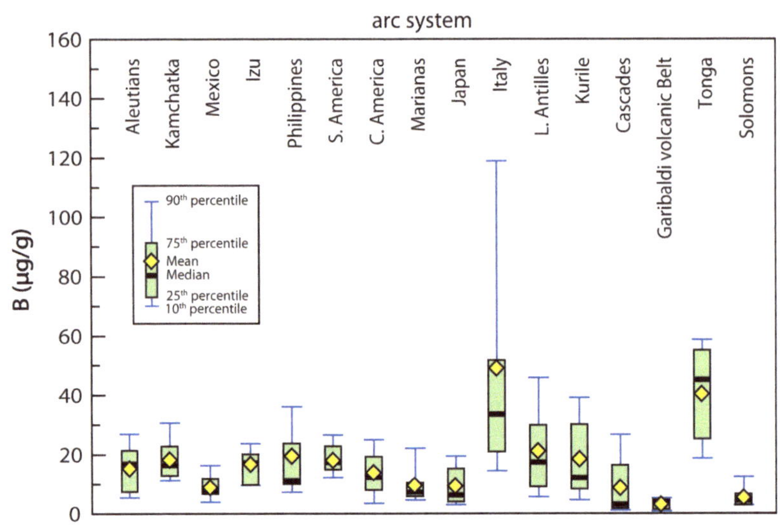

Fig. 9.6 Compilation of B concentrations in arc volcanic rock (Green and Savov, unpublished), showing that most arcs have high B abundances compared to primitive mantle (0.19 µg/g; Marschall et al. 2017) and ocean floor basalts (0.4–2.5 µg/g; Marschall et al. 2017). It also reveals that the hot and mafic arcs overlying relatively young slabs (S. Cascades and the Garibaldi Volcanic Field of the N. Cascades; Solomons) have the lowest B concentrations, most probably due to extensive shallow devolatilization (Savov et al. 2007; Hyndman et al. 2015)

Early studies of the boron isotopic composition of arc lavas interpreted the data as binary mixtures between a subduction fluid (with high $\delta^{11}B$) and mantle boron (Ishikawa and Nakamura 1994; Leeman 1996; Tonarini et al. 2001), with volcanic front magma sources being most strongly influenced by B-rich fluids. The discovery that the B isotopic composition of slab-derived fluids becomes progressively lighter with depth (Peacock and Hervig 1999; Rose et al. 2001; Rosner et al. 2003), consistent with high $\delta^{11}B$ fluid outfluxes in the shallow forearc of the Izu-Bonin-Mariana (IBM) arc system (Mottl et al. 2004; Savov et al. 2004), provide an alternative explanation for cross-arc $\delta^{11}B$ trends. However, the increasingly light slab was difficult to reconcile with the heavy B isotope ratios observed in many arcs if these fluids were originating strictly from the deeply subducted sediments and altered oceanic crust alone. This led to consideration of alternative fluid sources with heavy B isotope ratios, in particular serpentinites (Tonarini et al. 2007, 2011; Bouvier et al. 2010; Scambelluri and Tonarini 2012). The role of serpentinites is still somewhat controversial as key subduction zone tracers are not abundant in the serpentinites and it is therefore difficult to unequivocally link $\delta^{11}B$ and B variations with serpentinite dehydration. In addition, experimental studies of serpentinite dehydration behavior under sub-arc conditions are limited (Tenthorey and Hermann 2004; Padron-Navarta et al. 2010).

In the following sections we aim to further elucidate these issues based on available volcanic arc B isotope data from the literature, a database of which has been compiled for the purpose of this chapter.

9.5.1 Global Boron Isotopic Database from Volcanic Arcs

Our database contains about 130 arc volcanic rocks with high quality $\delta^{11}B$ data that have also been analyzed for their B concentrations, major and trace elements, and commonly used radiogenic isotopes, such as Sr and Nd. Our review is based on nearly 75 of these rocks with MgO >4 wt%, which include datasets from Kamchatka (Ishikawa et al. 2001), Kuriles (Ishikawa and Tera 1997), Japan (Moriguti et al. 2004), Izu-Bonin (Ishikawa and Nakamura 1994; Straub and Layne 2002), the Tonga-Kermadec arc (Leeman et al. 2017) and the Marianas (Ishikawa and Tera 1999). The rest of the $\delta^{11}B$ studies are from three continental arcs: the Cascades (Leeman et al. 2004; note melt inclusions from this arc reported by Walowski et al. 2015), the Andes (Rosner et al. 2003; note melt inclusions from this arc reported in Jones et al. 2014), and Central America (El Salvador; Tonarini et al. 2007), and from an Atlantic intra-oceanic volcanic arcs, the South Sandwich arc (Tonarini et al. 2011). We excluded the Aeolian Island arc from our discussion as these samples show evidence of considerable crustal assimilation (Tonarini et al. 2001) and no reliable subduction parameter data are available (Syracuse and Abers 2006). The B isotope ratios reported and reviewed here are independent from the variations in the loss on ignition (LOI), which argues against modification of B isotope signals by secondary, post eruption modifications/alterations.

Before we attempt to make any generalizations about controls on B isotope variations in volcanic arc settings, we want to point out some surprising features of the published $\delta^{11}B$ database (and accompanying trace element and isotope data). First, although B abundance data are available for most volcanic arcs (see Fig. 9.6), the amount of mafic arc volcanic rocks (MgO >4 wt%) with high quality $\delta^{11}B$ data is limited. Although the B isotopes are largely unaffected by shallow fractionation processes, the same cannot be said for trace elements, and it is therefore that we base our review primarily on primitive arc rocks.

Second, there is a bias in the global arc volcanic B database towards intra-oceanic volcanic arcs, which are studied preferentially to avoid influence from continental crust. Just in the last few years the Izu-Bonin region of the W. Pacific has been drilled during four different expeditions, recovering hundreds of meters of core. Accordingly, many elemental and isotope "Subduction

Factory" studies, including the systematics of fluid-mobile elements like boron, have been conducted at these tectonic settings and roughly 75% from the reported global arc $\delta^{11}B$ data are derived from the Western Pacific region.

Finally, it is worth noting that still no high quality B isotope data is available for long arc segments (including Alaska, Aleutians, Bismarck-New Britain, Colombia, Panama, Guatemala, Mexico, Central America, Southern Chile, New Zealand, Vanuatu, Aegean and Bismarck), and that as of yet no published data exists for slab melts (i.e., adakites sensu stricto) and boninites.

9.5.2 Variations of Boron Isotope Signatures with Subducting Slab Parameters

Boron is an incompatible element and, therefore, shows elevated concentrations in all arc volcanic rocks with respect to the upper mantle (<0.1 µg/g; Fig. 9.6). In addition, B abundances in arc volcanic rocks show evidence of extensive fluid-rock interactions in magma sources, as originally proposed in the pioneering paper by Ryan and Langmuir (1993). The range in B in all arc rocks is 1.3–36.6 µg/g, with an average B concentration of ca. 12 µg/g, the highest values coming from the Aeolian and Tonga arcs (Fig. 9.6). This compares to ca. 0.4–2.5 µg/g for MORB (Marschall et al. 2017). The mafic volcanic arcs that are formed above relatively young, hot subducted slabs (Cascades, Solomon arc) have the lowest B concentrations, probably due to early (forearc) devolatilization reactions and associated B loss (Savov et al. 2007; Hyndman et al. 2015).

Boron isotope variations in arc rocks show a dramatically large range of ca. 25‰ ($\delta^{11}B$ = −9‰ to +16‰) with an average $\delta^{11}B$ value of +4.1 ± 6.2‰ for rocks with MgO >4 wt%, and only slightly lower $\delta^{11}B$ value of +3.2 ± 6.2‰ when including evolved arc rocks (andesites to rhyolites; MgO <4 wt%). Compared to the $\delta^{11}B$ value of −7.4 ± 2.6‰ for ocean floor basalts (Marschall et al. 2017), arc volcanics have consistently more B and more variable (and on average higher) $\delta^{11}B$ values (Fig. 9.7).

Boron isotopic compositions of arc volcanics show significant correlations with subduction zone geometry (listed in Table 9.1). Complex/multi-variable relationships between convergent margin geometry and B isotope geochemistry exist (Fig. 9.8). Figure 9.8a shows a negative correlation between $\delta^{11}B$ and the trench–arc front distance (Leeman et al. 2004; Marschall et al. 2007; Pabst et al. 2012; this chapter). This suggests that arc volcanic fronts that are furthest away from the trench experience a smaller impact from the subducting slab fluids as these fluids tend to be released primarily in close proximity to the trench (Savov et al. 2007;

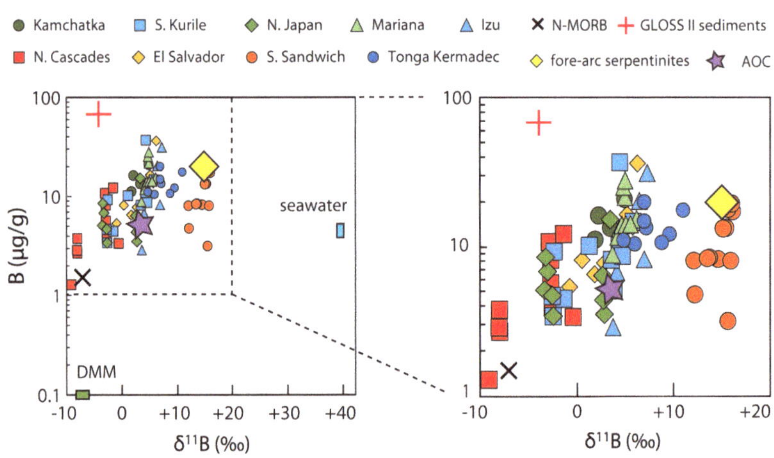

Fig. 9.7 Variation of $\delta^{11}B$ values of arc volcanic rocks with B concentrations, only including rocks with MgO >4 wt%

Table 9.1 Subduction parameters of arcs in boron isotope database

Arc system	Approximate location	Depth to slab (km)	Arc—trench distance (km)	Slab dip (°)	Plate velocity (mm/y)	Age (Ma)	Φ/100 km[a]	Sediment pile (subducted) (km)	Upper plate
N. Cascades	122°W 42°N	95	325	21	35	9	1.2	2.0 (0.4)	Continental
El Salvador	90°W 14°N	92–132	180–185	55–63	67–71	18	10–12	0.5 (0.3)	Continental
Kamchatka	159°E 54°N	82–99	210–240	51–53	74–78	89–104	52–63	0.6 (0.3)	Continental
S. Kurile	152°E 47°N	72–111	175–545	32–47	72–79	107–115	45–62	0.6 (0.3)	Continental
Honshu	141°E 39°N	96–141	285–365	30–32	80–82	132	54–57	0.6 (0.3)	Continental
Izu	140°E 33°N	107–148	195–310	38–53	40–48	130–138	41–67	0.4 (0.2)	Oceanic
Marianas	146°E 18°N	117–135	215–240	58–61	19–21	149–152	25–27	0.4 (0.2)	Oceanic
Tonga-Kermadec	176°W 25°S	123–171	166–196	52–56	66–166	106–109	57–143	0.4 (0.2)	Oceanic
S. Sandwich	26 W 58°S	79–147	120–155	59–68	2–7	31–75	1–5	0.4 (0.2)	Oceanic

[a]Φ thermal parameter (vertical descent rate x age of slab; Kirby et al. 1991). Subduction parameters from Syracuse and Abers (2006) and Syracuse et al. (2010)

Hyndman et al. 2015). Accordingly, due to the lack of significant slab B input, the subarc mantle signature will be dominant and associated arc volcanics tend to have MORB-like $\delta^{11}B$ values (e.g., Cascades; Leeman et al. 2004). This is confirmed by trace-element variations, as hot slabs that dehydrate early, like the one beneath the Cascades, tend to have the smallest B/Ce ratios (0.02–0.2) compared to other arcs. This is in contrast to arcs that have volcanic fronts proximal to the subduction trench and thus experience the effects of voluminous early fluid loss to a much lesser extent (e.g., S. Sandwich; Tonarini et al. 2011).

Additional evidence for slab geometry control of the $\delta^{11}B$ signatures of volcanic arc magmas is apparent from $\delta^{11}B$ variations with slab dip angle. Arcs overlying flat slabs (such as the Kurile and Cascades arcs) show consistently more negative $\delta^{11}B$ values than arcs overlying steep slabs (such as the S. Sandwich arc; Fig. 9.8b). Surprisingly, there is no clear correlation between B isotopic variation of the arc rocks and the thermal parameter Φ of the subduction system, where Φ = subducting plate velocity × its age × the sine of its dip angle (Kirby et al. 1991), with high values indicating cold slabs. For example, both the Cascades and South Sandwich arcs have "hot" slabs (Table 9.1), yet these show extremely contrasting $\delta^{11}B$ signatures (Fig. 9.8c). Although the slab dips can also be affected by differences in the convergence rate and/or the presence of topographic anomalies like aseismic ridges (e.g., near Ecuador), the described relationships may indicate that correlations of $\delta^{11}B$ with slab dip and trench distance are perhaps not due solely to subduction zone thermal structure. We also note there is no correlation between depth of the slab top beneath volcanoes and their $\delta^{11}B$ signatures (Fig. 9.8d).

We hypothesize that flat slabs reside much longer and thus dehydrate more efficiently under the forearcs, stripping the boron from the slabs and extensively fractionate the slab residues. The slab residues that continue with the subduction process toward the subarc mantle are so low in boron concentrations that,

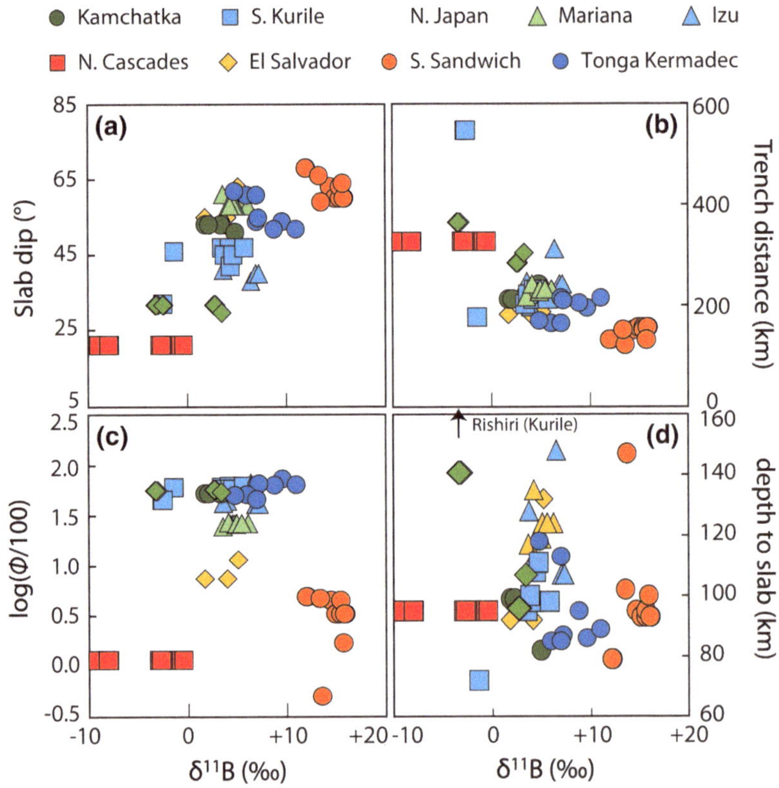

Fig. 9.8 Variation of $\delta^{11}B$ values of arc volcanic rocks with key convergent margin parameters: **b** the distance of the volcanic arc front relative to the subduction trench; **a** the steepness of the subducting slabs as reflected by their dip angles, **c** the thermal parameter Φ of the subduction systems and **d** depth of slab top beneath volcano. See Table 9.1 for convergent margin parameters

although highly fractionated, they play an insignificant role in governing the arc front magmatic signatures; hence the similarity between the samples from the classical hot arcs (Cascades) and MORB. Thus, $\delta^{11}B$ signatures in arc volcanic rocks appear to reflect a unique combination of thermal structure of subduction zones, the depths of occurrence and the geometry of the metamorphic dehydration zones near the subducting slabs. As the slab geometry controls the thermal structure of a subduction zone, which in turn determines where dehydration reactions take place and where fluids and thus B is released into the overlying mantle wedge, this ultimately impacts the gradual decrease in B abundances and $\delta^{11}B$ values with increasing slab depth (see Sect. 9.2). Covariation of B with key major and trace elements and radiogenic isotopes should therefore take into consideration the impact of the geometry of each volcanic arc or relevant arc segment.

9.5.3 Variations of Boron Isotope Signatures with Geochemical Proxies

Here we explore co-variations of boron isotopes with other geochemical parameters to highlight the potential use of $\delta^{11}B$ values in support of key tectonic/subduction zone geometry insights that are usually derived from seismology, numerical modeling or trace elements and radiogenic isotope studies. The majority of the samples in our database are basalts and basaltic-andesites, and predictably the MgO or the SiO_2 content of these rocks does not correlate with $\delta^{11}B$ (Fig. 9.9a, b), as B isotope ratios are insensitive to crystal fractionation due to the low B concentrations in phenocryst assemblages (e.g., Jones et al. 2014). In fact, the only correlations between the global $\delta^{11}B$ in arcs and their major element concentrations are the negative correlations of $\delta^{11}B$ with Na_2O (Fig. 9.9c) and, to a lesser extent, TiO_2 (Fig. 9.9d) and P_2O_5

(not shown). These particular major elements are enriched in hot (and often more alkaline) subduction zones with young slabs (Cascades, C. America) and those tend to have lighter $\delta^{11}B$ values at similar SiO_2 or MgO contents.

Boron isotopes appear to be positively correlated with Sc and negatively correlated with widely used subduction zone tracers like Sr, Zr and Nb (see discussion of B/Nb ratios below). Consequently in arcs with hot slabs, which have low Sc whereas Sr, Zr and Nb tend to be elevated (abundance patterns typical for subducted basaltic crust), $\delta^{11}B$ is low (Fig. 9.10). It seems that the low $\delta^{11}B$ in these arcs is related to the involvement of extensively dehydrated and nearly B-free (and increasingly ^{11}B-depleted) sediment and/or modified altered oceanic crust (AOC) additions (see Sect. 9.2), leaving the ambient mantle to dominate B isotope signatures. This is also supported by the low B/Ce ratios in these same samples (see below). Although it is tempting to then explain the high $\delta^{11}B$ of arc rocks by these sourcing slabs that have undergone less dehydration and therefore being richer in B as well as having higher $\delta^{11}B$ signatures, trace element and isotope systematics do not support such a scenario. The most commonly used sediment tracers in subduction zones, the radiogenic isotope ratios of Sr, Nd, Hf and Pb, do not co-vary with $\delta^{11}B$ in the global (mafic) arc dataset (Fig. 9.11a). An exception is $^{87}Sr/^{86}Sr$, which appears to be somewhat elevated in the arc rocks with high $\delta^{11}B$, although there is considerable scatter (Fig. 9.11b). Indeed, mixing between MORB and GLOSS II sedimentary inputs clearly fails to explain the $\delta^{11}B$, Sr and Nd isotope variations in our dataset (Fig. 9.11). Instead, arc volcanics appear to define an array between MORB and forearc-modified serpentinites from the Marianas (ODP Legs 125 and 195; Benton et al. 2004; Savov et al. 2005, 2007; Fig. 9.11 and 9.12).

Thus, it appears that the $\delta^{11}B$ values of arc rocks are controlled by the involvement of non-sedimentary, non-crustal but yet ultimately slab-derived material. The only such available

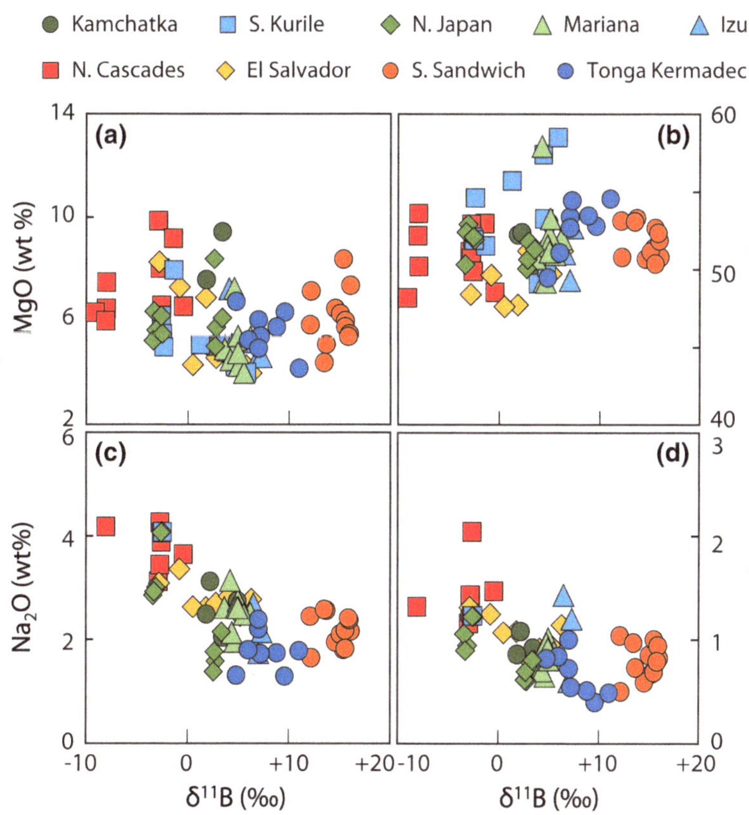

Fig. 9.9 **a–d** $\delta^{11}B$ ratios in arc volcanic rocks with MgO >4 wt% contrasted against major element concentrations; **a** $\delta^{11}B$ versus MgO; **b** $\delta^{11}B$ versus SiO_2; **c** $\delta^{11}B$ versus Na_2O; **d** $\delta^{11}B$ versus MgO. Note that while Izu, and South Sandwich arcs are similar in their MgO and SiO_2, these tend to have dramatically different Na_2O and TiO_2 contents, perhaps highlighting the importance of the degree of slab melting involved. $\delta^{11}B$ are usually lower in hot arcs (e.g., Cascades), where it is indistinguishable from MORB. Symbols are the same as for Fig. 9.7. Only rocks with MgO >4 wt% are shown here

source underneath the volcanic arc front are serpentinites, either as subducted forearc-modified serpentinites (Hattori and Guillot 2003; Savov et al. 2005; 2007; Pabst et al. 2011), serpentinite mélanges overlaying the subducting slabs (Snyder et al. 2004; Spandler et al. 2008; Marschall and Schumacher 2012; Martin et al. 2016), or hydrated oceanic mantle forming the lower part of the subducting slab (Hacker 2008; Spandler and Pirard 2013; Konrad-Schmolke et al. 2016). These have been shown to dramatically impact arc magma $\delta^{11}B$ compositions and yet show little other effects to the radiogenic isotopes and the major and most trace elements (Tonarini et al. 2007, 2011; Marschall and Schumacher 2012; Scambelluri and Tonarini 2012).

Additional evidence for sources of B can be derived from systematic $\delta^{11}B$ variations with trace-element ratios widely used in subduction zone studies. In Fig. 9.12 we show the $\delta^{11}B$ values of arc rocks against Nb/B and Zr/B ratios, respectively. These ratios are particularly useful as they contrast an element with highly fluid mobile behavior (B) against elements with a highly fluid-immobile nature (Zr, Nb; Noll et al. 1996; Leeman et al. 2004; Tonarini et al. 2011). Collectively, these ratios have been repeatedly shown to vary across arcs, decreasing behind the volcanic front, with gradually diminishing volumes of available slab fluids (and thus B) with depth (Ishikawa and Tera 1997; Tonarini et al. 2001; Leeman et al. 2004).

The same relationship is shown amongst volcanic arcs worldwide (Fig. 9.12): the $\delta^{11}B$ values of arc rocks show negative correlations with Nb/B and Zr/B ratios. The rocks with low $\delta^{11}B$ values (and high Nb/B) are those from hot and shallow arcs, consistent with a decreased role of subduction fluids in a mantle-dominated signal, as for these subduction systems (i.e., Cascades), boron is effectively lost in the forearc region (Mottl et al. 2004; Savov et al. 2004).

Samples representing high $\delta^{11}B$ arcs are rich in boron (have low Nb/B ratios) and these are hard to explain if we only consider sediments and AOC as possible boron sources. The reason is

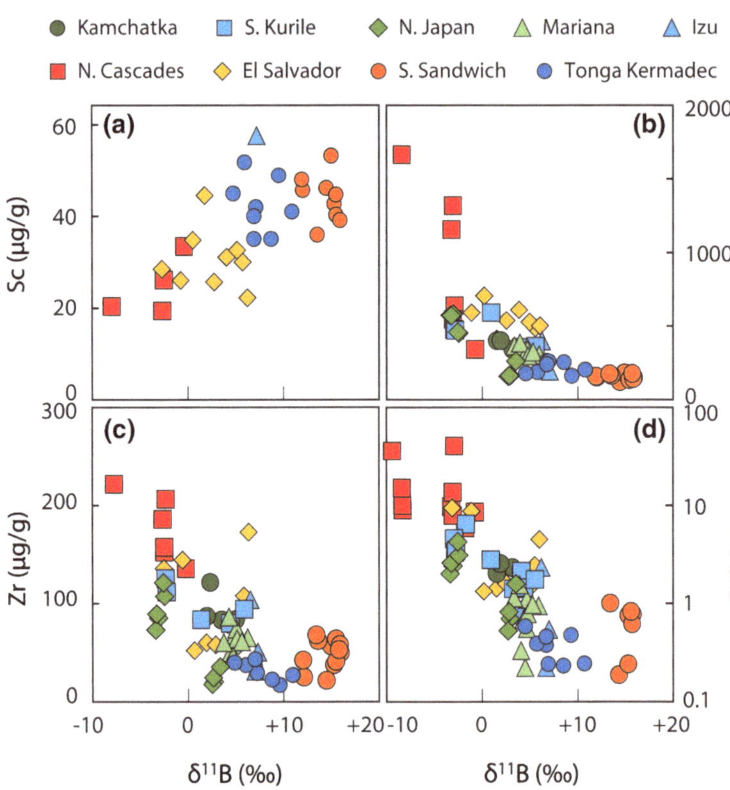

Fig. 9.10 **a–d** $\delta^{11}B$ values in arc volcanic rocks with MgO >4 wt% contrasted against key trace-element concentrations widely used in subduction zone studies: **a** $\delta^{11}B$ versus Sc; **b** $\delta^{11}B$ versus Sr; **c** $\delta^{11}B$ versus Zr; **d** $\delta^{11}B$ versus Nb. Note that the scale for Nb is logarithmic. Symbols the same as for Fig. 9.7

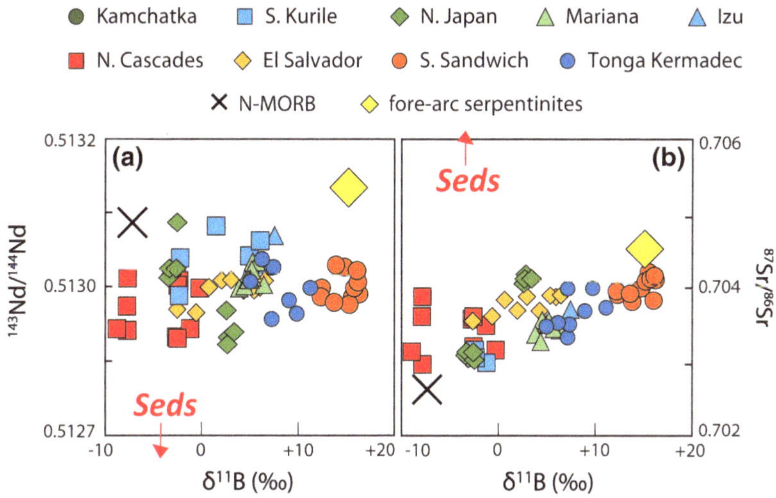

Fig. 9.11 δ^{11}B ratios in arc volcanic rocks with MgO >4 wt% contrasted against radiogenic isotope ratios of ^{143}Nd/^{144}Nd (**a**) and ^{87}Sr/^{86}Sr (**b**) widely used in arc geochemistry. Symbols are the same as on Fig. 9.7. Also shown are values for N-MORB (*black cross* x; Smith et al. 1995; Marschall et al. 2017), average GLOSS II subducted sediments (red cross +; Plank 2014) and forearc-serpentinized subarc mantle peridotites (*yellow diamond*; Benton et al. 2004; Savov et al. 2004; 2005; 2007)

Fig. 9.12 a–d: δ^{11}B variations in arc volcanic rocks with MgO >4 wt% with key trace element ratios widely used in arc geochemistry. **a** δ^{11}B versus La/Sm ratios. **b** δ^{11}B versus B/Ce ratios. **c** δ^{11}B versus Nb/B ratios. **d** δ^{11}B versus Zr/B ratios. Note that all ratio scales are logarithmic. See Fig. 9.11 for data sources

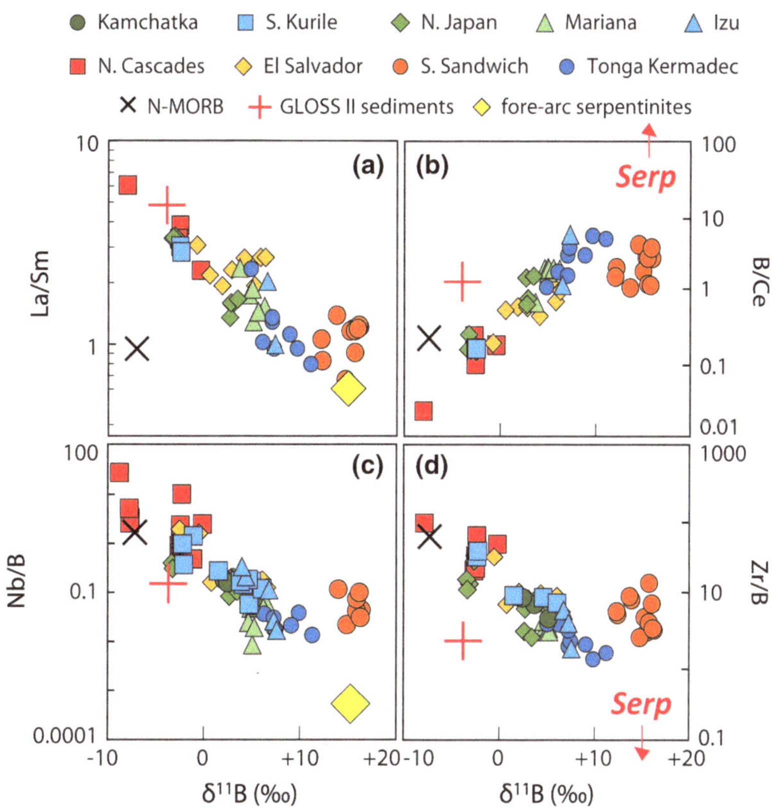

that excess boron (high B/Ce and low Nb/B or Zr/B ratios) can only be slab-sourced and yet the arcs with the most extreme B enrichments show the least subduction imprint in most of their other geochemical characteristics, i.e., are rather similar or closest to MORB in terms of Sc, Sr and Zr contents and La/Sm (Fig. 9.12a). These signatures are in fact similar to those of serpentinized peridotites from the Izu-Bonin-Mariana forearc, which have La/Sm ~ 0.6 and B/Ce ratios of ~ 1500 or more (Fig. 9.12a, b). We therefore repeat our earlier hypothesis that forearc modified serpentinites are the most likely source of these B-rich fluids in volcanic arcs, as these have not only high $\delta^{11}B$, but also very low Nb/B ($\sim 10^{-3}$) and Zr/B ($\sim 10^{-2}$) ratios. We note that Zr/B variations have been proposed to trace the influence of serpentinites from the subducted oceanic fracture zones (Manea et al. 2014), whereas Jones et al. (2014) attributed elevated $\delta^{11}B$ values during the Miocene in rocks from the South Central Chile Arc to subduction of a strongly serpentinized Juan Fernández Ridge. An important role for serpentinites in subduction zone fluid recycling was also inferred by Scambelluri and Tonarini (2012) based on high $\delta^{11}B$ of exhumed formerly subducted serpentinites and the lack of other lithologies that could explain the high $\delta^{11}B$ of many arc volcanoes.

Boron isotope systematics may give additional clues as to the origin of these serpentinites. For example, forearc-modified serpentine-dominated mélanges have been shown to be enriched in isotopically heavy boron (Benton et al. 2001; Savov et al. 2004, 2005, 2007; Tonarini et al. 2007, 2011), whereas deeper parts of the mantle wedge beyond the forearcs tend to have much lower $\delta^{11}B$ (Cannaò et al. 2015; Martin et al. 2016). On the other hand, abyssal serpentinites, and by inference hydrated oceanic mantle, may also have high $\delta^{11}B$ (Spivack and Edmond 1987; Boschi et al. 2008, 2013; Vils et al. 2009; Harvey et al. 2014b). We conclude that more work combining B isotopes with other isotopic and trace element tracers is needed to further constrain the contribution of various serpentinite sources.

9.6 Deep B Recycling

Although >90% of subducted B is recycled back to the surface in subduction zones (Savov et al., 2007), a small amount of B may be recycled into the deep mantle. As mentioned in Sect. 9.2.2, sediments may contain phengite and/or tourmaline, both of which are stable at high T and may survive in the slab to depths beyond the volcanic arc. The eclogitized mafic part of the slab may still contain several μg/g B at depths beyond the arc, again mostly hosted by phengite, but its B isotopic composition is poorly constrained. Relatively high amounts of B with heavy B isotopic signatures have been reported from (partially) dehydrated serpentinites. For example, high B contents (10–20 μg/g) were reported in metamorphic olivine and Ti-clinohumite in chlorite harzburgite from Cerro del Almirez as well as in dehydration veins from Erro-Tobbio (Scambelluri et al. 2004; De Hoog et al. 2014). Although serpentinite dehydration processes are complex and may involve interaction with external fluids (Harvey et al. 2014a), metamorphic olivine from Erro-Tobbio was shown to have high $\delta^{11}B$ (+17 to +23‰) similar to that of the serpentinite protolith (De Hoog et al. 2014), which strongly suggests that olivine may be a significant sink for isotopically heavy B derived from serpentine (antigorite) breakdown.

Furthermore, in the oceanic mantle part of very cold subduction slabs (550–660 °C at 160 km depth) a new high-pressure hydrous phase, Phase A, may form upon the breakdown of serpentine (Rüpke et al. 2004; Komabayashi et al. 2005; Hacker 2008). The behavior of fluid-mobile elements such as B during this antigorite breakdown reaction is yet unknown. However, only few subducting slabs are cold enough for Phase A to form (northeast Japan,

Aleutian and possibly Kamchatka and Kuril; Komabayashi et al. 2005).

Ultimately these subducted materials will participate in mantle convection and may contribute to magma sources of oceanic island basalt (OIB) volcanism. The presence of deeply recycled materials in OIB has long been suspected based on radiogenic isotopes (Sr, Nd, Pb, Hf, Os; Hofmann 2014), and has also been suggested based on halogen signatures of MORB and OIB (Kendrick et al. 2017), but indisputable proof has remained elusive. Boron isotopes may provide such evidence but the data has proven difficult to obtain as B isotopes of OIB magmas are very sensitive to assimilation of hydrothermally altered material that can be trapped *en route* to the surface (Fig. 8.2; Marschall 2017). In addition, the very low B contents of the erupted materials have hampered attempts to detect the small variations in B isotopic composition of different mantle domains (see Marschall 2017 for more discussion). Additional work is needed to detect recycled B in OIB and put further constraints on the amount of B recycled beyond volcanic arcs into the deep mantle.

9.7 Outstanding Issues and Future Work Needed

Although great advances have been made over the last decades in our understanding of how subduction zones work, what the sources of fluids are, how elements are transferred from slab to magma source, and how subduction geometries affect these, many important outstanding questions still remain. Here we summarize several topics that we suggest need addressing in future research related to the use of boron and boron isotopes as subduction zone tracers.

1. **What is the significance of serpentinites in the mantle beneath oceanic crust**? Geochemical is no clear correlation between B dehydration of serpentinite in the oceanic mantle may be an important source of fluids, and is perhaps even required to explain melting of subducted crust (Spandler and Pirard 2013; Konrad-Schmolke and Halama 2014; Freymuth et al. 2016; Konrad-Schmolke et al. 2016). However, the ability of fluids to penetrate that deep into the oceanic crust/mantle has recently been disputed and geophysical evidence can be interpreted by porosity (Korenaga 2017). In addition, the slab-bending mechanism of hydration of the lower slab crust and mantle (Ranero et al. 2003) has unresolved issues, such as the buoyancy of slabs if they are extensively hydrated (they become difficult to subduct) and importantly, the lack of evidence for volumetrically important seawater penetration even in highly fractured and deformed oceanic crust (e.g., in IODP Site 1256, the deepest oceanic crust sites show <5% alteration and MORB-like Sr and Pb isotope compositions; Höfig et al. 2014 and references therein). Proposed future deep IODP drilling efforts aims to resolve some of those issues.
2. **How can we discriminate between different serpentinite origins**? As discussed earlier, it is possible that two types of B outfluxes sourced from serpentinites co-exist in subduction zones: (1) one from the mechanical downdrag of forearc serpentinites and their diapiric rise toward the arcs (Savov et al. 2005; Marschall and Schumacher 2012) and (2) one from the late dehydration of the serpentinized mantle section of the slabs themselves (Konrad-Schmolke et al. 2011; Konrad-Schmolke and Halama 2014). Although data for widely used radiogenic isotope systems is limited for serpentinite, it has been shown that these can vary significantly in abyssal serpentinites due to interaction with seawater (Frisby et al. 2016). Boron isotopic compositions of various serpentinite sources ('cold-nose' forearc, mantle wedge, oceanic serpentinites) are still poorly constrained, relying on small datasets, and clearly overlap (Sect. 9.2.4). Generally, a correlation of high $\delta^{11}B$ with radiogenic $^{87}Sr/^{86}Sr$ ratios is observed (Savov et al. 2007; Vils et al. 2009; Harvey et al. 2014b). Finding and evaluating serpentinites with

isotopically very heavy boron and highly radiogenic strontium may help us better address the true range of $\delta^{11}B$ in abyssal settings. The latter is very important if we are to use the differences in B concentrations and the $\delta^{11}B$ values in serpentinites from abyssal versus forearc (subduction-related) settings to discriminate between types of serpentinites contributing to boron outfluxes across arcs. We expect that a combination of radiogenic (Sr, Nd, Hf) and B isotope studies of serpentinites has the potential to pinpoint some of the ultimate sources of (isotopically heavy) boron at various depths under the arcs. It follows that additional $\delta^{11}B$ data is much needed for evaluation and quantification of the role of serpentinites in arc magmas.

3. **Role of subduction erosion of the 'cold nose' fore arc**. Related to point (2) above, the extremely high $\delta^{11}B$ values of the S. Sandwich Arc have been explained by subduction erosion of the forearc (Tonarini et al. 2011). As such high $\delta^{11}B$ values are not observed elsewhere, this could be used as an argument that subduction erosion of the forearc is volumetrically insignificant in other arcs. We note that the S. Sandwich Arc has the lowest plate velocity of all convergent margins on Earth. Further exploring the relationship between boron isotopes and subduction parameters may give new insights into fore-arc recycling in subduction zones.
4. **What happens to B-enriched olivine during further subduction**? Metamorphic olivine formed during dehydration of serpentine may contain significant boron (Sect. 9.2.4), which limits the amount of B that could be transferred to arc magma sources. Since B in olivine is in trigonal coordination (Ingrin et al. 2014) favored by ^{11}B, as opposed to B in serpentine (tetrahedral), fluids derived from serpentinite may be isotopically lighter than the residue, which would have also have ramifications for the modeling of serpentinite-derived fluids in subduction zones. Finally, as B in the mantle is too low to permit a significant subduction input, it remains unclear what happens to B in metamorphic olivine during deeper subduction. More work is needed to better understand the fate of B in metamorphic minerals (olivine, pyroxenes, amphiboles, phlogopites, etc.).
5. **Slab melting**. Serpentinites and the fluids released from their dehydration are now often implicated in slab melting, which requires a high water flux (Van Keken et al. 2011; Spandler and Pirard 2013; Freymuth et al. 2016). However, no boron or B isotope studies exist to explore that link. We note a remarkable lack of B isotope data for slab melts ('true' adakites).
6. **Chlorite**. The subduction interface is too hot for serpentinite to exist and therefore chlorite, which has a higher T stability, may play an important role in transporting water and potentially B in deeper parts of the subduction zone (Spandler et al. 2008; Van Keken et al. 2011; Marschall and Schumacher 2012). However, chlorite is thought to be poor in B (Marschall et al. 2006a) compared to amphibole and phengite, but limited data is available from ultramafic rocks. It is also unclear how B isotopes are affected by chloritization.
7. **Carbonate**. Carbonate-rich sediments can have high $\delta^{11}B$ (Sect. 9.2.2; Marschall 2017), but it is unclear what happens to this B during deep subduction. Mantle-derived carbonatites have been shown to have $\delta^{11}B$ values of −8 to +5‰ that correlate with radiogenic Sr isotopes pointing to subduction recycling of crustal carbon (Hulett et al. 2016). On the other hand, with the exception of the class IIb (blue) diamonds, diamonds have <0.1 µg/g boron (Gaillou et al. 2012), suggesting that carbon in diamonds is not derived from subducted carbonate. This is in agreement with recent experimental work, which indicates that significant amounts of carbonate can break down beneath the arc to form CO_2-rich fluids with trace-element signatures traditionally ascribed to AOC (Skora et al. 2015). On the other hand, as carbonates typically have isotopically heavy B (Simon et al. 2006), this may affect B isotope signature of arcs with a significant carbonate sediment

sequences (e.g., Central America, New Zealand; Fig. 9.3).

8. **Boron and boron isotopic composition of subducting sediments**. Although the trace element and radiogenic isotope signatures of the subducted sediment pile are well known (Plank 2014), their boron abundances are not well studied and the reported GLOSS II value for boron is indirectly derived based on knowledge of the element Li and the general trench sediment B/Li ratios of ~2. Boron isotope signatures of subducted sediments from many key arcs are completely missing and yet may contribute significantly to the outstanding issues of B and volatile sources and recycling across subduction zones.

Acknowledgements We thank the editors for their patience and persistence. Bill Leeman and Ralf Halama are thanked for their thoughtful reviews and comments. Stephen Turner and Ralf Halama contributed to the development of the volcanic arc B and B isotope and trace element database. IPS would like to thank DTM-Carnegie and particularly Fouad Tera for inspirational and career-influencing discussions on light element geochemistry. IPS was introduced to the world of B and B isotopes by Jeff Ryan and Sonia Tonarini, who are thanked for their support and mentoring over the years. IPS and CJH are grateful to Samuele Agostini and the staff of IGG at CNR-Pisa for their continuous support with often difficult B isotope measurements and unselfish advice, and for always being open to the ever growing B isotope science community.

References

Alt JC, Garrido CJ, Shanks WC, Turchyn A, Padron-Navarta JA, Sanchez-Vizcaino VL, Pugnaire MTG, Marchesi C (2012) Recycling of water, carbon, and sulfur during subduction of serpentinites: a stable isotope study of Cerro del Almirez. Spain Earth Planet Sci Lett 327:50–60

Angiboust S, Pettke T, De Hoog JCM, Caron B, Oncken O (2014) Channelized fluid flow and eclogite-facies metasomatism along the subduction shear zone. J Petrol 55(5):883–916

Bebout GE (1991) Field-based evidence for devolatilization in subduction zones—implications for arc magmatism. Science 251(4992):413–416

Bebout GE (2013) Metasomatism in subduction zones of subducted oceanic slabs, mantle wedges, and the slab-mantle interface. The Role of Fluids in Terrestrial and Extraterrestrial Processes, Metasomatism and the Chemical Transformation of Rock, pp 289–349

Bebout GE, Barton MD (2002) Tectonic and metasomatic mixing in a high-T, subduction-zone melange—insights into the geochemical evolution of the slab-mantle interface. Chem Geol 187(1–2):79–106

Bebout GE, Nakamura E (2003) Record in metamorphic tourmalines of subduction-zone devolatilization and boron cycling. Geology 31(5):407–410

Bebout GE, Ryan JG, Leeman WP (1993) B-Be systematics in subduction-related metamorphic rocks—characterization of the subducted component. Geochim Cosmochim Acta 57(10):2227–2237

Bebout GE, Ryan JG, Leeman WP, Bebout AE (1999) Fractionation of trace elements by subduction-zone metamorphism—effect of convergent-margin thermal evolution. Earth Planet Sci Lett 171(1):63–81

Bebout GE, Bebout AE, Graham CM (2007) Cycling of B, Li, and LILE (K, Cs, Rb, Ba, Sr) into subduction zones: SIMS evidence from micas in high-P/T metasedimentary rocks. Chem Geol 239(3–4):284–304

Benton LD, Ryan JG, Tera F (2001) Boron isotope systematics of slab fluids as inferred from a serpentine seamount Mariana forearc. Earth Planet Sci Lett 187 (3–4):273–282

Benton LD, Ryan JG, Savov IP (2004) Lithium abundance and isotope systematics of forearc serpentinites, Conical Seamount, Mariana forearc: insights into the mechanics of slab-mantle exchange during subduction. Geochem Geophys Geosyst 5:Q08J12

Boschi C, Dini A, Fruh-Green GL, Kelley DS (2008) Isotopic and element exchange during serpentinization and metasomatism at the Atlantis Massif (MAR 30 degrees N): insights from B and Sr isotope data. Geochim Cosmochim Acta 72(7):1801–1823

Boschi C, Bonatti E, Ligi M, Brunelli D, Cipriani A, Dallai L, D'Orazio M, Fruh-Green GL, Tonarini S, Barnes JD, Bedini RM (2013) Serpentinization of mantle peridotites along an uplifted lithospheric section, Mid Atlantic Ridge at 11 degrees N. Lithos 178:3–23

Bouvier AS, Métrich N, Deloule E (2008) Slab-derived fluids in the magma sources of St. Vincent (Lesser Antilles Arc): volatile and light element imprints. J Petrol 49(8):1427–1448

Bouvier AS, Metrich N, Deloule E (2010) Light elements, volatiles, and stable isotopes in basaltic melt inclusions from Grenada, Lesser Antilles: inferences for magma genesis. Geochem Geophys Geosyst 11: Q09004

Cagnioncle AM, Parmentier EM, Elkins-Tanton LT (2007) Effect of solid flow above a subducting slab on water distribution and melting at convergent plate boundaries. J Geophys Res Solid Earth 112:B9 Artn B09402

Cann JR, McCaig AM, Yardley BWD (2015) Rapid generation of reaction permeability in the roots of black smoker systems, Troodos ophiolite. Cyprus Geofluids 15(1–2):179–192

Cannaò E, Agostini S, Scambelluri M, Tonarini S, Godard M (2015) B, Sr and Pb isotope geochemistry of high-pressure Alpine metaperidotites monitors fluid-mediated element recycling during serpentinite dehydration in subduction melange (Cima di Gagnone, Swiss Central Alps). Geochim Cosmochim Acta 163:80–100

Cannaò E, Scambelluri M, Agostini S, Tonarini S, Godard M (2016) Linking serpentinite geochemistry with tectonic evolution at the subduction plate-interface: the voltri massif case study (Ligurian Western Alps, Italy). Geochim Cosmochim Acta 190:115–133

Cannat M (1993) Emplacement of mantle rocks in the sea-floor at midocean ridges. J Geophys Res Solid Earth 98(B3):4163–4172

De Hoog JCM, Hattori K, Jung H (2014) Titanium- and water-rich metamorphic olivine in high-pressure serpentinites from the Voltri Massif (Ligurian Alps, Italy): evidence for deep subduction of high-field strength and fluid-mobile elements. Contrib Mineral Petrol 167(3):Art 990

Debret B, Andreani M, Godard M, Nicollet C, Schwartz S, Lafay R (2013) Trace element behavior during serpentinization/de-serpentinization of an eclogitized oceanic lithosphere: a LA-ICPMS study of the Lanzo ultramafic massif (Western Alps). Chem Geol 357:117–133

Deschamps F, Guillot S, Godard M, Chauvel C, Andreani M, Hattori K (2010) In situ characterization of serpentinites from forearc mantle wedges: timing of serpentinization and behavior of fluid-mobile elements in subduction zones. Chem Geol 269(3–4):262–277

Deschamps F, Guillot S, Godard M, Andreani M, Hattori K (2011) Serpentinites act as sponges for fluid-mobile elements in abyssal and subduction zone environments. Terra Nova 23(3):171–178

Dilek Y, Furnes H (2011) Ophiolite genesis and global tectonics: geochemical and tectonic fingerprinting of ancient oceanic lithosphere. Geol Soc Am Bull 123(3–4):387–411

Domanik KJ, Holloway JR (1996) The stability and composition of phengitic muscovite and associated phases from 5.5 to 11 Gpa: implications for deeply subducted sediments. Geochim Cosmochim Acta 60 (21):4133–4150

Domanik KJ, Hervig RL, Peacock SM (1993) Beryllium and boron in subduction zone minerals—an ion microprobe study. Geochim Cosmochim Acta 57 (21–22):4997–5010

England PC, Katz RF (2010) Melting above the anhydrous solidus controls the location of volcanic arcs. Nature 467(7316):700–703

Foster GL, Pogge von Strandmann PAE, Rae JWB (2010) Boron and magnesium isotopic composition of seawater. Geochem Geophys Geosyst 11:Q08015

Foustoukos DI, Savov IP, Janecky DR (2008) Chemical and isotopic constraints on water/rock interactions at the Lost City hydrothermal field, 30 degrees N Mid-Atlantic Ridge. Geochim Cosmochim Acta 72 (22):5457–5474

Freymuth H, Ivko B, Gill JB, Tamura Y, Elliott T (2016) Thorium isotope evidence for melting of the mafic oceanic crust beneath the Izu arc. Geochim Cosmochim Acta 186:49–70

Frisby C, Bizimis M, Mallick S (2016) Hf-Nd isotope decoupling in bulk abyssal peridotites due to serpentinization. Chem Geol 440:60–72

Fryer P (2012) Serpentinite mud volcanism: observations, processes, and implications. Annu Rev Mar Sci 4:345–373

Gaillou E, Post JE, Rost D, Butler JE (2012) Boron in natural type IIb blue diamonds: chemical and spectroscopic measurements. Am Mineral 97(1):1–18

Garth T, Rietbrock A (2014) Order of magnitude increase in subducted H_2O due to hydrated normal faults within the Wadati-Benioff zone. Geology 42(3):207–210

Gonfiantini R, Tonarini S, Groning M, Adorni-Braccesi A, Al-Ammar AS, Astner M, Bachler S, Barnes RM, Bassett RL, Cocherie A, Deyhle A, Dini A, Ferrara G, Gaillardet J, Grimm J, Guerrot C, Krahenbuhl U, Layne G, Lemarchand D, Meixner A, Northington DJ, Pennisi M, Reitznerova E, Rodushkin I, Sugiura N, Surberg R, Tonn S, Wiedenbeck M, Wunderli S, Xiao YK, Zack T (2003) Intercomparison of boron isotope and concentration measurements. Part II: evaluation of results. Geostandard Newslett 27 (1):41–57

Grove TL, Till CB, Lev E, Chatterjee N, Medard E (2009) Kinematic variables and water transport control the formation and location of arc volcanoes. Nature 459 (7247):694–697

Hacker BR (2008) H_2O subduction beyond arcs. Geochem Geophys Geosyst 9:Q03001

Halama R, Konrad-Schmolke M, Sudo M, Marschall HR, Wiedenbeck M (2014) Effects of fluid–rock interaction on 40Ar/39Ar geochronology in high-pressure rocks (Sesia-Lanzo Zone, Western Alps). Geochim Cosmochim Acta 126:475–494

Harlow GE, Flores KE, Marschall HR (2016) Fluid-mediated mass transfer from a paleosubduction channel to its mantle wedge: evidence from jadeitite and related rocks from the Guatemala suture zone. Lithos 258–259:15–36

Harvey J, Garrido CJ, Savov I, Agostini S, Padron-Navarta JA, Marchesi C, Sanchez-Vizcaino VL, Gomez-Pugnaire MT (2014a) B-11-rich fluids in subduction zones: the role of antigorite dehydration in subducting slabs and boron isotope heterogeneity in the mantle. Chem Geol 376:20–30

Harvey J, Savov IP, Agostini S, Cliff RA, Walshaw R (2014b) Si-metasomatism in serpentinized peridotite: the effects of talc-alteration on strontium and boron isotopes in abyssal serpentinites from Hole 1268a, ODP Leg 209. Geochim Cosmochim Acta 126:30–48

Hattori KH, Guillot S (2003) Volcanic fronts form as a consequence of serpentinite dehydration in the forearc mantle wedge. Geology 31(6):525–528

Hermann J, Spandler CJ (2008) Sediment melts at sub-arc depths: an experimental study. J Petrol 49(4):717–740
Hervig RL, Moore GM, Williams LB, Peacock SM, Holloway JR, Roggensack K (2002) Isotopic and elemental partitioning of boron between hydrous fluid and silicate melt. Am Mineral 87(5–6):769–774
Höfig TW, Geldmacher J, Hoernle K, Hauff F, Duggen S, Garbe-Schonberg D (2014) From the lavas to the gabbros: 1.25 km of geochemical characterization of upper oceanic crust at ODP/IODP site 1256, eastern equatorial Pacific. Lithos 210:289–312
Hofmann AW (2014) 3.3 Sampling mantle heterogeneity through oceanic basalts: isotopes and trace elements. In: Holland HD, Turekian KK (eds) Treatise on Geochemistry, 2nd edn. pp 67–101
Hulett SRW, Simonetti A, Rasbury ET, Hemming NG (2016) Recycling of subducted crustal components into carbonatite melts revealed by boron isotopes. Nature Geosci 9(12):904–908
Hupers A, Kasemann SA, Kopf AJ, Meixner A, Toki T, Shinjo R, Wheat CG, You CF (2016) Fluid flow and water-rock interaction across the active Nankai Trough subduction zone forearc revealed by boron isotope geochemistry. Geochim Cosmochim Acta 193:100–118
Hyndman RD, McCrory PA, Wech A, Kao H, Ague J (2015) Cascadia subducting plate fluids channelled to fore-arc mantle corner: ETS and silica deposition. J Geophys Res Solid Earth 120(6):4344–4358
Ingrin J, Kovacs I, Deloule E, Balan E, Blanchard M, Kohn SC, Hermann J (2014) Identification of hydrogen defects linked to boron substitution in synthetic forsterite and natural oliyine. Am Mineral 99 (10):2138–2141
Ishikawa T, Nakamura E (1993) Boron isotope systematics of marine sediments. Earth Planet Sci Lett 117 (3–4):567–580
Ishikawa T, Nakamura E (1994) Origin of the slab component in arc lavas from across-arc variation of B and Pb isotopes. Nature 370(6486):205–208
Ishikawa T, Tera F (1997) Source, composition and distribution of the fluid in the Kurile mantle wedge: constraints from across-arc variations of B/Nb and B isotopes. Earth Planet Sci Lett 152(1–4):123–138
Ishikawa T, Tera F (1999) Two isotopically distinct fluid components involved in the Mariana arc: evidence from Nb/B ratios and B, Sr, Nd, and Pb isotope systematics. Geology 27(1):83–86
Ishikawa T, Tera F, Nakazawa T (2001) Boron isotope and trace element systematics of the three volcanic zones in the Kamchatka arc. Geochim Cosmochim Acta 65(24):4523–4537
Johnson MC, Plank T (1999) Dehydration and melting experiments constrain the fate of subducted sediments. Geochem Geophys Geosyst, Artn 1007
Jones RE (2014) Subduction zone processes and continental crust formation in the southern Central Andes: insights from geochemistry and geochronology. PhD Thesis. The University of Edinburgh, Edinburgh, School of GeoSciences, 523p
Jones RE, De Hoog JCM, Kirstein LA, Kasemann SA, Hinton R, Elliott T, Litvak VD, Eimf (2014) Temporal variations in the influence of the subducting slab on Central Andean arc magmas: evidence from boron isotope systematics. Earth Planet Sci Lett 408:390–401
Kawakami T (2001a) Boron depletion controlled by the breakdown of tourmaline in the migmatite zone of the Aoyama area, Ryoke metamorphic belt, southwestern Japan. Can Mineral 39:1529–1546
Kawakami T (2001b) Tourmaline breakdown in the migmatite zone of the Ryoke metamorphic belt. SW Japan J Metamorph Geol 19(1):61–75
Kendrick MA, Hemond C, Kamenetsky VS, Danyushevsky L, Devey CW, Rodemann T, Jackson MG, Perfit MR (2017) Seawater cycled throughout Earth/'s mantle in partially serpentinized lithosphere. Nature Geosci 10(3):222–228
King RL, Bebout GE, Grove M, Moriguti T, Nakamura E (2007) Boron and lead isotope signatures of subduction-zone melange formation: hybridization and fractionation along the slab-mantle interface beneath volcanic arcs. Chem Geol 239(3–4):305–322
Kirby SH, Durham WB, Stern LA (1991) Mantle phase-changes and deep-earthquake faulting in subducting lithosphere. Science 252(5003):216–225
Kodolanyi J, Pettke T (2011) Loss of trace elements from serpentinites during fluid-assisted transformation of chrysotile to antigorite—an example from Guatemala. Chem Geol 284(3–4):351–362
Komabayashi T, Hirose K, Funakoshi K, Takafuji N (2005) Stability of phase A in antigorite (serpentine) composition determined by in situ X-ray pressure observations. Phys Earth Planet Inter 151(3–4):276–289
Konrad-Schmolke M, Halama R (2014) Combined thermodynamic-geochemical modeling in metamorphic geology: boron as tracer of fluid-rock interaction. Lithos 208:393–414
Konrad-Schmolke M, Zack T, O'Brien PJ, Barth M (2011) Fluid migration above a subducted slab—thermodynamic and trace element modelling of fluid-rock interaction in partially overprinted eclogite-facies rocks (Sesia Zone, Western Alps). Earth Planet Sci Lett 311(3–4):287–298
Konrad-Schmolke M, Halama R, Manea VC (2016) Slab mantle dehydrates beneath Kamchatka—yet recycles water into the deep mantle. Geochem Geophys Geosyst 17(8):2987–3007
Korenaga J (2017) On the extent of mantle hydration caused by plate bending. Earth Planet Sci Lett 457:1–9
Kowalski PM, Wunder B, Jahn S (2013) Ab initio prediction of equilibrium boron isotope fractionation between minerals and aqueous fluids at high P and T. Geochim Cosmochim Acta 101:285–301
Lafay R, Deschamps F, Schwartz S, Guillot S, Godard M, Debret B, Nicollet C (2013) High-pressure serpentinites, a trap-and-release system controlled by metamorphic conditions: example from the piedmont zone of the western Alps. Chem Geol 343:38–54

Lécuyer C, Grandjean P, Reynard B, Albarède F, Telouk P (2002) $^{11}B/^{10}B$ analysis of geological materials by ICP–MS Plasma 54: application to the boron fractionation between brachiopod calcite and seawater. Chem Geol 186(1–2):45–55

Leeman WP (1996) Boron and other fluid-mobile elements in volcanic arc lavas: implications for subduction processes. In: Subduction Top to Bottom, pp 269–276

Leeman WP, Tonarini S, Chan LH, Borg LE (2004) Boron and lithium isotopic variations in a hot subduction zone—the southern Washington cascades. Chem Geol 212(1–2):101–124

Leeman WP, Tonarini S, Turner S (2017) Boron isotope variations in Tonga-Kermadec-New Zealand arc lavas: implications for origin of subduction components and mantle influences. Geochem Geophy Geosyst 18:1126–1162. doi:10.1002/2016GC006523

MacGregor J, Grew ES, De Hoog JCM, Harley SL, Kowalski PM, Yates MG, Carson CJ (2013) Boron isotopic composition of tourmaline, prismatine, and grandidierite from granulite facies paragneisses in the Larsemann Hills, Prydz Bay, East Antarctica: evidence for a non-marine evaporite source. Geochim Cosmochim Acta 123:261–283

Manea VC, Leeman WP, Gerya T, Manea M, Zhu G (2014) Subduction of fracture zones controls mantle melting and geochemical signature above slabs. Nat Commun 5:5095

Marschall HR (2017) Boron isotopes in the ocean floor realm and the mantle. In: Marschall HR, Foster GL (eds) Boron Isotopes—The Fifth Element, Advances in Isotope Geochemistry, vol 7, Springer, Heidelberg, 191–217

Marschall HR, Schumacher JC (2012) Arc magmas sourced from melange diapirs in subduction zones. Nat Geosci 5(12):862–867

Marschall HR, Altherr R, Ludwig T, Kalt A, Gmeling K, Kasztovszky Z (2006a) Partitioning and budget of Li, Be and B in high-pressure metamorphic rocks. Geochim Cosmochim Acta 70(18):4750–4769

Marschall HR, Ludwig T, Altherr R, Kalt A, Tonarini S (2006b) Syros metasomatic tourmaline: evidence for very high-delta B-11 fluids in subduction zones. J Petrol 47(10):1915–1942

Marschall HR, Altherr R, Rüpke L (2007) Squeezing out the slab—modelling the release of Li, Be and B during progressive high-pressure metamorphism. Chem Geol 239(3–4):323–335

Marschall HR, Altherr R, Kalt A, Ludwig T (2008) Detrital, metamorphic and metasomatic tourmaline in high-pressure metasediments from Syros (Greece): intra-grain boron isotope patterns determined by secondary-ion mass spectrometry. Contrib Mineral Petrol 155(6):703–717

Marschall HR, Altherr R, Gmeling K, Kasztovszky Z (2009a) Lithium, boron and chlorine as tracers for metasomatism in high-pressure metamorphic rocks: a case study from Syros (Greece). Mineral Petrol 95(3–4):291–302

Marschall HR, Korsakov AV, Luvizotto GL, Nasdala L, Ludwig T (2009b) On the occurrence and boron isotopic composition of tourmaline in (ultra) high-pressure metamorphic rocks. J Geol Soc London 166:811–823

Marschall HR, Wanless VD, Shimizu N, Pogge von Strandmann PAE, Elliott T, Monteleone B 2017 The boron and lithium isotopic composition of mid-ocean ridge basalts and the mantle. Geochim Cosmochim Acta 207:102–138

Martin C, Flores KE, Harlow GE (2016) Boron isotopic discrimination for subduction-related serpentinites. Geology 44(11):899–902

Moran AE, Sisson VB, Leeman WP (1992) Boron depletion during progressive metamorphism—implications for subduction processes. Earth Planet Sci Lett 111(2–4):331–349

Moriguti T, Shibata T, Nakamura E (2004) Lithium, boron and lead isotope and trace element systematics of quaternary basaltic volcanic rocks in northeastern Japan: mineralogical controls on slab-derived fluid composition. Chem Geol 212(1–2):81–100

Morris JD, Leeman WP, Tera F (1990) The subducted component in Island-Arc Lavas—constraints from Be isotopes and B-Be systematics. Nature 344(6261):31–36

Mottl MJ (1992) Pore waters from serpentinite seamounts in the Mariana and Izu-Bonin forearcs, Leg 125: evidence for volatiles from the subducting slab. In: Fryer P, Pierce J, Stokking LB et al (eds) Proc. ODP, Sci, Results, pp 373–385

Mottl MJ, Wheat CG, Fryer P, Gharib J, Martin JB (2004) Chemistry of springs across the Mariana forearc shows progressive devolatilization of the subducting plate. Geochim Cosmochim Acta 68(23):4915–4933

Nakamura E, Ishikawa T, Birck JL, Allegre CJ (1992) Precise boron isotopic analysis of natural rock samples using a boron mannitol complex. Chem Geol 94 (3):193–204

Nakano T, Nakamura E (2001) Boron isotope geochemistry of metasedimentary rocks and tourmalines in a subduction zone metamorphic suite. Phys Earth Planet Inter 127(1–4):233–252

Noll PD, Newsom HE, Leeman WP, Ryan JG (1996) The role of hydrothermal fluids in the production of subduction zone magmas: evidence from siderophile and chalcophile trace elements and boron. Geochim Cosmochim Acta 60(4):587–611

Pabst S, Zack T, Savov IP, Ludwig T, Rost D, Vicenzi EP (2011) Evidence for boron incorporation into the serpentine crystal structure. Am Mineral 96(7):1112–1119

Pabst S, Zack T, Savoy IP, Ludwig T, Rost D, Tonarini S, Vicenzi EP (2012) The fate of subducted oceanic slabs in the shallow mantle: insights from boron isotopes and light element composition of metasomatized blueschists from the Mariana forearc. Lithos 132:162–179

Padron-Navarta JA, Hermann J, Garrido CJ, Sanchez-Vizcaino VL, Gomez-Pugnaire MT (2010) An experimental investigation of antigorite

dehydration in natural silica-enriched serpentinite. Contrib Mineral Petrol 159(1):25–42
Padrón-Navarta JA, López Sánchez-Vizcaíno V, Garrido CJ, Gómez-Pugnaire MT (2011) Metamorphic record of high-pressure dehydration of antigorite serpentinite to chlorite harzburgite in a subduction setting (Cerro del Almirez, Nevado-Filábride Complex, Southern Spain). J Petrol 52(10):2047–2078
Palmer MR (1991) Boron-isotope systematics of Halmahera Arc (Indonesia) lavas—evidence for involvement of the subducted slab. Geology 19(3):215–217
Palmer MR, Spivack AJ, Edmond JM (1987) Temperature and pH controls over isotopic fractionation during adsorption of boron on marine clay. Geochim Cosmochim Acta 51(9):2319–2323
Pawley A (2003) Chlorite stability in mantle peridotite: the reaction clinochlore + enstatite = forsterite + pyrope + H_2O. Contrib Mineral Petrol 144(4):449–456
Peacock SM (1990) Fluid processes in subduction zones. Science 248(4953):329–337
Peacock SM (1993) The importance of blueschist-eclogite dehydration reactions in subducting oceanic crust. Geol Soc Am Bull 105(5):684–694
Peacock SM (2001) Are the lower planes of double seismic zones caused by serpentine dehydration in subducting oceanic mantle? Geology 29(4):299–302
Peacock SM, Hervig RL (1999) Boron isotopic composition of subduction-zone metamorphic rocks. Chem Geol 160(4):281–290
Penniston-Dorland SC, Kohn MJ, Manning CE (2015) The global range of subduction zone thermal structures from exhumed blueschists and eclogites: rocks are hotter than models. Earth Planet Sci Lett 428:243–254
Plank T (2014) 4.17 The chemical composition of subducting sediments. In: Holland H, Turekian KK (eds.) Treatise on Geochemistry, 2nd Edn. pp 607–629
Plank T, Langmuir CH (1998) The chemical composition of subducting sediment and its consequences for the crust and mantle. Chem Geol 145(3–4):325–394
Poli S, Schmidt MW (2002) Petrology of subducted slabs. Annu Rev Earth Pl Sci 30:207–235
Portnyagin M, Manea VC (2008) Mantle temperature control on composition of arc magmas along the Central Kamchatka depression. Geology 36 (7):519–522
Ranero CR, Morgan JP, McIntosh K, Reichert C (2003) Bending-related faulting and mantle serpentinization at the Middle America trench. Nature 425 (6956):367–373
Reynard B (2013) Serpentine in active subduction zones. Lithos 178:171–185
Rose EF, Shimizu N, Layne GD, Grove TL (2001) Melt production beneath Mt. Shasta from boron data in primitive melt inclusions. Science 293(5528):281–283
Rosner M, Meixner A (2004) Boron isotopic composition and concentration of ten geological reference materials. Geostand Geoanal Res 28(3):431–441
Rosner M, Erzinger J, Franz G, Trumbull RB (2003) Slab-derived boron isotope signatures in arc volcanic rocks from the Central Andes and evidence for boron isotope fractionation during progressive slab dehydration. Geochem Geophy Geosyst Artn 9005
Rüpke LH, Morgan JP, Hort M, Connolly JAD (2004) Serpentine and the subduction zone water cycle. Earth Planet Sci Lett 223(1–2):17–34
Ryan JG, Langmuir CH (1993) The Systematics of boron abundances in young volcanic-rocks. Geochim Cosmochim Acta 57(7):1489–1498
Ryan JG, Morris J, Tera F, Leeman WP, Tsvetkov A (1995) Cross-arc geochemical variations in the kurile arc as a function of slab depth. Science 270 (5236):625–627
Saffer DM, Kopf AJ (2016) Boron desorption and fractionation in subduction zone fore arcs: implications for the sources and transport of deep fluids. Geochem Geophys Geosyst 17(12):4992–5008
Sanchez-Valle C, Reynard B, Daniel I, Lecuyer C, Martinez I, Chervin JC (2005) Boron isotopic fractionation between minerals and fluids: new insights from in situ high pressure-high temperature vibrational spectroscopic data. Geochim Cosmochim Acta 69 (17):4301–4313
Savov IP, Tonarini S, Ryan J, Mottl MJ (2004) Boron isotope geochemistry of serpentinites and porefluids from Leg 195, Site 1200, S.Chamorro Seamount, Mariana forearc region (abstract). International Geological Congress, Florence, Italy
Savov IP, Ryan JG, D'Antonio M, Kelley K, Mattie P (2005) Geochemistry of serpentinized peridotites from the Mariana Forearc Conical Seamount, ODP Leg 125: implications for the elemental recycling at subduction zones. Geochem Geophys Geosyst 6 Q04J15
Savov IP, Ryan JG, D'Antonio M, Fryer P (2007) Shallow slab fluid release across and along the Mariana arc-basin system: insights from geochemistry of serpentinized peridotites from the Mariana fore arc. J Geophys Res Solid Earth 112(B9):B09205
Scambelluri M, Tonarini S (2012) Boron isotope evidence for shallow fluid transfer across subduction zones by serpentinized mantle. Geology 40(10):907–910
Scambelluri M, Muntener O, Hermann J, Piccardo GB, Trommsdorff V (1995) Subduction of water into the mantle—history of an Alpine peridotite. Geology 23 (5):459–462
Scambelluri M, Muntener O, Ottolini L, Pettke TT, Vannucci R (2004) The fate of B, Cl and Li in the subducted oceanic mantle and in the antigorite breakdown fluids. Earth Planet Sci Lett 222(1):217–234
Scambelluri M, Pettke T, Cannaò E (2015) Fluid-related inclusions in Alpine high-pressure peridotite reveal trace element recycling during subduction-zone dehydration of serpentinized mantle (Cima di Gagnone, Swiss Alps). Earth Planet Sci Lett 429:45–59
Schmidt MW (1996) Experimental constraints on recycling of potassium from subducted oceanic crust. Science 272(5270):1927–1930
Schmidt MW, Poli S (1998) Experimentally based water budgets for dehydrating slabs and consequences for

arc magma generation. Earth Planet Sci Lett 163(1–4):361–379
Schmidt MW, Vielzeuf D, Auzanneau E (2004) Melting and dissolution of subducting crust at high pressures: the key role of white mica. Earth Planet Sci Lett 228 (1–2):65–84
Simon L, Lecuyer C, Marechal C, Coltice N (2006) Modelling the geochemical cycle of boron: implications for the long-term delta B-11 evolution of seawater and oceanic crust. Chem Geol 225(1–2):61–76
Skora S, Blundy JD, Brooker RA, Green ECR, De Hoog JCM, Connolly JAD (2015) Hydrous phase relations and trace element partitioning behaviour in calcareous sediments at subduction-zone conditions. J Petrol 56(5):953–980
Smith HJ, Spivack AJ, Staudigel H, Hart SR (1995) The boron isotopic composition of altered oceanic crust. Chem Geol 126(2):119–135
Smith HJ, Leeman WP, Davidson J, Spivack AJ (1997) The B isotopic composition of arc lavas from Martinique. Lesser Antilles Earth Planet Sci Lett 146 (1–2):303–314
Snyder GT, Savov IP, Muramatsu Y (2004) Iodine and boron in Mariana serpentine mud volcanoes (ODP 125 and 195): implications for fore-arc processes and subduction recycling. In: Shinohara M, Salisbury MH, Richter C (eds) proc. ODP, Sci. Results, vol 195
Spandler C, Pirard C (2013) Element recycling from subducting slabs to arc crust: a review. Lithos 170:208–223
Spandler C, Hartmann J, Faure K, Mavrogenes JA, Arculus RJ (2008) The importance of talc and chlorite "hybrid" rocks for volatile recycling through subduction zones; evidence from the high-pressure subduction melange of New Caledonia. Contrib Mineral Petrol 155(2):181–198
Spivack AJ, Edmond JM (1987) Boron isotope exchange between seawater and the oceanic crust. Geochim Cosmochim Acta 51(5):1033–1043
Spivack AJ, Palmer MR, Edmond JM (1987) The sedimentary cycle of the boron isotopes. Geochim Cosmochim Acta 51(7):1939–1949
Straub SM, Layne GD (2002) The systematics of boron isotopes in Izu arc front volcanic rocks. Earth Planet Sci Lett 198(1–2):25–39
Straub SM, Layne GD (2003) Decoupling of fluids and fluid-mobile elements during shallow subduction: evidence from halogen-rich andesite melt inclusions from the Izu arc volcanic front. Geochem Geophys Geosyst 4 Artn 9003
Syracuse EM, Abers GA (2006) Global compilation of variations in slab depth beneath arc volcanoes and implications. Geochem Geophys Geosyst Artn Q05017
Syracuse EM, van Keken PE, Abers GA (2010) The global range of subduction zone thermal models. Phys Earth Planet In 183(1–2):73–90
Tatsumi Y (1989) Migration of fluid phases and genesis of basalt magmas in subduction zones. J Geophys Res-Solid 94(B4):4697–4707
Tenthorey E, Hermann J (2004) Composition of fluids during serpentinite breakdown in subduction zones: evidence for limited boron mobility. Geology 32 (10):865–868
Tera F, Brown L, Morris J, Sacks IS, Klein J, Middleton R (1986) Sediment incorporation in island-arc magmas—inferences from Be-10. Geochim Cosmochim Acta 50(4):535–550
Till CB, Grove TL, Withers AC (2012) The beginnings of hydrous mantle wedge melting. Contrib Mineral Petrol 163(4):669–688
Toki T, Higa K, Shinjo R (2015) Data report: boron isotope ratios in interstitial waters from sites C0021 and C0022. In: Strasser M, Dugan B, Kanagawa K, Moore GF, Toczko S, Maeda L, and the Expedition 338 Scientists (eds.) Proceedings of the integrated ocean drilling program. Integrated Ocean Drilling Program, Yokohama, vol 338
Tonarini S, Pennisi M, Leeman WP (1997) Precise boron isotopic analysis of complex silicate (rock) samples using alkali carbonate fusion and ion-exchange separation. Chem Geol 142(1–2):129–137
Tonarini S, Leeman WP, Ferrara G (2001) Boron isotopic variations in lavas of the Aeolian volcanic arc, South Italy. J Volcanol Geothermal Res 110(1–2):155–170
Tonarini S, Agostini S, Doglioni C, Innocenti F, Manetti P (2007) Evidence for serpentinite fluid in convergent margin systems: the example of El Salvador (Central America) arc lavas. Geochem Geophys Geosyst 8:Q09014
Tonarini S, Leeman WP, Leat PT (2011) Subduction erosion of forearc mantle wedge implicated in the genesis of the South Sandwich Island (SSI) arc: evidence from boron isotope systematics. Earth Planet Sci Lett 301(1–2):275–284
Trommsdorff V, Sanchez-Vizcaino VL, Gomez-Pugnaire MT, Muntener O (1998) High pressure breakdown of antigorite to spinifex-textured olivine and orthopyroxene SE Spain. Contrib Mineral Petrol 132(2):139–148
Ulmer P, Trommsdorff V (1995) Serpentine stability to mantle depths and subduction-related magmatism. Science 268(5212):858–861
Van Hinsberg VJ, Henry DJ, Marschall HR (2011) Tourmaline: an ideal indicator of its host environment. Can Mineral 49(1):1–16
Van Keken PE, Kiefer B, Peacock SM (2002) High-resolution models of subduction zones: implications for mineral dehydration reactions and the transport of water into the deep mantle. Geochem Geophys Geosystems 3 Artn 1056
Van Keken PE, Hacker BR, Syracuse EM, Abers GA (2011) Subduction factory: 4. Depth-dependent flux of H_2O from subducting slabs worldwide. J Geophys Res Solid Earth 116 Artn B01401
Vannucchi P, Sage F, Phipps Morgan J, Remitti F, Collot J-Y (2012) Toward a dynamic concept of the subduction channel at erosive convergent margins with implications for interplate material transfer. Geochem Geophys Geosyst 13(2):Q02003

Vils F, Tonarini S, Kalt A, Seitz HM (2009) Boron, lithium and strontium isotopes as tracers of seawater-serpentinite interaction at Mid-Atlantic ridge, ODP Leg 209. Earth Planet. Sci. Lett. 286(3–4):414–425

Vils F, Muntener O, Kalt A, Ludwig T (2011) Implications of the serpentine phase transition on the behaviour of beryllium and lithium-boron of subducted ultramafic rocks. Geochim Cosmochim Acta 75(5):1249–1271

Wada I, Behn MD, Shaw AM (2012) Effects of heterogeneous hydration in the incoming plate, slab rehydration, and mantle wedge hydration on slab-derived H_2O flux in subduction zones. Earth Planet Sci Lett. 353:60–71

Walowski KJ, Wallace PJ, Hauri EH, Wada I, Clynne MA (2015) Slab melting beneath the Cascade Arc driven by dehydration of altered oceanic peridotite. Nat Geosci 8(5):404–408

Wunder B, Meixner A, Romer RL, Wirth R, Heinrich W (2005) The geochemical cycle of boron: constraints from boron isotope partitioning experiments between mica and fluid. Lithos 84(3–4):206–216

Yamaoka K, Ishikawa T, Matsubaya O, Ishiyama D, Nagaishi K, Hiroyasu Y, Chiba H, Kawahata H (2012) Boron and oxygen isotope systematics for a complete section of oceanic crustal rocks in the Oman ophiolite. Geochim Cosmochim Acta 84:543–559

Yamaoka K, Matsukura S, Ishikawa T, Kawahata H (2015) Boron isotope systematics of a fossil hydrothermal system from the Troodos ophiolite, Cyprus: water-rock interactions in the oceanic crust and subseafloor ore deposits. Chem Geol 396:61–73

You CF, Chan LH, Spivack AJ, Gieskes JM (1995) Lithium, boron, and their isotopes in sediments and pore waters of Ocean drilling program site-808, Nankai trough—implications for fluid expulsion in accretionary prisms. Geology 23(1):37–40

You CF, Spivack AJ, Gieskes JM, Martin JB, Davisson ML (1996) Boron contents and isotopic compositions in pore waters: a new approach to determine temperature induced artifacts—geochemical implications. Mar Geol 129(3–4):351–361

Boron Isotopes in the Continental Crust: Granites, Pegmatites, Felsic Volcanic Rocks, and Related Ore Deposits

10

Robert B. Trumbull and John F. Slack

Abstract

Boron is an incompatible lithophile element that is readily transported by granitic melts and hydrous fluids and therefore is concentrated in the continental crust relative to the mantle. The isotopic composition of boron in crystalline rocks of the continental crust (e.g., metamorphic and igneous lithologies) varies over a wide range of −20 to +10‰, depending on the B-isotope composition of the protoliths and on fractionation effects caused by phase transitions (metamorphic devolatilization reactions, fluid exsolution from magmas). Studies of progressive metamorphism and anatexis show that the behavior of boron and its isotopes depends heavily on the presence or absence of B-retentive minerals like tourmaline. In general, boron is prone to loss during devolatilization reactions, and metamorphic fluid preferentially removes the heavier isotope, but growth of tourmaline can minimize or prevent these effects. A new compilation of over 250 B-isotope analyses from about 90 localities of felsic igneous rocks in the continental crust shows a first-order distinction in composition between I-type magmas (subduction-related having meta-igneous sources) and S-type magmas (derived from metasedimentary rocks). Boron in I-type magmas is isotopically heavy (mean $\delta^{11}B = -2‰$, s.d. = 5) relative to unaltered MORB (mean $\delta^{11}B = -7‰$, s.d. = 1), presumably because of a greater contribution by subducted oceanic crust and pelagic sediments.

R.B. Trumbull (✉)
GFZ German Research Centre for Geosciences, Telegrafenberg, 14473 Potsdam, Germany
e-mail: bobby@gfz-potsdam.de

J.F. Slack
U.S. Geological Survey, National Center, MS 954, Reston, VA 20192, USA
e-mail: jfslack@usgs.gov

J.F. Slack
U.S. Geological Survey (Emeritus), Farmington, ME 04938, USA

H. Marschall and G. Foster (eds.), *Boron Isotopes*, Advances in Isotope Geochemistry,
https://doi.org/10.1007/978-3-319-64666-4_10

Boron in S-type granitic rocks has a much lighter isotopic signature (mean $\delta^{11}B = -11‰$, s.d. = 4). The latter corresponds to the commonly cited B-isotope value of −10‰ for continental crust, but because much of Earth's crust is derived from I-type magmas, its average B-isotope value is probably higher than previously thought. The dichotomy of B-isotope compositions in I- and S-type granitoids is also observed in their genetically related magmatic-hydrothermal ore deposits, as we demonstrate in a review of data from porphyry and Iron Oxide-Copper-Gold (IOCG) systems (I-type) and from Sn-W veins and granitic pegmatites (S-type). However, it is important to note that in all of these systems, there are significant and locally complex effects of isotopic fractionation due to magmatic fluid exsolution and to mixing of boron sourced from externally derived fluids.

10.1 Introduction

> "Boron is a quintessential element of the Earth's upper continental crust"
> *Edward S. Grew in: Elements (2015), 11/3, p. 162*

This quotation from a leading researcher on boron mineralogy and geochemistry is a good starting point for this chapter. The continental crust is the main home of boron on planet Earth for two reasons. First, the crust is built to a considerable extent from intermediate-composition to felsic igneous rocks and boron partitions preferentially into granitic melts, where it tends to remain and become concentrated during magma ascent, emplacement, and crystallization. Second, boron is a volatile element, readily transported by hydrous fluids, which are prevalent in and on the continents. As a consequence, important concentrations of boron and boron minerals are common in many types of hydrothermal ore deposits and in evaporitic borates, the latter being the most important commercial source of boron.

The continental crust is far from homogeneous and its constituent parts represent different stages in the rock cycle of erosion, sedimentation, metamorphism, melting, and magmatism. The focus of this chapter is on the volumetrically most important components of the continental crust, which are the intermediate-composition to felsic igneous rocks and their metamorphosed equivalents (Wedepohl 1995). We here review current knowledge about the concentrations of boron and fractionation of its isotopes during metamorphism and anatexis, describe what is known about boron isotope systematics in granitic magmas of I- and S-types, and finally show how B-isotope studies have contributed to a better understanding of the formation of diverse magmatic-hydrothermal ore deposits (e.g., rare-element pegmatites; Cu–Mo–Au porphyries; greisen-, and vein-type Sn-W mineralization). Not discussed in this chapter is the behavior of boron and its isotopes during subduction (see de Hoog and Savov 2017) except as it relates to the composition of juvenile magmas added to continental crust at magmatic arcs, nor do we consider surface processes of continental weathering, erosion, and sedimentation (see Gaillardet and Lemarchand 2017).

10.2 Boron in the Continental Crust

There are several sources of information on the chemical composition of continental crust as reviewed by Rudnick and Gao (2003). Their recommended values for average boron concentrations in the upper crust, lower crust, and bulk crust are 17 ± 8 µg/g, 2 µg/g, and ca. 11 µg/g, respectively. Only the upper crustal number is well enough constrained by geologic samples and mapped proportions of different lithologies for an uncertainty estimate. As shown by an extensive sampling of surface rocks in East

Table 10.1 Boron concentrations in crustal rocks and minerals

1a: Boron (μg/g) in crustal rocks: average bulk-rock values, with "n" in parentheses

Amphibolite	9	(189)	
Carbonate	8, 22	(2088)	Archean: 8, post-Archean: 22 μg/g
Pelitic rocks	38, 101	(1410)	Archean: 38, post-Archean: 101 μg/g
Tonalite-trondhemite-granodiorite	5	(1194)	
Intermediate felsic granulites	2	(136)	
Granites (undifferentiated)	7, 3	(662)	Archean: 7, post-Archean: 3 μg/g
Diorite	7	(260)	
Felsic volcanics	11	(972)	
Felsic metavolcanics	5	(41)	

1b: Boron (μg/g) in minerals from metapelitic rocks: average values with "n" in parentheses

Reference	1	2	3a	3b	3c
Muscovite	68 (10)	72 (4)	101 (7)	68 (5)	47 (17)
Biotite	16 (6)	10 (1)	–	–	3 (2)
Chlorite	5 (7)	4 (3)	16 (10)	3 (5)	6 (13)
Plagioclase	2 (9)	2 (4)	0.6 (2)	0.5 (2)	0.6 (1)
Epidote	2 (7)	2 (3)	–	–	–
Amphibole	–	11 (1)	–	–	–

1c: Boron (μg/g) in minerals from granitic rocks: average values with "n" in parentheses

Reference	1	2	3	4	5	6
Muscovite	24 (8)	15 (8)	140 (11)	–	–	–
Biotite	10 (10)	7 (3)	4 (3)	20 (3)	18	41
Chlorite	–	–	–	14 (3)	–	–
Plagioclase	13 (7)	5 (6)	8 (6)	9 (2)	13	13
K-feldspar	4 (7)	2 (2)	4 (9)	–	–	–

Averages of "n" samples from regional sampling in E. China (Gao et al. 1998)
1 Pelona metagraywacke, California: Leeman and Sisson (2002)
2 Catalina Schist metagraywacke, California: Leeman and Sisson (2002)
3 Sambabawa Belt: Nakamo and Nakamura (2001): *3a* chlorite zone, *3b* albite-biotite zone, *3c* oligoclase-biotite zone
1 Hauzenberg granite, Bavaria: Sauerer and Troll (1990)
2 Peña Negra pluton, Spain: (Pereira Gomez and Shaw (1997)
3 S-type granite and pegmatite, Ikaria, Greece: Hezel et al. (2011)
4 I-type granite, Ikaria, Greece: Hezel et al. (2011)
5 S-type Dadongshan granite, SE China: Zhao et al. (2015)
6 S-type Qianlishan granite, SE China: Zhao et al. (2015)

China by Gao et al. (1998), pelitic rocks (variable metamorphic grade) have by far the highest concentrations of boron (Table 10.1a) and this relates to their clay and mica contents. Studies summarized in Tables 10.1b, c show that the main boron-hosting minerals within pelitic metamorphic rocks and granites are sheet silicates, with boron concentrations in muscovite being highest. Tourmaline, having about 3 wt% B, is the predominant host of boron in rocks where present but in the general case, boron in felsic crystalline rocks is mainly hosted by micas. In felsic volcanic rocks containing a glassy groundmass, the highest boron concentration may reside in the glass (Schmitt and Simon 2004).

Boron concentrations in MORB and its depleted mantle source are 1.2 ± 0.14 μg/g and

0.19 ± 0.02 μg/g, respectively, according to recent work summarized by Marshall et al. (2016; see also Marschall 2017). The contrast by one or two orders of magnitude between crust and mantle boron concentrations implies a net transfer of boron from the mantle to the continental crust. The rate of accumulation of boron in continental crust over time is obviously related to the rate of crustal formation, for which an extensive literature exists (see review by Cawood et al. 2013). In a landmark early study on the B-isotope composition of the continental crust, Chaussidon and Albarede (1992) proposed that progressive sequestration of boron over time caused a change in the average isotopic composition of crustal boron. These authors produced a box model for boron exchange within the system composed of oceanic crust, continental crust, and seawater, assuming that crustal boron is not returned to the mantle and that neglible isotope fractionation occurs between mantle and juvenile crust. By combining mass-balance calculations for these three components with growth curves for continental crust, Chaussidon and Albarede (1992) predicted changes for average $\delta^{11}B$ values of the crust from −25 at 4000 Ma to −5‰ at present day. A more recent mass-balance model for the present-day Earth by Marschall et al. (2017) predicts a $\delta^{11}B$ value of −9.1 ± 2.4‰ in the continental crust. The exact values for average crustal composition are model-dependent, but the predicted trend toward higher $^{11}B/^{10}B$ ratios in the crust with time can be tested from the rock record. Chaussidon and Albarede (1992) were the first to attempt this test by analyzing tourmaline from rocks having ages of 3800 to 6 Ma. The authors attached significance to an inverse correlation of $^{11}B/^{10}B$ ratios with age for a subset of Li-rich tourmalines, as predicted by the model, but their full data set lacked systematic variation of $\delta^{11}B$ with time. Since that publication, many more studies on B-isotope compositions of tourmaline have been reported, including two for Middle to Early Archean rocks (Farber et al. 2015; Grew et al. 2015). These recent data serve to confirm what was already apparent from Chaussidon and Albarede (1992): namely, that crustal boron incorporated into magmatic tourmaline shows a range of $\delta^{11}B$ values, chiefly between −20 and −5‰, for all rock ages examined. There is no support in the rock record for a systematic trend to higher $\delta^{11}B$ values for continental crust with time, as was predicted by earlier crustal growth and boron sequestration models.

10.3 B-Isotope Systematics in Crustal Processes

Primary input to the continental crust of material from Earth's mantle takes place at convergent margins via subduction-related magmatism and to a lesser extent in intraplate settings. The most important examples of the latter type are the Large Igneous Provinces (LIPs) where huge volumes of mantle melting and magmatic additions to the crust are driven by mantle plumes. Continental growth via subduction has been dominant in the post-Archean, whereas the opposite is thought to be true for the Hadean and Archean. Much debate exists on when the switch occurred, but many workers have put the mark at ca. 3.5–3.0 Ga (e.g., Cawood et al. 2013; Griffin et al. 2014; Dhuime et al. 2015). The question is not without interest for the B-isotope story because, although mantle melting produces basalt in both cases (which may differentiate or re-melt to form granitic crust), the subduction-related basalts are more hydrous than MORB or LIP basalts and have higher concentrations of fluid-mobile elements that are derived from dehydration of the subducted slab (e.g., Wilson 1989). Boron is one of those elements and its isotopic composition depends on the relative contributions of boron derived from marine sediments (low $\delta^{11}B$) and altered oceanic crust (high $\delta^{11}B$). Mineral-fluid isotope partitioning during dehydration of the slab gives the slab-derived boron that is incorporated into arc magmas a lower isotopic composition compared with boron that enters the subduction zone. This phenomenon is well documented in both oceanic and continental arcs, as shown by parallel, cross-arc decreases in B/Nb ratios and $\delta^{11}B$ values of volcanic rocks (Ishikawa et al.

2001; Rosner et al. 2003; de Hoog and Savov 2017). By contrast, the basalts produced by anhydrous melting of the mantle in LIPs are expected to have relatively low boron concentrations and approximately the same B-isotope composition as their source (MORB and mantle $\delta^{11}B$ is uniform at $-7 \pm 1‰$; see Marschall et al. 2017).

The formation of juvenile crust is related to mantle melting and the ascent of basaltic magmas, with or without modification by subduction. Boron systematics related to magma genesis in the mantle are outside the scope of this chapter (but see chapters by Marschall; deHoog and Savov, this volume). Our concern here is with processes operating within the continental crust that rework and modify the juvenile magmatic input. We discuss how crustal boron and its isotopes are affected by metamorphism, partial melting, magmatic differentiation, and hydrothermal activity. Surface processes of erosion and sedimentation are discussed by Gaillardet and Lemarchand (2017).

10.3.1 Metamorphism and Partial Melting

The data compiled in Table 10.1a and the review by Leeman and Sisson (1996) show that, among common metamorphic rock types within the continental crust: amphibolite, metacarbonate, metagraywacke, and metapelite, the last is by far the most important carrier of boron. Pelitic rocks (shale, slate, mica schist) typically contain ca. 100 μg/g B, much more than the upper crustal average of 17 ± 8 μg/g (Rudnick and Gao 2003). It is also well established (e.g., Table 10.1b) that in the absence of accessory tourmaline and other more rare boro-silicates or borates, the main hosts of boron within these rocks are hydrous sheet silicates, chiefly muscovite (clay minerals in low-grade rocks). Therefore, prograde metamorphism in terranes having significant amounts of pelitic rocks may be expected to produce boron-rich metamorphic fluids by the dehydration of clay minerals and micas, and thus a progressive depletion of the rocks in boron. For example, Moran et al. (1992) determined boron concentrations in rocks from a range of metamorphic grades and found a negative correlation of boron concentration with metamorphic grade as expected. The low overall concentrations of boron in granulite-facies rocks (ca. 2 μg/g B; Table 10.1a) is logically attributed to dehydration loss. However, not all granulite-facies rocks are poor in boron. MacGregor et al. (2013) reported whole-rock concentrations of up to 2 wt% B in a high-temperature paragneiss from the Larsemann Hills, Antarctica. This locality contains an unusual assemblage of borosilicate minerals including tourmaline, prismatine, and grandidierite (see Grew 1996 for a review of borosilicates in metamorphic rocks, and MacGregor et al. (2013) for discussion of B-isotope fractionation among them). The survival of high boron contents in this case was attributed to the early and abundant formation of tourmaline, from which the other borosilicates formed at high metamorphic grade. A similar explanation was invoked by Slack et al. (1993) for the local abundance (~30–50 vol.%) of tourmaline in numerous rocks of the two-pyroxene granulite facies at Broken Hill, Australia. These relatively rare examples serve to make a general point: that the early formation of tourmaline or other stable boron-rich minerals can minimize or prevent the loss of boron during progressive metamorphism.

A very important aspect of the dehydration process is the strong boron isotopic fractionation between clays and micas on the one hand (IV-coordinated boron, rich in ^{10}B) and aqueous fluid on the other (III-coordinated B, rich in ^{11}B). Depending on temperature, the $\delta^{11}B$ value of a metamorphic fluid can be nearly 20‰ higher than that of the clays or micas within the coexisting rock (Palmer and Swihart 1996; Wunder et al. 2005; Kowalski and Wunder 2017). The expectation, then, is that prograde metamorphism will produce progressively boron-depleted and isotopically lighter rocks, but are these trends actually observed? In subduction systems, good evidence exists to support a model in which progressive dehydration of the subducting slab produces a loss of boron and a decrease in $\delta^{11}B$

of the solid residue (see Konrad-Scholke and Halama 2014). Case studies of boron concentrations and B-isotope variations in prograde metamorphic sequences show that although this general trend may hold, there is complexity caused by specific compositions of the protoliths and the stability of boron-hosting phases, especially sheet silicates, but also of tourmaline and other aluminous borosilicates. For example, Nakano and Nakamura (2001) studied a prograde metapelite sequence in the Sanbagawa belt of Japan and reported no change in bulk-rock boron concentrations or $\delta^{11}B$ values from greenschist to amphibolite facies (ca. 300–600 °C). These authors showed that prograde growth of tourmaline (modal abundance increased from 0.01 to 0.11 vol.%) was enough to effectively prevent boron loss from the rock, despite the metamorphic destruction of B-bearing sheet silicates. Importantly, the study employed in situ isotopic analysis by SIMS and found that zoned tourmaline grains in the high-grade samples have lower $\delta^{11}B$ values in the rims relative to the cores (Fig. 10.1). This isotopic pattern was attributed to the incorporation of boron into tourmaline following metamorphic release from mica (Nakano and Nakamura 2001). In a similar study but with a different result, Bebout and Nakamura (2003) determined that prograde metamorphism of the Catalina Schist (California) removed about 75% of the whole-rock boron from precursor shale. The high-grade rocks there (epidote-amphibolite facies) also contain tourmaline, for which SIMS data revealed "prograde zoning" as in the Sanbagawa belt, with lower $\delta^{11}B$ values occurring in crystal rims. However, tourmaline abundance in the Catalina Schist is much lower than in the Sanbagawa rocks, and hence was apparently insufficient to prevent boron loss to the metamorphic fluids (Bebout and Nakamura 2003). Tourmaline in the Catalina Schist also shows evidence of retrograde zoning, i.e., thin outer rims have high $\delta^{11}B$ values attributed to a late fluid ingress. Zoning of this type, which can plausibly be related to rehydration or retrograde metamorphism, has been reported in metamorphic tourmaline from other terranes including the Alps (Bebout and Nakamura 2003; Bebout et al. 2013), Syros in Greece (Marschall et al. 2008), and Greenland (Chaussidon and Appel 1997; Grew et al. 2015). In their review of tourmaline B-isotope variations in high-pressure metamorphic terranes, Marschall et al. (2009) gave a graphical overview of observed $\delta^{11}B$ zoning patterns, shown here in simplified form on Fig. 10.2. The authors distinguished on this plot three scenarios that can produce isotopic zonation, designated as A (boron released by prograde breakdown of mica), B (boron introduced by retrograde fluids), and C (boron present in detrital cores). These examples show that both prograde and retrograde zoning are commonly observed. Tourmaline has received the most attention in studies of boron systematics during metamorphism, but white micas can also be important. An example is the study by Halama et al. (2014) who combined in situ boron isotope analyses with in situ $^{40}Ar/^{39}Ar$ dating of phengite to reconstruct the fluid-rock interaction history of a metamorphosed subduction complex in the western Alps. In that study, phengite rims yielded consistently younger ages and lower B concentrations than cores, which was related to deformation and fluid interaction during exhumation. However, the B-isotopic composition of cores and rims overlap, which the authors attributed to a redistribution of boron without significant fractionation.

Romer and Meixner (2014) determined boron and lithium concentrations and isotope variations in a well-documented prograde sequence of volcano-sedimentary units from the Variscan orogen in Germany. The studied rocks comprise two protolith series (Phycodes Group and Frauenbach Group), which are exposed in several nappes, which allows analysis of equivalent rocks from very low-grade to eclogite-facies conditions. Readers are referred to the publication for details, particularly as relates to the Li isotope results, which are beyond the scope of this chapter. In short, the authors found a net loss of boron (by 30–60%) and lower $\delta^{11}B$ values in the higher grade rocks of the Phycodes Group. The Frauenbach rocks underwent a similar loss of boron during metamorphism (apart from some altered by late fluid ingress), but their

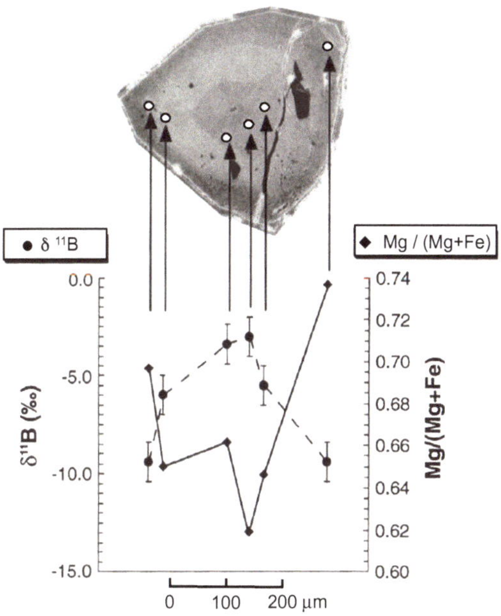

Fig. 10.1 Example of zoned tourmaline from metapelite in Sambagawa metamorphic belt of Japan (from Nakamo and Nakamura 2001). Core-to-rim decrease in $\delta^{11}B$ is attributed to loss of ^{11}B to fluid phase during prograde metamorphism

B-isotope compositions were less affected. In comparing the two protolith series, Romer and Meixner (2014) emphasized that the nature of B (and Li) variations during metamorphism depends on the specific protolith mineralogy, and to what extent volatile elements are sequestered in metamorphic minerals or lost to dehydration fluids. An important conclusion of that study is that "background" B-isotope variations of the protolith can be larger than any shift produced by prograde metamorphism (see also Kasemann et al. 2000).

Whether boron is concentrated in the anatectic melt relative to its solid source depends on partitioning among the melt and residual minerals that are stable during prograde metamorphism, which is a function of the melting reactions. Critically important, for example, is the extent to which dehydration and boron loss affected the rock prior to melting, and whether the rock contains tourmaline or other borosilicates (London et al. 1996; Acosta-Vigil et al. 2001; Kawakami and Ikeda 2003). Pereira Gomez and

Fig. 10.2 Compilation of $\delta^{11}B$ values in tourmaline from high-pressure metamorphic rocks (diagram modified from Marschall et al. 2009) indicates that isotopic zoning is common. *Arrows* show core-to-rim zoning trends; color schemes distinguish different types of zoning processes (A, B, C) invoked to explain the variations (see text)

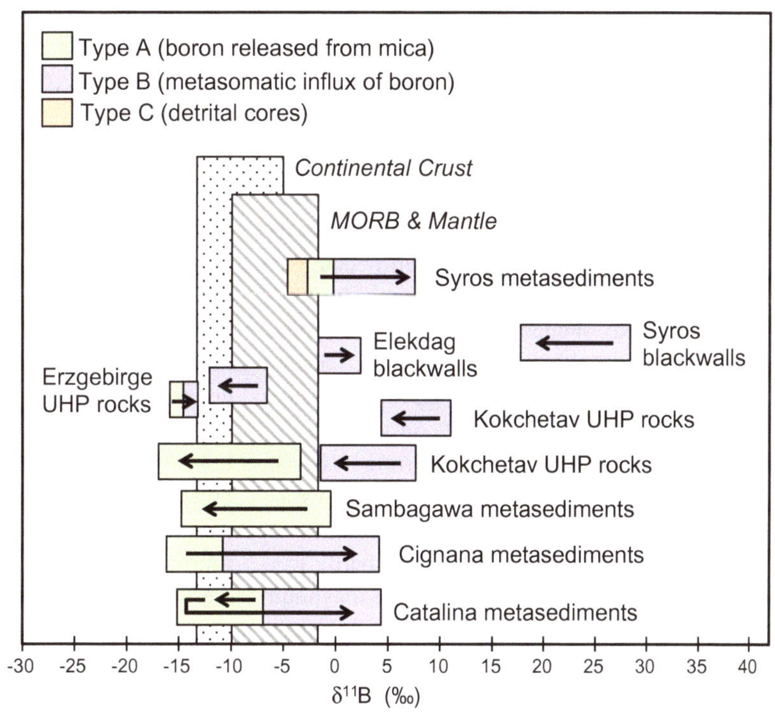

Shaw (1997) found that leucosomes in the Peña Negra anatectic complex (Spain) have lower boron contents than the unmelted protolith, attributing this to the formation of sillimanite in the restate, a mineral with a high affinity for boron (see also Acosta-Vigil et al. 2001). More generally, dehydration-melting reactions involving the breakdown of mica will release boron to the melt. Experimental and empirical evidence shows that tourmaline may be consumed during prograde metamorphism in water-saturated conditions because B-solubility in aqueous fluid increases with temperature, but tourmaline commonly persists until anatexis and then melts incongruently, giving up boron to the melt (London et al. 1996, 2012; Wolf and London 1997; Acosta-Vigil et al. 2003; London 2011). Only under special (dry) conditions, does tourmaline stably persist into granulite-facies rocks (Slack et al. 1993; MacGregor et al. 2013), but this outcome is apparently rare. The issue of ***boron concentration*** in partial melts is therefore a complex one, depending on metamorphic history, mineral assemblage, and compositional variables like water content. However, all available evidence suggests that ***boron isotopic*** composition is not greatly affected by the melting process. The fractionation effect of B-isotopic exchange between solids and a melt will depend on the degree to which boron in the respective phases has a different coordination number (B–O bond length) and on the temperature at which the exchange occurs. Based on experimental data and theoretical studies (Wunder et al. 2005; Kowalski and Wunder 2017), the complete transfer of boron from trigonal to tetrahedral coordination, which is approached for the system mica-fluid, will produce a shift on the order of 6‰ at 700 °C (with fluid preferring ^{11}B). In anatexis involving protoliths that contain varied B-bearing mineral hosts producing granitic melts in which boron coordination is only partly trigonal, one should expect less than the maximum 6‰ effect. In principle, considering the uncertainty of 1–2‰ for in situ B-isotope analyses by secondary ion mass spectrometry (SIMS) or laser-ablation inductively coupled mass spectrometry (LA-ICPMS), the isotopic effect of melting might be discernable. Commonly, however, the natural B-isotope variability in metasedimentary rocks and granites derived therefrom is as large, or larger, than the theoretical effect of melting alone. Evidence from field studies supports the prediction of minimal B-isotope fractionation between source rocks and the melt during partial melting. For example, Kasemann et al. (2000) found no significant difference in the B-isotope composition of tourmaline from leucosomes and melanosomes in migmatites from northwest Argentina, nor any systematic difference between the δ^{11}B values of migmatite and local S-type granites of Ordovician and Mesozoic age (Fig. 10.3). The authors concluded that the high-grade metamorphism and partial melting that occurred in this area had little or no systematic effect on the B-isotope composition of the rocks; and furthermore, that the boron contained in the Andean metamorphic basement is essentially recycled by partial melting and granite formation. MacGregor et al. (2013) also found overlapping δ^{11}B values of tourmaline from paragneiss and anatectic pegmatites in high-grade rocks in the Larsemann Hills, Antarctica. Trumbull et al. (2008) studied a tourmaline-bearing S-type granite from Namibia that was derived from partial melting of metapelites containing tourmaline-rich layers or "tourmalinites" (cf. Slack 1996). The overlapping B-isotopic composition of magmatic tourmaline in the granite and of metasedimentary tourmalines from the country rocks supports the notion of negligible fractionation during melting. Slack et al. (1993) reported similar B-isotopic ratios in granitic tourmalines and metapelitic country rocks in the Broken Hill district, Australia, suggesting a metasedimentary source for boron in the granitic magmas. Other cases of crustal boron recycling from metapelite source rocks to granites were proposed for tourmaline-bearing granites and pegmatites in western Spain by Pesquera et al. (2005) and for granites from Variscan massifs of central Germany by Romer et al. (2014).

To conclude, both natural case studies and experimental/theoretical results on isotope fractionation suggest that partial melting of rocks

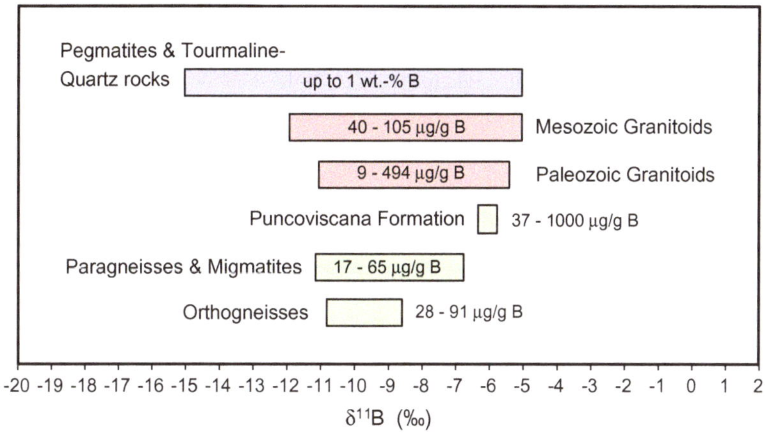

Fig. 10.3 Boron recycling during metamorphism, anatexis, and formation of S-type granites in central Andes basement (modified from Kasemann et al. 2000). Boron concentrations (whole-rock) and B-isotope ratios (tourmaline) from metamorphic rocks, migmatites, two generations of S-type granites (Paleozoic and Mesozoic) and related pegmatites, and hydrothermal quartz-tourmaline rocks show a complete overlap

will not produce a large shift in the B-isotopic composition of the participating phases. The implication of this model is that B-isotope data obtained from bulk rocks, magmatic minerals, and glass or melt inclusions can be used with some confidence to constrain the provenance of the magma source(s).

10.3.2 Boron in Granitic Magmas

London et al. (1996) and Dingwell et al. (1996) provided extensive reviews of empirical and experimental studies relating to the incorporation of boron in granitic magmas, important effects of boron on melt structure, behavior of boron during magmatic differentiation, and consequences of its eventual partitioning into a separate, late-magmatic fluid phase. Those reviews made only passing mention of boron isotope variations because at that time few studies had been done. After publication of the Reviews in Mineralogy volume on boron (Grew and Anovitz 1996), research on the ***geochemical*** behavior of boron and boron minerals in granites, especially with relevance to pegmatites and granite-related ore deposits, has expanded considerably. We make no effort to compile and review that literature here, but focus only on work that has addressed the behavior of boron isotopes in granitic systems. One point worth emphasizing, however, is the increased number of melt-inclusion studies that include analysis of boron concentrations (e.g., Audetat et al. 2000; Anderson et al. 2000; Schmitt et al. 2002; Badanina et al. 2004, 2010; Thomas 2002; Thomas et al. 2003; Savov et al. 2009; Wittenbrink et al. 2009; Jones et al. 2014). The melt inclusion data are important because the incompatible nature and volatility of boron means that its concentrations in solid granites are almost certainly much lower than in granitic melts (London et al. 1996). Furthermore, melt-inclusion studies have demonstrated that volatile-rich granitic melts can undergo liquid immiscibility before the loss of a hydrous fluid phase (e.g., Veksler and Thomas 2002; Thomas et al. 2003). This phase separation may play a significant role in the concentration of boron and other elements in some granitic segregations and pegmatites (Trumbull et al. 2008; Drivenes et al. 2015), although debate continues on the scale at which the effects of immiscibility are important (Thomas et al. 2012; London 2015).

Most overviews of B-isotope variations in natural reservoirs have grouped data for "granites and pegmatites" into a single category (e.g., Jiang and Palmer 1998; Marschall and Ludwig 2006; Marschall and Jiang 2011; Trumbull et al. 2013). This grouping of data can be misleading because of the well-established contrasts in magma sources and geologic settings of felsic magmas that crystallize as granite. For example, the classic division of granites into I-type (igneous source) and S-type (sedimentary source: Chappell and White 2001) is expressed in part by a contrast in their Sr- and O-isotope compositions, and it is logical to expect that the B-isotope compositions will also differ because of the difference in magma source. The I-type granites originate mostly from juvenile magmas produced at island arcs and Cordilleran-type active margins, so their B-isotope compositions should more strongly reflect the subduction-related mantle origin of magmas. In contrast, the S-type granites are derived by partial melting of pre-existing continental rocks that experienced weathering at the surface and contain abundant clay minerals and mica. The expectation of a source-related contrast in B-isotope composition between I- and S-type granites is confirmed by the compilation of published studies that follows (Fig. 10.4). We recognize that the diversity of granites in the continental crust is greater than the simple I- and S-type division can encompass (Frost et al. 2001; Clemens and Stevens 2012) plus there are complexities introduced by magma mixing, assimilation, and differentiation that can blur the distinction between these two major suites.

In this context it is worth mentioning a third group of granitic rocks (A-type: anorogenic, anhydrous, alkaline), which have a distinctive, peralkaline composition and distinctive tectonic association with continental rifting. However, there have been very few studies of the B-isotope compositions of A-type magmas. One exception is that of Tonarini et al. (2004, 2009), who reported a narrow range of $\delta^{11}B$ values from about −10 to −7‰ for an alkaline basalt-trachyte-phonolite series in the Campi Flegrei region of Italy. Those values are close to the average B-isotope composition of the mantle (−7 ± 1‰; Marschall et al. 2017), with an offset to lower values that can be attributed to the assimilation of continental crust. In another example, Kaliwoda et al. (2011) studied

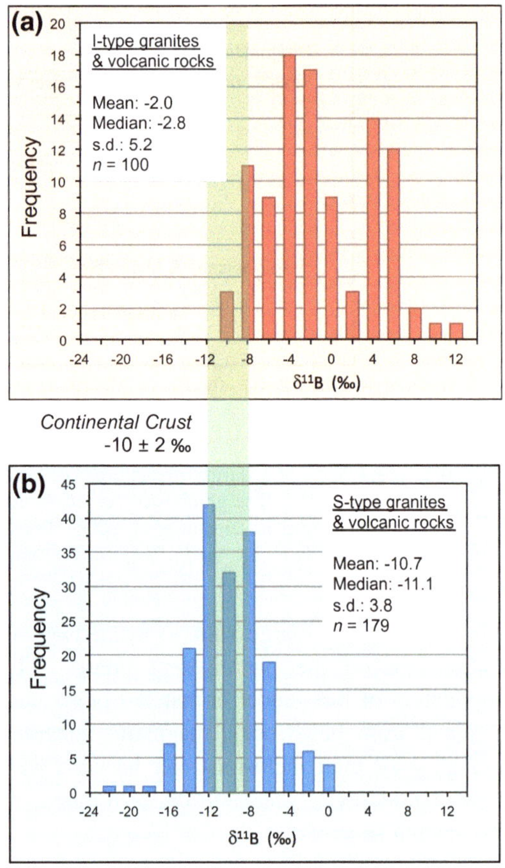

Fig. 10.4 Frequency histograms based on new compilation of B-isotope data from felsic intrusive and volcanic rocks of I-type (**a**) and S-type (**b**) affinity. Virtually all I-type data are from rock or glass analyses; about 85% of the S-type data are analyses of magmatic tourmaline, which differ negligibly from the melt composition values (1–3‰ heavier; see text). Vertical green bar shows the commonly cited range of $\delta^{11}B$ for continental crust (Marschall and Jiang 2011), which corresponds with peak of the S-type composition; value for bulk continental crust will be higher (see text). Data sources for histograms: Chaussidon and Albarede (1992), da Costa et al. (2014), Di Renzo et al. (2011), Drivenes et al. (2015), Duncan et al. (2014), Huang et al. (2016), Iveson et al. (2016), Jiang (2001), Jiang et al. (2008), Jiang and Palmer (1998), Jochum et al. (2006), Jones et al. (2014), Kasemann et al. (2000), Leeman et al. (2004), Ludwig et al. (2011), Marschall and Ludwig (2006), Matthews et al. (2003), Molnár et al. (2016), Palmer and Sturchio (1990), Pesquera et al. (2005), Romer et al. (2014), Rosner et al. (2003), Savov et al. (2009), Schmitt et al. (2002), Schmitt and Simon (2004), Siegel et al. (2016), Smith et al. (1997), Smith and Yardley (1996), Swihart and Moore (1989), Tonarini et al. (1998), Trumbull and Chaussidon (1999), Trumbull et al. (2008; 2013), Wittenbrink et al. (2009), Yang and Jiang (2012), Yang et al. (2015), Yavuz et al. (2011), Zhao et al. (2011; 2015)

nepheline syenite and related alkaline intrusive rocks from Illimaussaq, Greenland. Their SIMS analyses of amphibole, feldspar, and sodalite (sodalite being the main B-carrier in these rocks) showed considerable evidence of zoning and subsolidus boron redistribution, making interpretation of the observed range in $\delta^{11}B$ values (−20 to −5‰) difficult. The authors proposed an initial melt composition of −17‰, much lower than the mantle average, but because of compelling geochemical evidence that the Illimaussaq magmas are mantle derived, Kaliwoda et al. (2011) postulated a source component of deep-subducted oceanic crust depleted in ^{11}B.

10.3.2.1 Boron in I-Type Magmas

The juvenile continental crust, since at least Late Archean times, formed from magmas of mafic to intermediate composition derived by mantle melting at accretionary plate margins. B-isotope systematics related to subduction, slab dehydration, and partial melting of the overlying mantle wedge are treated by de Hoog and Savov (2017). Here we discuss progress in understanding the isotope geochemistry of boron in the continental products of subduction magmatism: the I-type granitic rocks and their volcanic equivalents. Boron concentrations in I-type granitic rocks are 10 µg/g or less (see Table 10.1a for tonalite-trondhjemite-granodiorite and diorite). However, since boron is highly incompatible in rock-forming minerals and partitions strongly into residual melt and/or fluid, the bulk-rock contents of boron are minimum estimates of the magmatic concentrations, especially if fluids were exsolved. More reliable estimates have come from in situ analysis of matrix glass and quartz-hosted melt inclusions in volcanic equivalents of these rocks, for which data suggest the parent melts had ca. 20–120 µg/g B, which is up to 10 times greater than the bulk-rock value (Anderson et al. 2000; Schmitt et al. 2002; Schmitt and Simon 2004; Savov et al. 2009; Wittenbrink et al. 2009; Jones et al. 2014).

A frequency histogram of B-isotope compositions for I-type granites and their volcanic equivalents is shown in Fig. 10.4a, based on 100 data points for 29 separate occurrences compiled from 13 publications. Most of these data represent in situ analysis by SIMS or LA-ICPMS of quartz-hosted melt inclusions and/or groundmass glass; included are some bulk-rock analyses, and two data points from SIMS analysis of tourmaline interpreted to be magmatic by the contributing authors. The total range in $\delta^{11}B$ values extends from −9 to +12‰. There is much detail in these publications on the causes of isotopic variations reported in the individual studies, to which interested readers are referred. The important point to make here is that 75% of the $\delta^{11}B$ values fall in the range of −4 to +6‰. Mean and median values are −2.0 (s.d. = 5.2) and −2.8‰, respectively. These B-isotope compositions clearly distinguish I-type granites from S-type granites, as discussed further below.

10.3.2.2 Processes Affecting B-Isotope Composition in I-Type Magmas: Differentiation, Assimilation, Degassing

Once boron is incorporated within arc magma by melting at source, its concentration is expected to increase with differentiation because no crystalline phases occur in metaluminous I-type magmas with which boron is compatible. For this reason, and because of the high temperatures involved (>700 °C), the B-isotopic composition of these types of magmas will be unaffected by differentiation. Assimilation of crustal materials by arc magmas could affect their B isotope ratios if the assimilant contains enough boron to offset the magma composition. The observations of systematic across-arc trends in boron concentrations and isotope ratios that correlate with depth to the Wadati-Benioff zone (e.g., Ishikawa and Nakamura 1994; Ishikawa et al. 2001; Rosner et al. 2003) indicate that variations at the source are preserved despite assimilation. Especially significant is the study of Rosner et al. (2003), who found regular across-arc trends in B-isotope ratios in Neogene volcanic rocks of the central Andes, where crustal thickness exceeds 50 km and where the Sr and Pb isotope ratios of the volcanics record considerable crustal contamination.

Exsolution and loss of an aqueous fluid phase from magma will potentially cause an isotopic shift in the residual magma. Fluid-melt fractionation of boron isotopes is not well understood, and this is a subject that cries out for more research. The only experimental study to date (Hervig et al. 2002) predicted fractionation factors that are higher than subsequent theoretical and empirical studies of B-isotope partitioning would seem to support (see discussion in Kowalski and Wunder 2017). For example, the experimental melt-fluid fractionation was −7.5‰ at 700 °C for rhyolite, which is about 1‰ more than a complete shift in boron coordination number (III–IV) at that temperature would produce (Wunder et al. 2005; Kowalski et al. 2013). This would imply that boron in the melt is entirely in four-fold coordination but this is inconsistent with a spectroscopic study by Tonarini et al. (2003). Experiments by Deegan et al. (2016) addressed the special case of carbonate assimilation by magma followed by CO_2 degassing. They argued that assimilation enhances the proportion of four-fold B coordination in the melt and that degassing at the assimilation site can produce locally strong B-isotope fractionation even at 1200 °C (>10‰). The experiments were dynamic and no equilibrium fractionation factor was derived so their geologic importance is unclear. More generally relevant is the empirical study of B-isotope compositions in rhyolite by Schmitt and Simon (2004). They concluded that degassing of rhyolite from Long Valley caldera (California) had no effect on B-isotope compositions because the same $\delta^{11}B$ values were found in quartz-hosted melt inclusions with 3–4 wt% H_2O as in the degassed matrix glass that contained only about 1/10th as much H_2O.

To conclude this section, evidence from current studies suggests that the B-isotope compositions of I-type granitic or volcanic rocks do not differ significantly from those of their parental melts even after geochemical evolution and late-stage degassing. Furthermore, and importantly, our compilation of B-isotope data (Fig. 10.4a) implies that the magma source for these igneous rocks is isotopically heavier than that of the average mantle. This interpretation is consistent with a subduction setting for I-type magma genesis, wherein boron released from slab dehydration is dominated by the isotopic signature of ^{11}B-rich altered oceanic crust and pelagic sediments.

10.3.2.3 Boron in S-Type Magmas

Characteristic of the S-type or "sediment-source-type" granitic magmas is a peraluminous bulk composition and other properties that relate to a substantial component of metasedimentary rocks in the magma source (Chappell and White 2001). Owing to the breakdown of mica and other hydrous silicates within their source(s), S-type granites tend to be relatively rich in water and other volatile elements (fluorine and boron, especially), which have the effect of reducing melt viscosity and extending the temperature range of crystallization (Dingwell et al. 1996; London et al. 1996). The accumulation of incompatible trace elements and volatiles in differentiated S-type magmas produces what the economic geology literature refers to as "fertile granites," which are spatially associated with pegmatite and greisen/vein-type mineralization having local ore grades of Sn, W, Li, Nb, Ta, Cs, or Be. Boron is among the incompatible elements that are concentrated during the formation and evolution of S-type granitic magmas, and thereby are transported into the upper crust or, in the case of rhyolites, to the surface. In peraluminous granites, the presence of tourmaline is the most conspicuous evidence of boron enrichment, where it may occur as dispersed accessory grains or in quartz-tourmaline clots and orbicules, both typically forming late in the crystallization sequence. Tourmaline is a common mineral in granitic pegmatites and also occurs as a product of boron metasomatism in contact zones surrounding granites and pegmatites, and in associated ore deposits (e.g., London et al. 1996; Drivenes et al. 2015). It is not surprising that studies of B-isotopes in S-type granites have been dominated by analysis of tourmaline because it is easily recognized in the field and in thin section, and is easily analyzed by microbeam techniques. Also, this chemically complex and widely stable mineral is a high-fidelity recorder of the changing geochemical environment in which it crystallized (van Hinsberg et al. 2011). Tourmaline is much less common in peraluminous rhyolites, but exceptions in southeast China and Ikaria, Greece, were described by Yang and Jiang (2012) and Baltatzis et al. (2009), respectively. Figure 10.4b shows the distribution of B-isotope compositions of S-type granites and volcanic equivalents, based on 179 analyses from 66 occurrences reported in 28 publications. About 140 of the analyses are of tourmaline that was judged by the contributing authors to be magmatic. For simplicity, we assume here that the $\delta^{11}B$ value of magmatic tourmaline is the same as the melt from which it crystallized. This is unlikely to be strictly true, but even at a low solidus temperature of 500 °C that is relevant for some evolved pegmatites, the expected tourmaline-melt fractionation is only ca. 3‰ (Trumbull et al. 2013; Siegel et al. 2016), so for most granites there should be little isotopic difference between tourmaline and the melt from which it crystallized. The total range in published $\delta^{11}B$ values for S-type granites is −20 to 0‰, with a strong peak between −12 and −8‰; mean and median values are −10.7 (s.d. = 3.8) and −11.1‰, respectively. A discussion of the range of compositions is outside the scope of this review.

We mentioned above that the mineral compositions and metamorphic history in the magma source ultimately control its initial boron concentration and B-isotope ratio, but processes acting near the end of crystallization can cause additional variations, as demonstrated by studies of late-stage zones in granites and granitic pegmatites described in the next section. Researchers seeking to know the "magmatic" signature of boron in a specific study might better analyze the less-differentiated samples from a granite suite to avoid potential ambiguity imparted by late-stage processes.

10.3.2.4 Implications for B-Isotope Composition of the Continental Crust

The frequency maximum of B-isotope ratios in S-type granites (Fig. 10.4b) corresponds very well with the value of −10‰ that is commonly taken as representative for continental crust as a whole (e.g., Marschall and Jiang 2011). However, the frequency maximum for I-type granites is around −3‰ (Fig. 10.4a) and since these rocks are volumetrically more abundant than S-type granites, one should expect the bulk crustal value to be considerably higher than −10‰. In fact, an independent estimate of the continental B-isotope composition based on mass-balance calculations of the continent-mantle-ocean system by Marschall et al. (2016) suggested a $\delta^{11}B$ value of -9.1 ± 2.4‰. This value seems low considering the dominance of I-type granitic rocks in the crust and their relatively high $\delta^{11}B$ values, but one must recall that it is based on boron mass-balance and that I-type granites have low boron concentrations compared to S-types and their metapelite sources.

10.3.3 Late-Stage Granites and Pegmatites: The Magmatic-Hydrothermal Transition

Research on the B-isotope systematics of granitic pegmatites, aplites, and related rocks has been driven by an interest in understanding the complex processes that operate at the end stage of magmatic evolution, and by the economic importance of some of these rocks as sources of rare-metal ores and gemstones. From the earliest studies of Swihart and Moore (1989) and Barth (1993) onward, B-isotope research on pegmatites has been dominated by analysis of tourmaline. On the other hand, much progress in understanding late-magmatic processes has come from studies of melt and fluid inclusions. For example, inclusion studies have demonstrated that some hydrous silicate melts crystallize well below 500 °C and that some show a complete transition from silicate melt to supercritical hypersaline fluids (e.g., London 1986, 2014; Thomas et al. 2000, 2003; Veksler and Thomas 2002; Sirbescu and Nabelek 2003; Peretyazhko et al. 2004). Boron concentrations in melt and fluid inclusions have been measured in several studies, but B-isotope data are essentially lacking, the only known exception being for melt inclusions in Sn-related porphyry intrusions in Bolivia (Wittenbrink et al. 2009) discussed below. Tourmaline continues to play the leading role in B-isotope research because it is easy to study, it constitutes the main boron sink in most systems, it grows under both late-magmatic and hydrothermal conditions, and its zoning provides evidence of changing geochemical conditions. A summary of tourmaline B-isotope studies from late-magmatic and hydrothermal tourmalines is given below (also see Kowalski and Wunder 2017), but we emphasize that tourmaline remains an indirect recorder of fluid compositions and that many more studies of boron and B-isotope variations in fluid/melt inclusions are needed.

Experimental data relevant to late-magmatic processes are also lacking. There is one experimental study of B-isotope fractionation between melt and fluid (Hervig et al. 2002) but inconsistencies with empirical and theoretical evidence suggest a need for revision (see discussion above of I-type magma degassing). Qualitatively, experimental studies and partitioning theory predict that an aqueous fluid exsolved from peraluminous silicate magma will be enriched in ^{11}B relative to the melt to some extent, depending on the temperature and, critically, on the proportion of trigonal and tetrahedral B–O complexes within the melt. The consequence of fluid exsolution for the composition of remaining melt, and for any tourmaline crystallizing from it, is a lowering of $\delta^{11}B$ values relative to the starting magma. In contrast, the exsolved fluid and any hydrothermal tourmaline forming from it will be isotopically heavier than the starting magma (Fig. 10.5). Trumbull et al. (2008, 2013), Drivenes et al. (2015), and Siegel et al. (2016) noted that crystallization of abundant tourmaline can also lower the $\delta^{11}B$ values of the residual melt, and that the formation of muscovite or other minerals having tetrahedral-coordinated boron can generate the opposite effect. The relative importance of crystallization versus fluid exsolution on the B-isotope composition of a residual melt thus depends on the specific conditions of mass balance and open versus closed-system behavior, which may vary on a small scale and with evolution of the system, especially for complex zoned pegmatites that typically contain near-monomineralic zones and replacement bodies (London 2014; Trumbull et al. 2013; Siegel et al. 2016).

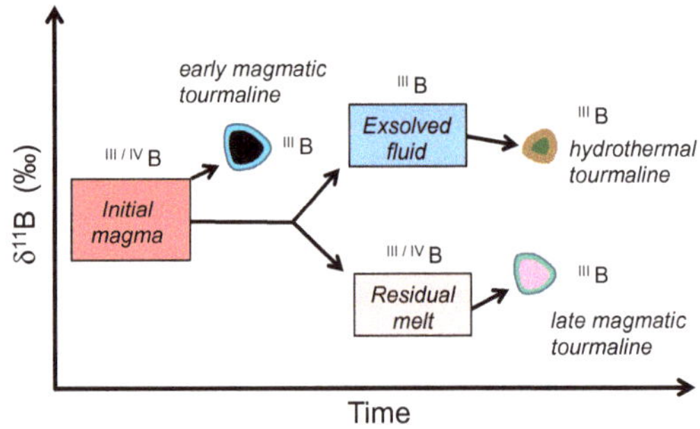

Fig. 10.5 Schematic diagram showing predicted effect of fluid exsolution from granitic magma on B-isotope composition of fluid phase, residual melt, and tourmaline (or other minerals) that form from them. Superscripts III and IV indicate boron coordination numbers in the phases. This diagram illustrates the potential use of B-isotope studies to distinguish the magmatic-hydrothermal transition in mineralized pegmatites and other late-stage products of granitic systems

10.3.3.1 B-Isotope Variations in Early- to Late-Magmatic Tourmaline

A large amount of B-isotope data has been obtained from granitic and pegmatitic tourmalines (Fig. 10.4b and sources therein), but only a handful of detailed studies document B-isotope variations in tourmaline from early- to late-formed units. Smith and Yardley (1996) found a decrease in $\delta^{11}B$ values of 3‰ in coarse-grained to fine-grained magmatic tourmaline from Cornish (England) granites. These data were obtained from sieved fractions of crushed rock, but the authors argued from petrographic observation that the fine-grained fraction represents late growth, and hence related the isotopic difference to loss of ^{11}B to an exsolved fluid. Matthews et al. (2003) also invoked fluid loss to explain a core-to-rim drop of 6‰ for zoned tourmaline in pegmatites from Naxos, Greece. The same trend and a similar magnitude of B-isotope zoning was found by Trumbull et al. (2008) in late-magmatic tourmaline from the Erongo granite, Namibia. However, results of other studies show that B-isotopic evolution in pegmatites can be more complex. For example, Tonarini et al. (1998) found no isotopic difference from root to tip of zoned tourmaline from Elba (Italy) pegmatites, despite evidence for fluid exsolution and a 300 °C temperature gradient during growth of the tourmaline. Trumbull and Chaussidon (1999) compared B-isotope data for magmatic tourmaline from the Sinceni granite, Swaziland, with those for four cogenetic aplite and pegmatite dikes. Three of the four dikes have the same $\delta^{11}B$ value as the granite, whereas one is ca. 5‰ lower, which the authors attibuted to loss of ^{11}B from the melt due to fluid exsolution.

The Borborema district in Brazil contains both barren and rare-element pegmatites that host gem-quality tourmaline. Trumbull et al. (2013) analyzed tourmaline from four mineralized and zoned pegmatites, one barren and unzoned pegmatite, and the parental granite. The granite and all five pegmatites yield a "main range" of $\delta^{11}B$ values from −17 to −9‰. The B-isotopic compositions of the barren and mineralized pegmatites overlap each other but the latter shows a greater isotopic range, typically about 6‰. Among the four zoned pegmatites, one displays a strong and consistent trend to lower $\delta^{11}B$ values in its later, inner and more evolved zones; the same trend was observed from core to rim of zoned tourmaline crystals. The shift to lower B-isotope values in late-formed tourmaline was explained by ^{11}B loss to an exsolved fluid and/or to crystallization of tourmaline in the outer zones. The other zoned pegmatites in that study display equally strong but inconsistent zoning patterns, with some values beyond the "main range" ($\delta^{11}B$ to +1.6‰) being attributed to ingress of externally derived ^{11}B-rich fluids. Trumbull et al. (2013) concluded that the factors controlling B-isotope evolution (crystallization, melt structure, fluid saturation and exsolution, mixing with external reservoirs) can vary locally and on a small scale within individual pegmatites from the same district.

Siegel et al. (2016) made a detailed study of tourmaline from the zoned, rare-element Varutrask pegmatite in Sweden. Their results document a two-stage evolution in B-isotopes, including an early trend toward higher $\delta^{11}B$ values from outer to intermediate zones that the authors attributed to crystallization of abundant mica, followed by a decrease of $\delta^{11}B$ in the most evolved inner zones and replacement bodies. The latter trend was suggested to result from the exsolution of a fluid enriched in ^{11}B after crystallization of the intermediate zones.

There have been other applications of B-isotopes in tourmaline related to the economic geology of gemstones. Galbraith et al. (2009) used B-isotopes and trace elements in tourmaline as a proximity indicator for emerald mineralization in western Canada. Ludwig et al. (2011) showed that the B- and Li-isotope ratios could serve as a provenance control to distinguish the highly valueable, Paraíba-type blue gem tourmalines of Brazil from those of other localities.

To conclude this section, we emphasize that case studies of B-isotope variations in late-stage granites and pegmatites have shown significant and diverse internal variations and zoning patterns that defy explanation by one simple model. There are many factors that influence isotope fractionation at the end stages of granite crystallization (e.g., modal abundance of minerals, variable melt structure, fluid separation, liquid immiscibility), combined with the relatively low solidus temperature of residual melts (<500 °C) that enhances fractionation. In complex pegmatites, these factors can and do vary on local scales (e.g., London 2014). Mineralization in rare-element (Sn, Ta–Nb, Cs, Li, Be) pegmatites tends to occur in the latest, most evolved inner zones and/or in replacement bodies that root within the inner zones and cut outer ones (Černý 2005; Linnen et al. 2012; London 2014). There is still controversy on how these bodies form and whether hydrothermal or magmatic processes are responsible for the high concentration of rare elements. The studies described here suggest that B-isotope data may be able to detect the isotopic shift caused by fluid exsolution (Fig. 10.5), and may also reveal the influence of external fluids. If valid, this approach would provide a major contribution to understanding ore formation. However, to do so will require resolving the question of fluid-melt-mineral isotopic fractionation, by means of experimental work and direct analyses of the melt/fluid phases occurring in nature.

10.3.4 Boron Isotopes in Magmatic-Hydrothermal Ore Deposits

Boron is an element readily mobilized by aqueous fluids, and many hydrothermal ore deposits are associated in one way or another with boron metasomatism, which is expressed most commonly in the form of tourmaline (e.g., Slack 1996; Slack and Trumbull 2011). The value of B-isotope studies for constraining fluid provenance and for recognizing multiple fluid sources in hydrothermal ore deposits has long been recognized (e.g., Palmer and Slack 1989), but the number of applications using this approach remained small until the 1990s with increased availability of microbeam techniques for in situ isotope analysis (SIMS and LA-ICPMS) and of experimentally determined constraints on fluid-mineral fractionation.

Here our focus is on magmatic-hydrothermal ore deposits more or less directly related to I- or S-type magmatism, but it is important to emphasize that B-isotope studies have been applied to many other types of ore deposits as well, owing to the fact that boron metasomatism and tourmaline formation are strikingly widespread. Examples include sediment-hosted volcanogenic massive sulfide (VMS) and sedimentary-exhalative (SEDEX) deposits (Palmer and Slack 1989; Slack et al. 1989, 1993; Jiang et al. 1999), different varieties of iron oxide-copper-gold (IOCG) deposits (Xavier et al. 2008; Trumbull et al. 2011; Tornos et al. 2012), and vein- and reef-type Au, W, and U deposits (Rowins et al. 1997; Jiang et al. 2002; Krienitz et al. 2008; Esmaeily et al. 2009; Garda et al. 2009; Pal et al. 2010; Yavuz et al. 2011; Baksheev et al. 2015; Büttner et al. 2016; Lambert-Smith et al. 2016). The ores and metallogenetic settings represented in these studies are too diverse for a summary here. Suffice it to say that B-isotope compositions in tourmaline have proven valuable in discerning the provenance of ore-bearing fluids and in demonstrating the presence and relative timing of multiple fluid sources (see Slack and Trumbull 2011 and references therein).

It has long been known that mineral deposits related to I- and S-type magmas have different metal associations, i.e., Cu–Fe–Au for I-types and Sn–W for S-types (Chappell and White 2001). Nevertheless, the style of mineralization may be the same for both (for example, porphyry-type deposits). Good examples are "paired" magmatic and mineral belts such as in the central Andes porphyry province where B-isotope studies have been especially informative. Wittenbrink et al. (2009) found major differences in boron concentration and $\delta^{11}B$ values of quartz-hosted melt inclusions in mineralized porphyry intrusions of the Chilean versus Bolivian parts of the province. Melt inclusions associated with the giant El Salvador porphyry Cu–Au deposit in Chile, within the main arc, have moderate boron concentrations (15–78 µg/g, mean 52 µg/g, s.d. 23, n = 10) and B-isotope compositions (−7 to +11.7‰, mean +0.6‰, s.d. 4.6, n = 10), within the typical range for I-type intrusions (Fig. 10.4a). In contrast, melt inclusions occurring within magmatic quartz from the Llallagua, Chorolque, and Cerro Rico Sn–Ag deposits in the "tin belt" of Bolivia have higher B contents (63–576 µg/g, mean 230 µg/g, s.d. 158, n = 10) and much lower $\delta^{11}B$ values (−16.1 to −7.4‰, mean −12.3‰, s.d. 2.7, n = 10), which are typical for S-type granites (Fig. 10.4b). This shift to more negative B-isotope ratios, higher boron contents, and a change in associated metals is consistent with peraluminous bulk compositions, more radiogenic Sr-isotope ratios, and higher O-isotope ratios of the S-type silicic granites from Cerro Rico and other intrusions of the Bolivian tin belt, linked by most workers to the presence of metasedimentary rocks in the basement beneath this region (Lehmann et al. 2000; Wittenbrink et al. 2009). For completeness, it should be mentioned that there is also a distinctive metal association in deposits related to A-type granites (Nb, Zr, REE, Y), but B-isotope studies of these rocks are so far lacking.

10.3.4.1 Ore Deposits with I-Type Association

The most common ore associated with I-type magmas is in porphyry Cu±Mo±Au deposits and related skarns. These occur worldwide, but to date B-isotope studies of porphyry deposits have been restricted to the Cordilleran belts of North and South America. The melt inclusion study of Wittenbrink et al. (2009) on the El Salvador porphyry Cu deposit of Chile was described above. All other published studies have involved analysis of B-isotopes in tourmaline. For example, Tornos et al. (2012) reported $\delta^{11}B$ values of −3 to −1‰ for tourmaline in the Copucha porphyry Cu deposit in Chile that fall in the range for I-type magmas (Fig. 10.4a). Iveson et al. (2016) found very similar values for tourmaline in the Margaret porphyry Cu-Mo deposit and in the associated Spirit Lake Pluton, California (+3.8 and −2.6 to +1.9‰, respectively). Unpublished analyses from the Potsdam SIMS laboratory of tourmaline from breccias in the Rio Blanco-Los Bronces porphyry Cu deposit in Chile also reveal isotopically heavy B (average +3‰), consistent with an I-type magma affinity. The Cordilleran provinces of Chile and the southwestern USA also contain breccia- and vein-hosted Cu±Au deposits that are spatially associated with I-type intrusions like the porphyry deposits, but are rich in iron oxides and therefore classified as IOCG type (Sillitoe 2003; Barton 2014). These include deposits near Copiapó and Taltal in northern Chile and some deposits in the Yerington district of Nevada. The genetic relation of these IOCG deposits to magmatism is controversial and B-isotope studies have provided important insights. For example, tourmaline from IOCG deposits at Taltal studied by Tornos et al. (2012) yield $\delta^{11}B$ values of −8 to +6‰ (one outlier at −10.5‰), which were attributed to magma-derived boron based on similarity to values for Chilean arc granites and to tourmaline (−3 to −1‰) from the nearby Copucha porphyry Cu deposit. Barton (2014) showed that tourmaline from IOCG deposits in the Copiapó area (Chile) and Yerington (USA) has essentially the same range of $\delta^{11}B$ values (−7.5 to +4.2‰ and −5 to +5‰, respectively). Barton (2014) acknowledged that the B-isotope data are permissive of a magmatic fluid source, but pointed out other features (e.g., alteration patterns) better explained by externally derived fluids. Examples where B-isotopes in tourmaline from IOCG deposits indicate non-magmatic or mixed fluids include the study by Xavier et al. (2008) from the Carajás district, Brazil, who inferred a marine boron source based on $\delta^{11}B$ values as high as +26‰, and Su et al. (2016) who suggested a mixed source of magmatic (low $\delta^{11}B$) and external (high $\delta^{11}B$) fluids in the Dahungshan IOCG deposit, southwest China.

10.3.4.2 Ore Deposits with S-Type Association

The B-isotope composition of subvolcanic porphyry stocks associated with Sn mineralization in Bolivia (Wittenbrink et al. 2009) was mentioned above. Classic examples of hydrothermal ore deposits spatially related to S-type granites are the greisen- and vein-hosted Sn–W deposits of Cornwall (England), which was the subject of an early B-isotope study by Smith and Yardley (1996). They reported $\delta^{11}B$ values of tourmaline from quartz-cassiterite veins (−10 to −8‰) and quartz-wolframite-cassiterite-Li mica veins (−7.3 to −5.2‰) at Cligga Head. Magmatic tourmaline in the Cligga Head granite has lower $\delta^{11}B$ values (−12.5 to −7‰), the difference being consistent with an isotopically heavier fluid exsolved from the granite. Drivenes et al. (2015) studied tourmaline from the Land's End granite in Cornwall and a quartz-tourmaline-cassiterite vein, reporting essential overlap in the B-isotope ratios of both but a tendency for lower values in the vein. Some studies of granite-greisen Sn and W–Sn vein systems in China have yielded similar results as the Cornish examples. For the Yidong granite in South China (Jiang 2001), a progression was observed to isotopically heavier tourmaline in the sequence of granite (−13.9‰) to greisen (−12.7 to −11.9‰) to tourmaline-cassiterite-sulfide veins (−11.4 to −11.2‰). Another example of this type is the Baotan Sn deposit (Zhang et al. 2014) in which granitic tourmaline has $\delta^{11}B$ values of −13.9 to −12.0‰, whereas hydrothermal tourmaline from quartz-cassiterite veins is isotopically slightly heavier (−13.0 to −9.5‰). For the Baiganhu W-Sn deposit (northwest China), in situ LA-ICPMS analysis of tourmaline from quartz veins (Zheng et al. 2016) revealed a relatively narrow range of $\delta^{11}B$ values from −12.9 to −7.9‰ and a lack of correlation with either grain color (brown versus blue-green) or crystal zones (core versus rim). No B-isotope data from the related monzogranite were reported. Hydrothermal tourmaline from Mo–W stockwork veins in a deposit in southeast Finland (Molnár et al. 2016) yields slightly lower B-isotope compositions than magmatic tourmaline in the same district (−17.2 to −12.2‰ vs −14.5 to −10.8‰, respectively). Duncan et al. (2014) compared tourmaline compositions from granitic pegmatites and various quartz-tourmaline veins in the Mount Isa district, Australia. Data for pegmatitic tourmalines and four out of five vein samples overlap in the range of −14.2 to −5.1‰, which was interpreted as a result of magmatic degassing during tourmaline crystallization. Tourmaline in the fifth vein has much higher $\delta^{11}B$ values (+6 and +11‰), which was attributed to an external fluid with isotopically heavy boron derived from evaporitic sediments. High Br/Cl ratios in fluid inclusions and unusually Al-poor and Fe-rich compositions of the tourmaline provided supporting evidence for this latter model (Duncan et al. 2014). As a last example, Huang et al. (2016) used B-isotope data to negate a magmatic fluid origin for vein-type Cu mineralization associated in time and space with a peraluminous granite in southwest China. They reported a sharp contrast in B-isotope compositions between tourmaline from the granite (−14.6 to −12.2‰) and from the veins and alteration selvages (−6.3 to +0.5‰).

10.4 Summary

We have shown that the isotopic composition of boron in metamorphic and igneous rocks of the continental crust varies over a wide range from −20 to +10‰. This range is related mainly to the protolith but fractionation during phase transitions also plays a role at low temperatures (<500 °C). In general, boron is prone to loss during metamorphic devolatilization reactions, which also causes a decrease in the $\delta^{11}B$ value. However, prograde metamorphism and anatexis have little effect on B-isotopes if boron-retentive minerals like tourmaline form in sufficient amount. Our new compilation of over 250 B-isotope analyses from about 90 localities of felsic igneous rocks, presented here, documents a first-order distinction between I-type and S-type magmas. Boron in I-type magmas is isotopically heavy (mean $\delta^{11}B$ = −2‰, s.d. = 5), even relative to unaltered MORB, which we attribute to the contribution of heavy boron from seawater-altered oceanic crust and pelagic sediments. Boron in S-type magmas has a light isotopic signature (mean $\delta^{11}B$ = −11‰, s.d. = 4), which derives from the preferential take-up of ^{10}B by clays and micas. This contrast between I- and S-type magmas requires some caution when discussing the composition of "crustal" boron and discriminating a magmatic versus non-magmatic origin for hydrothermal deposits. The commonly cited value of −10‰ for the continental crust is based on data from upper crustal rocks and unlikely to represent the bulk crust. Nevertheless, mass-balance considerations demand that the continental crust be isotopically lighter than the mantle because seawater, pelagic marine sediments, and altered oceanic crust all have B-isotope values that are higher than the mantle (Marschall et al. 2017). A mass-balance model for Earth's crust-mantle-ocean system that takes into account the reservoir masses and their composition by Marschall et al. (2017) predicted a $\delta^{11}B$ value for the bulk crust of −9.1‰ ± 2.4 versus −7‰ ± 1 for the mantle.

The contrasting B-isotope compositions of I- versus S-type granites and rhyolites also holds for the magmatic-hydrothermal ore deposits related to them, as documented mainly by the composition of tourmaline from these deposits. This fact is demonstrated in a review of data from porphyry and IOCG systems (I-type) and of Sn–W vein systems and granitic pegmatites (S-type). In principle, therefore, the B-isotope composition of hydrothermal tourmaline can help distinguish the nature of the genetically related igneous system. However, significant and locally complex effects of isotopic fractionation due to fluid-rock and fluid-magma interaction also need to be considered.

Finally, it is worth emphasizing that empirical studies of B-isotopes have partly out-paced experimental studies on boron complexation and isotopic fractionation behavior. For complex granitic pegmatites, in particular, there are a wealth of intriguing isotopic variations and zonations but a lack of fundamental experimental data with which to interpret them. Furthermore, the overwhelming emphasis of recent research on B-isotopes in continental rocks has been on tourmaline. That mineral has many excellent qualities and will remain important long into the future, but it can be expected that additional studies of boron and B-isotopes in other rock-forming minerals such as white mica, and in melt inclusions from various igneous rocks, will also yield important new insights.

Acknowledgements The authors wish to thank Horst Marschall for the invitation to contribute to this book and for his insightful comments on the manuscript. This chapter benefitted from reviews by Robert Ayuso and Robert Seal (both U.S. Geological Survey, Reston), Shao-Yong Jiang (China University of Geosciences, Wuhan), and Mathias Konrad-Schmolke (University of Gothenburg, Göteborg).

References

Acosta-Vigil A, Pereira MD, Shaw DM, London D (2001) Contrasting behavior of boron during crustal anatexis. Lithos 56:15–31

Acosta-Vigil A, London D, Morgan GB IV, Dewers TA (2003) Solubility of excess aluminum in hydrous granitic melts in equilibrium with peraluminous minerals at 700–800 °C and 200 MPa, and applications of the aluminum saturation index. Contrib Min Petrol 146:100–119

Anderson AT, Davis AM, Lu FQ (2000) Evolution of Bishop Tuff rhyolitic magma based on melt and magnetite inclusions and zoned phenocrysts. J Petrol 41:449–473

Audetat A, Guenther D, Heinrich CA (2000) Magmatic-hydrothermal evolution in a fractionating granite: a microchemical study of the Sn–W–F-mineralized Mole Granite (Australia). Geochim Cosmochim Acta 64:3373–3393

Badanina EV, Syritso LF, Volkova EV, Thomas R, Trumbull RB (2010) Composition of Li–F granite melt and its evolution during the formation of the ore-bearing Orlovka massif in eastern Transbaikalia. Petrology 18:131–157

Badanina EV, Veksler IV, Thomas R, Syritso LF, Trumbull RB (2004) Magmatic evolution of Li–F, rare-metal granites: a case study of melt inclusions in the Khangilay Complex, eastern Transbaikalia (Russia). Chem Geol 210:113–133

Baksheev IA, Prokofiev VY, Trumbull RB, Wiedenbeck M, Yapaskurt VO (2015) Geochemical evolution of tourmaline in the Darasun gold district deposits, Transbaikal region, Russia: evidence from chemical and boron isotope compositions in tourmaline. Min Deposita 50:125–138

Baltatzis E, Kostopoulos D, Godelitsas A, Zachariadis P, Papanikolaou D (2009) Pliocene tourmaline rhyolite dykes from Ikaria Island in the Aegean back-arc region: geodynamic implications. Geodinam Acta 22:189–199

Barth S (1993) Boron isotope variations in nature: a synthesis. Geologische Rund 82:640–651

Barton MD (2014) Iron oxide (−Cu–Au–REE–P–Ag–U–Co) systems. In: Scott SD (ed) Geochemistry of mineral deposits. Treatise on Geochemistry, Elsevier, Amsterdam, vol 13, pp 515–541

Bebout GE, Nakamura E (2003) Record in metamorphic tourmalines of subduction zone devolatilization and boron cycling. Geology 31:407–410

Bebout GR, Agard P, Kobayashi K, Moriguti T, Nakamura E (2013) Devolatilization history and trace element mobility in deeply subducted sedimentary rocks: evidence from Western Alps HP/UHP suites. Chem Geol 342:1–20

Büttner SH, Reid W, Glodny J, Wiedenbeck M, Chuwa G, Moloto T, Gucsik A (2016) Fluid sources in the Twangiza-Namoya gold belt (Democratic Republic of Congo): evidence from tourmaline and fluid compositions, and from boron and Rb–Sr isotope systematics. Prec Res 280:161–178

Cawood PA, Hawkesworth CJ, Dhuime B (2013) The continental record and the generation of continental crust. Geol Soc America Bull 125:14–32

Černý P (2005) The Tanco rare-element pegmatite deposit, Manitoba: regional context, internal anatomy, and global comparisons. In: Linnen RL, Samson IM (eds) Rare-element geochemistry and mineral deposits, GAC Short Course Notes 17, pp 127–158

Chappell BW, White AJR (2001) Two contrasting granite types: 25 years later. Austral J Earth Sci 48:489–499

Chaussidon M, Albarede F (1992) Secular boron isotope variations in the continental crust: an ion microprobe study. Earth Planet Sci Lett 108:229–241

Chaussidon M, Appel PWU (1997) Boron isotopic composition of tourmalines from the 3.8 Ga-old Isua supracrustals, West Greenland: implications on the $\delta^{11}B$ value of Early Archean seawater. Chem Geol 136:171–180

Clemens JD, Stevens G (2012) What controls chemical variation in granitic magmas? Lithos 134–135:317–329

da Costa LB, Mourão C, Récio C, Guimarães F, Antunes IM, Ramos JF, Barriga FJAS, Palmer MR, Milton JA (2014) Tourmaline occurrences within the Penamacor-Monsanto granitic pluton and host-rocks (central Portugal): genetic implications of crystal-chemical and isotopic features. Contrib Mineral Petrol 167:993. doi:10.1007/s00410-014-0993-7

Deegan FM, Troll VR, Whitehouse MJ, Jolis EM, Freda C (2016) Boron isotope fractionation in magma via crustal carbonate dissolution. Scientific Reports 6:30774. doi:10.1038/srep30774

De Hoog CJ, Savov IP (2017) Subduction zones, dehydration, metasomatism, mud and serpentinite volcanoes, and arc magmatism. In: Marschall HR, Foster GL (eds) Boron isotopes—The fifth element, Advances in Isotope Geochemistry, vol 7, Springer, Heidelberg, 219–250

Dhuime B, Wuestefeld A, Hawkesworth CJ (2015) Emergence of modern continental crust about 3 billion years ago. Nature Geosci 8:552–555

Di Renzo V, Arienzo I, Civetta L, D'Antonio M, Tonarini S, Di Vito MA, Orsi G (2011) The magmatic feeding system of the Campi Flegrei caldera: architecture and temporal evolution. Chem Geol 281:227–241

Dingwell DB, Pichavant M, Holtz F (1996) Experimental studies of boron in granitic melts. In: Grew ES, Anovitz LM (eds) Boron: mineralogy, petrology and geochemistry. Rev Mineral 33, Mineral Soc America, Washington, DC, pp 331–386

Drivenes K, Larsen RB, Mueller A, Borensen BE, Wiedenbeck M, Raanes MP (2015) Late-magmatic immiscibility during batholith formation: assessment of B isotopes and trace elements in tourmaline from the Land's End Granite, SW England. Contrib Min Petrol 169:56. doi:10.1007/s00410-015-1151-6

Duncan RJ, Buick IS, Kobayashi K, Wilde AR (2014) Chemical and stable isotopic characteristics of syn-tectonic tourmaline from the western fold belt, Mount Isa inlier, Queensland, Australia. Chem Geol 381:131–143

Esmaeily D, Trumbull RB, Haghnazar M, Krienitz M-S, Wiedenbeck M (2009) Chemical and boron isotopic composition of hydrothermal tourmaline from scheelite-quartz veins at Nezamabad, western Iran. Eur J Min 21:347–360

Farber K, Dziggel A, Trumbull RB, Meyer FM, Wiedenbeck M (2015) Tourmaline B-isotopes as tracers of fluid sources in silicified Palaeoarchean oceanic crust from the Mendon Formation, Barberton greenstone belt, South Africa. Chem Geol 417:134–147

Frost BR, Barnes CG, Collins WJ, Arculus RJ, Ellis DJ, Frost CD (2001) A geochemical classification for granitic rocks. J Petrol 42:2033–2048

Gaillardet J, Lemarchand D (2017) Boron isotopes in riverine systems and the weathering environment. In: Marschall HR, Foster GL (eds) Boron Isotopes—The Fifth Element, Advances in Isotope Geochemistry, vol. 7, Springer, Heidelberg, 165–189

Galbraith CG, Clarke DB, Trumbull RB, Wiedenbeck M (2009) Assessment of tourmaline compositions as an indicator of emerald mineralization at the Tsa da Glisza prospect, Yukon Territory, Canada. Econ Geol 104:713–731

Garda GM, Trumbull RB, Beljavskis P, Wiedenbeck M (2009) Boron isotope composition of tourmalinite and vein tourmalines associated with gold mineralization, Serra do Itaberaba Group, central Ribeira belt, SE Brazil. Chem Geol 264:207–220

Gao S, Luo T-C, Zhang B-R, Zhang H-F, Han Y-W, Zhao Z-D, Hu Y-K (1998) Chemical composition of the continental crust as revealed by studies in East China. Geochim Cosmochim Acta 62:1959–1975

Grew ES (1996) Borosilicates (exclusive of tourmaline) and boron in rock-forming minerals in metamorphic rocks. In: Grew ES, Anovitz LM (eds) Boron: mineralogy, petrology and geochemistry. Rev Mineral 33, Mineral Soc America, Washington, DC, pp 387–502

Grew ES, Anovitz LM, eds (1996) Boron: mineralogy, petrology and geochemistry. Rev Mineral 33, Mineral Soc America, Washington, DC, 864 p

Grew ES, Dymek RF, DeHoog JCM, Harley SL, Boak J, Hazen RM, Yates MG (2015) Boron isotopes in tourmaline from the ca. 3.7–3.8 Ga Isua supracrustal belt, Greenland: sources for boron in Eoarchean continental crust and seawater. Geochim Cosmochim Acta 163:156–177

Griffin WL, Belousova EA, O'Neill C, O'Reilly SY, Malkovets V, Pearson NJ, Spetsius S, Wilde SA (2014) The world turns over: Hadean-Archean crust-mantle evolution. Lithos 189:2–15

Halama R, Konrad-Schmolke M, Sudo M, Marschall HR, Wiedenbeck M (2014) Effects of fluid–rock interaction on $^{40}Ar/^{39}Ar$ geochronology in high-pressure rocks (Sesia-Lanzo zone, western Alps). Geochim Cosmochim Acta 126:475–494

Hervig RL, Moore GM, Williams LB, Peacock SM, Holloway JR, Roggensack K (2002) Isotopic and elemental partitioning of boron between hydrous fluid and silicate melt. Am Min 87:769–774

Hezel DC, Kalt A, Marschall HR, Ludwig T, Meyer H-P (2011) Major-element and Li, Be compositional evolution of tourmaline in an S-type granite-pegmatite system and its country rocks: an example from Ikaria, Aegean Sea, Greece. Can Min 49:321–340

Huang S-Q, Song Y-C, Hou Z-Q, Xue C-D (2016) Chemical and stable isotopic (B, H, and O) compositions of tourmaline in the Maocaoping vein-type Cu deposit, western Yunnan, China: constraints on fluid source and evolution. Chem Geol 439:173–188

Ishikawa T, Nakamura E (1994) Origin of the slab component in arc lavas from across-arc variation of B and Pb isotopes. Nature 370:205–208

Ishikawa T, Tera F, Nakazawa T (2001) Boron isotope and trace element systematics of the three volcanic zones in the Kamchatka arc. Geochim Cosmochim Acta 65:4523–4537

Iveson AA, Webster JD, Rowe MC, Neill OK (2016) Magmatic–hydrothermal fluids and volatile metals in the Spirit Lake pluton and Margaret Cu–Mo porphyry system, SW Washington, USA. Contrib Min Petrol 171, 20. doi:10.1007/s00410-015-1224-6

Jiang S-Y (2001) Boron isotope geochemistry of hydrothermal ore deposits in China: a preliminary study. Phys Chem Earth (A) 26:851–858

Jiang S-Y, Palmer MR (1998) Boron isotope systematics of tourmaline from granites and pegmatites: a synthesis. Eur J Min 10:1253–1265

Jiang S-Y, Palmer MR, Slack JF, Shaw DR (1999) Boron isotope systematics of tourmaline formation in the Sullivan Pb–Zn–Ag deposit, British Columbia, Canada. Chem Geol 158:131–144

Jiang S-Y, Palmer MR, Yeats CJ (2002) Chemical and boron isotopic compositions of tourmaline from the Archean Big Bell and Mount Gibson gold deposits, Murchison Province, Yilgarn craton, Western Australia. Chem Geol 188:229–247

Jiang S-Y, Radvanec M, Nakamura E, Palmer M, Kobayaski K, Zhao H-X, Zhao K-D (2008) Chemical and boron isotopic variations of tourmaline in the Hnilec granite-related hydrothermal system, Slovakia: constraints on magmatic and metamorphic fluid evolution. Lithos 106:1–11

Jochum K-P, et al (2006) MPI-DING reference glasses for in situ microanalysis: new reference values for element concentrations and isotope ratios. Geochem Geophys Geosyst 7:Q02008. doi:10.1029/2005GC001060

Jones RE, De Hoog JCM, Kirstein LA, Kasemann S, Hinton R, Elliot T, Litvak VD, EIMF (2014) Temporal variations in the influence of the subducting slab on central Andean arc magmas: evidence from boron isotope systematics. Earth Planet Sci Lett 408, 390–401

Kaliwoda M, Marschall HR, Marks MAH, Ludwig T, Altherr R, Markl G (2011) Boron and boron isotope systematics in the peralkaline Ilímaussaq intrusion (South Greenland) and its granitic country rocks: a record of magmatic and hydrothermal processes. Lithos 125:51–64

Kasemann S, Erzinger J, Franz G (2000) Boron recycling in the continental crust of the central Andes from the Paleozoic to Mesozoic, NW Argentina. Contrib Min Petrol 140:328–343

Kawakami T, Ikeda T (2003) Boron in metapelites controlled by the breakdown of tourmaline and retrograde formation of borosilicates in the Yanai area, Ryoke metamorphic belt, SW Japan. Contrib Min Petrol 145:131–150

Konrad-Schmolke M, Halama R (2014) Combined thermodynamic–geochemical modeling in metamorphic geology: boron as tracer of fluid–rock interaction. Lithos 208–209:393–414

Kowalski PM, Wunder B, Jahn S (2013) Ab initio prediction of equilibrium boron isotope fractionation between minerals and aqueous fluids at high P and T. Geochim Cosmochim Acta 101:285–301

Kowalski P, Wunder B (2017) Boron-isotope fractionation among solids-fluids-melts: experiments and atomic modeling. In: Marschall HR, Foster GL (eds) Boron Isotopes—The Fifth Element, Advances in Isotope Geochemistry, vol 7, Springer, Heidelberg, 33–70

Krienitz M-S, Trumbull RB, Hellmann A, Kolb J, Meyer FM, Wiedenbeck M (2008) Hydrothermal gold mineralization at the Hira Buddini gold mine, India: constraints on fluid sources and evolution from boron isotopic compositions of tourmaline. Min Deposita 43:421–434

Lambert-Smith JS, Rocholl A, Treloar PJ, Lawrence DM (2016) Discriminating fluid source regions in orogenic gold deposits using B-isotopes. Geochim Cosmochim Acta 194:57–76

Leeman WP, Sisson, VB (1996) Geochemistry of boron and its implications for crustal and mantle processes. In: Grew ES, Anovitz LM (eds) Boron: mineralogy, petrology and geochemistry. Rev Mineral 33, Miner Soc America, Washington, DC, pp 645–707

Leeman WP, Tonarini S, Chan LH, Borg LE (2004) Boron and lithium isotopic variations in a hot subduction zone—the southern Washington Cascades. Chem Geol 212:101–124

Lehmann B, Dietrich A, Heinhorst J, Metrich N, Mosbah M, Palacios C, Schneider H-J, Wallianos A, Webster J, Winkelmann L (2000) Boron in the Bolivian tin belt. Min Deposita 35:223–232

Linnen RL, Lichtervelde MV, Černý P (2012) Granitic pegmatites as sources of strategic metals. Elements 8:275–280

London D (1986) Magmatic-hydrothermal transition in the Tanco rare metal pegmatite: evidence from fluid inclusions and phase equilibrium experiments. Amer Min 71:376–395

London D (2011) Experimental synthesis and stability of tourmaline: a historical overview. Can Min 49:117–136

London D (2014) A petrologic assessment of internal zonation in granitic pegmatites. Lithos 184–187:74–104

London D (2015) Reply to Thomas and Davidson on "A petrologic assessment of internal zonation in granitic pegmatites" (London, 2014a). Lithos 212–215:469–484

London D, Morgan V1 GB, Wolf MB (1996) Boron in granitic rocks and their contact aureoles. In: Grew ES, Anovitz LM (eds) Boron: mineralogy, petrology and geochemistry. Rev Mineral 33, Mineral Soc America, Washington, DC, pp 299–330

London D, Morgan VIGB, Acosta-Vigil A (2012) Experimental simulations of anatexis and assimilation involving metapelite and granitic melt. Lithos 153:292–307

Ludwig T, Marschall HR, Pogge von Strandmann PAE, Shabaga BM, Fayek M, Hawthorne FC (2011) A secondary ion mass spectrometry (SIMS) re-evaluation of B and Li isotopic compositions of Cu-bearing elbaite from three global localities. Min Mag 75:2485–2494

MacGregor JR, Grew ES, DeHoog JCM, Harley SL, Kowalski PM, Yates MG, Carson CJ (2013) Boron isotopic composition of tourmaline, prismatine, and grandidierite from granulite facies paragneisses in the Larsemann Hills, Prydz Bay, East Antarctica: evidence for a non-marine evaporite source. Geochim Cosmochim Acta 123:261–283

Marschall HR (2017) Boron isotopes in the ocean floor realm and the mantle. In: Marschall HR, Foster GL (eds) Boron Isotopes—The Fifth Element, Advances in Isotope Geochemistry, vol 7, Springer, Heidelberg, 191–217

Marschall HR, Jiang S-Y (2011) Tourmaline isotopes: no element left behind. Elements 7:313–319

Marschall HR, Ludwig T (2006) Re-examination of the boron isotopic composition of tourmaline from the Lavicky granite, Czech Republic, by secondary ion mass spectrometry: back to normal. Geochem J 40:631–638

Marschall HR, Altherr R, Kalt A, Ludwig T (2008) Detrital, metamorphic and metasomatic tourmaline in high pressure metasediments from Syros (Greece): intra-grain boron isotope patterns determined by secondary-ion mass spectrometry. Contrib Min Petrol 155:703–717

Marschall HR, Korsakov AV, Luvizotto GL, Nasdala L, Ludwig T (2009) On the occurrence and boron isotopic composition of tourmaline in (ultra) high-pressure metamorphic rocks. J Geol Soc London 166:811–832

Marschall HR, Wanless VD, Shimizu N, Pogge von Strandmann PAE, Elliott T, Monteleone BD (2017) The boron and lithium isotopic composition of mid-ocean ridge basalts and the mantle. Geochim Cosmochim Acta 207:102–138

Mathews A, Pullitz B, Hamiel Y, Hervig RL (2003) Volatile transport during the transportation of anatectic melts: oxygen, boron and hydrogen stable isotope study on the metamorphic complex of Naxos, Greece. Geochim Cosmochim Acta 67:3145–3163

Molnár F, Mänttäri I, O'Brient H, Lahaye Y, Pakkanen L, Johanson B, Käpyaho A, Sorjonen-Ward P, Whitehouse M, Sakellaris G (2016) Boron, sulphur and copper isotope systematics in the orogenic gold deposits of the Archaean Hattu schist belt, eastern Finland. Ore Geol Rev 77:133–162

Moran AE, Sisson VB, Leeman WP (1992) Boron depletion during progressive metamorphism: implications for subduction processes. Earth Planet Sci Lett 111:331–349

Nakamo T, Nakamura E (2001) Boron isotope geochemistry of metasedimentary rocks and tourmalines in a subduction zone metamorphic suite. Phys Earth Planet Inter 127:233–252

Pal DC, Trumbull RB, Wiedenbeck M (2010) Chemical and boron isotope compositions of tourmaline from the Jaduguda U (–Cu–Fe) deposit, Singhbhum shear zone, India: implications for the source and evolution of the mineralizing fluid. Chem Geol 277:245–260

Palmer MR, Slack JF (1989) Boron isotopic composition of tourmaline from massive sulfide deposits and tourmalinites. Contrib Min Petrol 103:434–451

Palmer MR, Sturchio NC (1990) The boron isotope systematics of the Yellowstone National Park (Wyoming) hydrothermal system: a reconnaissance. Geochim Cosmochim Acta 54:2811–2815

Palmer MR, Swihart GH (1996) Boron isotope geochemistry: an overview. In: Grew ES, Anovitz LM (eds) Boron: mineralogy, petrology and geochemistry. Rev Mineral 33, Mineral Soc America, Washington, DC, pp 709–744

Pereira Gomez MD, Shaw DM (1997) Behaviour of boron in the generation of an anatectic complex: the Peña Negra complex, central Spain. Lithos 40:179–188

Peretyazhko IS, Zagorsky VY, Smirnov SZ, Mikhailov MY (2004) Conditions of pocket formation in the Oktyabrskaya tourmaline-rich gem pegmatite (the Malkhan field, central Transbaikalia, Russia). Chem Geol 210:91–111

Pesquera A, Torres-Ruiz J, Gil-Crespo PP, Jiang S-Y (2005) Petrographic, chemical and B-isotopic insights into the origin of tourmaline-rich rocks and boron recycling in the Martinamor antiform (central Iberian zone, Salamanca, Spain). J Petrol 46:1013–1044

Romer RL, Meixner A (2014) Lithium and boron isotopic fractionation in sedimentary rocks during metamorphism– the role of rock compositions and protolith mineralogy. Geochim Cosmochim Acta 128:158–177

Romer RL, Meixner A, Förster H-J (2014) Lithium and boron in late-orogenic granites—isotopic fingerprints for the source of crustal melts? Geochem Cosmochim Acta 131:98–114

Rosner M, Erzinger J, Franz G, Trumbull RB (2003) Slab-derived boron isotope signatures in arc volcanic rocks from the central Andes and evidence for boron isotope fractionation during progressive slab dehydration. Geochem Geophys Geosys. doi:10.1029/2002GC000438

Rowins SM, Groves DI, McNaughton NJ (1997) A reinterpretation of the role of granitoids in the genesis of Neoproterozoic gold mineralization in the Telfer dome, Western Australia. Econ Geol 92:133–160

Rudnick RL, Gao S (2003) Composition of the continental crust. In: Holland HD, Turrekian KK (eds) Treatise on geochemistry, Elsevier, Amsterdam, vol 3, pp 1–64

Sauerer A, Troll G (1990) Abundance and distribution of boron in the Hauzenberg (Bavaria) granite complex. Geochim Cosmochim Acta 54:49–55

Savov IP, Leeman WP, Lee CA, Shirey SB (2009) Boron isotopic variations in NW USA rhyolites: Yellowstone, Snake River Plain, eastern Oregon. J Volc Geother Res 188:162–172

Schmitt AK, Simon JI (2004) Boron isotopic variations in hydrous rhyolitic melts: a case study from Long Valley, California. Contrib Min Petrol 146:590–605

Schmitt AK, Kasemann S, Meixner A, Rhede D (2002) Boron in central Andean ignimbrites: implications for crustal boron cycles in an active continental margin. Chem Geol 183:333–347

Siegel K, Wagner T, Trumbull RB, Jonsson E, Matalin G, Wälle M, Heinrich CA (2016) Stable isotope (B, H, O) and mineral-chemistry constraints on the magmatic to hydrothermal evolution of the Varuträsk rare-element pegmatite (northern Sweden). Chem Geol 421:1–16

Sillitoe RH (2003) Iron oxide-copper-gold deposits: an Andean view. Min Deposita 38:787–812

Sirbescu MLC, Nabelek PI (2003) Crystallization conditions and evolution of magmatic fluids in the Harney Peak Granite and associated pegmatites, Black Hills, South Dakota—evidence from fluid inclusions. Geochim Cosmochim Acta 67:2443–2465

Slack JF (1996) Tourmaline associations with hydrothermal ore deposits. In: Grew ES, Anovitz LM (eds) Boron: mineralogy, petrology and geochemistry, Rev Mineral 33, Mineral Soc America, Washington, DC, pp 559–644

Slack JF, Trumbull RB (2011) Tourmaline as a recorder of ore-forming processes. Elements 7:321–326

Slack JF, Palmer MR, Stevens BPJ (1989) Boron isotope evidence for the involvement of non-marine evaporites in the origin of the Broken Hill ore deposits. Nature 342:913–916

Slack JF, Palmer MR, Stevens BPJ, Barnes RG (1993) Origin and significance of tourmaline-rich rocks in the Broken Hill district, Australia. Econ Geol 88:505–541

Smith HJ, Leeman WP, Davidson J, Spivack AJ (1997) The B isotopic composition of arc lavas from Martinique, Lesser Antilles. Earth Planet Sci Lett 146:303–314

Smith MP, Yardley BWD (1996) The boron isotopic composition of tourmaline as a guide to fluid processes in the southwestern England orefield: an ion microprobe study. Geochim Cosmochim Acta 60:1415–1427

Su Z-K, Zhao X-F, Li X-C, Zhou M-F (2016) Using elemental and boron isotopic compositions of tourmaline to trace fluid evolutions of IOCG systems: the world class Dahongshan Fe–Cu deposit in SW China. Chem Geol 441:265–279

Swihart GH, Moore PB (1989) A reconnaissance of the boron isotopic composition of tourmaline. Geochim Cosmochim Acta 53:911–916

Thomas R (2002) Determination of the H_3BO_3 concentration in fluid and melt inclusions in granite pegmatites by laser Raman microprobe spectroscopy. Amer Min 87:56–68

Thomas R, Davidson P, Beurlen H (2012) The competing models for the origin and internal evolution of granitic pegmatites in the light of melt and fluid inclusion research. Min Petrol 106:55–73

Thomas R, Foerster H-J, Heinrich W (2003) The behavior of boron in a peraluminous granite-pegmatite system and associated hydrothermal solutions: a melt and fluid-inclusion study. Contrib Min Petrol 144:457–472

Thomas R, Webster JD, Heinrich W (2000) Melt inclusions in pegmatite quartz: complete miscibility between silicate melts and hydrous fluids at low pressure. Contrib Min Petrol 139:394–401

Tonarini S, D'Antonio M, Di Vito MA, Orsi G, Carandente A (2009) Geochemical and B–Sr–Nd isotopic evidence for mingling and mixing processes in the magmatic system that fed the Astroni Volcano (4.1–3.8 ka) within the Campi Flegrei caldera (southern Italy). Lithos 107:135–151

Tonarini S, Dini A, Pezzotta F, Leeman WP (1998) Boron isotopic composition of zoned (schorl–elbaite) tourmalines, Mt Capanne Li–Cs pegmatites, Elba (Italy). Eur J Min 10:941–952

Tonarini S, Forte C, Petrini R, Ferrara G (2003) Melt/biotite $^{11}B/^{10}B$ isotopic fractionation and the boron local environment in the structure of volcanic glasses. Geochim Cosmochim Acta 67:1863–1873

Tonarini S, Leeman WP, Civetta L, D'Antonio M, Ferrara G, Necco A (2004) B/Nb and δ11B systematics in the Phlegrean volcanic district, Italy. J Volc Geother Res 133:123–139

Tornos F, Wiedenbeck M, Velasco F (2012) The boron isotope geochemistry of tourmaline-rich alteration in the IOCG systems of northern Chile: implications for a magmatic-hydrothermal origin. Min Deposita 47:483–499

Trumbull RB, Chaussidon M (1999) Chemical and boron isotopic compositions of magmatic and hydrothermal tourmalines from Sinceni granite-pegmatite system in Swaziland. Chem Geol 153:125–137

Trumbull RB, Krienitz M-S, Gottesmann B, Wiedenbeck M (2008) Chemical and boron-isotope variations in tourmalines from an S-type granite and its source rocks: the Erongo granite and tourmalinites in the Damara belt, Namibia. Contrib Min Petrol 155:1–18

Trumbull RB, Slack JF, Krienitz M-S, Belkin HE, Wiedenbeck M (2011) Fluid sources and metallogenesis in the Blackbird Co–Cu–Au–Bi–Y–REE district, Idaho, USA: insights from major-element and boron isotopic compositions of tourmaline. Can Min 49:225–244

Trumbull RB, Beurlen H, Wiedenbeck M, Soares DR (2013) The diversity of B-isotope variations in tourmaline from rare-element pegmatites in the Borborema Province of Brazil. Chem Geol 352:47–62

van Hinsberg VJ, Henry DJ, Dutrow BL (2011) Tourmaline as a petrologic forensic mineral: a unique recorder of its geologic past. Elements 7:327–332

Veksler IV, Thomas R (2002) An experimental study of B-, P- and F-rich synthetic granite pegmatite at 0.1 and 0.2 GPa. Contrib Min Petrol 143:673–683

Wedepohl KH (1995) The composition of the continental crust. Geochim Cosmochim Acta 59:1217–1232

Wilson BM (1989) Igneous petrogenesis, a global tectonic approach. Unwin Hyman, London, p 466

Wittenbrink J, Lehmann B, Wiedenbeck M, Wallianos A, Dietrich A, Palacios C (2009) Boron isotope composition of melt inclusions from porphyry systems of the central Andes: a reconnaissance study. Terra Nova 21:111–118

Wolf MB, London D (1997) Boron in granitic magmas: stability of tourmaline in equilibrium with biotite and cordierite. Contrib Min Petrol 130:12–30

Wunder B, Meixner A, Romer RL, Wirth R, Heinrich W (2005) The geochemical cycle of boron: constraints from boron isotope partitioning experiments between mica and fluid. Lithos 84:206–216

Xavier RP, Wiedenbeck M, Trumbull RB, Dreher AM, Monteiro LVS, Rhede D, de Araújo CEG, Torresi I (2008) Tourmaline B-isotopes fingerprint marine evaporites as the source of high salinity ore fluids in iron oxide-copper-gold deposits, Carajás mineral province (Brazil). Geology 36:743–746

Yang S-Y, Jiang S-Y (2012) Chemical and boron isotopic composition of tourmaline in the Xiangshan volcanic–intrusive complex, southeast China: evidence for boron mobilization and infiltration during magmatic–hydrothermal processes. Chem Geol 312–313:177–189

Yang S-Y, Jiang S-Y, Palmer MR (2015) Chemical and boron isotopic compositions of tourmaline from the Nyalam leucogranites, south Tibetan Himalaya: implication for their formation from B-rich melt to hydrothermal fluids. Chem Geol 419:102–113

Yavuz F, Jiang S-Y, Karakaya N, Karakaya MC, Yavuz R (2011) Trace-element, rare-earth element and boron isotopic compositions of tourmaline from a vein-type Pb–Zn–Cu–U deposit, NE Turkey. Inter Geol Rev 53:1–24

Zhang ST, Ma DS, Lu JJ, Zhang RQ, Gao SY (2014) Chemical and boron isotopic composition of tourmaline in Baotan tin deposit, northern Guangxi, South China. Acta Geol Sinica (English edn) 88(Supl 2):485–486

Zhao K-D, Jiang S-Y, Nakamura E, Moriguti T, Palmer MR, Yang S-Y, Dai B-Z, Jiang Y-H (2011) Fluid–rock interaction in the Qitianling granite and associated tin deposits, South China: evidence from boron and oxygen isotopes. Ore Geol Rev 43:243–248

Zhao K-D, Jiang S-Y, Nakamura E, Moriguti T, Wei H-Z (2015) A preliminary study on boron isotope fractionation of major rock-forming minerals in granite. Acta Petrol Sin 31:740–746 (Chinese with English abstract)

Zheng Z, Deng X-H, Chen H-J, Yue S-W, Dong L-H, Qu X, Chen Y-J (2016) Fluid sources and metallogenesis in the Baiganhu W–Sn deposit, east Kunlun, NW China: insights from chemical and boron isotopic compositions of tourmaline. Ore Geol Rev 72: 1129–1142

The Cosmochemistry of Boron Isotopes

11

Ming-Chang Liu and Marc Chaussidon

Abstract

The boron elemental abundances and isotopic compositions in the universe and constituents (stars, interstellar medium and the Solar System material) within it have important implications for the astrophysical origins of this element. Astronomical observations and laboratory analysis have revealed that despite a significant difference in boron abundances among different objects, the $^{11}B/^{10}B$ ratio of 4 appears to be ubiquitous (within measurement uncertainties) across the Galaxy. Galactic Cosmic Ray (GCR) spallation, which yields $^{11}B/^{10}B = 2.5$, cannot have been the sole source of B; another mechanism that favors the production of ^{11}B over that of ^{10}B must have operated over the Galactic timescale. However, how exactly the Galaxy, interstellar medium, and the Solar System acquired the $^{11}B/^{10}B$ ratio of 4 remains poorly understood.

Keywords

Boron isotopes · Meteorites · Nucleosynthesis

11.1 Basics of Boron Isotopes

11.1.1 Boron Nucleosynthesis

Boron (B) is the fifth element (atomic number 5) in the periodic table, and has two naturally occurring and stable isotopes, ^{10}B (19.9%) and ^{11}B (80.1%). The elemental abundance of boron

M.-C. Liu (✉)
Department of Earth, Planetary, and Space Sciences, UCLA, 595 Charles Young Drive E, Los Angeles, CA, USA
e-mail: mcliu@ucla.edu

M. Chaussidon
Institut de Physique du Globe de Paris (IPGP), CNRS, Paris, France
e-mail: chaussidon@ipgp.fr

H. Marschall and G. Foster (eds.), *Boron Isotopes*, Advances in Isotope Geochemistry,
https://doi.org/10.1007/978-3-319-64666-4_11

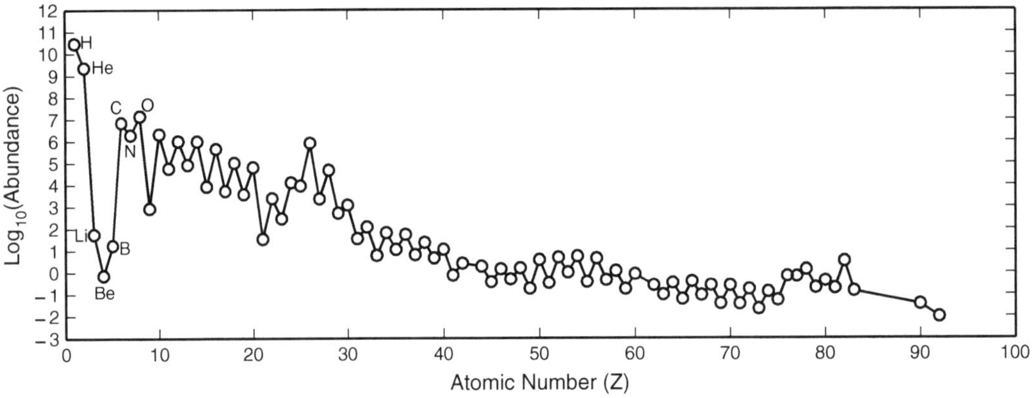

Fig. 11.1 The cosmic abundance of elements. Boron (along with lithium and beryllium) is several orders of magnitude depleted compared to its neighbor elements, such as H, He, C, N, and O

in the Solar System (the "cosmic abundance") determined from the measurements of bulk CI chondritic meteorites is 17–21 per million Si atoms (B/Si = 17–21 × 10^{-6}, Anders and Grevesse 1989; Zhai and Shaw 1994; Lodders 2003). This abundance, along with that of lithium and beryllium, is several orders of magnitude lower than its neighbor elements (helium, carbon, nitrogen and oxygen) (Anders and Grevesse 1989; Lodders 2003) (Fig. 11.1). This is because boron (as well as lithium and beryllium) is destroyed in stellar interiors during hydrogen burning, once the temperature goes beyond several million Kelvin[1] (e.g., Burbidge et al. 1957). A theory that the presence of boron in the universe primarily involves non-thermal nuclear reactions between high-energy (>GeV) particles (protons and alphas) in Galactic Cosmic Rays (GCRs) and C, N and O nuclei in the interstellar medium (ISM) was first proposed by Reeves et al. (1970) and such a spallation process is now quantitatively well understood after many years of studies (e.g., Prantzos 2012). The fact that the boron (lithium and beryllium as well) abundance in GCRs is much more enriched compared to its abundance in the Solar System by several orders of magnitude provides strong observational support for this hypothesis (e.g., Simpson 1983; Prantzos 2007). Based on the cross sections for the $^{12}C(p,x)^{10,11}B$, $^{14}N(p, x)^{10,11}B$ and $^{16}O(p, x)^{10,11}B$ reactions (Fig. 11.2), spallation produces ^{10}B and ^{11}B with an isotopic ratio of $^{11}B/^{10}B \sim 2.5$ (Meneguzzi et al. 1971). Minor isotopes of C, N and O contribute very little to the production of ^{10}B and ^{11}B due to their low abundances. This theoretical prediction is consistent with B isotope measurements of present-day GCRs that arrived at Earth by the ISEE 3 spacecraft, which revealed $^{11}B/^{10}B$ = 2.22–2.89, depending on the energy regime measured (Krombel and Wiedenbeck 1988; Gibner et al. 1992). The low cosmic abundance of boron is a result of poor efficiency of spallation reactions.

Although GCR spallation is an important pathway to synthesize boron isotopes, this mechanism has been shown to fail to account for the Solar System abundance of ^{11}B (e.g., Meneguzzi et al. 1971). The average $^{11}B/^{10}B$ in the Solar System obtained from the measurements of CI chondritic meteorites, whose bulk chemical composition most resembles that of the solar photosphere, is 4.04 (e.g., Agyei and McMullen 1968; Zhai et al. 1996), much higher than the GCR-derived $^{11}B/^{10}B$ ratio. Boron in the diffuse interstellar clouds is also enriched in ^{11}B ($^{11}B/^{10}B = 3.4 \pm 0.7$) compared to GCRs (Federman et al. 1996; Lambert et al. 1998) (Table 11.1). The ratio difference between the Solar System (and interstellar gas) and GCRs

[1]The thermal nuclear reactions that destroy B isotopes are $^{10}B(p,\alpha)^{7}Be(e^{-},\nu)^{7}Li(p,\alpha)^{4}He$ and $^{11}B(p,\alpha)^{8}Be \rightarrow {}^{4}He + {}^{4}He$ (Burbidge et al. 1957).

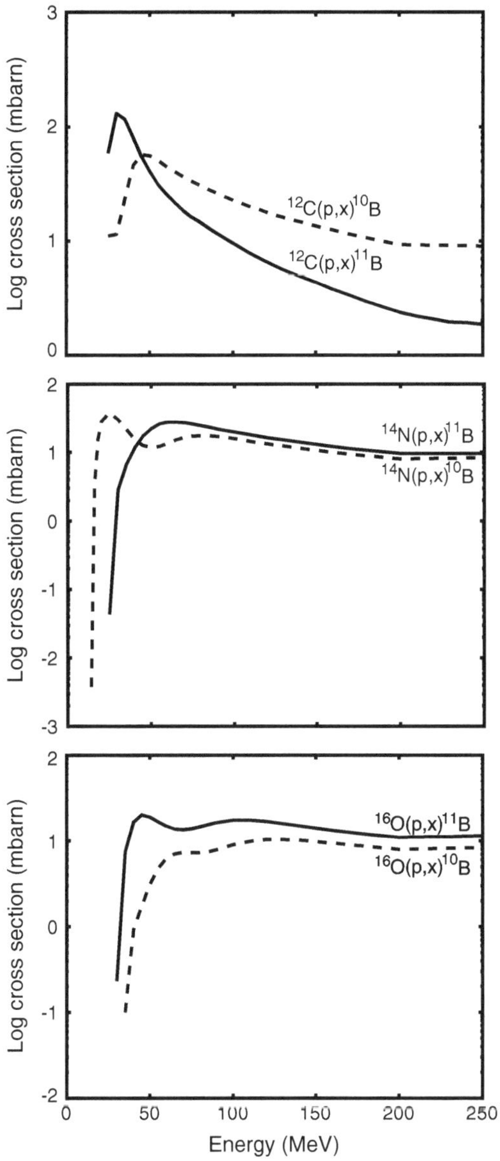

Fig. 11.2 Calculated cross-sections for the main spallation reactions that account for the production of boron isotopes: **a** $^{12}C(p, x)^{10,11}B$; **b** $^{14}N(p, x)^{10,11}B$; **c** $^{16}O(p, x)^{10,11}B$. Calculations were performed with the TALYS code (Koning et al. 2007)

calls for an additional source of ^{11}B. Two nucleosynthetic mechanisms for ^{11}B have been considered to account for the higher-than-GCR $^{11}B/^{10}B$ ratios observed in the Solar System and elsewhere:

(1) Low-energy cosmic ray spallation: This hypothesis stemmed from the discovery by Bloemen et al. (1994) who first reported the detection of intense gamma-ray emission from the Orion complex in the 3–7 MeV range by the COMPTEL telescope. These observed lines at 4.44 and 6.13 MeV originated from de-excitation of ^{12}C and ^{16}O nuclei ejected from supernovae when they collide with the surrounding gas (inverse spallation), which is molecular and atomic hydrogen. Such levels of gamma-ray emission suggested the existence of abundant carbon and oxygen in the low-energy part of GCRs (Ramaty et al. 1979; Bloemen et al. 1994). Detailed numerical calculations have shown that spallation of hydrogen by the low energy C and O nuclei may favor the production of ^{11}B over ^{10}B, and the resulting $^{11}B/^{10}B$ ratio varies from ~ 8 to ~ 3 depending on the energy, assumed spectral shape, and compositions of incident particles and interstellar medium (e.g., Cassé et al. 1995; Ramaty et al. 1996). Therefore, the Solar System $^{11}B/^{10}B$ ratio can be understood as a result of mixing of boron produced by high-energy and low-energy GCRs (Cassé et al. 1995; Chaussidon and Robert 1995; Ramaty et al. 1996). However, it should be pointed out that observational support for such a model is still lacking, mainly due to the effects of solar modulation that make constraints on the GCR flux below 1 GeV highly uncertain. In addition, the result of Bloemen et al. (1994) was later found to have been affected by the instrumental background that was underestimated (Bloemen et al. 1999). After properly correcting for the background, no convincing detection of gamma-ray can be claimed, indicating a much lower (if any) flux of low-energy GCR components. To date, the production of boron by low-energy spallation remains hypothetical.

(2) Neutrino-induced spallation (ν-process) of ^{12}C in Type II supernovae: This ν-process is

Table 11.1 Boron isotopic compositions of the constituents in the Galaxy

Location	$^{11}B/^{10}B$	Source(s) of B	Measurement methods
Interstellar medium	3.4 ± 0.7	GCRs, low energy spallation and/or ν-process	HST-GHRS observations of B II line at 1362 Å
Nearby stars	3.7–4.7	GCRs, low energy spallation and/or ν-process	HST-GHRS and HST-STIS observations of B III line at 2065.8 Å
Solar system	4.04	GCRs, low energy spallation and/or ν-process	Isotopic analysis of chondritic meteorites
GCRs	2.22–2.89	GCRs	ISEE 3 spacecraft

predicted to produce a significant amount of ^{11}B in the C-rich shell via $^{12}C(\nu, \nu'n)^{11}C(e^+$ decay$)^{11}B$ and $^{12}C(\nu, \nu'p)^{11}B$ reactions; very little or no ^{10}B is synthesized by this process (Woosley et al. 1990; Woosley and Weaver 1995). The ^{11}B abundance can build up over the Galactic timescale and a major portion of the Solar System's ^{11}B could have been derived from this source. However, the yields of ^{11}B by this process strongly depend on the total energy and temperatures of neutrinos involved (e.g., Yoshida and Kajino 2005). One could constrain these neutrino characteristics by using Galactic Chemical Evolution (GCE) models that include cosmic ray spallation to match the Solar System B isotopic abundance, but these calculation results are model-dependent and sensitive to input parameters, which require detailed understandings of physics of supernova explosion (e.g., Yoshida et al. 2005).

In addition to the aforementioned mechanisms that result in direct production of B isotopes, a trace amount of ^{10}B can also be made from the decay of a radioactive isotope ^{10}Be ($t_{1/2}$ = 1.39 Myr, Korschinek et al. 2010), which is also synthesized during spallation of C, N, and O by energetic projectiles. The GCR-derived ^{10}Be contributes very little ^{10}B to the total Solar System B inventory due to its low abundance. McKeegan et al. (2000) estimated a steady-state $^{10}Be/^{9}Be$ ratio of $\sim 8 \times 10^{-6}$ in the interstellar medium based on the known cosmic abundance of ^{9}Be, an assumed production rate of ^{9}Be and a $^{10}Be/^{9}Be$ production ratio of 0.1 in the Galaxy by GCRs. However, it has been demonstrated through large ^{10}B excesses (up to a 40% enrichment relative to the chondritic abundance) in meteoritic Ca–Al-rich Inclusions (CAIs), the oldest datable solids in the Solar System with a radiometric age of 4.567 Gy (e.g., Connelly et al. 2012), that ^{10}Be existed in the early Solar System at high abundances, with $^{10}Be/^{9}Be = (3–9) \times 10^{-4}$. It should be pointed out that large ^{10}B excesses, hence high inferred $^{10}Be/^{9}Be$ ratios, can only be found in phases enriched in Be and depleted in B, such as CAIs (Fig. 11.3) or refractory hibonite ($CaAl_{12}O_{19}$) and grossite ($CaAl_4O_7$), so that one would not be able to see any resolvable radiogenic ^{10}B excess in a gas/solid of solar composition, which has Be/B = 0.04 (e.g., the solar nebula or bulk chondritic meteorites). Although the decay of ^{10}Be is not a major source of ^{10}B in the Solar System, the $^{10}Be/^{9}Be$ ratios inferred in CAIs have important implications for irradiation conditions in the solar nebula, and such irradiation has left observable traces in B isotopes in the early Solar System solids. This topic will be discussed in detail later in the text.

11.1.2 Cosmochemistry of Boron

Chemically, B is a moderately volatile element. Its 50% condensation temperature, which is a measure of volatility and defined as the temperature at which 50% of the element would condense into a solid from a gas of solar composition at a given pressure, has been estimated to be

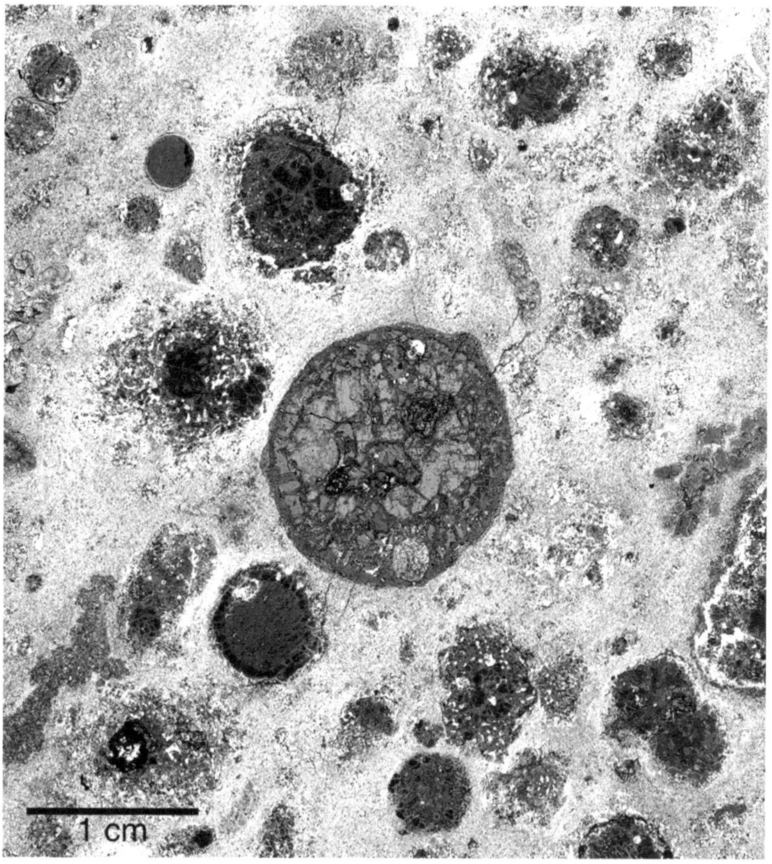

Fig. 11.3 Backscattered electron image of a slab of the Allende meteorite. The round-shaped object in the center of the slab is a Ca–Al-rich inclusion (CAI). Other smaller round inclusions are chondrules

within the range of 910–970 K at 10 Pa based on the B elemental abundance in CI chondrites (as a result of volatility-dependent fractionation in the solar nebula) and thermodynamic calculations (e.g., Zhai 1995; Lauretta and Lodder 1997). Thermodynamically, the majority of B would condense as $CaB_2Si_2O_8$ in solid solution with anorthite ($CaAl_2Si_2O_8$) (Lauretta and Lodder 1997); however, this theoretical prediction has not been verified by laboratory measurements of meteorite samples primarily because of pervasive terrestrial boron contamination which obscures the original distribution of B in meteoritic components. Most chondritic meteorites analyzed so far (carbonaceous and ordinary chondrites) have bulk B abundances ranging from 0.3 to 1.4 μg/g (B/Si = 7×10^{-6} to 3.4×10^{-5}). The majority of B resides in the matrices (Zhai and Shaw 1994), but no specific B host phases have been identified.

The idea of B condensing at a specific temperature in the solar nebula has been called into question by Chaussidon and Robert (1997) based on the B/Si and $^{11}B/^{10}B$ variations measured in a chondrule of the Semarkona chondrite, an unequilibrated, primitive ordinary chondrite, reported in Chaussidon and Robert (1995). Spot analysis revealed that the chondrule's $\delta^{11}B$[2] spans from −40 to +40‰, 87% of data being however clustered between −40 and +10‰, and is positively correlated with the B/Si ratios (8×10^{-6}–2×10^{-5}), and that 15–50% of B in Semarkona is contained in chondrules. Given that chondrules are formed by high temperature processing (>1200 K, e.g., flash heating by shockwaves; see Morris et al. 2012) followed by

[2] $\delta^{11}B = (R_{sp}/R_{std} - 1) \times 1000$, where R_{sp} and R_{std} are the $^{11}B/^{10}B$ ratio of the sample and standard, respectively. In Chaussidon and Robert (1995), R_{std} = 4.04558 is used (NBS951).

rapid cooling (100–1000 K/h, e.g., Hewins 1997), they should have lost most of their volatile elements, such as Na, K and B. Evaporative loss of B would lead to enrichments of ^{11}B in an evaporation residue, thus a negative correlation between B/Si and $\delta^{11}B$ should be expected, in clear contrast to the observation in Chaussidon and Robert (1995). In addition, the 80‰ variation in $\delta^{11}B$ is too large to have originated from isotopic fractionation processes associated with evaporation (or condensation) of chondrules. Chaussidon and Robert (1997) argued that the meteoritic observations of B (concentrations and isotopic compositions) are most consistent with B-bearing grains that condensed in the Solar System's prenatal molecular cloud and carried anomalous $^{11}B/^{10}B$ due to low energy spallation being incorporated into the solar nebula. In this context, condensation of B from the nebular gas at a unique temperature may not be the main mechanism for chondritic meteorites to acquire the observed B abundances. However, a similar study carried out on chondrules from various meteorites, including Semarkona, by Hoppe et al. (2001), did not find as large $\delta^{11}B$ variations and as high B contents in chondrules as did Chaussidon and Robert (1995, 1998), arguing against the hypothetical presence of B-bearing inclusions. More details about B isotopic heterogeneities in chondrules and their implications will be addressed later in the text.

11.2 Boron Isotopes in the Galaxy

The evolution of the B abundance and isotopic composition in the Galaxy is strongly controlled by its nucleosynthetic mechanisms. As stated in the previous section, nucleosynthesis of B largely involves non-thermal nuclear reactions between relativistic GCR protons and CNO nuclei in the ISM. Theoretical modeling shows that GCR spallation alone fails to account for the Solar System's $^{11}B/^{10}B$ ratio due to the underproduction of ^{11}B. Other sources of ^{11}B must be invoked (see above), but the extent to which these nucleosynthetic processes supplement the Solar System with ^{11}B remains poorly understood. Since B nucleosynthesis occurs in, or is closely associated with, the ISM, it is important to know the interstellar and stellar B elemental abundances and isotopic ratios and then compare them with those in the Solar System to obtain better understandings of the origins of boron isotopes.

11.2.1 Boron Abundances in Stars

The boron abundance in stellar atmospheres can be inferred by observing the B I resonance lines (2496.771 and 2497.723 Å) for F and G (solar-type) stars, and B II resonance line (1362 Å) for hotter and larger stars (A and B stars). In recent studies, the use of the BIII resonance line (2065.8 Å) has been favored over that of BII in estimating B abundances in stars because the BII line in B stars is week and prone to blending with nearby absorption lines, while the BIII line is stronger and less blended. Moreover, the large isotopic shift at this wavelength also allows for the possibility of directly measuring the $^{11}B/^{10}B$ ratio (e.g., Proffitt et al. 1999).

The first determinations of B abundances, expressed in the form of B/H (as opposed to B/Si determined from measurements of meteorites), in a number of type A and B stars in the solar neighborhood (including Vega and Sirius) were performed by using the *Copernicus* satellite to observe the B II resonance line (Praderie et al. 1977; Boesgaard and Heacox 1978). These stars were found to have a wide range of B/H ratios, from $<5 \times 10^{-12}$ (Sirius) to 2×10^{-10} after correcting for LTE (local thermodynamic equilibrium) effects. This range is broadly consistent with B derived from GCR spallation, which predicts a B/H ratio of $\sim 3 \times 10^{-10}$ (Meneguzzi et al. 1971). A later study by Cunha et al. (1997) of the BII line in four type-B stars in the Orion association with the *Hubble Space Telescope* Goddard High Resolution Spectrograph

(HST-GHRS) also revealed LTE-corrected B/H values similar to those found by Boesgaard and Heacox (1978), but these abundance ratios were significantly revised upward to $B/H = 5^{+2.9}_{-1.9} \times 10^{-10}$ after applying corrections for non-LTE effects. Cunha et al. (1997) also reanalyzed the LTE-abundances provided by Boesgaard and Heacox (1978) for non-LTE effects and obtained a mean value of $B/H = 4^{+1.1}_{-0.9} \times 10^{-10}$. These stars appear to have B abundances close to what was obtained for the Solar System based on the analysis of chondritic meteorites ($B/H = (6 \pm 0.6) \times 10^{-10}$, e.g., Zhai and Shaw 1994; Lodders 2003) and on the observations of solar photosphere ($B/H = (4–5) \times 10^{-10}$, e.g., Kohl et al. 1977; Cunha and Smith 1999; Grevesse et al. 2007; Asplund et al. 2009). However, more recent work, observing the BIII line at 2065.8 Å with HST-GHRS and HST-STIS (Space Telescope Imaging Spectrograph), has yet again changed the stellar B abundances (including two of the stars, HD886 and HD35299, that were studied in Cunha et al. 1997) to a level at least a factor of two lower than the Solar System abundance (e.g., Proffitt et al. 1999; Proffitt and Quigley 2001; Venn et al. 2002). Abundances deduced from the observation of B III line appear to be more robust because of better signal-to-noise ratios and less corrections for blending. It should be noted that several mechanisms have been proposed to account for B depletion in stars (e.g., Venn et al. 2002) so that it is not yet clear if these stars formed with sub-solar B abundances or certain processes (e.g., rotational mixing) during stellar evolution depleted B (Proffitt and Quigley 2001). Despite the difference in B abundances between the solar and stellar values, Proffitt et al. (1999) estimated the $^{11}B/^{10}B$ ratio in HD886 and HD35299, for the first time, to be $4.7^{+0.8}_{-0.6}$ and $3.7^{+0.8}_{-0.6}$, respectively, which are indistinguishable from the average Solar System $^{11}B/^{10}B = 4.04$ (Zhai 1995) within uncertainties. This result implies that in addition to GCR spallation, stars in the solar neighborhood must have also acquired additional ^{11}B from other source(s).

11.2.2 Boron in the Interstellar Medium

The first solid detection of interstellar B was reported by Meneguzzi and York (1980) by measuring absorption from the B II λ1362 resonance line in the direction of κ Orionis with the *Copernicus* satellite. An interstellar abundance of $B/H = (1.5 \pm 0.7) \times 10^{-10}$ was inferred, which is in agreement with the best-estimated stellar values at that time ($B/H = (1–2) \times 10^{-10}$; see above) deduced by Praderie et al. (1977) and Boesgaard and Heacox (1978) who used the same instrument. However, later studies with the HST-GHRS found that boron is significantly depleted in the interstellar diffuse clouds relative to disk stars by at least a factor of 2, with the B/H ratios $<1 \times 10^{-10}$ (e.g., Federman et al. 1993; Jura et al. 1996; Federman et al. 1996; Lambert et al. 1998). Revisions of interstellar boron abundances have been made through observations with more advanced instrumentations, such as HST-STIS, which yielded an average value of $B/H = (2.5 \pm 0.6) \times 10^{-10}$ (Howk et al. 2000; Ritchey et al. 2011). With this precision, it is evident that the boron abundances in the interstellar medium are lower than the Solar System (meteoritic) value of $(6 \pm 0.6) \times 10^{-10}$ (Zhai and Shaw 1994; Lodders 2003). However, one should also note that the aforementioned interstellar B abundances were derived from the measurements of the gas phase, and should only be taken as a lower limit because some boron could be incorporated into interstellar dust grains and cannot be seen spectroscopically (e.g., Jura et al. 1996; Howk et al. 2000). There could still be other reasons for boron depletion in the interstellar medium, for example, the interstellar B abundance was low to begin with (Jura et al. 1996), but at this moment, it is still not possible to favor one particular reason over the others.

The isotopic ratios of boron shed more light than the elemental abundance on the origins of this element. The interstellar $^{11}B/^{10}B$ ratio was determined at 3.4 ± 0.7 (Federman et al. 1996; Lambert et al. 1998), in agreement with those in the Solar System (Zhai et al. 1996) and stellar atmospheres (Proffitt et al. 1999). For a long time, solar $^{11}B/^{10}B$ was considered anomalous due to some local processes. The results obtained from the ISM imply that $^{11}B/^{10}B \sim 4$ appears to be a general feature in the local solar neighborhood (including stars and molecular clouds). This requires that nucleosynthesis mechanism(s) producing B with high $^{11}B/^{10}B$ ratios (>solar $^{11}B/^{10}B$) operate over the Galactic timescale to bring GCR-derived $^{11}B/^{10}B$ up to $\sim$4. The two aforementioned nucleosynthetic pathways, low-energy cosmic ray spallation and the ν-process, are possibilities, but assessing the proportion to which each source contributed ^{11}B-rich components to the solar neighborhood remains a challenging task.

11.3 Boron Isotopes in the Early Solar System

The boron isotopic studies of meteoritic material is an important research topic as they allow for a precise determination of $^{11}B/^{10}B$ of the Solar System and its rocky constituents, which can be used to constrain the sources of boron isotopes and the irradiation environment inside or around the solar nebula. Large isotopic variations caused by various processes have been found to exist among different objects formed at the very beginning of the Solar System history. In this section, we first introduce the laboratory techniques used to analyze boron isotopes in solid matter and then summarize the boron isotope distributions in chondritic meteorites and their components. Implications of boron isotopes for early Solar System processes will be discussed.

11.3.1 Mass Spectrometry for B Isotope Measurements

(i) Secondary Ion Mass Spectrometry (SIMS): SIMS is an in situ surface analysis technique that can be applied to obtain chemical and isotopic compositions of solid material (Foster et al. 2017). Analysis of boron isotopes has been extensively performed on different types of SIMS instruments by many cosmochemistry and geochemistry research groups. Here is a brief summary of the SIMS technique. A polished, gold-coated sample is sputtered with an $^{16}O^-$ primary ion beam of 10–50 nA (depending on the concentration of boron) to generate sufficient count rates of $^{10}B^+$ and $^{11}B^+$. A key step in SIMS boron isotope analysis, especially for cosmochemistry, is to minimize surface boron contamination; therefore sputtering of the sample surface prior to analysis ("presputtering") is always performed until the boron signals stop dropping rapidly. The mass resolution (M/ΔM) is usually set at ca. 2000 so that isobaric interferences (e.g., hydrides, $^{30}Si^{3+}$, $^{40}Ca^{4+}$) can be separated from the peaks of interest. Secondary ions can be collected by magnetic peak switching onto an electron multiplier (CAMECA ims-series ion probes; e.g., McKeegan et al. 2000; Chaussidon et al. 2006; Liu et al. 2009) or by multiple EMs simultaneously under a static magnetic field (CAMECA NanoSIMS; Liu et al. 2010). In general, SIMS data need to be corrected for matrix effects as samples of different chemical compositions have different instrumental mass fractionation. Boron isotopes appear to be an exception: it has been shown that boron matrix effects are at the permil level for a wide range of silicate glass compositions, with the exception of NBS 61x glass series, and are generally negligible compared to the counting statistics (a few permil or larger) in cosmochemical applications due to low boron concentrations in the samples (e.g., Chaussidon et al. 1997; Marschall and Monteleone 2015).

(ii) Thermal Ionization Mass Spectrometry (TIMS): The boron isotopic ratios can be determined in natural material by thermal ionization mass spectrometry (TIMS) (Foster et al. 2017). To achieve high precision and high accuracy, boron isotopes are often measured as $Cs_2BO_2^+$ by positive-ion TIMS

(P-TIMS) because of the small relative mass difference between $^{133}Cs_2^{10}B^{16}O_2^+$ and $^{133}Cs_2^{11}B^{16}O_2^+$ causing small degrees of instrumental mass fractionation. A reproducibility of ±0.1‰ has been constantly reported in the literature; however, one major drawback is large sample consumption (~100–3000 ng of boron is needed, e.g., Xiao and Wang 1998; Zhai 1995; Ishikawa and Nagaishi 2011; Foster et al. 2017), which can sometimes be a limiting factor when it comes to dealing with quantity-limited meteorite samples.

Developments on achieving reasonably high precision with less sample consumption have been made for P-TIMS in the past several years. Ishikawa and Nagaishi (2011) presented a methodology involving preheating samples and filaments at 240 °C to stabilize the chemical species of boron on a filament and to improve the ionization efficiency of $Cs_2BO_2^+$. This new method enabled high precision analysis without consuming large quantities of samples, and an external reproducibility of ±0.1‰ for 50–100 ng of boron and of ±0.2‰ for 10 ng of boron can be achieved. Such a terminal precision and amounts of samples needed would be suitable for analysis of meteorite material.

Measurements of boron as BO_2^- by negative-ion TIMS (N-TIMS) require much less material (~as small as a few picograms of boron) due to better ionization yields, but the data are usually associated with larger uncertainties (~0.7–2‰) (Foster et al. 2017). This is mainly because much larger instrumental mass fractionation (due to lighter masses compared to $Cs_2BO_2^+$) cannot be properly corrected for. An analytical protocol developed by Shen and You (2003), in which they measured not only $^{10,11}B^{16}O_2^-$ (mass 42 and 43) but also $^{11}B^{16}O^{18}O^-$ (mass 45) to characterize the instrumental mass fractionation with $^{18}O/^{16}O$ and correct for it, improved the analytical reproducibility by as much as tenfold. The resulting precision of ±0.19–0.47‰ (2σ) from a few ng of boron is comparable to, or only slightly worse than, the typical precision obtained with P-TIMS (see above), but the required amount of sample in each measurement is much less. This method is applicable to the studies with quantity-limited samples, such as looking for radiogenic ^{10}B excesses in CAIs due to the decay of short-lived ^{10}Be (see below), which was present in the early Solar System (e.g., McKeegan et al. 2000). However, because of high laboratory background (contamination) and low B abundances (<ppm) in CAIs, no successful attempts have been reported.

11.3.2 B Isotopes in Chondritic Meteorites

Boron isotopic ratios in bulk chondritic meteorites have been fairly well studied in the past four decades. The first analysis of $^{11}B/^{10}B$ in chondritic meteorites was performed by Shima (1962, 1963) by using TIMS, with the reported $\delta^{11}B$ values ranging from −55 to −35‰. These chondritic values were found to be too light in later studies (e.g., Agyei and McMullen 1968); reasons for such differences remain unknown. It has now been well established that most chondritic meteorites are characterized by a narrow range of $\delta^{11}B$ values from −10 to +7‰, except the Mokoia CV3 chondrite, whose $\delta^{11}B$ = + 19‰ (Fig. 11.4, left; Agyei and McMullen 1968; Chaussidon and Robert 1995; Zhai et al. 1996). It should be pointed out that bulk meteorite samples contain various amounts of CAIs, chondrules and fine-grained matrix material (e.g., primitive dust, secondary products, presolar grains) so that the resulting $\delta^{11}B$ value strongly relies on what is included in an analysis. The CI chondrites, in which matrix material comprises ~95% by volume (Scott and Krot 2014), have an average $\delta^{11}B$ ~ −3.3 ± 0.6‰ ($^{11}B/^{10}B$ = 4.04) (Zhai et al. 1996). In contrast, ordinary chondrites only have 10–15% (by volume) matrix but 60−80% chondrules, and show a larger $\delta^{11}B$ variation (−7 to +7‰), although the average ($\delta^{11}B$ = + 1.8 ± 1‰, Zhai et al. 1996) still falls close to the CI value. The origin of the large B isotopic variations observed in each chondrite group is not fully understood. The

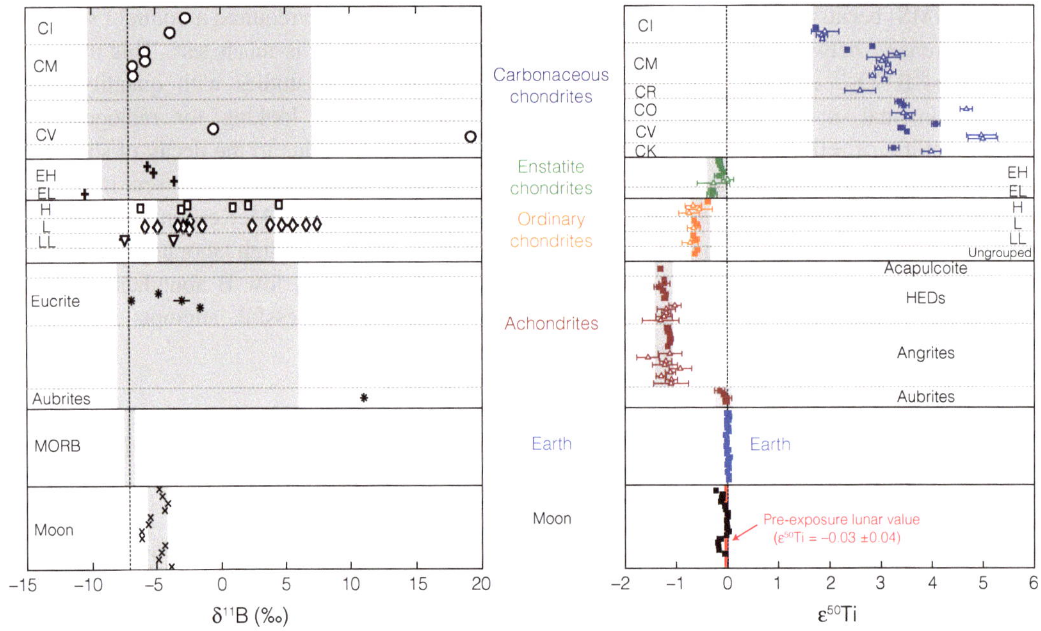

Fig. 11.4 (*Left*) Boron isotopic variations seen in bulk chondritic and planetary materials. *Data source* Zhai et al. (1996). The variation within each chondrite group is not yet fully understood. *Gray boxes* represent 1 sigma standard deviations for each group. (*Right*) Ti isotopes in various bulk planetary materials are shown for comparison (Zhang et al. 2012; *gray boxes* represent 2 sigma standard errors for each group.). Each group of planetary materials is characterized by a distinct ^{50}Ti value; in contrast, δ^{11}B variations among different groups of chondritic and planetary materials are less obvious (figure reproduced with kind permission from Nature Geoscience)

Earth mantle, as represented by the mid ocean ridge basalts (MORB; see Marschall 2017) has δ^{11}B in general at the low end of the range present in chondrites. Large B isotope fractionations are known to occur between fluids and B-bearing minerals at low temperature. Such processes, either nebular or parent body, might in part account for the B isotopic variations present in chondrites. This would not however explain the apparent systematic shift of bulk Earth towards a lighter B isotopic composition relative to chondrites. Contrary to, for instance Ti isotopes in which different groups of planetary materials show distinct isotopic compositions (Fig. 11.4, right; Zhang et al. 2012), there is no clear isotopic variations for B at such scale. Different groups of chondrites, despite having quite different proportions of constituents, have average δ^{11}B very similar to one another, suggesting a high degree of homogeneity (within a couple of permil) of the boron isotope distribution in the solar nebula, and that ^{11}B/^{10}B = 4.04 (δ^{11}B = −0.9‰) could best represent the average Solar System value.

11.3.3 B Isotopes in CAIs and Implications for Early Solar System Irradiation

Efforts in determining the boron isotopic compositions in Ca–Al-rich Inclusions (CAIs) are mainly to look for the fossil records of short-lived ^{10}Be ($t_{1/2}$ = 1.39 Myr), a radionuclide exclusively produced by energetic particle spallation. Given the fact that beryllium is refractory but boron is moderately volatile and that ^{10}Be decays to ^{10}B, CAIs are expected to be enriched in beryllium and depleted in boron, and the former existence of ^{10}Be in CAIs would appear in the form of radiogenic excesses of ^{10}B ($\equiv\delta^{10}B^*$).

The existence of ^{10}Be in the early Solar System was first demonstrated by McKeegan et al. (2000) using a large-radius ion microprobe (CAMECA ims-1270), where large excesses in ^{10}B (up to $\delta^{11}B^* \sim -400‰$) were found to be in correlation with Be/B ratios in melilite crystals of an Allende CAI, indicating an initial $^{10}Be/^{9}Be$ ratio of 9×10^{-4}. Subsequent boron isotope studies, all by SIMS, of different types of CAIs from various CV3 chondrites (Allende, Efremovka, Vigarano, Leoville, and Axtell), individual refractory hibonite ($CaAl_{12}O_{19}$) grains from the CM2 chondrite Murchison and grossite ($CaAl_4O_7$)-rich refractory inclusions from the CH/CB chondrite Isheyevo for ^{10}Be abundances carried out by others (Sugiura et al. 2001; Marhas et al. 2002; MacPherson et al. 2003; Liu et al. 2009, 2010; Gounelle et al. 2013; Srinivasan and Chaussidon 2013) showed a range of $^{10}Be/^{9}Be$ between $\sim 3 \times 10^{-4}$ and $\sim 1 \times 10^{-2}$.

From all the studies, it appears that the appearance of ^{10}Be was ubiquitous in the solar nebula (or at least in the region where refractory inclusions formed). However, this widespread distribution could not have been derived from the steady-state production by GCRs in the solar nebula because the predicted $^{10}Be/^{9}Be$ ratio by this process is too low ($\sim 8 \times 10^{-6}$; McKeegan et al. 2000). Instead, the high yet heterogeneous abundances recorded in refractory inclusions can be well understood as a result of in situ production due to intense irradiation by the proto-Sun (e.g., McKeegan et al. 2000; Gounelle et al. 2006; Liu et al. 2009, 2010; Srinivasan and Chaussidon 2013); the variability in the initial $^{10}Be/^{9}Be$ ratios simply reflects fluctuations in the fluences of energetic projectiles that the irradiated targets experienced. However, due to the nature of such mass spectrometric measurements (low B contents ($\ll$µg/g) in the analyzed spots), large analytical uncertainties ($\sim$10–30%) always accompany the inferred ^{10}Be abundances (as much as 50% if data were obtained from small-geometry SIMS instruments, such as CAMECA ims-4f and ims-6f, e.g., Sugiura et al. 2001; Marhas et al. 2002; MacPherson et al. 2003), and therefore would allow for an apparent uniformity at the level of $^{10}Be/^{9}Be \sim (6 \pm 3)$ 10^{-4}. This has led to a model in which ^{10}Be came into the early Solar System as trapped GCR components by the presolar molecular cloud core (Desch et al. 2004). Because the spatial scale of the molecular cloud core is so much larger than that of the Solar System, ^{10}Be would be homogeneously distributed in the solar nebula. However, with much improved analytical precision by using more advanced instrumentations (e.g., CAMECA ims-1270, ims-1280 and NanoSIMS), newer data (e.g., Chaussidon et al. 2006; Liu et al. 2009, 2010; Wielandt et al. 2012; Srinivasan and Chaussidon 2013; Gounelle et al. 2013) have shown that although ^{10}Be existed in all types of meteoritic refractory inclusions (where high Be/B ratios can be obtained), its inferred initial abundance could vary by as much as a factor of 30 (a factor of 4 if the extremely high value of $^{10}Be/^{9}Be = 1 \times 10^{-2}$ from Gounelle et al. (2013) is excluded), contradicting one major prediction by the trapping model. Another model involving the Solar System's parental cloud was proposed by Tatischeff et al. (2014) based on the common occurrence of ^{10}Be, although the details are largely different from those in Desch et al. (2004). Tatischeff et al. (2014) considered irradiation of the presolar molecular cloud by freshly accelerated cosmic rays escaped from an isolated supernova remnant so that the Solar System (and refractory inclusions) formed with background ^{10}Be at a level of 3×10^{-4}. This model also allows for additional in situ solar energetic particle spallation to explain the abundances higher than this background value. However, this model invokes complicated astrophysical considerations and is difficult to test. Discussing which model most likely accounts for the presence and abundance of ^{10}Be in the early Solar System is beyond the scope of this paper. Interested readers are encouraged to refer to the cited papers for more details.

Given the fact that ^{10}Be and boron isotopes are synthesized by spallation of ^{16}O (regardless of the location where irradiation occurs), processes that produced ^{10}Be should also yield collateral effects on boron isotopes. Interestingly, many of the ^{10}Be isochrons are accompanied by an elevated $^{10}B/^{11}B$ ratio (0.2477–0.2576) relative to the chondritic value (0.2476) as the intercept (Fig. 11.5), indicating that these refractory solids (CAIs and hibonite) formed with supra-chondritic boron isotopic compositions. Liu et al. (2010) considered two scenarios in the context of irradiation in the solar nebula: irradiation of refractory solids and irradiation of solar gas, with the primary goal of reconstructing the observed Be-B isotopic results in refractory inclusions. In the case of irradiation of refractory solids, the irradiation targets were assumed to be Be-bearing but B-free because of the difference in volatility of the two elements. At the end of irradiation, the irradiated solids should carry ^{10}Be and pure spallogenic B, with $^{10}B/^{11}B \sim 0.4$, the theoretical production value by spallation determined from the ratio of cross sections of $^{16}O(p, x)^{10,11}B$ (e.g., Yiou et al. 1968). To arrive at $^{10}Be/^{9}Be = (3–9) \times 10^{-4}$ and $^{10}B/^{11}B = 0.2477–0.2576$ seen in refractory inclusions, it is required that irradiated solids suffered diffusive loss of spallogenic boron (because boron is unstable in refractory phases at high temperatures) followed by isotopic exchange with a chondritic gas/reservoir. Only ~2–4% of the pure spallogenic B was preserved. All this must have taken place before a significant amount of ^{10}Be decayed.

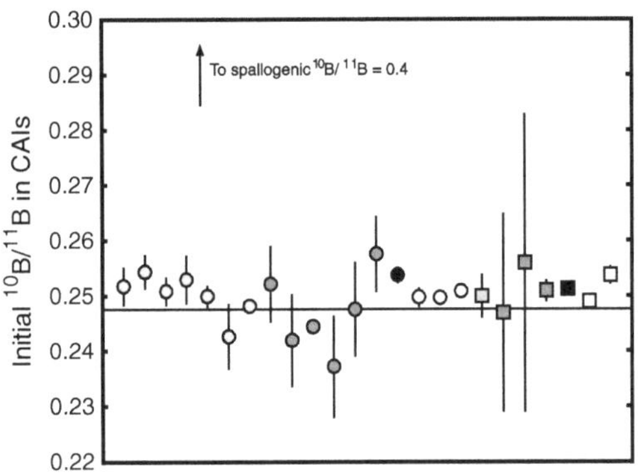

Fig. 11.5 The initial $^{10}B/^{11}B$ ratios inferred from $^{10}Be–^{10}B$ isochrons for various CAIs. Many CAIs are characterized by a supra-chondritic $^{10}B/^{11}B$ ratio, implying the existence of a spallogenic B component in the reservoir(s) where CAIs formed. The horizontal solid line represents the chondritic $^{10}B/^{11}B = 0.2476$. Data are taken from Wielandt et al. (2012) (*light-gray-filled circles*), MacPherson et al. (2003), (*dark-gray-filled circles*), Chaussidon et al. (2006) (*black circle*), Srinivasan and Chaussidon (2013) (*open circles*), McKeegan et al. (2001) (*light-gray-filled square*), Sugiura et al. (2001) (*dark-gray-filled squares*), Liu et al. (2009, 2010) (*black square*), and Gounelle et al. (2013) (*open squares*)

In the second scenario, irradiation of nebular gas of solar composition (e.g., Lodders 2003) was followed by condensation of precursors that ended up forming CAIs or hibonite. The solar gas is initially characterized by chondritic $^{10}B/^{11}B$, and Be-B isotopes were produced by spallation of ^{12}C, ^{14}N and ^{16}O. With the total fluence needed to reproduce $^{10}Be/^{9}Be$ at the observed levels in CAIs and hibonite, the resulting $^{10}B/^{11}B$ is essentially the same as the initial value, as there is too much B in the gas to be significantly affected by spallation reactions. If one were to elevate $^{10}B/^{11}B$ in the gas to a supra-chondritic level similar to the inferred values in refractory solids, an overproduction of $^{10}Be/^{9}Be$, ~30–100 times higher, would result. Such a high abundance has not been seen in any of the meteoritic samples (The high $^{10}Be/^{9}Be = 1 \times 10^{-2}$ found by Gounelle et al. (2013) is because of the low indigenous ^{9}Be content, not because of an extremely high proton fluence; see that paper for details). From this view, it is less likely that irradiation of solar gas plays a major role in making up the ^{10}Be inventory in the early Solar System.

Srinivasan and Chaussidon (2013) also performed mixing calculations similar to those in Liu et al. (2010), yet slightly different in the details, for the gas and refractory solid components in the solar nebula to explain the observed ^{10}Be abundances and $^{10}B/^{11}B$ variations in CAIs and hibonites. They considered different dust/gas ratios and different mixing fractions between irradiated gas and solids, and arrived at a conclusion that the current Be-B dataset are most consistent with the derivation of a major fraction of ^{10}Be in CAIs from solid precursors, and with the supra-chondritic $^{10}B/^{11}B$ ratios being a result of mixing between spallogenic B in the solids and normal B in the nebular gas. Such an addition of "normal" B is in line with CAI petrography and bulk chemistry: the abundances of moderately volatile elements (e.g., B, Na, and Li) in many CAIs with or without petrographically obvious alteration are not as depleted as would be expected in high-temperature condensates/recrystallized solids.

For other models concerning the origin of ^{10}Be, it is more difficult to evaluate the boron isotopic composition that would have ended up in the CAI-forming region, because of complicated astrophysical processes involved. It is important for any model which aims to explain the observed ^{10}Be abundances in the early Solar System to also consider the supra-chondritic ^{10}B/^{11}B ratios.

11.3.4 B Isotopes in Chondrules

Chondrules are the dominant constituent of most chondrites, ~60–80% by volume in ordinary chondrites and 5–60% in carbonaceous chondrites (Scott and Krot 2014). Searching for ^{10}Be in chondrules has been attempted (e.g., Sugiura 2001), but chondrules with high enough Be/B to resolve ^{10}B excesses appear to be extremely rare. Sugiura (2001) analyzed an anorthite-rich chondrule from the Yamato (CO 3.0) chondrite and found a fair correlation between ^{10}B excesses and Be/B, the latter of which is as high as 10. The best-fit isochron yielded a ^{10}Be/^{9}Be ratio of $(7.2 \pm 5.8) \times 10^{-4}$ (2σ) for this chondrule. Despite the large analytical errors and subjective removal of chondrule data points to get a good isochron, this value is consistent with those inferred from CAIs. This has an important implication for the early Solar System irradiation environment. To really settle the ^{10}Be abundances recorded in chondrules, more high quality work on other chondrules needs to be performed.

In most cases where chondrules have low Be/B ratios ($\ll$1), one can study the intrinsic ^{11}B/^{10}B variations within and among chondrules. As previously introduced, isotopic studies with SIMS by Chaussidon and Robert (1995, 1998) revealed that chondrules from various chondritic meteorites exhibit a large heterogeneity in δ^{11}B, spanning from −40 to +40‰, with 87% of data being clustered between −40 and +10‰, whereas chondrites' δ^{11}B vary only within ±10‰ (Zhai et al. 1996). Moreover, the variations were found to be positively correlated with B/Si of the spots analyzed. Early Solar System irradiation of chondrule precursors could not have produced such a large variation since chondrules are too B rich (similar to the case of irradiation of nebular gas), and if it could have, only negative δ^{11}B would have resulted, unless irradiation energy regime was only restrained to be at MeV level. Isotopic fractionation associated with evaporation during chondrule formation (if any) could have enriched ^{11}B in the residue but one should expect a negative correlation between δ^{11}B and the boron contents. This led Chaussidon and Robert (1995, 1998) to argue for the existence of ^{11}B-rich grains of molecular cloud origin, which were incorporated into the solar nebula during its formation and mixed with GCR-produced B (enriched in ^{10}B) that the Solar System formed with. This ^{11}B-rich component would be consistent with the production by hypothetical low-energy cosmic ray irradiation of interstellar hydrogen (and helium), although astronomical evidence for low-energy spallation in the interstellar medium has disappeared (Bloemen et al. 1999). It should be noted that the ν-process from Type II supernovae could have also led to the same ^{11}B enrichment, but evidence for presolar grains of supernova origin (e.g., SiC of Type X; Zinner 2014) carrying large ^{11}B anomalies is still very marginal (Fujiya et al. 2011).

A similar study of chondrules by Hoppe et al. (2001) yielded results in clear contrast to those in Chaussidon and Robert (1995, 1998). Sixteen chondrules of different textures from various meteorites (Allende, Murchison, Semarkona, Chainpur, Dhajala, Bjurbole, and Parsa) studied in this work show no significant variation in δ^{11}B outside their experimental errors of a few permil, and their boron isotopic compositions agree with those reported for bulk chondrites. Isotopic heterogeneities within individual chondrules are limited within ±20‰ (associated with 10–30‰ errors), as opposed to ±40‰ reported for carbonaceous and ordinary chondrites in

Chaussidon and Robert (1995, 1998). It was also found that the boron abundances in chondrules measured by Hoppe et al. (2001) are as much as 2 orders of magnitude lower than those analyzed by Chaussidon and Robert (1995, 1998), and are independent of $\delta^{11}B$. Although impossible to conclusively rule out the existence of chondrules carrying heterogeneities on the order of a few percent with anomalously high B contents, Hoppe et al. (2001) argued that such chondrules must be rare and that any potential boron isotope heterogeneities in the early Solar System caused by different nucleosynthetic processes (e.g., high-energy GCRs, low-energy GCRs and the ν-process) did not survive the chondrule formation process to a measurable degree.

Given the controversial results reported by two groups of researchers, it requires further analytical efforts to better understand which population best represents the overall boron isotopic compositions of chondrules. This will shed light on the origin of Solar System boron (see above), especially if large positive $\delta^{11}B$ values associated with high B concentrations turn out to be a common signature in chondrules.

11.4 Concluding Remarks

Astronomical observations and laboratory analyses have yielded a wealth of data on the abundances and isotopic ratios of boron in the Galaxy. However, the main question remains unsolved: how did the Galaxy (and the constituents within it) acquire its boron isotopes that have a $^{11}B/^{10}B$ ratio of $\sim$4? In addition to spallation nucleosynthesis of B by Galactic Cosmic Rays that produces $^{11}B/^{10}B$ of 2.5, another mechanism that favors the production of ^{11}B over that of ^{10}B must have operated over the Galactic timescale. Low-energy spallation and ν-process have been considered to be potential contributors of ^{11}B-rich components, but so far we are still ignorant as to which mechanism plays a more dominant role.

Another question that is poorly understood with regard to boron isotopes in the Solar System is that how much of the boron isotope anomalies was produced by energetic particle spallation by the protoSun? This process is known to have occurred in the very beginning of the Solar System history based on the fossil records of ^{10}Be, but the level of irradiation, and thus the resulting production of B isotopes, is not well constrained, especially if part of ^{10}Be came in as an irradiation product in the parental molecular cloud. High precision B isotope measurements of chondrules and ^{10}Be-free (potentially due to a lack of irradiation?) refractory inclusions will certainly help shed light on this question.

The average $\delta^{11}B$ of each chondrite group seems to be consistent with that of the CI chondrites, suggesting a high degree of homogeneity of B isotope distribution (within a couple of permil) in the Solar System. However, the bulk B isotopic composition could vary by as much as $\sim$20‰ from one chondrite to another, the reasons for which are not fully understood. Parent body alteration could affect the B isotopic composition of a chondrite, but its effects need to be evaluated. Early Solar System irradiation of meteoritic components before they were incorporated into parent bodies could potentially account for (some of) the $\delta^{11}B$ variability in chondrites, but more in-depth calculations are required to quantify the degree of heterogeneity that this process might have caused.

Acknowledgements We thank Dr. Horst Marschall for his invitation to write this review paper and his comments on the first and final versions of the manuscript. Constructive comments from two reviewers, Drs. Glenn MacPherson and Don Burnett, greatly improved the presentation of this paper.

References

Agyei EK, McMullen CC (1968) A study of the isotopic abundance of boron from various sources. Canadian J Earth Sci 5:921–927

Anders E, Grevesse N (1989) Abundances of the elements—meteoritic and solar. Geochim Cosmochim Acta 53:197–214

Asplund M, Grevesse N, Sauval AJ, Scott P (2009) The chemical composition of the sun. Ann Rev Astro Astrophys 47:481–522

Bloemen H, Wijnands R, Bennett K, Diehl R, Hermsen W, Lichti G, Morris D, Ryan J, Schonfelder V,

Strong AW, Swanenburg BN, de Vries C, Winkler C (1994) COMPTEL observations of the Orion complex: evidence for cosmic-ray induced gamma-ray lines. Astro Astrophys 281:L5–L8

Bloemen H, Morris D, Knödlseder J, Bennett K, Diehl R, Hermsen W, Lichti G, van der Meulen RD, Oberlack U, Ryan J, Schönfelder V, Strong AW, de Vries C, Winkler C (1999) The revised COMPTEL orion results. Astrophys J Lett 521:L137–L140

Boesgaard AM, Heacox WD (1978) The abundance of boron in B- and A-type stars. Astrophys J 226:888–896

Burbidge EM, Burbidge GR, Fowler WA, Hoyle F (1957) Synthesis of the elements in stars. Rev Modern Phys 29:547–650

Cassé M, Lehoucq R, Vangioni-Flam E (1995) Production and evolution of light elements in active star-forming regions. Nature 373:318–319

Chaussidon M, Robert F (1995) Nucleosynthesis of ^{11}B-rich boron in the pre-solar cloud recorded in meteoritic chondrules. Nature 374:337–339

Chaussidon M, Robert F (1997) Comment on "Boron cosmochemistry II: boron nucleosynthesis and condensation temperature" by M. Zhai, Meteor Planet Sci, p 32

Chaussidon M, Robert F (1998) ^{7}Li/ ^{6}Li and ^{11}B/ ^{10}B variations in chondrules from the Semarkona unequilibrated chondrite}. Earth Planet Sci Lett 164:577–589

Chaussidon M, Robert F, Mangin D, Hanon P, Rose EF (1997) Analytical procedures for the measurement of boron isotope compositions by ion microprobe in meteorites and mantle rocks. Geostand Newslett 21: 7–17

Chaussidon M, Robert F, McKeegan KD (2006) Li and B isotopic variations in an Allende CAI: evidence for the in situ decay of short-lived ^{10}Be and for the possible presence of the short-lived nuclide ^{7}Be in the early solar system. Geochim Cosmochim Acta 70:224–245

Connelly JN, Bizzarro M, Krot AN, Nordlund Å, Wielandt D, Ivanova MA (2012) The absolute chronology and thermal processing of solids in the solar protoplanetary disk. Science 338:651–655

Cunha K, Smith VV (1999) A determination of the solar photospheric boron abundance. Astrophys J 512:1006–1013

Cunha K, Lambert DL, Lemke M, Gies DR, Roberts LC (1997) Boron abundances of B stars of the orion association. Astrophys J 478:211–224

Desch SJ, Connolly HC Jr, Srinivasan G (2004) An interstellar origin for the beryllium-10 in calcium-rich, Aluminum-rich inclusions. Astrophys J 602:528–542

Federman SR, Lambert DL, Cardelli JA, Sheffer Y (1996) The boron isotope ratio in the interstellar medium. Nature 381:764–766

Federman SR, Sheffer Y, Lambert DL, Gilliland RL (1993) Detection of boron, cobalt, and other weak interstellar lines toward ζ Ophiuchi. Astrophys J 413: L51–L54

Foster GL, Marschall HR, Palmer MR (2017) Boron isotope analysis of geologic materials. In: Marschall HR, Foster GL (eds) Boron isotopes—the fifth element, vol 7. Advances in isotope geochemistry. Springer, Heidelberg (Chapter 2)

Fujiya W, Hoppe P, Ott U (2011) Hints for neutrino-process boron in presolar silicon carbide grains from supernovae. Astrophys J Lett 730:L7

Gibner PS, Mewaldt RA, Schindler SM, Stone EC, Webber WR (1992) The isotopic composition of cosmic-ray B, C, N, and O—evidence for an overabundance of O-18. Astrophys J Lett 391:L89–L92

Gounelle M, Shu FH, Shang H, Glassgold AE, Rehm KE, Lee T (2006) The irradiation origin of beryllium radioisotopes and other short-lived radionuclides. Astrophys J 640:1163–1170

Gounelle M, Chaussidon M, Rollion-Bard C (2013) Variable and extreme irradiation conditions in the early solar system inferred from the initial abundance of ^{10}Be in isheyevo CAIs. Astrophys J Lett 763:L33

Grevesse N, Asplund M, Sauval AJ (2007) The solar chemical composition. Space Sci Rev 130:105–114

Hewins RH (1997) Chondrules. Ann Rev Earth Planet Sci 25:61

Hoppe P, Goswami JN, Krähenbühl U, Marti K (2001) Boron in chondrules. Meteor Planet Sci 36:1331–1343

Howk JC, Sembach KR, Savage BD (2000) The abundance of interstellar boron. Astrophys J 543:278–283

Ishikawa T, Nagaishi K (2011) High-precision isotopic analysis of boron by positive thermal ionization mass spectrometry with sample preheating. J Anal At Spectrom 26:359–365

Jura M, Meyer DM, Hawkins I, Cardelli JA (1996) The interstellar boron abundance toward orion. Astrophys J 456:598–601

Kohl JL, Parkinson WH, Withbroe GL (1977) The solar boron abundance. Astrophys J Lett 212:L101–L104

Koning AJ, Hilaire S, Dujivestijn MC (2007) TALYS-1.0. In: Bersillon O, Gunsing F, Bauge E, Jacqmin R, Leray S (eds) Proceedings of the conference on nuclear data for science and technology, pp 211–214

Korschinek G, Bergmaier A, Faestermann T, Gerstmann UC, Knie K, Rugel G, Wallner A, Dillmann I, Dollinger G, Lierse von Gostomski C, Kossert K, Maiti M, Poutivtsev M, Remmert A (2010) A new value for the half-life of ^{10}Be by heavy-ion elastic recoil detection and liquid scintillation counting. Nucl Inst Meth Phys Res B 268:187–191

Krombel KE, Wiedenbeck ME (1988) Isotopic composition of cosmic-ray boron and nitrogen. Astrophys J 328:940–953

Lambert DL, Sheffer Y, Federman SR, Cardelli JA, Sofia UJ, Knauth DC (1998) The ^{11}B/^{10}B ratio of local interstellar diffuse clouds. Astrophys J 494:614–622

Lauretta DS, Lodders K (1997) The cosmochemical behavior of beryllium and boron. Earth Planet Sci Lett 146:315–327

Liu M-C, McKeegan KD, Goswami JN, Marhas KK, Sahijpal S, Ireland TR, Davis AM (2009) Isotopic records in CM hibonites: implications for timescales of mixing of isotope reservoirs in the solar nebula. Geochim Cosmochim Acta 73:5051–5079

Liu M-C, Nittler LR, Alexander CMO, Lee T (2010) Lithium-beryllium-boron isotopic compositions in meteoritic hibonite: Implications for origin of ^{10}Be and early Solar System irradiation. Astrophys J 719: L99–L103

Lodders K (2003) Solar system abundances and condensation temperatures of the elements. Astrophys J 591:1220–1247

MacPherson GJ, Huss GR, Davis AM (2003) Extinct ^{10}Be in type A calcium-aluminum-rich inclusions from CV chondrites. Geochim Cosmochim Acta 67:3165–3179

Marhas KK, Goswami JN, Davis AM (2002) Short-lived nuclides in hibonite grains from murchison: evidence for solar system evolution. Science 298:2182–2185

Marschall HR, Monteleone BD (2015) Boron isotope analysis of silicate glass with very low boron concentrations by secondary ion mass spectrometry. Geostand Geoanal Res 39:31–46

Marschall HR (2017) Boron isotopes in the ocean floor realm and the mantle. In: Marschall HR, Foster GL (eds) Boron isotopes—The fifth element, advances in isotope geochemistry, vol 7. Springer, Heidelberg

McKeegan KD, Chaussidon M, Robert F (2000) Incorporation of short-lived ^{10}Be in a calcium-aluminum-rich inclusion from the allende meteorite. Science 289:1334–1337

McKeegan KD, Chaussidon M, Krot AN, Robert F, Goswami JN, Hutcheon ID (2001) Extinct radionuclide abundances in Ca, Al-rich inclusions from the CV chondrites Allende and Efremovka: a search for synchronocity. In: LPSCXXXII, #2175 (abstract)

Meneguzzi M, York DG (1980) Detection of interstellar boron in front of kappa Orionus. Astrophys J Lett 235: L111–L114

Meneguzzi M, Audouze J, Reeves H (1971) The production of the elements Li, Be, B by galactic cosmic rays in space and its relation with stellar observations. Astro Astrophys 15:337–359

Morris MA, Boley AC, Desch, SJ, Athanassiadou T (2012) Chondrule formation in bow shocks around eccentric planetary embryos. Astrophys J 752:27 (17 pp)

Praderie F, Milliard B, Pitois ML, Boesgaard AM (1977) The abundance of boron in VEGA and Sirius. Astrophys J 214:130–139

Prantzos N (2007) Origin and evolution of the light nuclides. Space Sci Rev 130:27–42

Prantzos N (2012) Production and evolution of Li, Be, and B isotopes in the Galaxy. Astro Astrophys 542: A67

Proffitt CR, Quigley MF (2001) Boron abundances in early B stars: results from the B III resonance line in IUE data. Astrophys J 548:429–438

Proffitt CR, Jönsson P, Litzén U, Pickering JC, Wahlgren GM (1999) Goddard high-resolution spectrograph observations of the B III resonance doublet in early B stars: abundances and isotope ratios. Astrophys J 516:342–348

Ramaty R, Kozlovsky B, Lingenfelter RE (1979) Nuclear gamma-rays from energetic particle interactions. Astrophys J Suppl 40:487–526

Ramaty R, Kozlovsky B, Lingenfelter RE (1996) Light isotopes, extinct radioisotopes, and gamma-ray lines from low-energy cosmic-ray interactions. Astrophys J 456:525

Reeves H, Fowler WA, Hoyle F (1970) Galactic cosmic ray origin of Li, Be and B in stars. Nature 226:727–729

Ritchey AM, Federman SR, Sheffer Y, Lambert DL (2011) The abundance of boron in diffuse interstellar clouds. Astrophys J 728:70 (37 pp)

Scott ERD, Krot AN (2014) 1.2—chondrites and their components. In: Davis AM (ed) Meteorites and cosmochemical processes. Treatise of geochemistry, vol 1, 2nd edn, pp 65–137

Shen JJ, You CF (2003) A 10-fold improvement in the precision of boron isotopic analysis by negative thermal ionization mass spectrometry. Anal Chem 75:1972–1977

Shima M (1962) Boron in meteorites. J Geophys Res 67:4521–4523

Shima M (1963) Geochemical study of boron isotopes. Geochim Cosmochim Acta 27:911–913

Simpson JA (1983) Elemental and isotopic composition of the galactic cosmic rays. Ann Rev Nucl Part Sci 33:323–382

Srinivasan G, Chaussidon M (2013) Constraints on ^{10}Be and ^{41}Ca distribution in the early solar system from ^{26}Al and ^{10}Be studies of efremovka CAIs. Earth Planet Sci Lett 374:11–23

Sugiura N (2001) Boron isotopic compositions in chondrules: anorthite-rich chondrules in the Yamato 82094 (CO_3) Chondrite. In: LPSCXXXII, #1277 (abstract)

Sugiura N, Shuzou Y, Ulyanov A (2001) Beryllium-boron and aluminum-magnesium chronology of calcium-aluminum-rich inclusions in CV chondrites. Meteor Planet Sci 36:1397–1408

Tatischeff V, Duprat J, de Séréville N (2014) Light-element nucleosynthesis in a molecular cloud interacting with a supernova remnant and the origin of beryllium-10 in the protosolar nebula. Astrophys J 796:124 (20 pp)

Venn KA, Brooks AM, Lambert DL, Lemke M, Langer N, Lennon DJ, Keenan FP (2002) Boron abundances in B-type stars: a test of rotational depletion during main-sequence evolution. Astrophys J 565:571–586

Wielandt D, Nagashima K, Krot AN, Huss GR, Ivanova MA, Bizzarro M (2012) Evidence for multiple sources of ^{10}Be in the early solar system. Astrophys J Lett 748:L25

Woosley SE, Weaver TA (1995) The evolution and explosion of massive stars. II. Explosive hydrodynamics and nucleosynthesis. Astrophys J Suppl 101:181–235

Woosley SE, Hartmann DH, Hoffman RD, Haxton WC (1990) The nu-process. Astrophys J 356:272–301
Xiao YK, Wang L (1998) Effect of NO_3^- on the isotopic measurement of boron. Int J Mass Spec 178:213–220
Yiou F, Baril M, de Citres JD, Fontes P, Gradsztajn E, Bernas R (1968) Mass-spectrometric measurement of lithium, beryllium, and boron isotopes produced in ^{16}O by high-energy protons, and some astrophysical implications. Phys Rev 166:968–974
Yoshida T, Kajino T (2005) Supernova neutrino process and its impact on the galactic chemical evolution of the light elements. Nucl Phys A 758:35–38
Yoshida T, Kajino T, Hartmann DH (2005) Constraining the spectrum of supernova neutrinos from ν-process induced light element synthesis. Phys Rev Lett 94:231101
Zhai M (1995) Boron cosmochemistry. Part II: Boron nucleosynthesis and condensation temperature. Meteoritics 30:733–737
Zhai M, Shaw DM (1994) Boron cosmochemistry. Part 1: Boron in meteorites. Meteoritics 29:607–615
Zhai M, Nakamura E, Shaw DM, Nakano T (1996) Boron isotope ratios in meteorites and lunar rocks. Geochim Cosmochim Acta 60:4877–4881
Zhang J, Dauphas N, Davis AM, Leya I, Fedkin A (2012) The proto-Earth as a significant source of lunar material. Nat Geosci 5:251–255
Zinner E (2014) Presolar grains. In: Davis AM (ed) Meteorites and cosmochemical processes. Treatise of geochemistry, vol 1, 2nd edn, pp 181–213

MIX
Papier aus verantwortungsvollen Quellen
Paper from responsible sources
FSC® C105338

If you have any concerns about our products,
you can contact us on
ProductSafety@springernature.com

In case Publisher is established outside the EU,
the EU authorized representative is:
Springer Nature Customer Service Center GmbH
Europaplatz 3, 69115 Heidelberg, Germany

Printed by Libri Plureos GmbH
in Hamburg, Germany